2ND EDITION

CYCLONE

COMPLETE JUNIOR CYCLE GEOGRAPHY

STACY KENNY & ANDREW HORAN
Consultant Author: **CHARLES HAYES**

Gill Education
Hume Avenue
Park West
Dublin 12

www.gilleducation.ie

Gill Education is an imprint of M.H. Gill & Co.
© Stacy Kenny and Andrew Horan 2023
ISBN: 978 0 7171 95275

All rights reserved. No part of this publication may be copied, reproduced or transmitted in any form or by any means without written permission of the publishers or else under the terms of any licence permitting limited copying issued by the Irish Copyright Licensing Agency.

Design and layout: Liz White Designs

Project editor: Ciara McNee

Illustrations: Keith Barrett, Derry Dillon and Andrii Yankovskyi

At the time of going to press, all web addresses were active and contained information relevant to the topics in this book. Gill Education does not, however, accept responsibility for the content or views contained on these websites. Content, views and addresses may change beyond the publisher or authors' control. Students should always be supervised when reviewing websites.

The authors and publisher have made every effort to trace all copyright holders. If, however, any have been inadvertently overlooked, we would be pleased to make the necessary arrangement at the first opportunity.

Photo permissions: Please see p. 448

Ordnance Survey Ireland Permit No. 9275
© Ordnance Survey Ireland/Government of Ireland

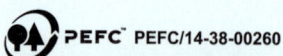

Acknowledgements

I would like to thank Andy and Charlie for being such a wonderful team to work with.

Thank you to all of the many wonderful people in Gill for bringing this book to life.

Stacy Kenny

I would like to thank Stacy and Charlie for all their help and advice while writing this wonderful 2nd edition of *Cyclone*. A big 'Thank you!' to the entire Gill team for making any task we set seem easy.

Andrew Horan

Contents

Introduction vi
Introduction to Assessments viii

1. Dynamic Earth 1
   1.1 Our Moving Continents 1
   1.2 Structure of the Earth 3
   1.3 Convection Currents 4
   1.4 Plate Tectonics 5
   Exam Focus 7

2. Volcanic Activity 9
   2.1 Think Like a Volcanologist 9
   2.2 Volcanic Mountains 10
   2.3 The Life Cycle of a Volcano 13
   2.4 The Pacific Ring of Fire 14
   2.5 The Economic and Social Impacts of Volcanoes ... 15
   Exam Focus 20
   CBA 1 Geography in the News 21

3. Earthquakes 23
   3.1 Distribution of Earthquakes 23
   3.2 Formation of an Earthquake 25
   3.3 Tsunami Waves 26
   3.4 Measuring Earthquakes 27
   3.5 Economic and Social Impacts of Earthquakes in Nepal and Chile 28
   3.6 Reducing Earthquake Damage 33
   Exam Focus 35

4. Fold Mountains 38
   4.1 Formation of Fold Mountains 38
   4.2 Distribution and Periods of Fold Mountain Formation 39
   4.3 Formation of the Dublin–Wicklow Mountains 40
   4.4 Economic and Social Impacts of Fold Mountains ... 41
   Exam Focus 43

5. Rocks 45
   5.1 Uses of Rock 45
   5.2 The Rock Cycle 46
   5.3 Distribution of Rocks in Ireland 48
   5.4 Rock Groups 49
   5.5 Human Interaction with Rocks: Environmental and Economic Consequences 54
   Exam Focus 59

6. Natural Energy Resources 61
   6.1 What Are Energy Resources? 61
   6.2 Exploitation of a Non-Renewable Resource: Oil ... 63
   6.3 The Search for Hydrocarbons in Irish Waters 65
   6.4 Exploitation of a Non-Renewable Resource: Peat ... 67
   6.5 Environmental Consequences of Non-Renewable Energies: Acid Rain 69
   6.6 Renewable Energy in Ireland 70
   Exam Focus 75

7. Geographical Skills: Ordnance Survey Maps 77
   7.1 Ordnance Survey Maps 78
   7.2 How to Measure Distances on a Map 79
   7.3 Directions on an OS Map 82
   7.4 Locating Places: The National Grid 83
   7.5 Identifying Symbols on an OS Map 85
   7.6 Altitude (Height) on an OS Map 88
   7.7 Drawing Sketch Maps 92
   7.8 Reading the Physical Landscape 94
   7.9 Reading the Human Landscape 97
   7.10 Settlement on an OS Map 98
   7.11 Large Scale Maps 106
   Exam Focus 110

8. Weathering 115
   8.1 The Landscape 115
   8.2 Denudation 116
   8.3 Sustainability and Tourism in the Burren 123
   Exam Focus 124

9. Mass Movement 126
   9.1 Mass Movement 126
   9.2 Types of Mass Movement 127
   Exam Focus 132

10. Rivers 134
   10.1 The Journey of a River 135
   10.2 The Stages of a River 136
   10.3 The Processes of a River 137
   10.4 Landforms of a Youthful River 139
   10.5 Landforms of a Mature River 141
   10.6 Landforms of an Old River 143
   10.7 Rivers and Human Interaction 146
   Exam Focus 150

Contents

11. The Sea 152
- 11.1 Ireland's Coastline 153
- 11.2 Coastal Processes 153
- 11.3 Landforms of Sea Erosion 156
- 11.4 Landforms of Sea Deposition 160
- 11.5 People and the Sea: Managing Surface Processes 164
- 11.6 Coastal Management 166
- Exam Focus 169
- CBA 1 Geography in the News 171

12. Glaciation: The Work of Moving Ice 173
- 12.1 Glaciation 174
- 12.2 Processes of Glacial Erosion 175
- 12.3 Transport and Deposition by Moving Ice 180
- 12.4 Glaciation and People 186
- Exam Focus 188

13. Measuring and Forecasting Weather 189
- 13.1 Weather Forecasting 190
- 13.2 Forecasting the Weather 193
- CBA 2 My Geography Moment 206
- Exam Focus 209
- CBA 2 My Geography Moment 211

14. Severe Weather 213
- 14.1 What Is Severe Weather? 213
- 14.2 Tropical Storms 214
- Exam Focus 220

15. Global Climates 222
- 15.1 What Is a Climate? 222
- 15.2 Classifying Global Climate Types 223
- 15.3 Ireland's Climate 227
- Exam Focus 233

16. Climate Change 235
- 16.1 How Do We Know That Our Climate Is Changing? 235
- 16.2 Changes in Global Climate 237
- 16.3 Climate Change and Human Activity 237
- 16.4 Implications and Effects of Climate Change 241
- 16.5 Solutions: A Climate Agreement 244
- Exam Focus 245
- CBA 1 Geography in the News 247

17. Soils 249
- 17.1 What Is Soil? 249
- 17.2 How Soil Is Formed 251
- 17.3 Soil Profiles 251
- 17.4 Ireland's Soil Types 253
- Exam Focus 255
- CBA 2 My Geography Moment 256

18. Population 264
- 18.1 World Population 264
- 18.2 Population Change and Density 266
- 18.3 The Population Cycle 267
- 18.4 Factors Influencing Population Change 268
- 18.5 Population Structure 269
- Exam Focus 275

19. Migration 277
- 19.1 People on the Move 277
- 19.2 Why People Migrate: Push and Pull Factors 279
- 19.3 Forced Migration 280
- 19.4 Individual Migration 282
- 19.5 Organised Migration 284
- 19.6 Globalisation and Migration 286
- Exam Focus 288

20. Geographical Skills: Aerial Photographs, Charts, Graphs and Infographics 290
Part 1: Aerial Photographs 291
- 20.1 Aerial Photographs 291
- 20.2 Examining Aerial Photographs: Colour and Shape 296
- 20.3 Urban Settlement: Towns and Cities on Aerial Photographs 298
- 20.4 Comparing OS Maps with Aerial Photographs 304
- Exam Focus 305

Part 2: Charts, Graphs and Infographics 312
- 20.5 Line Graphs 312
- 20.6 Pie Charts 313
- 20.7 Bar Charts 314
- 20.8 Infographics 314

21. Rural and Urban Settlement in Ireland 316
- 21.1 What Is Settlement? 316
- 21.2 The Physical Landscape 318
- 21.3 Historical Factors 321
- 21.4 The Importance of Dublin City 324
- Exam Focus 326

22. Urban Change in Dublin 327
- 22.1 Urban Change 327
- 22.2 Population Change in Dublin 328
- 22.3 The Effects of Urban Change 332
- Exam Focus 339
- CBA 2 My Geography Moment 340

23. Primary Economic Activities 341
- 23.1 The Physical Landscape 341
- 23.2 Introducing Economic Activities 344
- 23.3 What Are Primary Economic Activities? ... 345
- 23.4 The Physical Landscape: Farming and Forestry ... 346
- 23.5 The Physical Landscape: Fishing 350
- Exam Focus 352

24. Exploitation of Natural Resources 354
- 24.1 Natural Resources 355
- 24.2 What Is Sustainable Exploitation? 355
- 24.3 Sustainable Exploitation of Water: Egypt and Ireland ... 356
- 24.4 Sustainable Exploitation of Fish: Ireland ... 360
- 24.5 Sustainable Exploitation of Forestry: Ireland and Malaysia ... 364
- 24.6 Sustainable Exploitation of Soil: Urbanisation and Desertification ... 367
- Exam Focus 370

25. Secondary Economic Activities 374
- 25.1 What Are Secondary Economic Activities? ... 375
- 25.2 Location of Industry 376
- 25.3 Types of Industry 379
- 25.4 Change Over Time in the Location of Industry ... 383
- Exam Focus 384
- CBA 2 My Geography Moment 386

26. Tertiary Economic Activities 387
- 26.1 What Are Tertiary Economic Activities? ... 387
- 26.2 Tourism, the Physical Landscape and Transport ... 388
- 26.3 Tourism in Ireland 393
- 26.4 Sustainable Tourism 399
- Exam Focus 400

27. Economic Development and Inequality ... 402
- 27.1 Categories of Economic Development ... 402
- 27.2 How We Measure Economic Development ... 405
- 27.3 Causes of Unequal Economic Development ... 407
- 27.4 Developing an Economy 411
- Exam Focus 412

28. Human Development and Development Assistance ... 414
- 28.1 What Is Human Development? 414
- 28.2 Human Development Aid 417
- 28.3 Ireland's Bilateral Aid 419
- 28.4 Advantages and Disadvantages of Aid ... 423
- 28.5 Technology and Human Development ... 424
- Exam Focus 425

29. Life Chances in a Developed and a Developing Country ... 426
- 29.1 Developed and Developing Countries ... 427
- 29.2 Healthcare in Ireland and Nigeria 429
- 29.3 Education Opportunities in Ireland and Nigeria ... 430
- 29.4 Gender Equality for Young People in Ireland and Nigeria ... 432
- 29.5 Employment Opportunities in Ireland and Nigeria ... 433
- Exam Focus 436

30. Globalisation 438
- 30.1 What Is Globalisation? 438
- 30.2 Types of Globalisation 439
- 30.3 Some Consequences of Globalisation ... 442
- Exam Focus 446

Acknowledgements 448

Introduction

Welcome to *Cyclone*!

Welcome to Junior Cycle Geography – a wonderful subject that will allow you to make better sense of the world around you and to answer lots of the questions you hear on the news each day. Questions like: Why does Ireland experience the weather it does? Why is there such demand for housing in cities? What inspired so many big television and movie producers to shoot scenes in Ireland? How can we protect the planet and all its species?

Geography is also a very useful subject. The skills you develop in class, and through using the *Cyclone* textbook and Skills Book will serve you well throughout your life. The ability to recognise patterns and processes, to analyse data, to read maps and graphs, to debate and present and to live sustainably, among other things, are all fostered in the Junior Cycle Geography course.

Cyclone is an engaging and innovative textbook and suite of resources that will guide you on your journey through Junior Cycle Geography. It will help you to meet all the requirements of the specification and is written fully in line with the Junior Cycle Framework.

Cyclone puts *you* at the centre of the learning experience. As you make your way through the textbook and Skills Book, you will deepen your knowledge of different topics, while also developing the important skills of a geographer. The real-life examples and active learning methods will make for an exciting learning experience. We hope you truly enjoy this fascinating subject!

Using *Cyclone*

Contents

The Junior Cycle Geography specification emphasises a non-linear, integrated approach to teaching topics. Every effort has been made to interconnect learning outcomes and strands in each chapter.

Chapter headings have been colour-coded by strand, for easy reference. However, as many chapters deal with topics from across the different strands, this colour-coding is not an exact science. Colours have been designated according to the primary focus of the chapter, as follows:

STRAND 1	Exploring the Physical World
STRAND 2	Exploring How We Interact with the Physical World
STRAND 3	Exploring People, Place and Change
Geographical Skills	

Geographical Skills

There will be opportunities to develop your geographical skills in each chapter, and notably through using the *Cyclone Skills Book*. Chapters 7 and 20 are devoted explicitly to teaching geographical skills, such as reading and using maps and aerial photos, and developing graphicacy by interpreting and creating charts and graphs.

Introduction

Learning Outcomes and Learning Intentions

The relevant Learning Outcomes are highlighted at the start of each chapter, to assist planning.

The Learning Outcomes are then broken down into student-friendly Learning Intentions to help focus lessons. These list what you should be able to do when you have completed the chapter. You should refer back to them regularly when studying.

Key Skills activities

The book includes regular 'Key Skills' activities for integrating active teaching strategies, peer assessment and group work, and developing the eight key skills.

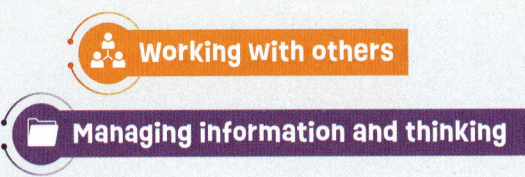

Classroom-Based Assessments

Geography in the News and My Geography Moment will introduce you to the Classroom-Based Assessments.

Skills Book

It is intended that you use this textbook in conjunction with the *Cyclone Skills Book*. The exercises in the Skills Book allow you to develop your skills, apply knowledge gained and revise and reinforce key terms.

Exam Focus

At the end of each chapter, you will complete the Exam Focus questions. These are sample exam questions that will focus your learning and help you to prepare for your Junior Cycle exam.

The questions have been carefully selected to help you hone your skills, knowledge and understanding of the topics. Exam hints and sample answers ready for you to complete are provided to help you to achieve your best.

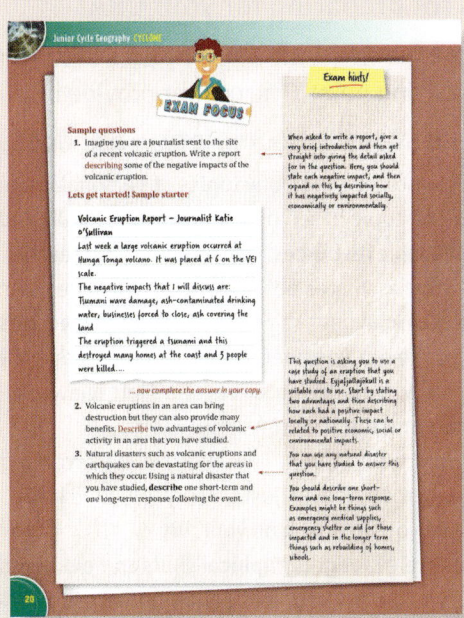

Introduction to Assessments

Classroom-Based Assessment 1: Geography in the News

What is it? An investigation of a recent geographical event, based on a media source

When does it take place? Second term of 2nd Year

Classroom-Based Assessment 2: My Geography Moment

What is it? An investigation of geographical aspect(s) in a local area

When does it take place? First term of 3rd Year

Why are CBAs important? Achievements in your Classroom-Based Assessments will be recorded on your Junior Cycle Profile of Achievement (JCPA). One of the following descriptors will be used: Exceptional; Above Expectations; In Line with Expectations; Yet to Meet Expectations.

Assessment Task

What is it? A written report that will assess your ability to reflect on your geographical thinking and skills, and to evaluate new knowledge or understanding. The report also assesses how your appreciation of Geography has been influenced by doing the CBA.

When does it take place? After CBA 2 has been completed

Why is it important? The Assessment Task is worth 10% of the final marks awarded to you for Geography by the State Examinations Commission.

Tips for preparing for your assessment task
Practise reflecting through completing the CBA activities, which have reflection built into them. Your teacher's resource book also includes activities and worksheets that will help you to develop the vocabulary you need for reflection.

Final Assessment

What is it? A common-level exam, set and marked by the State Examinations Commission. It will be no more than two hours in length.

When does it take place? June of 3rd Year

Why is it important? This exam is worth 90% of the final marks awarded to you for Geography by the State Examinations Commission.

Tips for preparing for your final assessment
The *Cyclone Skills Book* includes lots of different questions that will help you to prepare for the final assessment. By completing these exercises, you will revise new terms and apply your knowledge and skills in different contexts. All exam questions will be based on the Learning Outcomes, which are highlighted at the start of each chapter.

Tips for preparing for your CBAs

Use the Geography in the News and My Geography Moment sections of your textbook and Skills Book as a starting point. They might provide inspiration for your investigations.

Eleven steps for completing a successful CBA *(tick these off as you achieve them)*:

1. Brainstorm Geography **topics** of **personal interest**. ☐
2. Pick a topic based on your **interests, skills and available resources. Consult** with your teacher and with classmates before making a final decision. ☐
3. **Research** your topic using a **range of methods** (you might consider using the internet and the library, conducting surveys and interviews or carrying out experiments). ☐
4. **Work with others.** This might involve working in a group or even just having someone else give you feedback on your work. ☐
5. Use relevant **geographical skills** and **elements**. Try to incorporate maps and photos, use a variety of scales and consider the issue of sustainability. ☐
6. **Analyse** all available **data and research** before making judgements. ☐
7. **Organise your information** and decide how you will present it. ☐
8. **Communicate** the results of your investigation **clearly** and **in an engaging way**. ☐
9. **Explain** why your investigation is of particular interest and why it is **significant**. ☐
10. Draw and outline **conclusions** from your investigation. ☐
11. **Reflect** on your work. Think about what you learned and the skills you used and developed. ☐

Learning Outcome: 1.1

1

Dynamic Earth

Learning Intentions

You will be able to:
- Explain the term 'continental drift'
- Name and describe the layers of the earth
- Explain what tectonic plates are, and why they move
- Describe what happens when plates separate, collide or slide past each other.

key words

- Pangaea
- Continental drift
- Crust
- Molten
- Semi-molten
- Magma
- Mantle
- Convection currents
- Core
- Constructive
- Destructive
- Oceanic plates
- Continental plates
- Subduction
- Conservative

1.1 Our Moving Continents

The earth did not always look like it does today. Over 250 million years ago, the continents were all connected as one large landmass, known as **Pangaea**. A German scientist, Alfred Wegener, studied the movement of the continents. In 1912, he proposed that, over time, the continents broke apart and moved. He called this process **continental drift**.

1. This image shows the position of our continents 90 million years ago. Can you identify any familiar locations? Why do you think that our planet looks quite different to this today?

Junior Cycle Geography CYCLONE

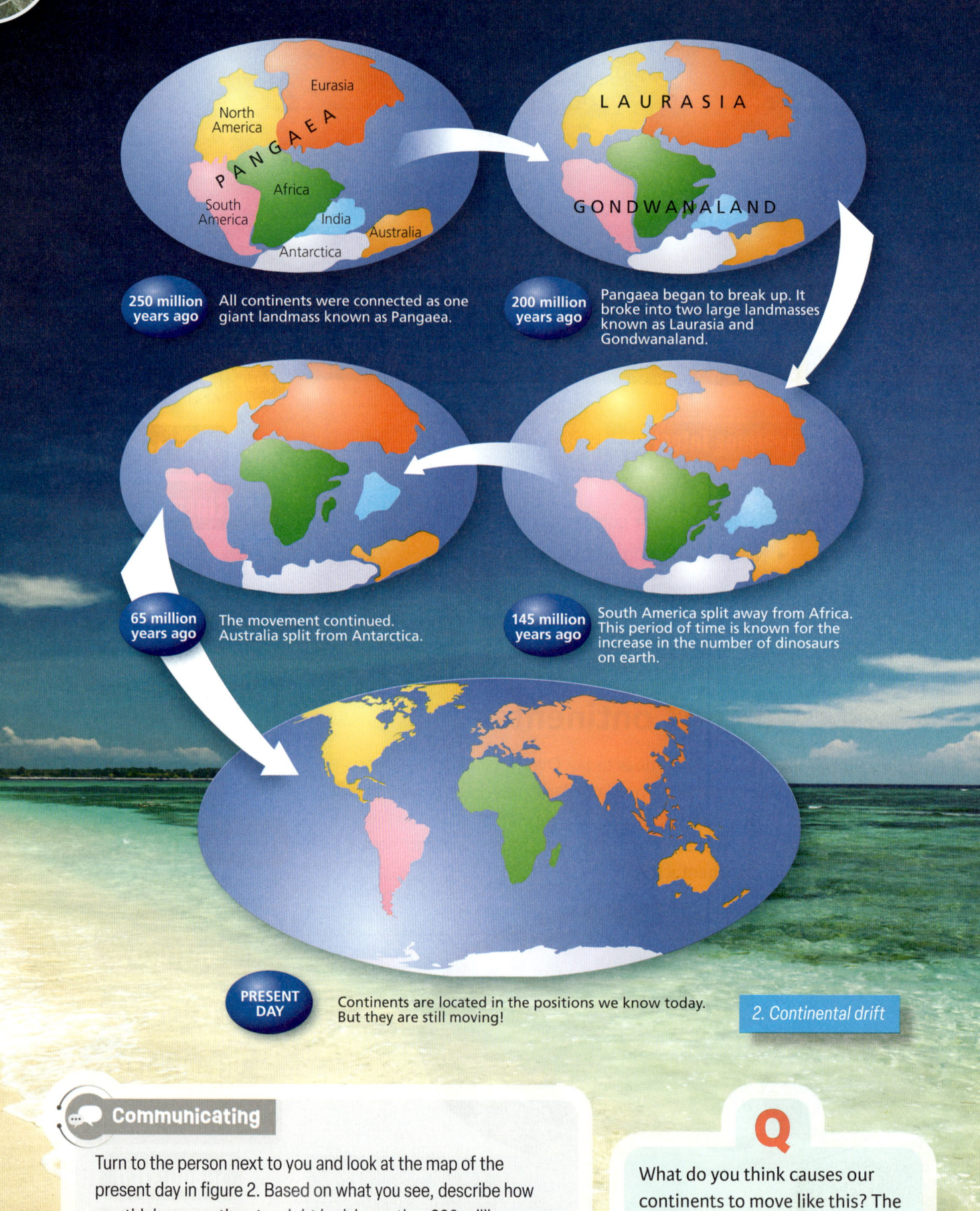

250 million years ago All continents were connected as one giant landmass known as Pangaea.

200 million years ago Pangaea began to break up. It broke into two large landmasses known as Laurasia and Gondwanaland.

145 million years ago South America split away from Africa. This period of time is known for the increase in the number of dinosaurs on earth.

65 million years ago The movement continued. Australia split from Antarctica.

PRESENT DAY Continents are located in the positions we know today. But they are still moving!

2. Continental drift

Communicating

Turn to the person next to you and look at the map of the present day in figure 2. Based on what you see, describe how you think our continents might look in another 200 million years.

Q What do you think causes our continents to move like this? The answer lies deep within our earth.

1.2 Structure of the Earth

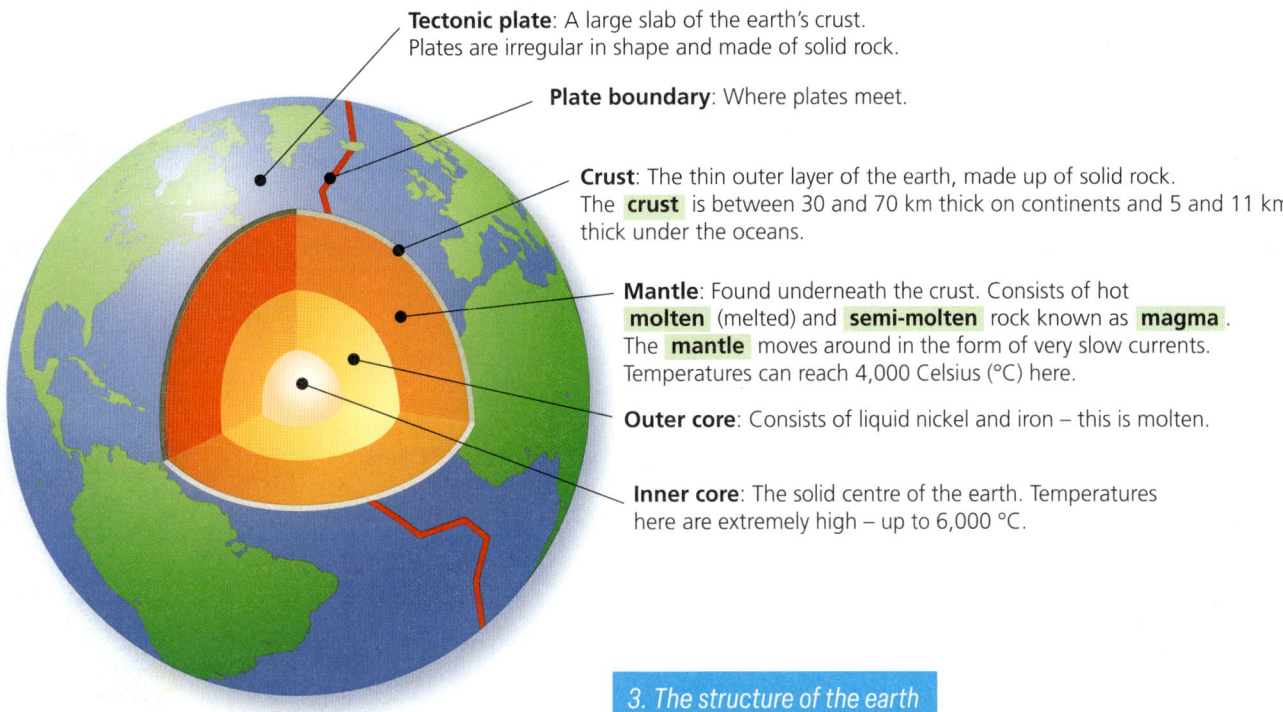

Tectonic plate: A large slab of the earth's crust. Plates are irregular in shape and made of solid rock.

Plate boundary: Where plates meet.

Crust: The thin outer layer of the earth, made up of solid rock. The crust is between 30 and 70 km thick on continents and 5 and 11 km thick under the oceans.

Mantle: Found underneath the crust. Consists of hot molten (melted) and semi-molten rock known as magma. The mantle moves around in the form of very slow currents. Temperatures can reach 4,000 Celsius (°C) here.

Outer core: Consists of liquid nickel and iron – this is molten.

Inner core: The solid centre of the earth. Temperatures here are extremely high – up to 6,000 °C.

3. The structure of the earth

The Earth's Crust

The earth is made up of several different layers. The outer layer is known as the crust and is made of solid rock. The crust ranges from 5 to 70 km in depth and is split into a number of pieces known as plates. Some plates have oceans on top and are known as oceanic crust. This is dense, thinner and mainly made of basalt rock. Other plates have land on top of them. This is known as continental crust and it is less dense, thicker and mainly made up of granite. The point where plates meet is called a plate boundary.

FUN FACT!
Below Central Valley in California, US, the crust is only 20 km thick, but under the Himalayas it has a thickness of up to 70 km. Underneath the ocean, it has a thickness of 5–11 km.

The Himalayas

Junior Cycle Geography CYCLONE

1.3 Convection Currents

The earth's plates float on semi-molten (half-melted) magma in the mantle. It is the movement of the magma that causes plates to collide, separate or slide past each other. Within the mantle, magma moves in circular motions called **convection currents**. The plates move in relation to each other due to these convection currents.

1. The earth's **core** heats the magma above, causing it to rise slowly upwards towards the crust.
2. As the magma rises, it then begins to cool.
3. This cooling results in the magma becoming heavier and sinking back down towards the core.
4. The cycle repeats continuously, creating a circular movement of magma (convection currents) in the mantle.
5. These convection currents cause the plates to move very slowly. The plates are dragged apart (separate), are pushed together (collide) or slide past each other.

Imagine heating water in a pot on a hob. As the water comes to the boil, the wooden blocks move apart due to the convection currents moving.

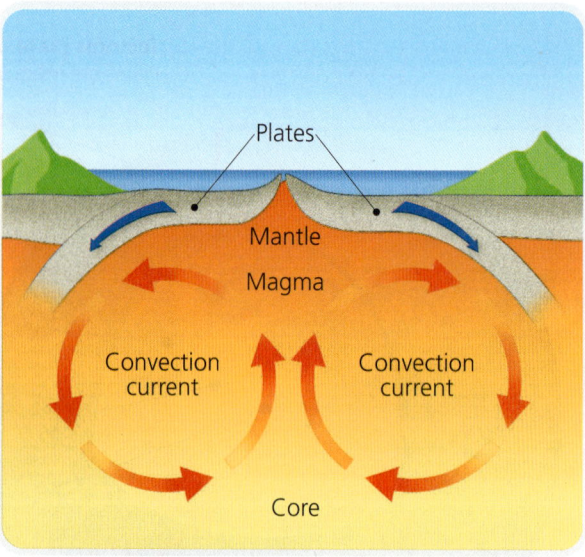

4. Convection currents

> **Managing information and thinking**
>
> Look back at figure 2 (p. 2). Thinking about how convection currents cause plates to move, can you identify two continents that were pulled apart due to convection currents in the mantle? Write them down and check your answer with the person next to you.

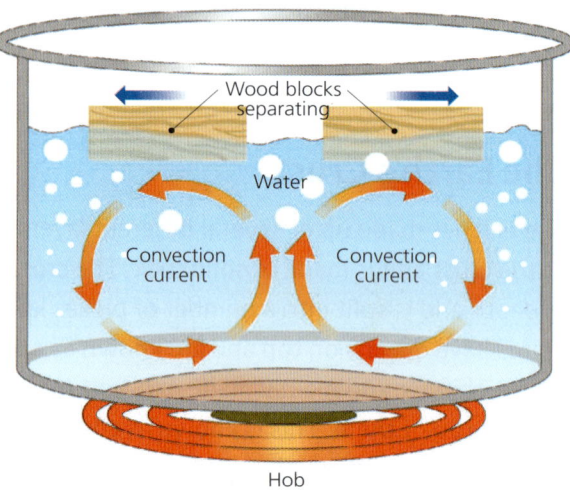

5. How convection currents work in a pot of boiling water

Question Time

1. What is continental drift?
2. Draw and label a diagram to show the layers of the earth. On your diagram, make sure to show the following:
   Crust, Mantle, Inner core, Outer core, Plate boundary, Plate
3. Explain what a plate boundary is.
4. With the aid of a labelled diagram, explain how convection currents cause plates to move.

Exam hint!

In Q4, 'with the aid of a labelled diagram' means draw a diagram and write a written explanation of what is being asked.

1.4 Plate Tectonics

Plate tectonics is the theory that the earth is made up of large, moving plates of solid rock. This map shows the world's main plates, their boundary lines and the direction in which they move.

FUN FACT! Juan de Fuca Plate is one of the smallest of the earth's tectonic plates, while the Pacific Plate is the largest.

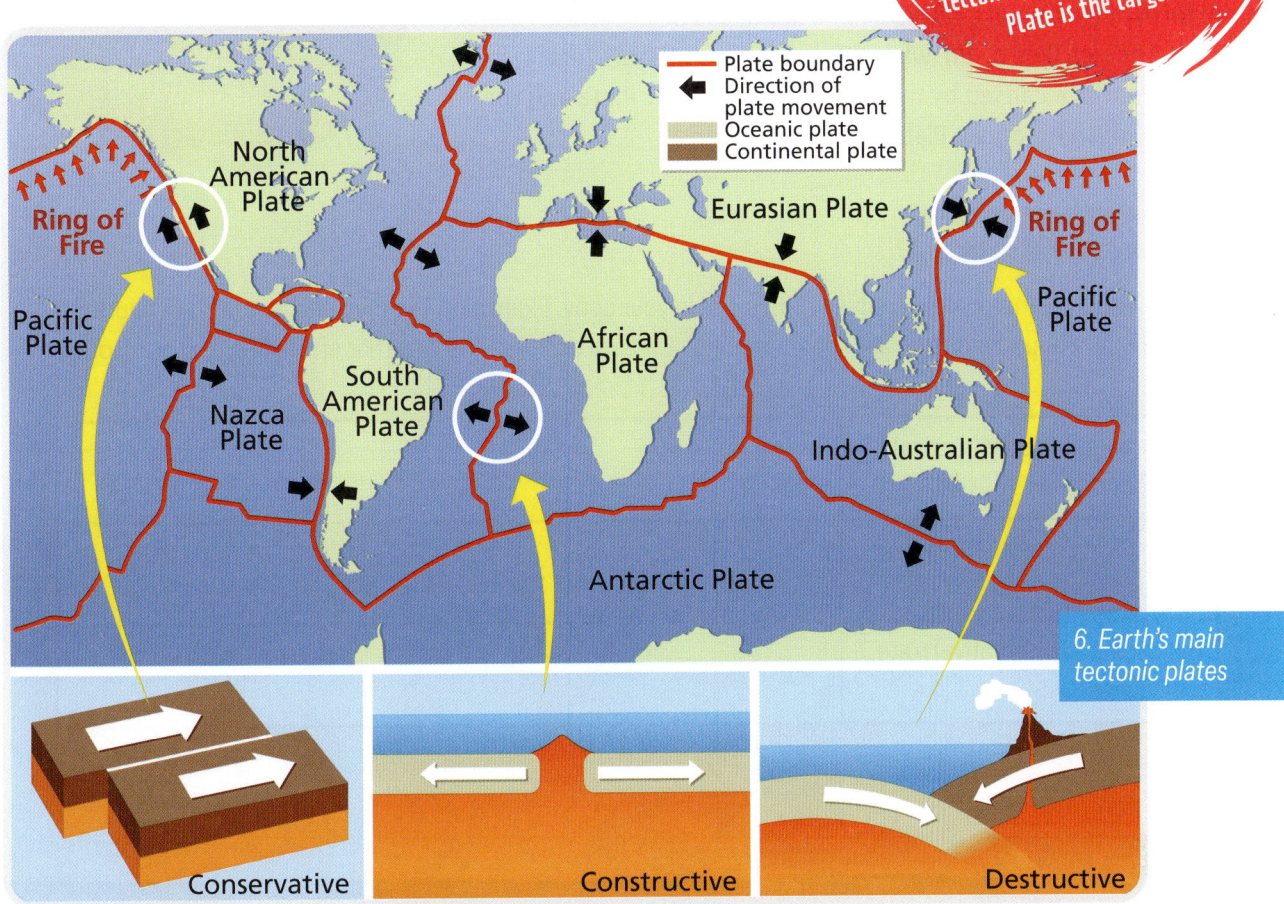

6. Earth's main tectonic plates

Separating Plates (Constructive plate boundary)

At a **constructive** plate boundary, two plates separate, and magma wells up to fill the space between. The magma cools and becomes solid, which forms a new crust. Volcanic islands, volcanic mountains and mid-ocean ridges form wherever plates separate. The Mid-Atlantic Ridge (see p. 11) is an example of where two plates are separating – here the North American and Eurasian Plates are pulling apart.

Colliding Plates (Destructive plate boundary)

At a **destructive** plate boundary, plates collide, and the crust is destroyed. When an **oceanic plate** collides with a **continental plate**, the heavier oceanic plate is forced under the lighter continental plate (**subduction**) and down into the hot mantle below. This causes part of the oceanic plate to melt and part of the continental plate to buckle. These processes form fold mountains and volcanic mountains. For example, the Andes Mountains have formed where the Nazca Plate and South American Plate collide.

Exam hint!
Remember: constructive = to construct/build new crust and destructive = crust is destroyed.

Sliding Plates (Conservative plate boundary)

At a **conservative** plate boundary, two plates slide past each other, and their edges sometimes become locked together. As pressure builds up, the edge of one plate may snap or jolt forward suddenly. Great waves of energy are then released, causing the earth to tremble in the form of earthquakes. This occurs along the San Andreas Fault in North America (see p. 25), where there is a deep fissure (crack) in the earth's surface caused by the sliding of the Pacific Plate and the North and South American Plates.

FUN FACT! Plates move at an average of 2–5 cm per year, which is about as fast as your fingernails grow!

Working with others

Working with a partner, create a table like this one in your copy. Examining the map and diagrams in figure 6, name two plate boundaries that are separating, two that are colliding and two that are sliding past each other.

Two separating (constructive) plate boundaries	1. 2.
Two colliding (destructive) plate boundaries	1. 2.
Two sliding (conservative) plate boundaries	1. 2.

Question Time

1. Name the plate boundary where Ireland is located.
2. Name the three types of plate boundaries.
3. (a) Draw a labelled diagram of each type of plate boundary. On your diagrams, try to use the following terms:
   Plate, Convection currents, Magma, Direction of movement, Constructive, Destructive, Conservative
   (b) Copy and complete the table below into your copy.

Plate Boundary	Colliding (Destructive)	Separating (Constructive)	Sliding (Conservative)
Features associated with this boundary			
Example of where this boundary is found			

1. Dynamic Earth

Exam hints!

Q1 is sample exam-style short question. This type of question requires a brief answer. You may be presented with a diagram such as this and be asked to name, provide or place the labels in the correct place.

Sample questions

1. **Examine** the image below showing the internal structure of the earth.

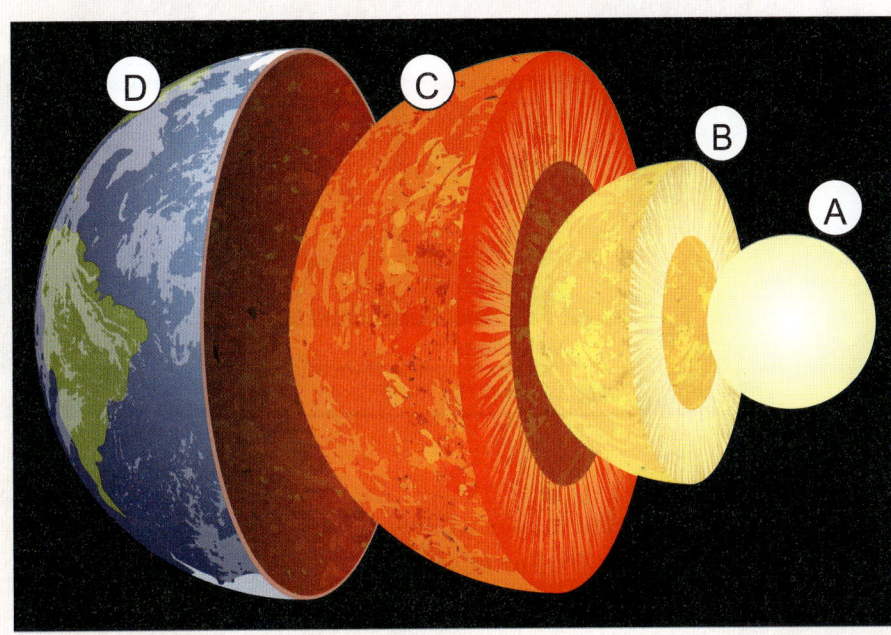

(a) **Name** the layers of the earth A to D as shown on the diagram above.
(b) **What** is the name of the process responsible for the movement of the plates?
(c) **What** term is given to the type of boundary where two plates slide alongside one another?

2. Review the map and answer the questions below.

Map showing tectonic plate boundaries with volcanoes, hot spots (present locations), and earthquake zones. Labels include San Andreas Fault, Pacific Ocean, Atlantic Ocean, Indian Ocean, and Equator.

(a) The area marked A on the map is where two plates are colliding. True or false?

(b) Using your pencil, **draw** arrows to show the direction of plate movement along the San Andreas Fault.

(c) **Describe** the pattern of tectonic activity along the San Andreas Fault.

(d) **Describe** how plates move at conservative plate boundaries.

(e) **Explain** why volcanic eruptions and earthquakes can occur at destructive plate boundaries. You may use evidence from the map to support your answer.

Exam hints!

The action verb 'Describe' requires you to build a detailed picture, using words and diagrams if appropriate.

You can use words and diagrams to show how plates are moving at this type of boundary.

In part (e), you are told you can use map evidence. Use the key to locate this type of boundary and name it, explaining what you find there. Use your geographical knowledge to explain what happens at a destructive boundary — look back to figure 6 on p. 5 to help you.

Learning Outcomes: 1.1, 2.1, 2.8, 2.9

2

Volcanic Activity

Learning Intentions

You will be able to:
- Describe and explain the formation of a volcano
- Describe why and where volcanic activity occurs
- Name the stages in the life cycle of a volcano
- Describe the economic and social impacts of the volcanic eruption in Iceland and how the country responded to a natural disaster.

- Volcanologist
- Vent
- Eruption
- Magma chamber
- Cone
- Crater
- Active
- Dormant
- Extinct

2.1 What Volcanologists Do

A **volcanologist** is a scientist who studies volcanoes. Volcanologists study volcanic rock, ash and gases. They monitor changes in the shape of volcanoes to try to understand why and how they erupt, and to predict when and where they might erupt in the future.

Working with others

    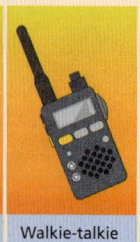    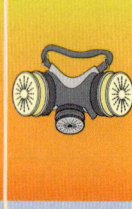

Thermal suit | Video camera | Rock hammer | Digital camera | Walkie-talkie | Heat-resistant gloves | Fire helmet | Metal bucket | Gas mask

1. Equipment used by volcanologists

The images above show some of the many different pieces of equipment used by a volcanologist when working at a volcano:

Thermal suits, video cameras, rock hammers, digital cameras, walkie-talkies, heat-resistant gloves, fire helmets, metal buckets and gas masks.

Working with a partner, jot down what you think the use is for each instrument mentioned.

2.2 Volcanic Mountains

Most volcanoes are cone-shaped mountains. They start out as a crack in the earth's crust.

FUN FACT!
The word 'volcano' comes from the name of the Roman God of fire, Vulcan.

The Formation of Volcanic Mountains

 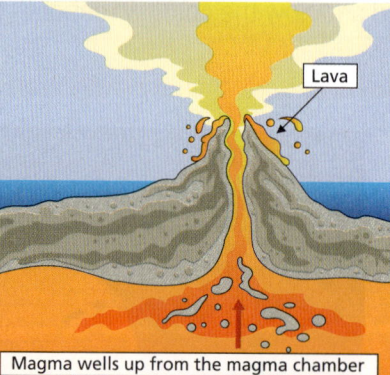 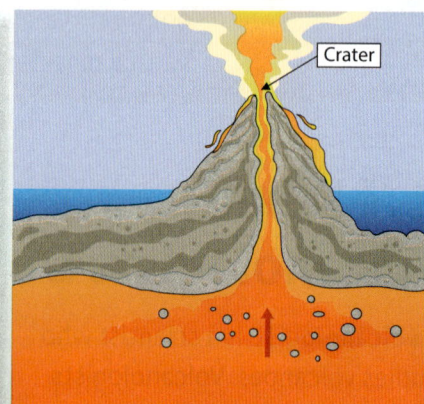

When two plates collide or separate, magma can rise from the mantle to fill the space that opens in the crust – the **vent**. We call this an **eruption**, and it can be violent due to the great pressure that the magma is under in the mantle.

When magma wells up from the **magma chamber** and reaches the surface, it is known as lava. The lava then cools and hardens.

After many eruptions, layers of lava and ash build up around the vent, and a **cone**-shaped mountain is formed with a **crater** at the top from which the lava flows.

2. The formation of a volcanic mountain

The Structure of a Volcano

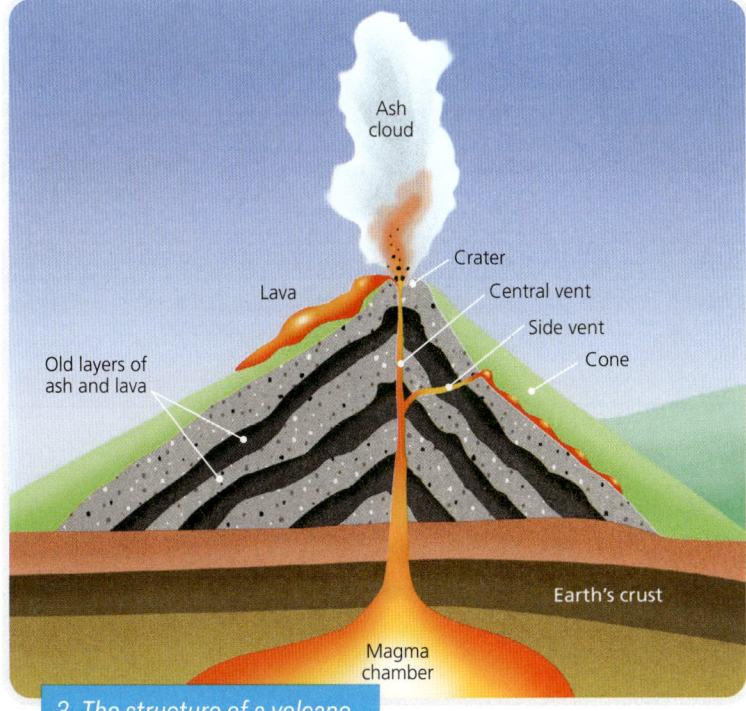

3. The structure of a volcano

Crater: this is the mouth of the volcano, where ash, lava, hot rocks and gases escape from.

Central vent: this is connected to the magma chamber below. It is the main passageway for magma to travel to the earth's surface.

Magma chamber: this contains the molten material below the surface.

Lava: this is the name given to the molten material once it reaches the surface of the earth.

Cone: this is the cone-shaped hill formed over time when the layers of ash, lava and volcanic material build up.

Distribution of Volcanoes

As the plates of the earth's crust separate or collide at constructive or destructive plate boundaries, volcanoes can occur.

At Constructive Plate Boundaries

- At constructive boundaries, plates are separating and pulling apart. Magma can rise to the surface through the crust, where it begins to cool and harden.
- When plates separate under the ocean, lava can pour out along cracks on the earth's surface, forming a ridge. This lava can build up and become what is called a mid-ocean ridge.
- In the middle of the Atlantic Ocean, there is a narrow chain of mountains called the Mid-Atlantic Ridge. Here, the Eurasian and North American Plates are slowly moving apart. As magma broke through the overlying crust, it formed this long ridge of underwater (submarine) mountains and volcanoes.
- Over time, some of these rose above the surface of the sea to form volcanic islands. Iceland is an example of a volcanic island.

The Mid-Atlantic Ridge

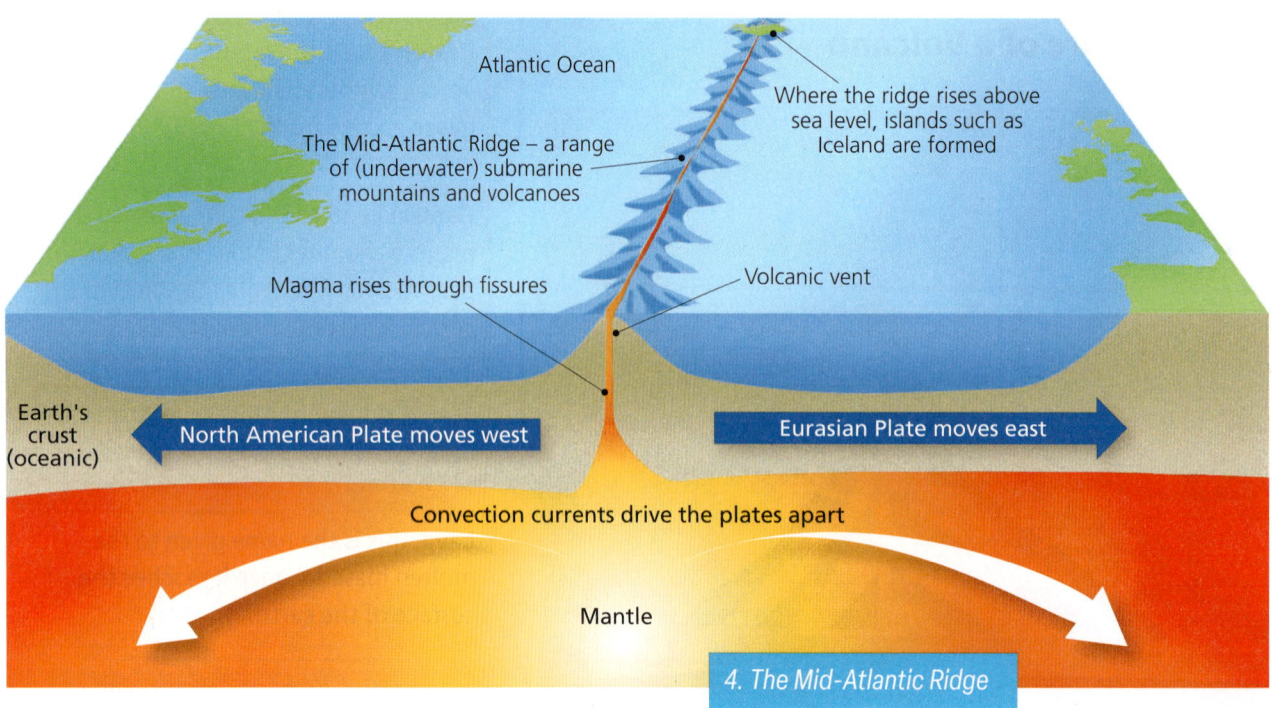

4. The Mid-Atlantic Ridge

At Destructive Plate Boundaries

- At destructive boundaries, plates are colliding with each other.
- When an oceanic plate collides with a continental plate, the heavier oceanic plate is forced down into the mantle below. The oceanic plate begins to melt.
- As it melts, magma can rise up to the surface through cracks in the crust above.
- The magma will continue to erupt and a volcanic mountain can form at the surface.
- Mount Saint Helens in the US is an example of this.

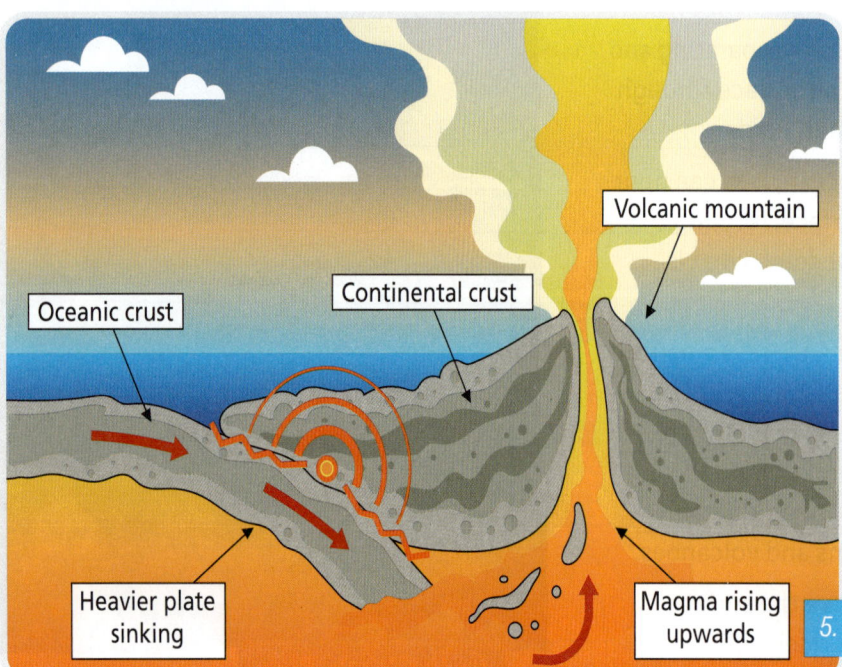

5. Volcanic activity at a destructive boundary

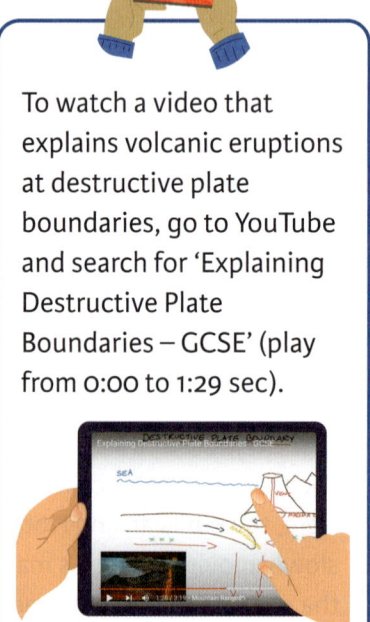

To watch a video that explains volcanic eruptions at destructive plate boundaries, go to YouTube and search for 'Explaining Destructive Plate Boundaries – GCSE' (play from 0:00 to 1:29 sec).

2. Volcanic Activity

📁 Managing information and thinking

Can you correctly identify the parts labelled A and B on the photograph of the volcanic mountain below?

2.3 The Life Cycle of a Volcano

Volcanoes go through three different stages.

6. The life cycle of a volcano

1. An **active** volcano is one that has erupted in the past 10,000 years. **Example:** Mount Etna in Italy

2. A **dormant** volcano has not erupted in the past 10,000 years but is expected to erupt again in the future. **Example:** Kilimanjaro in Tanzania

3. An **extinct** volcano is one that is not expected to erupt again. **Example:** Slemish, Co. Antrim

Junior Cycle Geography CYCLONE

> **Being creative**
>
> With a partner, devise a plan showing how you would use recycled materials to create a model of a volcano. What materials would you use? Decide how you would construct each part of the volcano, so that it would show clearly what it is.

Question Time

1. Explain briefly each of the following terms: (a) volcanologist, (b) crater, (c) cone, (d) magma.
2. Draw and label a diagram of the structure of a volcano. On your sketch, you must show and label the following items:
   Magma chamber, Central vent, Cone, Crater, Ash cloud, Layers of ash and lava, Side vent
3. Name a volcanic island found on a constructive plate boundary.
4. With the aid of a labelled diagram, describe how plate movement at destructive boundaries can lead to the formation of volcanoes.
5. Name two active volcanoes and state where they can be found.
6. Describe the life cycle of a volcano, using diagrams if you wish.
7. What does the fact that Ireland once had active volcanoes like Slemish tell us about Ireland's past location and continental drift?

2.4 The Pacific Ring of Fire

Figure 7 shows that active volcanoes and earthquakes occur near the meeting of plate boundaries. The Pacific Ring of Fire is a horseshoe-shaped area containing many plate boundaries. Because of the constant movements of plates, this region is home to almost 75% of the earth's active volcanoes and is where almost 90% of the world's earthquakes occur.

1. Why do you think this area is called the Pacific Ring of Fire?
2. Explain why volcanoes are generally found at constructive and destructive plate boundaries.
3. Thinking about what you have learned so far, and referring to figure 6 on p. 5, why do you think Ireland does not have any active volcanoes?

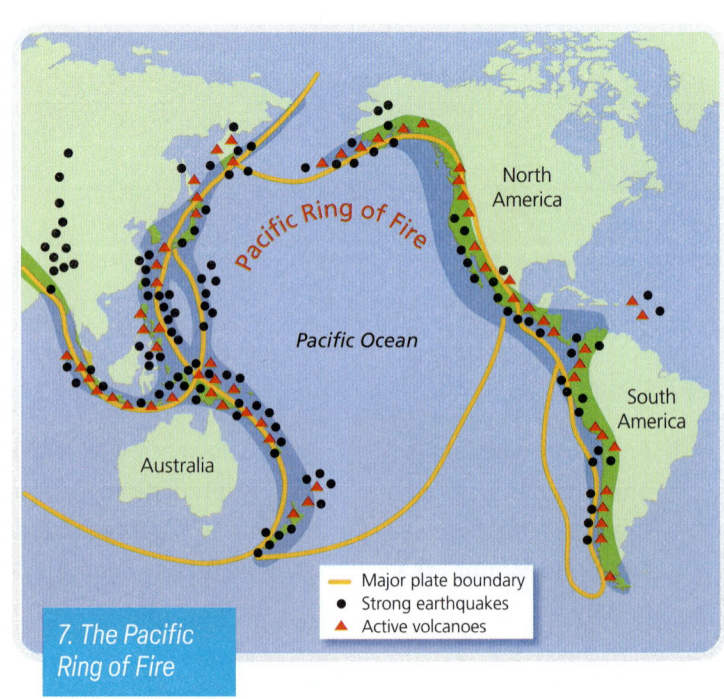

7. The Pacific Ring of Fire

2. Volcanic Activity

 For a video on plate boundaries and the Pacific Ring of Fire, go to YouTube and search for 'Plate Tectonics – Smithsonian's National Museum of Natural History' (4:28).

 On your smart device, download one of the free volcano and earthquake monitoring apps. Over the next week, track every volcanic eruption or earthquake using the app. Record at least two of these in your copy. Look up the location and find out what plate boundaries are located close by. Compare your record with your classmates'.

2.5 The Economic and Social Impacts of Volcanoes

Case Study: The Eyjafjallajökull Volcano in Iceland

Iceland is a volcanic island located on the Mid-Atlantic Ridge. Here, convection currents are pulling the North American and Eurasian Plates apart at a constructive plate boundary. This has resulted in many volcanoes, including Eyjafjallajökull. It is a volcano that is covered by an ice cap.

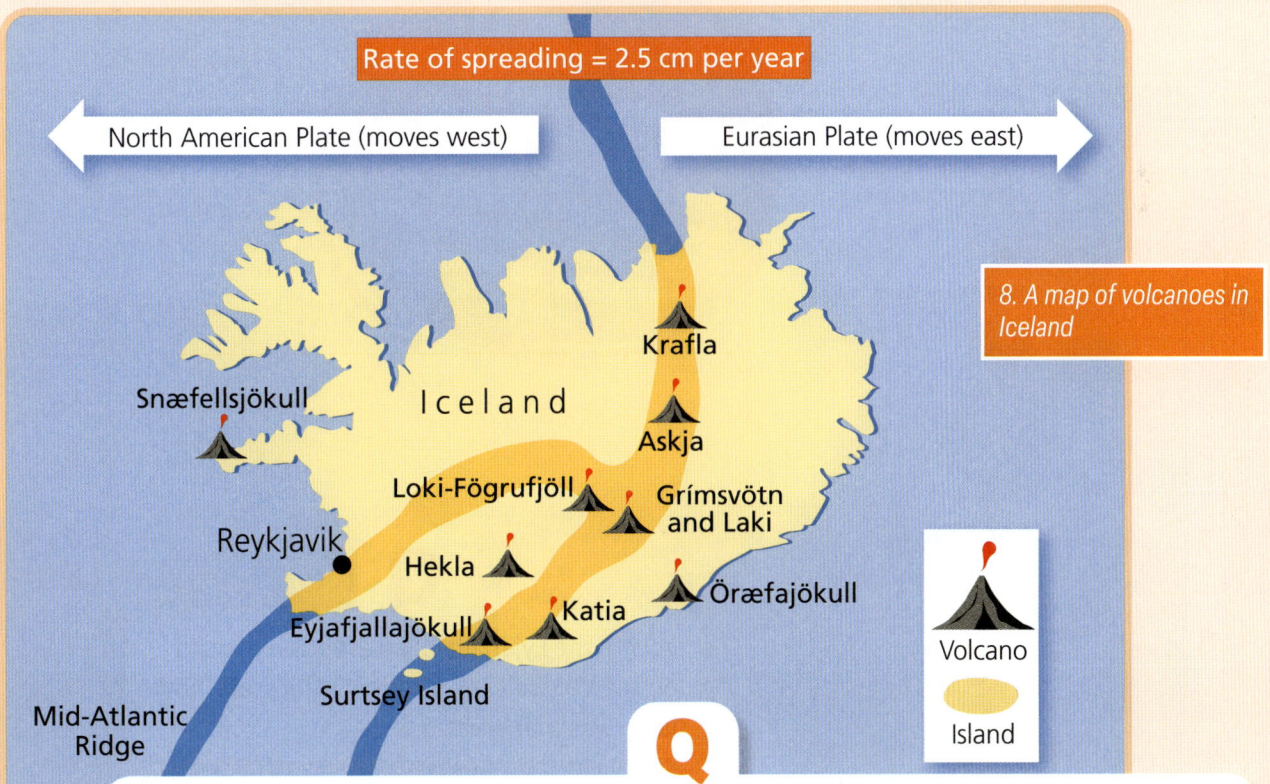

8. A map of volcanoes in Iceland

Q

1. Locate the Eyjafjallajökull volcano on the map. Where is it in relation to Iceland's capital city?
2. There are a number of active volcanoes in Iceland. How do you think the people living in the surrounding areas have adapted to life near a volcano?

Junior Cycle Geography CYCLONE

On 14 April 2010, Eyjafjallajökull had a violent and explosive eruption. The volcano had been dormant for 180 years. Very little lava came from the volcano, but huge amounts of ash were sent many kilometres into the air. The ash was then caught by winds and blown south-eastwards towards Europe. The eruption registered as a 3 to 4 on the VEI scale (see Fun Fact! below for an explanation). The eruption had many economic (financial) and social (affecting people) impacts.

Q Using an atlas or Google Maps if necessary, name the countries that are either fully or partially covered by ash cloud.

9. The span of the ash cloud caused by the eruption of Eyjafjallajökull

VEI		Examples
small non-explosive	1	2006, Raoul Island
	2	2010, Sinabung
moderate	3	2011, Nabro
large	4	2010, Eyjafjallajökull
	5	1980, Mount Saint Helens
very large	6	1883, Krakatoa
	7	1815, Tambora
	8	640,000 BCE, Yellowstone

FUN FACT! Volcanoes have a scale called the Volcanic Explosivity Index (VEI). The index gives us a way to measure volcanic eruptions. The scale goes from 0 to 8, with 0 being the weakest and 8 being the strongest.

Q Locate the Eyjafjallajökull eruption of 2010 on the VEI scale. According to the scale, how large was that eruption?

BCE = Before Common Era and is the equivalent of BC
CE = Common Era and is the equivalent of AD

2. Volcanic Activity

 Communicating

Take two minutes to look at the images A, B and C here and to think about what is happening in each. Then, working in pairs, discuss each image. Describe what you see and debate what might be happening in the images and the reasons why. Be prepared to explain your reasoning to your teacher.

Negative Impacts of the Volcanic Eruption

Social Impact

Contaminated water supply

Ash from the eruption polluted the local water. Farmers near the volcano could not let their animals drink from the streams. Farmers were only allowed to re-enter the hazard zone for two-hour periods each day so they could check on their cattle.

Evacuation of locals

Five hundred farmers and their families had to leave their homes. Locals and rescuers had to wear face masks to avoid choking.

Health concerns

A few weeks after the eruption, it was reported that exposure to the ash fall was associated with breathing difficulties for some locals. Many parents living close to the eruption reported that their children were worried or anxious following the experience.

Economic Impact

Travel chaos

Over eight days, 100,000 flights to and from Europe were cancelled because of poor visibility and the fear that ash would clog up the engines of aeroplanes. Twenty countries closed their airspace. This cost airlines and businesses about €145 million per day.

Loss of earnings

The president of the World Bank estimated that African countries lost over $65 million as a result of the airspace shutdown.

Countries, such as Kenya, that sell fresh fruit, vegetables and flowers to Europe were affected. Fresh produce is transported by plane so it can arrive quickly. The flight ban meant that this produce could not be sold. Over 1 million flowers went unsold in the first two days of the eruption.

> Can you think of two reasons why tourists would like to visit areas that have volcanoes?

Positive Impacts of the Volcanic Eruption

Social Impact

Increased tourism and employment

After the eruption, a campaign led to a big increase in the number of foreign tourists. In 2008, Iceland was visited by fewer than 500,000 people. In 2016, the number had increased to 1.8 million.

Reduced unemployment

The development of tourism as a result of the volcanic eruption created jobs.

Reduced carbon emissions

According to the Environmental Transport Association, the grounding of European flights prevented approximately 2.8 million tonnes of carbon dioxide entering the atmosphere.

Economic Impact

Income generated

A new Eyjafjallajökull visitor centre is generating income for the local community.

Fertile soil

Lava and ash are rich in nutrients, making the soils in volcanic areas very fertile and good for agricultural use.

Plankton bloom

Ash from the eruption was deposited into the North Atlantic Ocean. The ash contained iron, which was very good for the growth of a type of plant plankton. These are eaten by fish but they also take carbon dioxide out of the atmosphere and produce oxygen.

Response to the Natural Disaster

Short Term

- **Evacuation**: The area surrounding the eruption was evacuated. An emergency alert was issued and those living in the zone deemed most at risk were told to leave their farms and homes.
- **Emergency aid**: The Red Cross organised volunteers to help people who were impacted by the eruption. They provided food, emergency shelter, counselling and support.

Long Term

- **Planning and preparation:** Plans had been in place since 2002 to prepare for an eruption at Eyjafjallajökull. Locals were able to attend local community meetings to hear directly from the government officials about what they should do and how they should prepare for an eruption. Education about the hazards of volcanic eruptions and how to evacuate is very important in keeping people safe.
- **Repairs:** Areas impacted by heavy ash fall, lava flows, contaminated water supplies or loss of services all need to be correctly reconnected and checked following an eruption. This took a number of weeks following the eruption.

Volcanic Eruptions and Sustainability

Geothermal Energy

The many volcanoes in Iceland can be a source of heat and electricity. Geothermal energy (heat from the ground) is used to heat 90% of homes in Iceland. This source of energy is very efficient and saves households a lot of money.

Iceland uses its geothermal energy in the following ways:

- For heating public spaces and houses
- To generate electricity
- To heat up the country's many geothermal spas and pools
- To warm up streets so they do not get slippery in winter
- For greenhouses, so they can grow organic fruits and vegetables.

Go to YouTube and search for 'Power of Iceland – Iceland Academy' to watch a video explaining how geothermal energy is used in Iceland (1:37).

Go to YouTube and search for 'Eyjafjallajökull case study – Discover the World Education' to learn more about the impact of the volcanic eruption on Iceland (15:09). Look out for another positive impact of the volcano not discussed here.

Question Time

1. Why does Iceland have so many active volcanoes?
2. How long had Eyjafjallajökull been dormant before it erupted in 2010?
3. Describe two positive social impacts of the Eyjafjallajökull eruption.
4. Describe two negative economic impacts of the eruption.
5. What is geothermal energy and state two ways that it is used in Iceland.

Junior Cycle Geography **CYCLONE**

EXAM FOCUS

Sample questions

1. Imagine you are a journalist sent to the site of a recent volcanic eruption. Write a report **describing** some of the negative impacts of the volcanic eruption.

Lets get started! Sample starter

> **Volcanic Eruption Report – Journalist Katie O'Sullivan**
>
> Last week a large volcanic eruption occurred at Hunga Tonga volcano. It was placed at 6 on the VEI scale.
>
> The negative impacts that I will discuss are: Tsunami wave damage, ash-contaminated drinking water, businesses forced to close, ash covering the land
>
> The eruption triggered a tsunami and this destroyed many homes at the coast and 3 people were killed....

... now complete the answer in your copy.

2. Volcanic eruptions in an area can bring destruction but they can also provide many benefits. **Describe** two advantages of volcanic activity in an area that you have studied.

3. Natural disasters such as volcanic eruptions and earthquakes can be devastating for the areas in which they occur. Using a natural disaster that you have studied, **describe** one short-term and one long-term response following the event.

Exam hints!

When asked to write a report, give a very brief introduction and then get straight into giving the detail asked for in the question. Here, you should state each negative impact, and then expand on this by describing how it has negatively impacted socially, economically or environmentally.

Q2 is asking you to use a case study of an eruption that you have studied. Eyjafjallajökull is a suitable one to use. Start by stating two advantages and then describing how each had a positive impact locally or nationally. These can be related to positive economic, social or environmental impacts.

You can use any natural disaster that you have studied to answer Q3.

You should describe one short-term and one long-term response. Examples might be things such as emergency medical supplies, emergency shelter or aid for those impacted and in the longer term things such as rebuilding of homes, schools.

2. Volcanic Activity

CBA 1 Geography in the News

VOLCANIC HAZARDS AT TONGARIRO

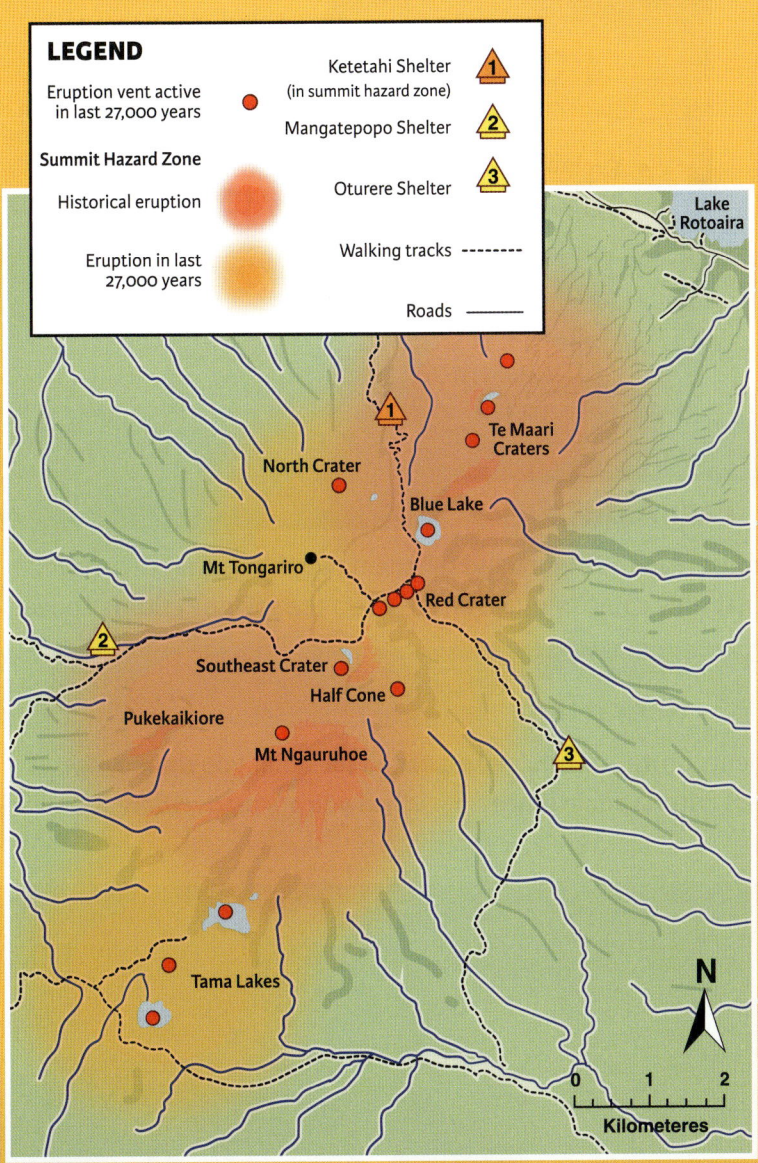

What to do!

When a volcanic eruption begins:
- You should make your way to safety by coming down the mountain and moving out of the hazard zones
- Keep out of valleys and try to stay on higher land on the ridges
- Keep away from the eruption vent
- Wait for the search and rescue team to inform you that is it safe to return to your home.

Volcanic hazards

Social Impacts
- Habitats may be damaged or destroyed by lava and ash in the hazard zones.
- People living in hazard zones will be evacuated and may be unable to return to their homes for long periods until damage can be assessed.

Environmental Impacts
- Vegetation and animal life may be badly impacted within the hazard zone.
- Ash can travel large distances by wind. We advise that N95 protective face masks are worn by all in the aftermath of the eruption.
- Ash and lava will provide natural fertilisation in time for the areas in which they are deposited.

Lava Flows
- Lava flows of molten rock are very hot.

Ash Fall
- Any place on this map is at risk from ash fall in an eruption – this will obscure vision and make it hard to breath, but is non-lethal.

You are a volcanologist working for the government at Mount Tongariro on the North Island of New Zealand. You must write an email to the local county council to report on recent activity at Mount Tongariro and give advice on how locals should prepare for any future eruptions.

This is an individual task.

Junior Cycle Geography CYCLONE

I Must

- Introduce myself and outline the recent activity on Mount Tongariro
- Outline which areas are likely to be impacted greatest by a volcanic eruption
- Describe three things that locals should be advised to do when an eruption occurs. You can use information provided in the graphic alongside your own knowledge to complete this.

I Should

- Explain one environmental and one social impact that could occur if the volcano erupts.

I Could

- Include images of the volcano, and additional interesting facts or figures.

Peer Assessment

Swap your email with a classmate. You must read their work and tell them two things that they have done very well, and suggest one thing that they could improve on based on the success criteria above.

Redrafting

Taking on board your classmate's comments on your work, make any changes that you think will improve your letter, reviewing the success criteria again if need be. When you are ready, show your work to your teacher.

Learning Outcomes: 1.1, 2.1, 2.8

Earthquakes

3

Learning Intentions

You will be able to:
- Describe the occurrence and distribution of earthquakes
- Explain how an earthquake forms
- Outline how earthquakes are measured
- Describe how a tsunami wave occurs
- Describe the economic and social impacts of earthquakes in Chile and Nepal
- Describe the response to these natural disasters
- Explain ways to reduce the damage caused by earthquakes.

key words

- Friction
- Focus
- Shock waves
- Tremor
- Epicentre
- Seismologist
- Fault
- Tsunami
- Seismograph
- Richter scale
- Magnitude
- Moment magnitude scale
- Aftershock

3.1 Distribution of Earthquakes

An earthquake is the sudden shaking of the earth caused by the movement of the earth's crust.

Most earthquakes occur along the edges of large plates, at destructive or conservative boundaries where they are colliding or sliding past each other (see pp. 5–6). As the plates push against each other, they create great **friction**. This causes energy to build up between the plates. The plates then slip at a point in the crust called the **focus**. **Shock waves**, also known as **tremors** or seismic waves, spread out from the **epicentre**, which is the point on the earth's surface above the focus.

Seismologists are scientists who study earthquakes. They use specialised equipment to measure activity in earthquake zones. They can then use this information to predict (estimate) when and where an earthquake may occur.

Junior Cycle Geography CYCLONE

👍 Staying well

With the person next to you, examine these images A–D and then answer the following questions:

A — Before

After

FUN FACT! An average earthquake lasts around one minute.

B

Q

1. Describe what is happening or has happened in photographs A, B and C.
2. How do you think locals responded to damage shown in each photograph A–C?
3. Dr Lucy Jones, shown in photograph D, is a retired seismologist. What do you think are two benefits of seismologists studying earthquakes?

C

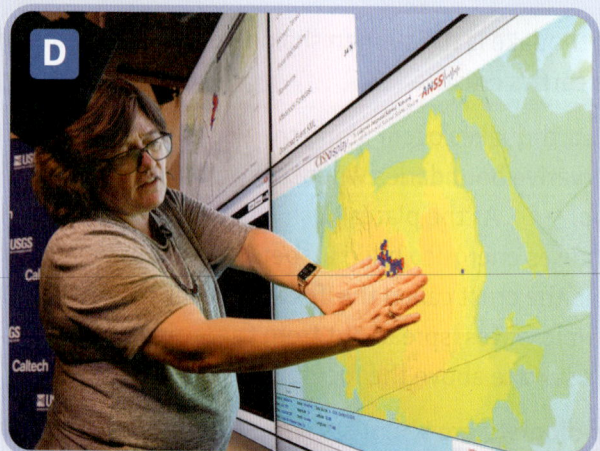
D

3.2 Formation of an Earthquake

Figure 1 shows the features and impacts of an earthquake.

1. **Focus**: This is where an earthquake begins beneath the earth's surface.
2. **Shock waves**: Also called seismic waves or tremors, these spread out in circles from the focus.
3. **Epicentre**: This is the point at the earth's surface that is directly above the focus of the earthquake. The strength of the earthquake is greatest here.
4. **Fault**: This is a large crack in the earth's crust where plates get stuck and energy builds up between them.

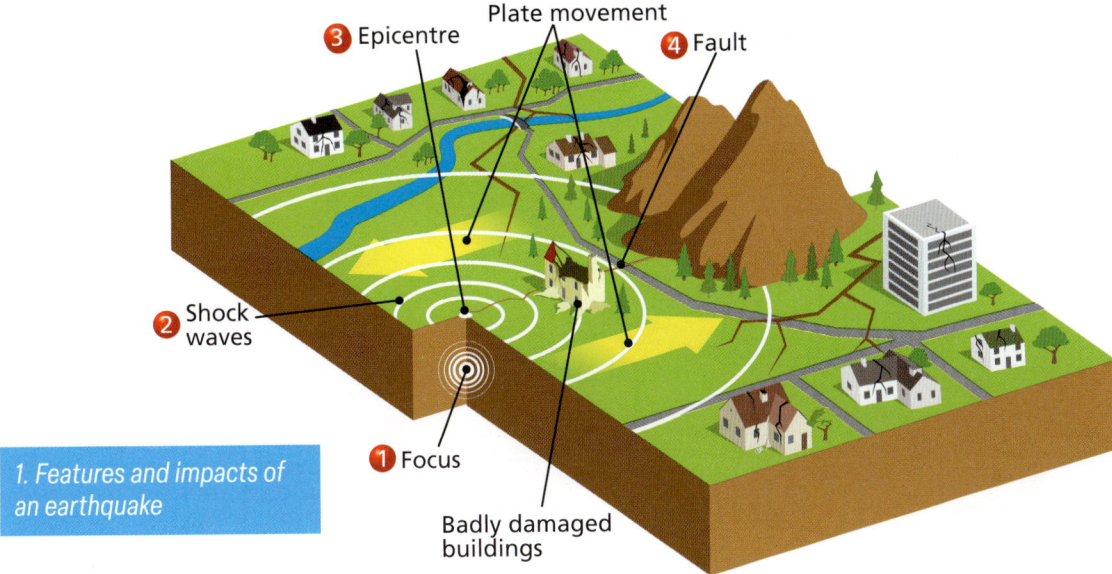

1. Features and impacts of an earthquake

The San Andreas Fault

The San Andreas Fault in California is a large crack in the earth's crust which marks part of a plate boundary. The slow-moving North American Plate is sliding in the same direction as the faster-moving Pacific Plate. The plates are always moving, but where they touch each other, they often lock together. Eventually, one plate will jump forward suddenly or slip or break, causing an earthquake.

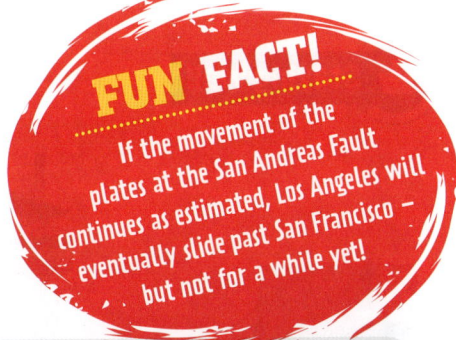

FUN FACT! If the movement of the plates at the San Andreas Fault continues as estimated, Los Angeles will eventually slide past San Francisco – but not for a while yet!

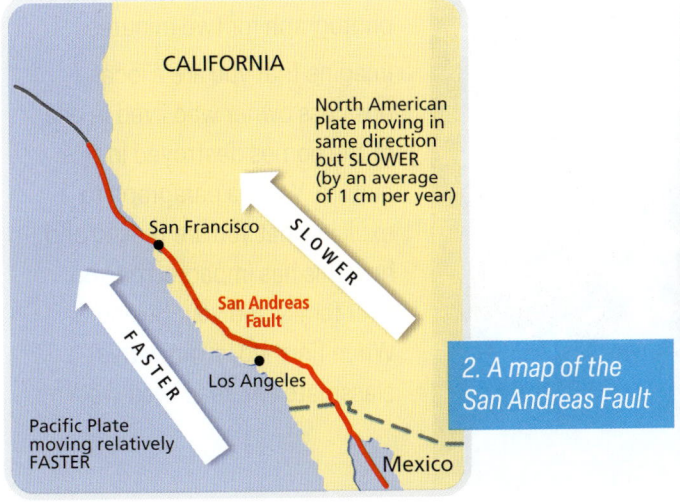

2. A map of the San Andreas Fault

An aerial photo of the San Andreas Fault

3.3 Tsunami Waves

When an earthquake occurs under the sea, it can trigger a **tsunami**. A tsunami is a huge tidal wave that moves very quickly towards the shore. These waves can cause a huge amount of damage along the coastline and can result in many deaths. See figure 3 for how a tsunami happens.

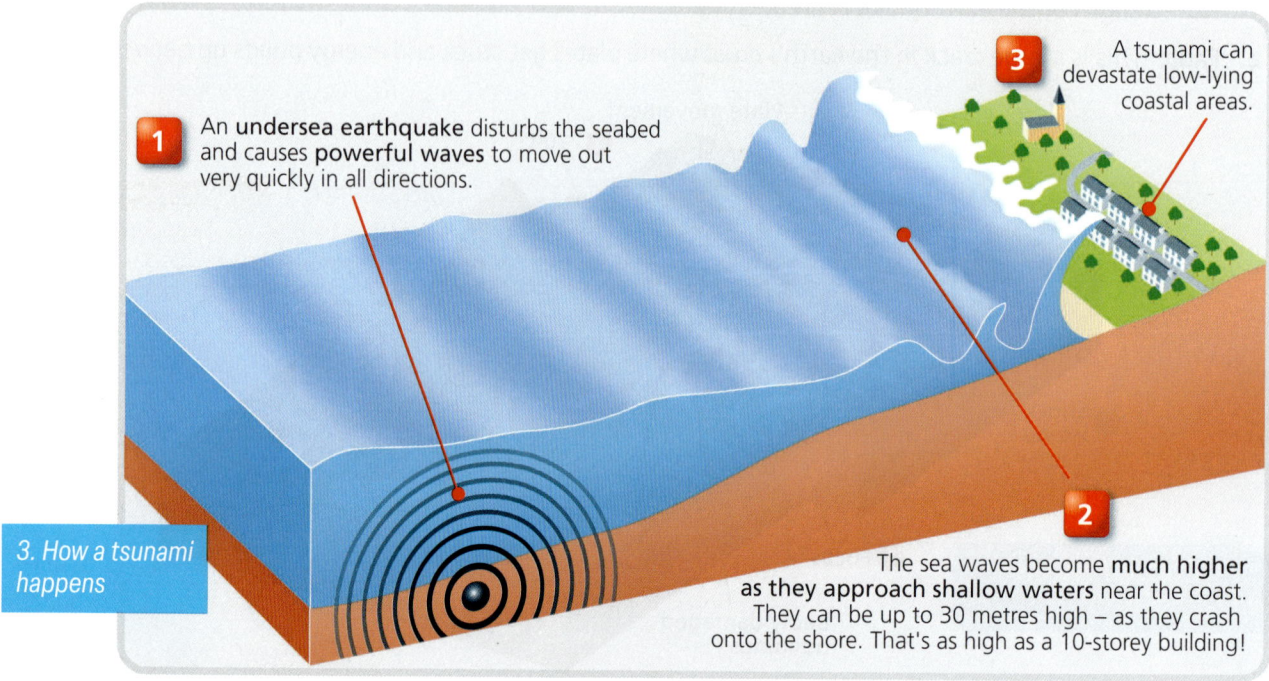

3. How a tsunami happens

Staying well

On 11 March 2011 the Great Sendai Earthquake, with a magnitude of 9, struck off the coast of Japan, triggering a series of large tsunami waves.

First, locate Japan using an atlas or Google Maps. Then study this photograph for two minutes.

Imagine that you are a local business owner who lived in one of the homes destroyed in this picture. Write a paragraph to explain what you are feeling and how this has impacted on you, your family and your business. When finished, swap with a classmate and read their work.

3.4 Measuring Earthquakes

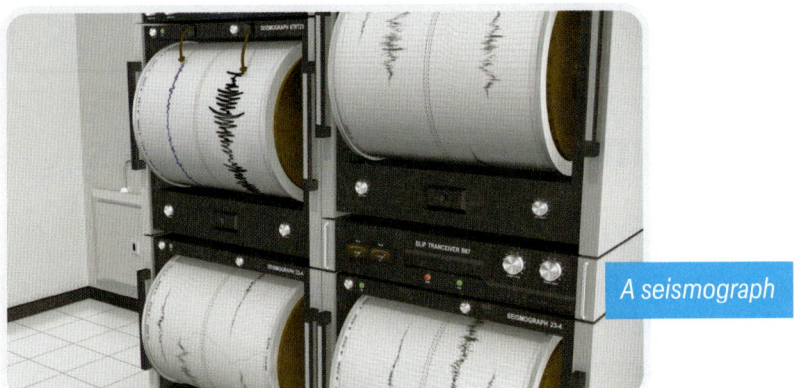

A seismograph

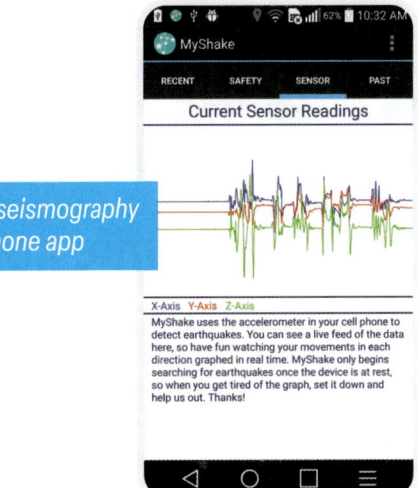

A seismography phone app

Seismograph	The instrument used to measure and record an earthquake and how long it lasted. Smart phone apps mean we can now carry a **seismograph** in our pockets!
Richter scale	The **Richter scale** was the original scale used to measure the **magnitude** (strength) of an earthquake. This scale goes from 1 to 8. Each one-unit increase means that the earthquake is 10 times more powerful than one at the previous unit. It may be unreliable for larger earthquakes.
Moment magnitude scale	The **moment magnitude scale** (MMS) is now used to measure the magnitude of earthquakes. It has replaced the Richter scale because it is more reliable for larger earthquakes. The scale goes from 1 to 10.
Modified Mercalli Scale	This scale is based on eyewitness accounts of the shaking experienced during the earthquake. The scale goes from I to X (1 to 10 in roman numerals) – the higher the number the larger the damage seen.

FUN FACT! The 1960 Chile earthquake was measured as 8.5 on the Richter scale. Scientists revised this using the moment magnitude scale, and the size was updated to magnitude 9.5. This is the largest earthquake ever recorded since seismographs were invented.

Question Time

1. Examine the diagram of the earthquake and explain each of the key words labelled.
2. On what type of plate boundaries do earthquakes typically occur?
3. Explain the formation of a tsunami wave using a diagram to support your answer.
4. Name the piece of equipment used to measure and record an earthquake.
5. Explain the difference between the Richter scale and the moment magnitude scale.

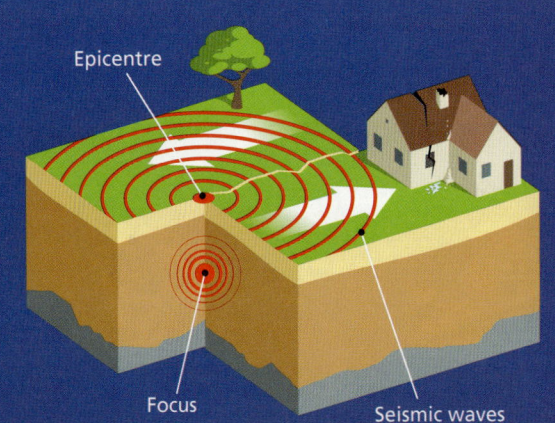

Using a smart device, investigate seismography apps that you can download for free. Then try using your device as a seismograph to record any tremors in your classroom or home. If your seismograph picks up any larger tremors, try to investigate why.

3.5 Economic and Social Impacts of Earthquakes in Nepal and Chile

Earthquakes can cause great damage and loss of life in the regions where they strike. The structural damage, economic loss and numbers injured or killed can depend greatly on whether the country affected is wealthy and more economically developed or poorer and less economically developed.

We will examine the impacts of earthquakes in Chile and in Nepal. Chile is considered to be more economically developed than Nepal.

Case Study: Nepal Earthquake, 2015

Earthquakes are not uncommon in Nepal. Nepal is a developing country, which means it has less money and fewer resources to prepare and plan for the potential damage caused by earthquakes.

Date of earthquake	25 April 2015
Magnitude	7.8 on the moment magnitude scale
Location	The epicentre was located about 80 km away from Nepal's capital city, Kathmandu.
Earthquake depth	The earthquake was close to the surface – the focus was only 15 km below ground. The closer an earthquake is to the surface, the more damage it tends to cause.
Aftershocks	Two very large **aftershocks** occurred within an hour of the main earthquake. They measured 6.6 and 6.7 on the moment magnitude scale.

3. Earthquakes

📁 Managing information and thinking

1. Name the two plates that have a boundary in this region.
2. What type of plate boundary is this: constructive, destructive or conservative? (Look back at figure 6 on p. 5 to help you.)
3. Some aftershocks can cause more damage than the stronger, original earthquake. Can you suggest a reason for this?

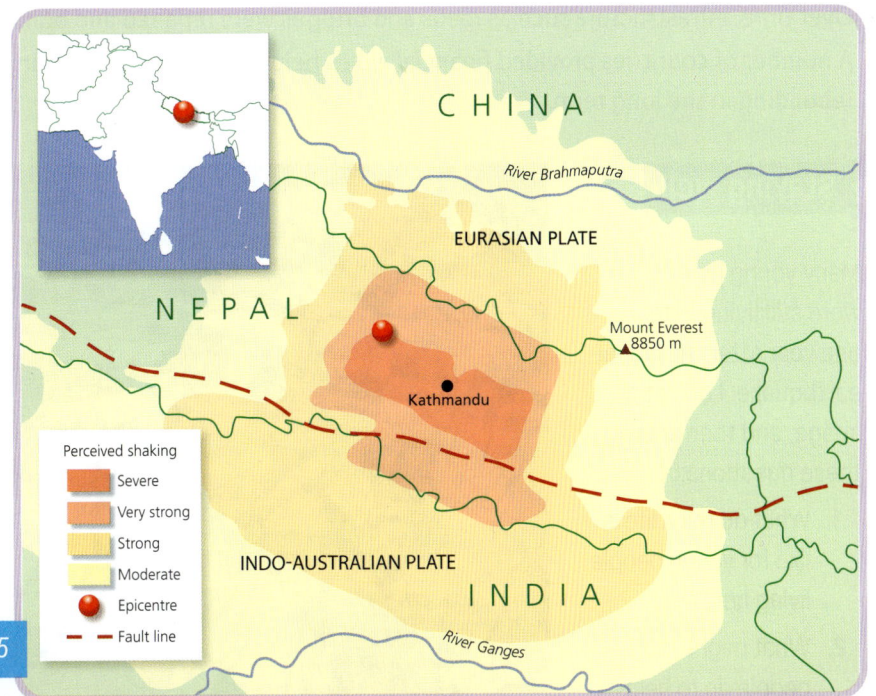

4. The Nepal earthquake, 2015

Social Impact

Over 8,600 people died, and almost 17,000 were injured.

Electricity, water supplies, sewage systems and communication networks were all affected.

Up to 90% of health clinics and schools in some areas were left unusable.

The earthquake caused an avalanche on Mount Everest, which killed 19 people.

The earthquake produced landslides that affected villages and parts of the city of Kathmandu.

Economic Impact

The Nepalese government estimated the cost of the damage at almost €9 billion.

Schools needed to be rebuilt. About 1 million children were left without schools to attend.

3.7 million people needed emergency aid (help), including food, water, medical supplies and tents.

Response to the Natural Disaster

Short Term

- Almost 300,000 people had to migrate (move) from Kathmandu to seek shelter and supplies.
- Short-term aid – such as water, food, tents and clothes – was sent from other countries to help the people impacted by the earthquake. Agencies such as the Red Cross and Oxfam provided support.

Junior Cycle Geography **CYCLONE**

Long Term

- The earthquake triggered landslides and many areas needed to be cleared of this material.
- Over time, infrastructure such as roads and bridges were repaired and buildings were rebuilt.
- A number of countries provided financial aid to help with the large costs involved of rebuilding in the long term.

👍 Staying well

Many young people were still living in tents one year after the Nepal earthquake. Look at the image, and then answer these questions:

1. What do think life is like for young people living here?
2. What might people do to help themselves stay well?

One year after the Nepal earthquake

Being numerate

Referring to the infographic, answer these questions:

1. How many people were displaced?
2. How much were the economic losses?
3. How many schools were damaged by the earthquake?
4. How many tents were dispatched?

NEPAL 2015 EARTHQUAKE

- 8,622 dead
- 16,808 injured
- 2.8 million people displaced
- $10 billion economic losses
- 491,620 buildings fully damaged
- 269,253 buildings partially damaged
- 7,532 schools damaged

RESCUE AND RELIEF RESPONSE

- 66,069 Nepalese army
- 44,629 Nepalese police
- 21,812 armed police force
- 1,571,978 tonnes of food dispatched
- 703,234 tents dispatched
- 48 international medical teams
- 101,182 people treated

3. Earthquakes

Case Study: Chile Earthquake, 2010

Chile is in a very active earthquake zone and has experienced some of the largest earthquakes ever recorded. Due to its long history of earthquakes, Chile has developed strict building codes to help prepare and limit damage.

Date of earthquake	27 February 2010
Magnitude	8.8 on the moment magnitude scale
Location	The epicentre was located over 300 km from the capital city of Santiago.
Earthquake depth	The focus was located 35 km below the surface in the Pacific Ocean.
Aftershocks	Hundreds of smaller aftershocks were felt in the weeks after the earthquake.
Tsunami	As the earthquake occurred at sea, the force pushed the seabed upwards and caused a tsunami wave to move across the Pacific Ocean at up to 700 km per hour.

Social Impact

More than 500 people died – 150 because of the tsunami wave – and 12,000 were injured.

Twelve million people, or three-quarters of the population of the country, were in areas that felt strong tremors.

Water, electricity and communication systems were badly affected.

Many coastal towns were devastated by tsunami waves. In other areas, many roads were blocked by debris, and bridges collapsed.

Economic Impact

The economic cost was $30 billion.

The government reported that 441,000 houses were severely damaged or completely destroyed.

In some areas, the limited availability of food and fuel led to widespread looting (stealing), which resulted in arrests.

The tourism industry was hit hard after the earthquake, as many hostels and hotels received cancellations.

Managing information and thinking

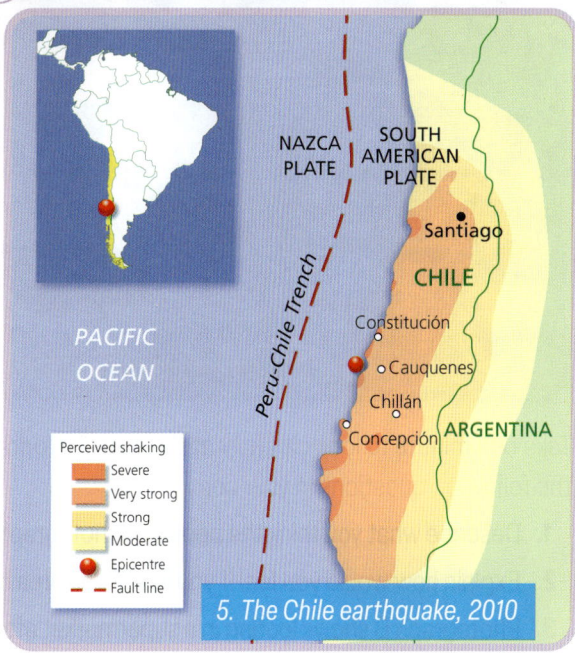

5. The Chile earthquake, 2010

1. What two plates can you identify in this region?
2. Where would you expect the greatest damage from this earthquake to have occurred? Explain why.

Junior Cycle Geography CYCLONE

Response to the Natural Disaster

Short Term
- The Chilean president arranged for retailers to distribute necessary items free of charge in the days after the earthquake.
- Some 10,000 Chilean troops were dispatched to help with the recovery and to keep order.
- Aid was provided by the United Nations and from the United States, the European Union and several Asian countries.

Long Term
Four months after the earthquake, more than 50,000 homes had been built and grants provided to many people who had lost their homes. Roads and airports that had been damaged were repaired.

Staying well

Take one minute to look carefully at the two photographs. Then, working with a classmate, write the answers to the following questions in your copy:

1. Describe what you think the people in photograph A are thinking and feeling.
2. Explain how would you feel if you lived in an area hit by a huge earthquake.
3. Photograph B shows looting in a supermarket after the earthquake. What is your opinion on this?

Digital: Geography in the News

Go to YouTube and search for 'Deadly quake strikes Chile – CSB' to watch a news report on the earthquake in Chile (2:06).

3. Earthquakes

Q

1. Compare and contrast the two earthquakes in Nepal and Chile using the table below. Copy the table in your copy, and fill it in.

Country	Nepal	Chile
Magnitude of earthquake		
Name of plate boundaries		
Number of people killed and injured		
Social impacts	1. 2.	1. 2.
Economic impacts	1. 2.	1. 2.
Response	Short term: Long term:	Short term: Long term:

2. What evidence is there to suggest that Chile is richer and more economically developed than Nepal?

3.6 Reducing Earthquake Damage

There are a number of ways that we can try to reduce the damage caused by earthquakes.

1. **Predicting where an earthquake might occur**: Scientists study historical records of when and where earthquakes have occurred at different plate boundaries. They can then predict where another earthquake might occur. Using monitoring equipment and lasers, and observing the behaviour of animals and levels of certain gases, scientists can gather information on any changes that are occurring. While this provides valuable data, scientists cannot be certain when earthquakes will occur.

2. **Planning for earthquakes**: Using maps of regions most at risk of earthquakes, planners can make decisions on which areas are unsuitable for the construction of certain buildings. For example, they would not build a hospital in an area that is at high risk of an earthquake.

3. **Construction of earthquake-safe buildings**: Buildings and bridges can be designed and built to resist the shaking that occurs during an earthquake. An example of an earthquake-proof building is the Torre Mayor skyscraper in Mexico City.

Junior Cycle Geography CYCLONE

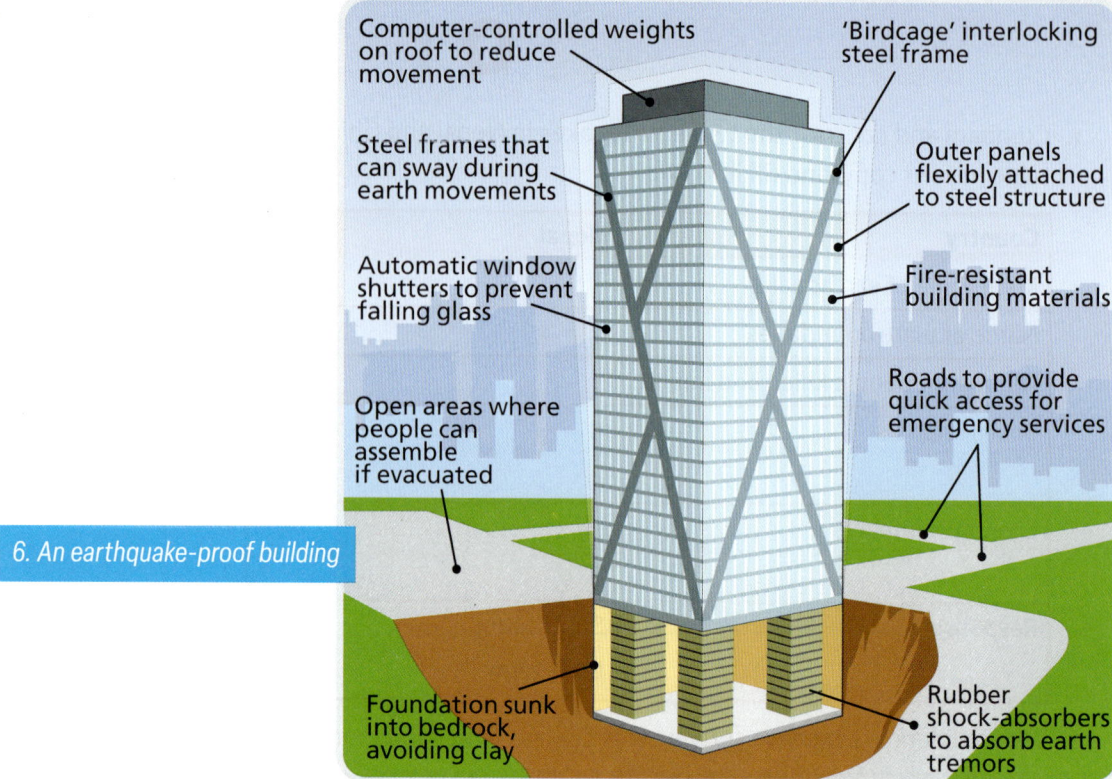

6. An earthquake-proof building

Being creative

Having examined figure 6 above, redesign your school or home to withstand the effects of a magnitude 8 earthquake. Your answer should include sketches of the new building(s) and a written explanation of the measures taken to withstand the shaking that takes place during the earthquake.

4. **Preparing**: In countries prone to earthquakes, many schools and businesses carry out earthquake drills, similar to fire drills, in order to prepare and be confident of what they should do if an earthquake occurs. Drop-Cover-Hold On is one such method to keep people safe.

DROP where you are, onto your hands and knees. This position protects you from being knocked down and reduces your chances of being hit by falling or flying objects.

COVER your head and neck with one arm and hand.
- If a sturdy table or desk is nearby, crawl underneath for shelter.
- If no shelter is nearby, crawl next to an interior wall.
- Stay on your knees; bend over to protect vital organs.

HOLD ON until the shaking stops.
- Under shelter: hold on to it with one hand; be ready to move with your shelter if it shifts.
- No shelter: hold on to your head and neck with both arms and hands.

3. Earthquakes

EXAM FOCUS

Sample questions

1. In your copy, **fill in** the blanks in the paragraph below. Use the words provided in the box. You do not need to use all of the words in the box.

   an earthquake   core   a volcano
   a fold mountain   mantle   subduction
   conservative   tectonic plates   magma

   ### Exam hint!
   When presented with a question that requires you to fill in the blanks, make sure you attempt each part and do not leave any blank spaces. The answers are given in the box, so based on your knowledge of the topic, give it your best attempt.

   The earth is made up of three layers consisting of the crust, _____ and _____. The earth's crust is made up of many plates that fit together like a jigsaw. These plates are called _____ _____. The place where these plates meet is called a plate boundary. Plates that are moving apart are called divergent or constructive plate boundaries. A landform associated with this boundary is _____ _____. Plates that are colliding are known as convergent or destructive plate boundaries. Where one plate is pushed under another, _____ occurs. _____ plate boundaries are plates that are sliding past each other. This may cause a build-up of pressure resulting in _____ _____.

   (NCCA assessment items, Sample 4, Q1)

2. In 2010 two earthquakes occurred of similar magnitude and similar depth and close to built-up areas. One of these earthquakes occurred in Christchurch, New Zealand, the other in Port-au-Prince, Haiti. Haiti experienced devastating results compared to New Zealand, which experienced mild damage. Haiti is located on an island in the Caribbean. It is classified as a low-income country. It is densely populated with many of its population living in extreme poverty. New Zealand is an island located in the Pacific Ocean. It is classified as a high-income country. Standard of living is very high in New Zealand.

Christchurch

Haiti

35

(i) **Explain** three factors that may have caused the effects to be felt worse in Haiti.

(NCCA assessment items, Sample 4, Q6)

Lets get started! Sample starter

> The three factors that I will discuss are:
> 1. High population density
> 2.
> 3.
> As Haiti has a high population density, there will be a large number of people living in each square kilometre. As the earthquake occurred in a built-up area, this means there was a large number of people who could be injured during the event. It also means that many people may need to seek medical support and the facilities may be unable to cope with large numbers ...

... now complete the answer in your copy.

(ii) **Outline** two short-term and two long-term supports that local communities will require in the aftermath of a significant earthquake.

(iii) **Describe** two ways that local communities can limit the impact that an earthquake has on their area.

Exam hints!

The action verb 'Explain' requires you to make something clear by giving details about it.

Start by stating the three factors that may have caused the effects to be felt worse in Haiti. Next, take each factor individually and explain it clearly. Below, the first one has been completed for you.

The action verb 'outline' requires you to set out the main points or ideas.

Start by naming the two short-term supports that communities will need. Next, outline why they will need each short-term support. Then do the same for long-term supports.

In (iii), you must describe two ways to limit the impact of an earthquake in an area. Use the suggestions on p. 33 to support your answer.

3. Read the news article below and answer the questions that follow.

Earthquake: Two dead after Japan hit by tremor

A powerful earthquake of magnitude 7.4 hit north-east Japan on 16th March 2022, leaving two people dead and 160 injured.

Two million homes suffered power outage and a train derailed, but none of its passengers were hurt. Passengers and staff were trapped for four hours on board before being able to escape. A number of people across north-eastern Japan were hurt by falling objects or in falls.

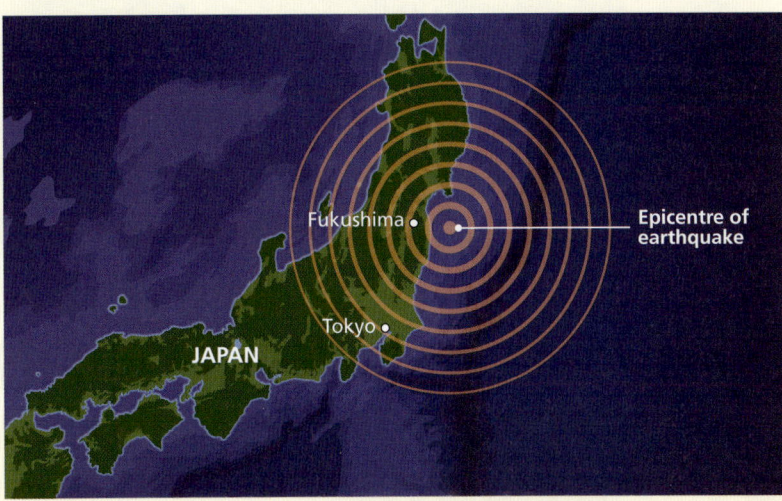

Immediately after the event, an advisory warning for tsunami waves was issued. People were also urged to remain vigilant over the risk of mudslides.

Electricity to more than 2.2 million homes was temporarily cut but by mid-morning power had been restored to most places, the local electricity provider said. Many businesses had to temporarily close down.

Buildings in Tokyo shook and swayed for more than two minutes. Japan's Air Self-Defense Force sent fighter jets to assess the damage to the area.

Exam hint!

When presented with a news article in the exam, read it carefully and highlight any key words. The information required for the initial questions will typically be found within the article. The longer questions will test your geographical knowledge of the topic and generally carry more marks.

(a) **What** was the magnitude of the earthquake that occurred in Japan on 16 March 2022?

(b) **Where** was the epicentre of the earthquake located?

(c) **List** one social and one economic impact of the earthquake for the area impacted.

(d) According to the article, 'An advisory warning for tsunami waves was issued'. **Describe**, with the aid of a labelled diagram, how a tsunami wave forms.

(e) Imagine you are a journalist sent to visit the site of an earthquake that you have studied.

   (i) **Name** the earthquake studied and state where and when it occurred.

   (ii) **Write** a brief report describing some of the negative impacts of the earthquake.

   (iii) **Outline** the response to this natural disaster.

4 Fold Mountains

Learning Outcomes: 1.1, 2.1, 2.9

Learning Intentions

You will be able to:

- Describe, with the aid of a diagram, how fold mountains are formed
- Explain why fold mountains are found in certain locations
- Describe the social and economic impacts of fold mountains in the Alps.

- Anticline
- Syncline
- Caledonian
- Armorican
- Alpine
- Folding

4.1 Formation of Fold Mountains

Fold mountains are found on plate boundaries where two plates have collided. As the two plates collide, the heavier oceanic plate sinks into the mantle below. The lighter continental plate is forced upwards, with the layers of rock in the crust becoming folded. The long ridges, known as **anticlines**, are the mountains. The downfolds (valleys) in between the ridges are known as the **synclines**.

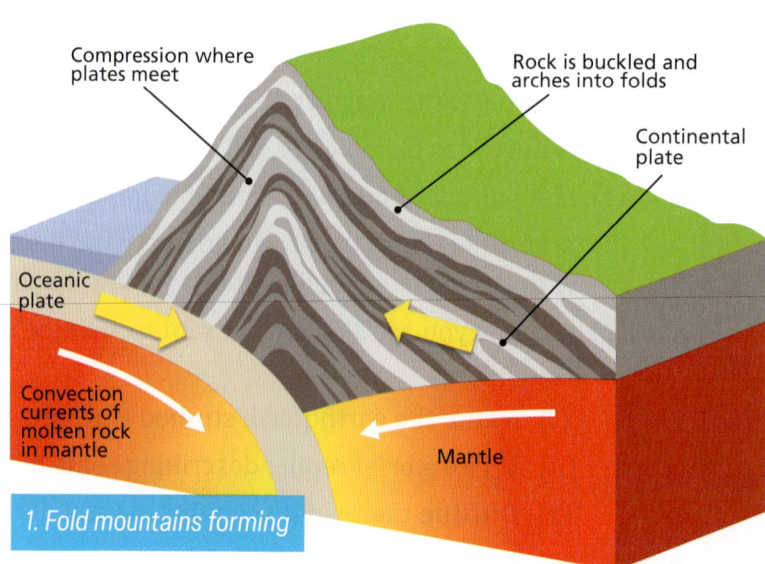

1. Fold mountains forming

4. Fold Mountains

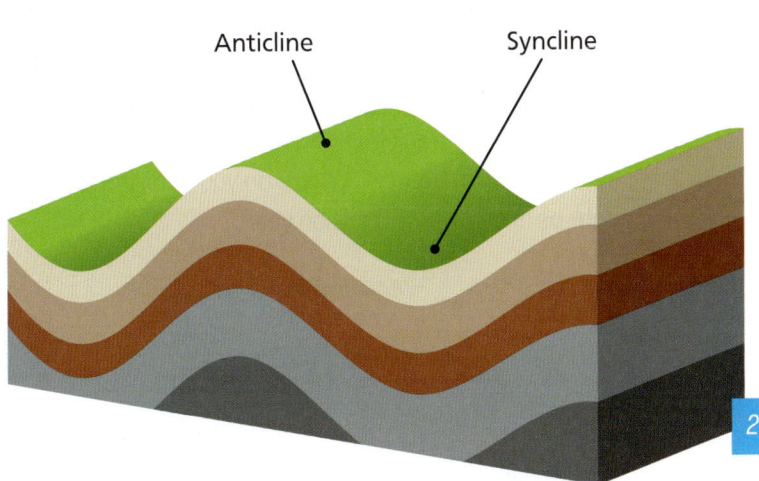

To watch an animation on fold mountain formation, go to YouTube and search for 'Mountain building animation – Tectonics Observatory' (0:16).

2. Anticlines and synclines

4.2 Distribution and Periods of Fold Mountain Formation

What evidence in this photo tells us that this rock has been folded?

There have been three major periods of fold mountain formation: the **Caledonian** period, the **Armorican** period and the **Alpine** period.

1. Caledonian **folding** occurred 400 million years ago. In Ireland, the Dublin–Wicklow Mountains were formed during this period. Mountains formed during this period of folding have been worn down over time.

2. Armorican folding occurred 250 million years ago. Many Munster fold mountains were formed during this time, for example, MacGillycuddy's Reeks in Kerry and the Galtee Mountains in Tipperary. These mountain ranges were once as high as the Alps but have been worn down over time.

3. Alpine folding started 35 million years ago and is still taking place today. These are the youngest and highest fold mountains in the world. They are so high because they have not yet been worn down by weathering. They include the Rockies in North America, the Himalayas in Asia and the Alps in Europe.

Managing information and thinking

Using an atlas or Google Maps, locate and identify the mountain ranges labelled 1–5 in figure 3.

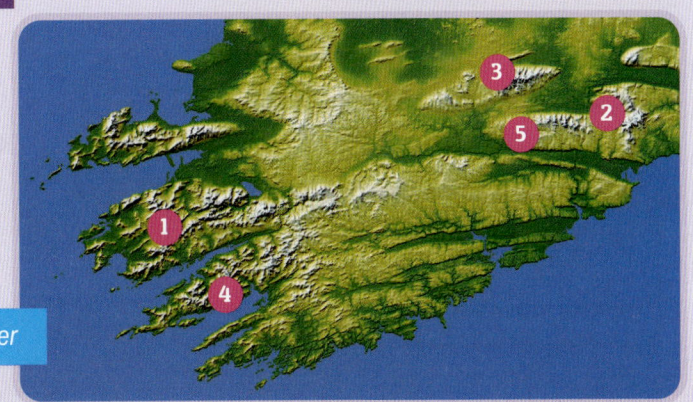

3. Distribution of fold mountains in Munster

Junior Cycle Geography CYCLONE

4.3 Formation of the Dublin–Wicklow Mountains

Ireland was once located in a very different geographical position to where it is today. One half of Ireland was connected to a landmass near North America, and the other half was connected to a landmass further south. Continental drift (as explained in Chapter 1) pulled these continents apart.

During the later Caledonian folding period, the two continents collided again and the two parts of 'Ireland' finally fused together. This collision resulted in the formation of the Dublin–Wicklow fold mountains.

Question Time

1. Outline how fold mountains are formed.
2. Draw a labelled diagram of a fold mountain. On your sketch, show and label the following: (a) Anticline, (b) Syncline, (c) Folds.
3. Name the three major periods of fold mountain formation.
4. During which folding period were the Dublin–Wicklow Mountains formed?
5. Name one range of fold mountains found in Ireland and one found outside of Ireland.

To watch an animation on the movement of the landmass of Ireland, go to YouTube and search for 'Ireland on the move – Wilson Jim' (1:05).

Communicating

Working in pairs, track Ireland's movement in the images shown in figure 4. Where do you think Ireland might be in another 100 million years?

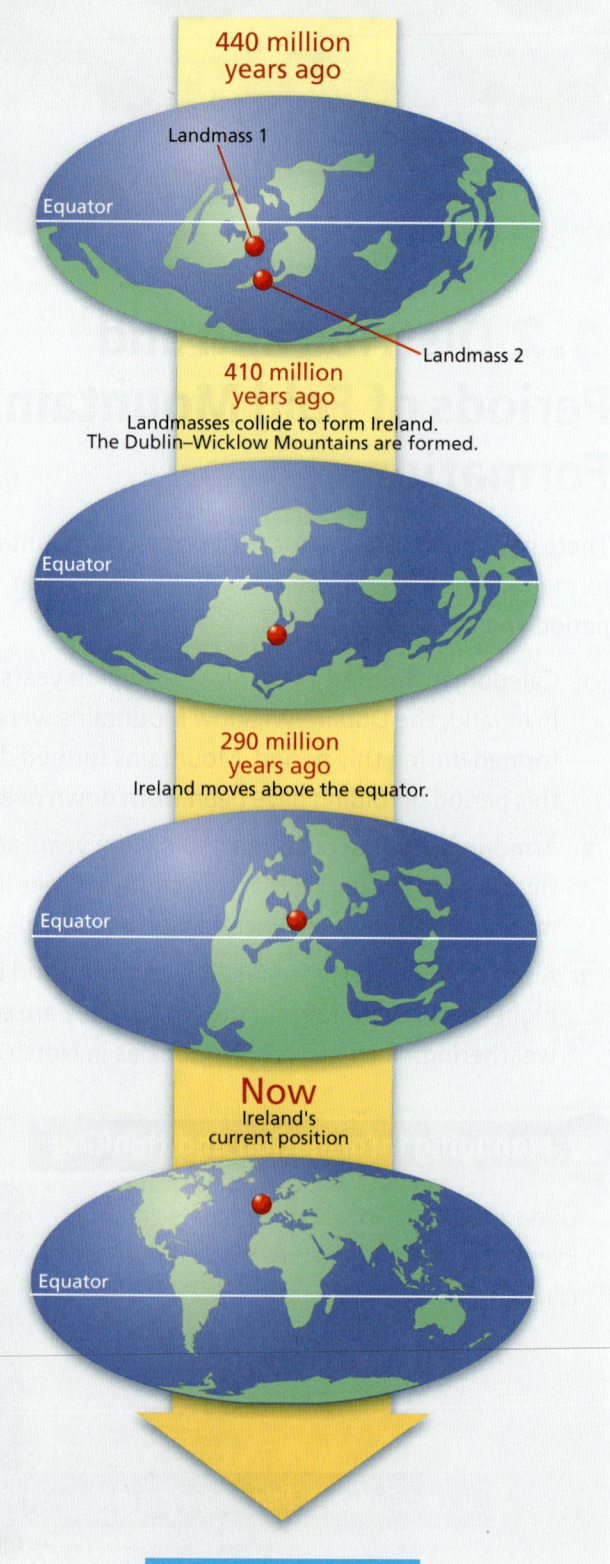

4. Ireland on the move

4.4 Economic and Social Impacts of Fold Mountains

Case Study: The Alps

The Alps is a long row of mountains that stretches across seven European countries. They formed 35 million years ago during the Alpine folding period, when the African and Eurasian Plates collided.

Managing information and thinking

Look at the map in figure 5. Identify the countries the Alps occupy. Which of these countries is not a member of the EU?

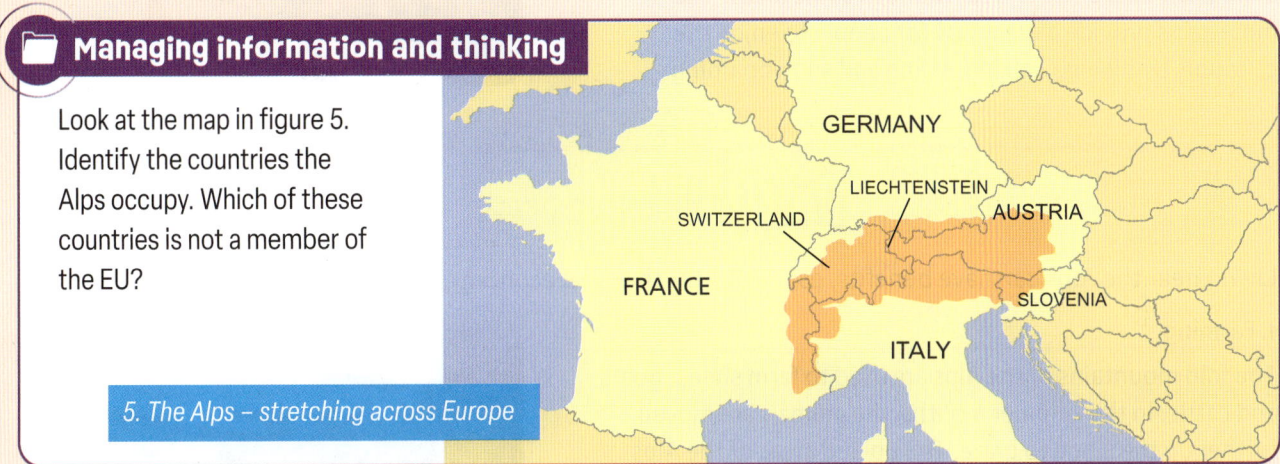

5. The Alps – stretching across Europe

Life in the Alps

Social Impact

Despite many positive social impacts, life in the mountains can still present some challenges.

1. Accessibility

Many parts of the Alps are difficult to access as slopes can be very steep. Road-building is expensive. Living in upland areas can be challenging as they can experience colder temperatures and higher levels of snow, rainfall and winds. As a result, fewer people live there.

2. Farming

Dairy farmers are very busy during summer months milking cows and making cheese. Hay is also made to prepare for the cold winter months. During winter, farming is very difficult.

→

Junior Cycle Geography CYCLONE

Social Impact

3. Avalanches
Devastating avalanches happen when snow and ice suddenly come loose and crash down mountain slopes. The impact of climate change on avalanches is being investigated. Scientists are concerned by increased avalanche activity.

4. Recreational activities
Skiing can take place all year round in some areas. The mountains provide a great way for people to enjoy the outdoors by hiking, walking, skiing and snowboarding.

Economic Impact

Living near the Alps can have benefits, especially for the local economy.

1. Farming
The fold mountains of the Alps are used to farm sheep, goats and cattle. In the summer months, the animals are moved further up the slope to graze. The upland areas of the Alps provide the ideal landscape for this type of farming.

Soil can be thin and poor quality, allowing only for rough-grazing, which can make farming difficult or unprofitable.

2. Hydroelectric power
The water from melting snow is used to generate hydroelectric power (where moving water is used to create electricity). Water is a sustainable energy resource as it can be used over and over again and does not negatively impact on the local environment. It is good for both the economy and the environment.

3. Tourism
Many people visit the Alps to ski or snowboard in winter, or hike or climb in summer. Visitors will stay in hotels, spend money in restaurants and buy local produce. The income generated by tourism can provide a better standard of living for locals.

Q

1. Name the two plates that collided to form the Alps.
2. How do fold mountains impact on settlements and farming in the Alps?
3. Describe one positive and one negative impact of fold mountains for those living in the Alps.

4. Fold Mountains

Sample questions

1. Write these statements into your copy and circle the correct answer.

   (a) An example of fold mountains in South America are the Andes / Himalayas.

   (b) Fold mountains are formed at a constructive / destructive plate boundary.

   (c) The mountains of Munster formed during the Alpine / Armorican foldings.

   You may find short-style questions such as this in the exam. Carefully read the statement and decide which term is the correct answer. Make sure to circle the answer that you think is correct.

2. In your copy, write the correct labels from the box below for the diagram showing fold mountain formation. The first one has been completed for you.

   Oceanic crust   Anticline   Continental crust
   Magma   Syncline   Plate sinking

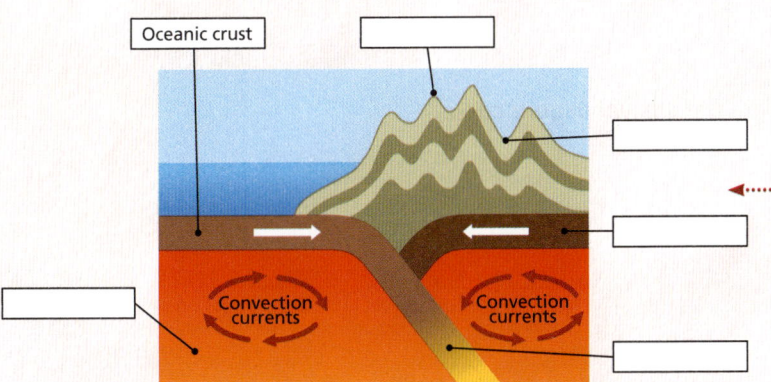

Labelling a diagram is a commonly asked question in the exam. As the terms are all provided above, be sure to carefully study the diagram and write the correct term in the spaces. If you make a mistake, simply place a line through the incorrect answer and write the correct one above it.

3. (i) **Name** an example of a range of fold mountains located outside of Ireland that you have studied.

   (ii) **Draw** a labelled diagram to show the formation of fold mountains at a colliding plate boundary. Label each of the following on your diagram:

   - Two colliding plates
   - Mantle
   - Convection currents
   - Fold mountains.

   (2022 SEC exam paper, Q8 (b) (i) & (ii))

   In part (ii), you are asked to draw and label the diagram. Marks will be awarded for the accuracy of your diagram. Marks are also given for each label correctly placed on your diagram.

4. **Outline** the three main periods of folding: Caledonian, Armorican and Alpine.

Exam hints!

43

5. **Describe** how fold mountains have had a positive social and economic impact on a region that you have studied.

Let's get started! Sample starter

> The fold mountain range that I have studied is the Alps. This is a mountain range that stretches across seven European countries and was formed 35 million years ago during the Alpine folding period.
>
> A positive social impact of the fold mountains here is that they provide a great recreational area for thousands of people every year. Skiing can take place all year round in many areas in the Alps. The mountains provide people with a scenic outdoor area, with hiking, walking and climbing activities to stay fit.
>
> A positive economic impact of the fold mountains here is ...

... now complete the answer in your copy.

Exam hints!

For each period of folding, state when it occurred, where we can see examples and what they may look like today.

For this question, you should state the fold mountain region that you have studied. Next, name a positive social impact and then develop your answer by describing how fold mountains have a positive social impact. Follow the same structure as in the sample for a positive economic impact and complete this in your copy.

Learning Outcomes: 1.2, 2.2

5

Rocks

Key words

- Mineral
- Compressed
- Inorganic
- Igneous
- Sedimentary
- Metamorphic
- Rock cycle
- Intrusive
- Extrusive
- Crystals
- Strata
- Deposits
- Permeable
- Soluble
- Natural resources
- Drilling
- Quarrying
- Shaft mining

Learning Intentions

You will be able to:
- Describe how each type of rock changes into another type as it moves through the rock cycle
- Name the three rock groups and give examples of each
- Explain the formation of each rock group
- Describe how we extract rocks and use them, and evaluate the environmental, economic and social consequences of rock exploitation
- Evaluate the environmental, economic and social consequences of mining in Galmoy Mine.

5.1 Uses of Rock

As you stand on the surface of the earth, underneath your feet is rock. Rocks can differ in their colour, texture (how they feel), hardness and **mineral** content. They can be made of a single mineral or many minerals **compressed** (crushed) together. Minerals are **inorganic** materials, meaning that they do not come from an animal or a plant. Rocks make up our landscapes and provide us with many essential resources.

Junior Cycle Geography CYCLONE

Q Make a list of anything not referred to in figure 1 that you use or see every day that may be made from a rock or mineral.

WHERE DOES IT COME FROM?

Glass: Made from sand (which is made up of quartz). Also comes from **glass sand**.

Slate: A rock of many layers. Used to make roof slates.

Brick: Can come from **clay** to make **red** or **yellow brick**, or can come from **limestone**. Used in building houses.

Oil or Petroleum: Used to fuel cars.

Coal: Used to make heat by burning.

Silica: Used in making electronic items like computers.

Iron ore: Used in making **steel** pipes, pots and pans.

Marble: Used in columns and fireplaces.

Copper: Extracted from **copper ore** and used in water pipes.

Aggregate: Crushed rock of various types, including **basalt** and **limestone**. Used in building roads.

1. Uses of rock

5.2 The Rock Cycle

Based on their origin or how they were formed, rocks are divided into three groups: **igneous**, **sedimentary** and **metamorphic**. Rocks are constantly changing in a process known as the **rock cycle**.

To watch a video on the rock cycle, go to YouTube and search for 'Rocks (Colin Stuart)' (3:41).

46

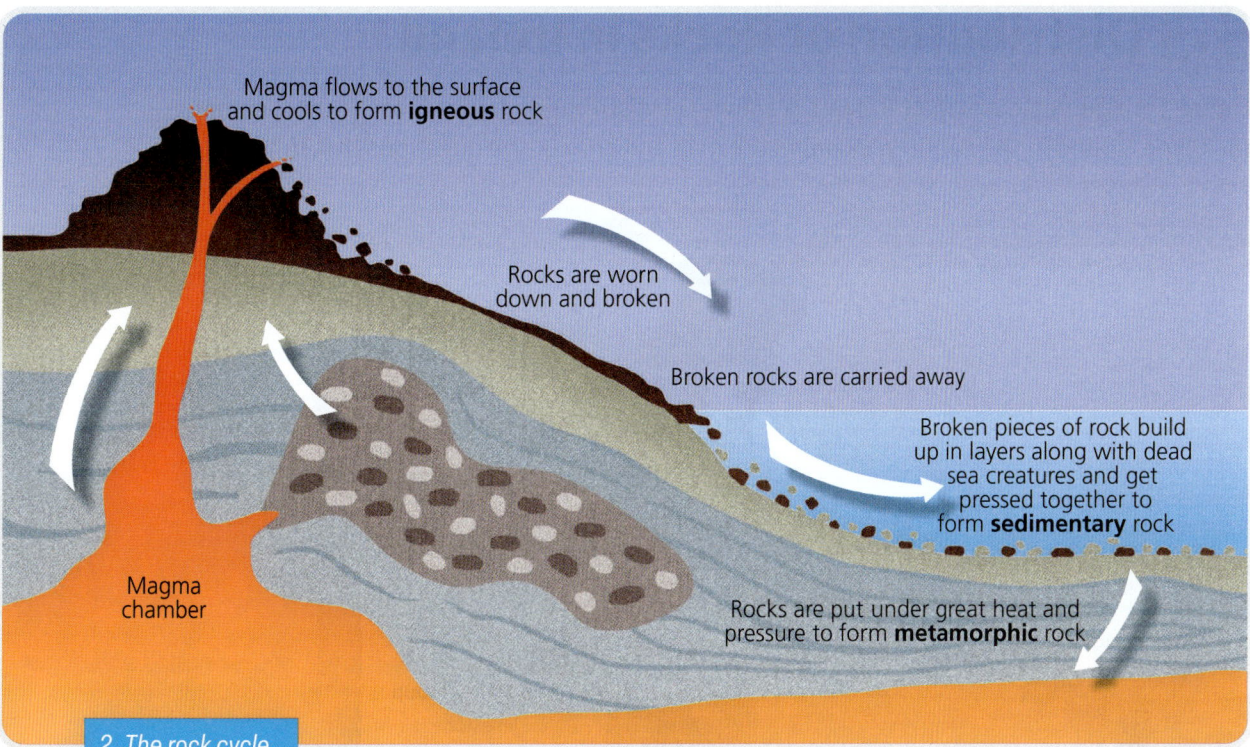

2. The rock cycle

1. Magma rises from the mantle. As it does, some may become trapped below the surface and begin to cool very slowly. This slowly cooled molten material is known as **intrusive** igneous rock. Granite is an example of this type of rock. The magma that reaches the surface flows outwards, and once it hits the open air it begins to cool and forms **extrusive** igneous rock. Basalt is an example of this type of rock.

2. Rocks on the surface of the earth are worn down over time. Some broken pieces are dropped in layers on the sea floor. Here, they become pressed and cemented together, mixing with dead sea creatures to form sedimentary rock. Examples include sandstone and limestone.

3. Metamorphic rocks are formed when igneous or sedimentary rocks are put under great heat or pressure. This changes the rock into a new, harder rock. Examples include marble and quartzite.

Question Time

1. What are rocks made of?
2. List three uses of rocks in everyday life.
3. Copy and complete this table:

Rock group	Description of formation	Examples

Junior Cycle Geography CYCLONE

5.3 Distribution of Rocks in Ireland

Figure 3 is a geological map of Ireland. It shows the distribution (spread) of the most common rock types found in the country. Examine the map and answer the questions that follow.

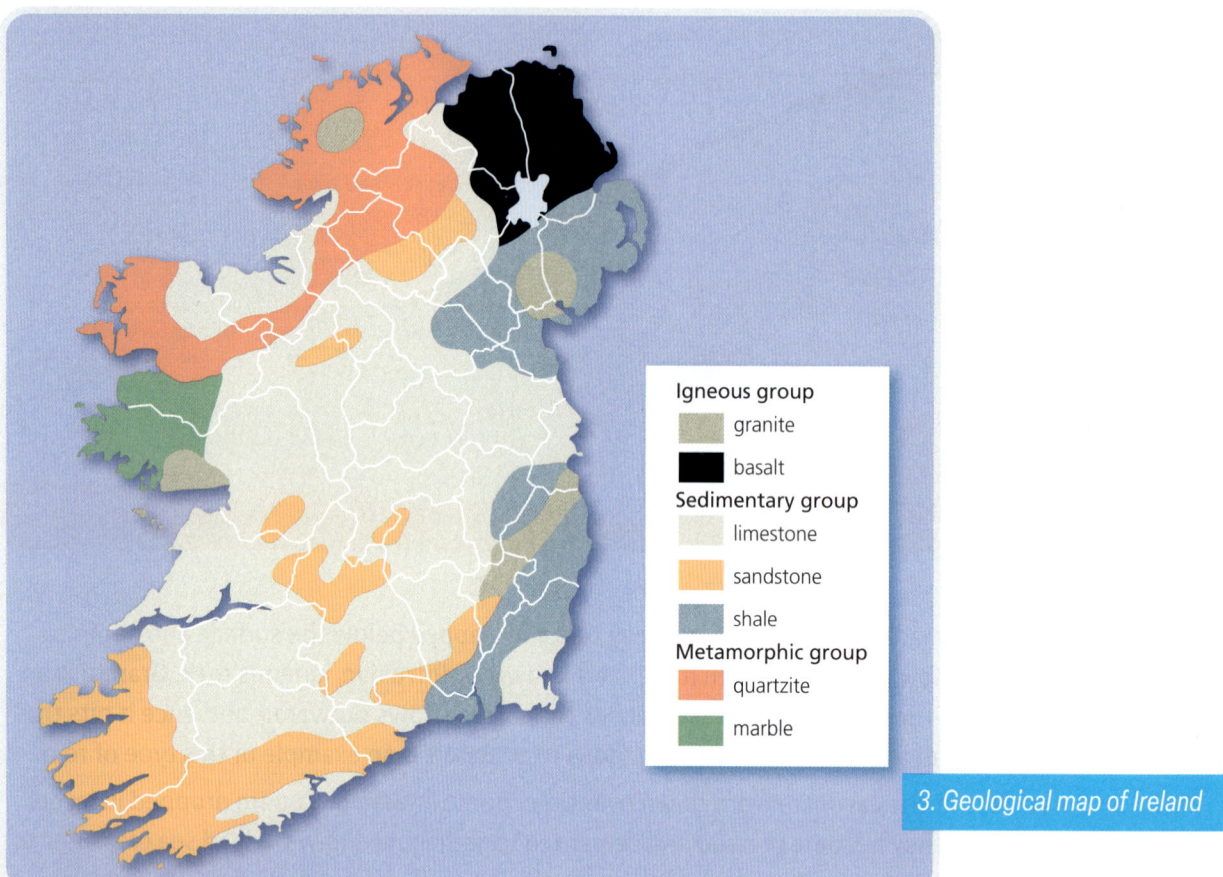

3. Geological map of Ireland

Question Time

1. What is the most common rock type found in Ireland?
2. Approximately what percentage of Ireland is made up of this most common rock type?
3. Name the rock type found in your county.
4. Copy the table below and fill it in by estimating what percentage of Ireland is covered by each rock listed. Once complete, compare your figures with a classmate's.

Rock type	Estimated percentage cover of Ireland	Rock type	Estimated percentage cover of Ireland
Granite		Sandstone	
Basalt		Quartzite	
Limestone		Marble	

5.4 Rock Groups

Igneous Rocks

As shown in figure 4, igneous rocks are formed from the magma found in the mantle just below the crust. When magma is forced up into the crust and becomes trapped, it begins to cool very slowly. Rocks formed in this way are known as intrusive igneous rock. Igneous rocks that form when lava reaches the earth's surface are called extrusive. As the lava spreads out on the surface and meets the air, it cools and hardens very quickly.

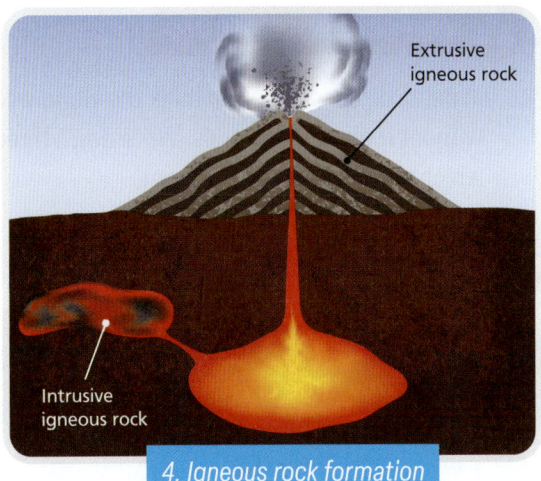

4. Igneous rock formation

Granite

Formation	Granite is an intrusive igneous rock. As it cooled very slowly over millions of years, the minerals in the magma formed very large **crystals**. These crystals include quartz, mica and feldspar. Granite is a hard rock and has a rough texture. Granite is multi-coloured.
How it is used	It is used for building monuments and gravestones, kitchen countertops and fireplaces.
Location	It can be found in the Dublin–Wicklow and Mourne Mountains.

Granite

Granite countertop

Basalt

Formation	Basalt is an extrusive igneous rock. Lava cools very quickly when it reaches the surface. Because it cooled so fast, basalt has very small crystals that can only be seen using a microscope. Basalt is grey to black in colour. It is very hard and has a smooth texture.
How it is used	Crushed basalt is used in road-building.
Location	Basalt can be found in the Antrim–Derry plateau. The Giant's Causeway in Antrim consists of columns of basalt.

The world-famous Giant's Causeway in Antrim

FUN FACT! The Giant's Causeway is Northern Ireland's most popular tourist destination.

Junior Cycle Geography CYCLONE

> **Working with others**
>
> In pairs, discuss what you think the world would be like without rocks and minerals. Write down three ways in which things would be different.

To watch a video on the formation of the Giant's Causeway, go to YouTube and search for 'How the Giant's Causeway was formed (Alistair Hamill)' (1:04).

Question Time

1. Explain why large crystals form in granite and small crystals form in basalt.
2. Igneous rocks have many uses. Can you give some examples?
3. Name one igneous rock and explain how it was formed.

Sedimentary Rocks

Sedimentary rocks make up about three-quarters of the rocks at the earth's surface. They are made up of pieces (sediments) of other rocks and the remains of dead plants and animals. Over time, these formed layers (**strata**) that became compressed and cemented together to form rock.

Sandstone

Formation	Sandstone was formed when large amounts of sand were worn away from the earth's crust. This sand was then carried away by wind or rivers and deposited on the beds of rivers and lakes or on the seafloor. The grains of sand built up and the **deposits** became compressed and cemented together. Sandstone is typically yellow/brown to red in colour.
How it is used	Sandstone can be used to make tiles or monuments.
Location	It can be found in MacGillycuddy's Reeks, Co. Kerry.

Q Sandstone is laid down in layers called strata. Can you see these layers in this image?

Limestone

Formation Limestone is formed when the remains of dead sea creatures, shells and fish bones pile up on the bed of the sea. Over millions of years, under great pressure, these remains become compressed and cemented together, forming strata. Some remains are preserved in the rock as fossils. Limestone is a **permeable** rock, which means that water can pass through it. Limestone is also **soluble**, meaning that it dissolves in water. It contains the mineral calcium carbonate.

How it is used Limestone is used as a base for building roads, as well as in monuments. It is also used to create agricultural lime used in farming.

Location It is found in the Burren, Co. Clare, and is the most common rock type in Ireland.

Fossils in limestone

Q What fossils can you identify?

Question Time

1. Explain the following words: (a) strata, (b) permeable, (c) soluble.
2. Name two uses of sandstone.
3. Describe the formation of one sedimentary rock.

Managing information and thinking

Animals can benefit from having limestone in their diet so it is often added to their feed. Find out why.

Metamorphic Rocks

Metamorphic rocks began life as igneous or sedimentary rock. When these two rock groups come into contact with great heat (from magma) or great pressure (due to folding), they can be changed into completely new rocks known as metamorphic rocks.

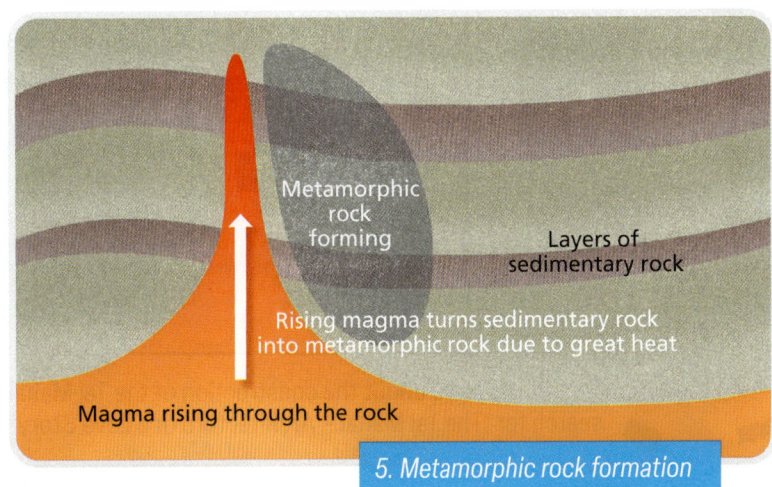

5. Metamorphic rock formation

Junior Cycle Geography CYCLONE

The Taj Mahal

FUN FACT!
The Taj Mahal in India is made entirely of marble. Construction of the Taj Mahal took around 20 years.

Marble

A marble fireplace

Formation	Marble is a metamorphic rock. It is formed when limestone is placed under great heat and pressure. In its pure form, marble is white, but it can also be green, red or black. Marble has a hard texture but can shatter if dropped. Marble is a smooth rock.
How it is used	Marble is used in headstones, fireplaces, monuments, sculptures and tiles.
Location	White marble can be found on Rathlin Island, green marble in Connemara, red marble in Cork and black marble in Kilkenny.

 To find out more about the construction of the Taj Mahal and why it was built, go to YouTube and search for 'The Taj Mahal – Architecture of a Love Story' (4:41).

5. Rocks

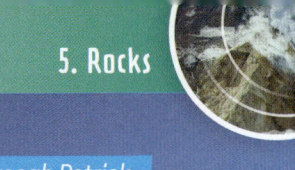

Croagh Patrick

For a fun, interactive video on rocks, go to YouTube and search for 'The Rock Cycle – Sedimentary, Metamorphic, Igneous – Learning Made Fun' (6:39).

Quartzite

Formation	Quartzite is formed when sandstone comes under great heat and pressure. This is usually during periods of folding, when the sandstone comes into contact with magma within the crust. Quartzite is smooth, grey to white in colour and extremely hard in texture.
How it is used	It is used for surfacing of roads, in watches and in glass.
Location	Quartzite can be found on the top of some mountains, such as the Sugarloaf in Co. Wicklow and Croagh Patrick in Co. Mayo.

Quartzite

Question Time

1. Name the two processes that are involved in the formation of metamorphic rock.
2. (a) Select one metamorphic rock, and using a diagram/diagrams, describe how it was formed.
   (b) State two uses of your chosen metamorphic rock.

5.5 Human Interaction with Rocks: Environmental and Economic Consequences

Rocks provide us with important **natural resources** (resources that come from nature and that humans can use).

How We Use Natural Resources

- Oil is used as a fuel source and to produce petrol, tar and plastic products.
- Natural gas is used in the heating of homes and for cooking.
- Coal is burned to generate heat in our homes and in industry.

- Gravel, stone and sand are used as building materials.
- They are used in concrete and some decorative stones are used on the front of buildings.

- Gold is used in jewellery and in the medical, electronics and automotive industries.
- Diamonds, metals and precious stones are all used in jewellery.

- Metal ores such as copper and lead are used in pipes and batteries.
- Copper is used in the plumbing industry.
- Zinc is used to rust-proof cars.

These resources can be mined or extracted from the earth in three different ways.

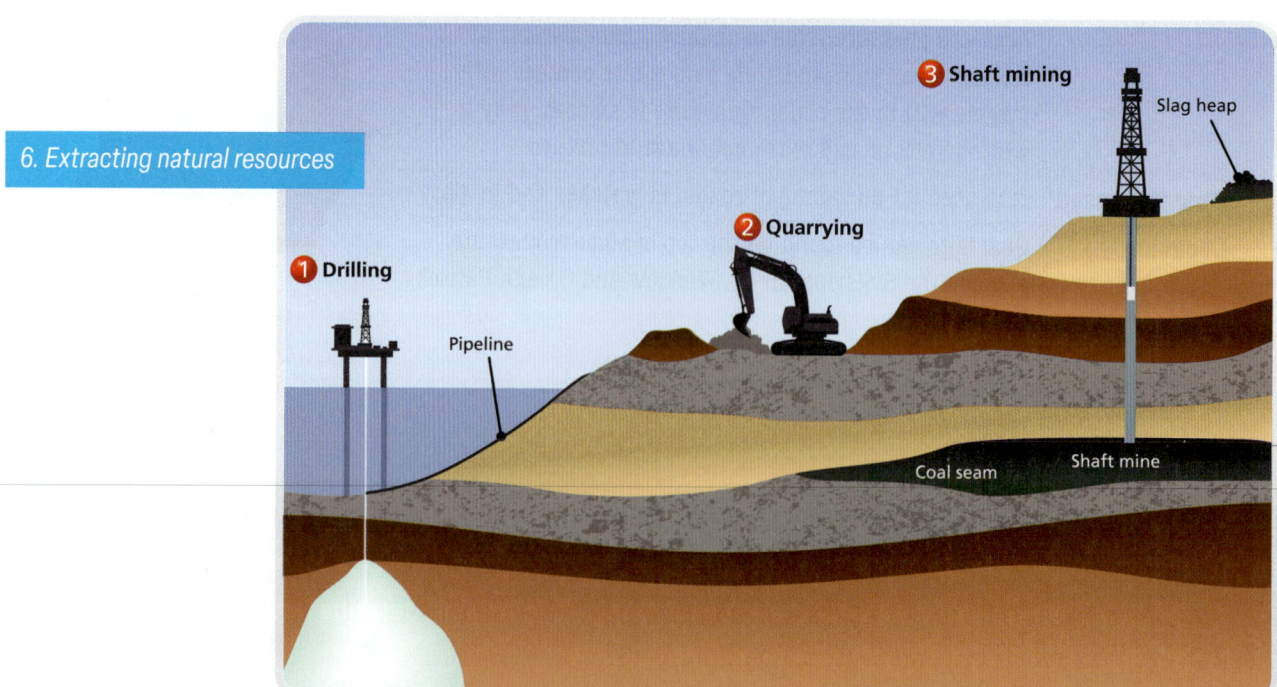

6. Extracting natural resources

1. Drilling

Drilling is the process of boring down into the earth's crust to extract oil or gas. This can happen on land or out at sea. Reserves of fossil fuels such as oil and gas are located underground. When drilling takes place, they are pumped to the surface. Oil and gas are used as fuel sources in cars, homes and industry.

Environmental Impact

Onshore and offshore oil spills can cause damage to the local environment. Birds and animals can ingest (swallow) the oil, or it can cover their bodies, resulting in damage to the animal or death. Oil spills can destroy the natural plant and animal life in the area of the spill.

Social Impact

Oil and gas exploration can create employment. However, this must be weighed up against the potential risks involved such as accidents or oil spills.

Economic Impact

The Corrib Oil and Gas Field project off the Mayo coast has had investment of almost €4.3 billion from three companies. At full production, Corrib has the potential to provide up to 60% of Ireland's gas needs and is expected to supply fuel until 2032.

2. Quarrying

Quarrying is the process of removing rock, sand, gravel or minerals from the ground. Also known as open-cast mining, quarrying happens at or near the surface of the earth. The materials extracted are used by the construction industry to build roads and buildings.

Environmental Impact

Quarrying produces large amounts of dust, which can cause air pollution.

Social Impact

Quarrying can be a very noisy process, from the actual breaking of the rock to trucks coming and going. This results in noise pollution for those living close to a quarry.

Economic Impact

Quarrying has an important impact on the Irish economy. There are approximately 10,000 people employed in quarrying in Ireland. Quarrying and mining in Ireland generate over €1 billion for the Irish economy.

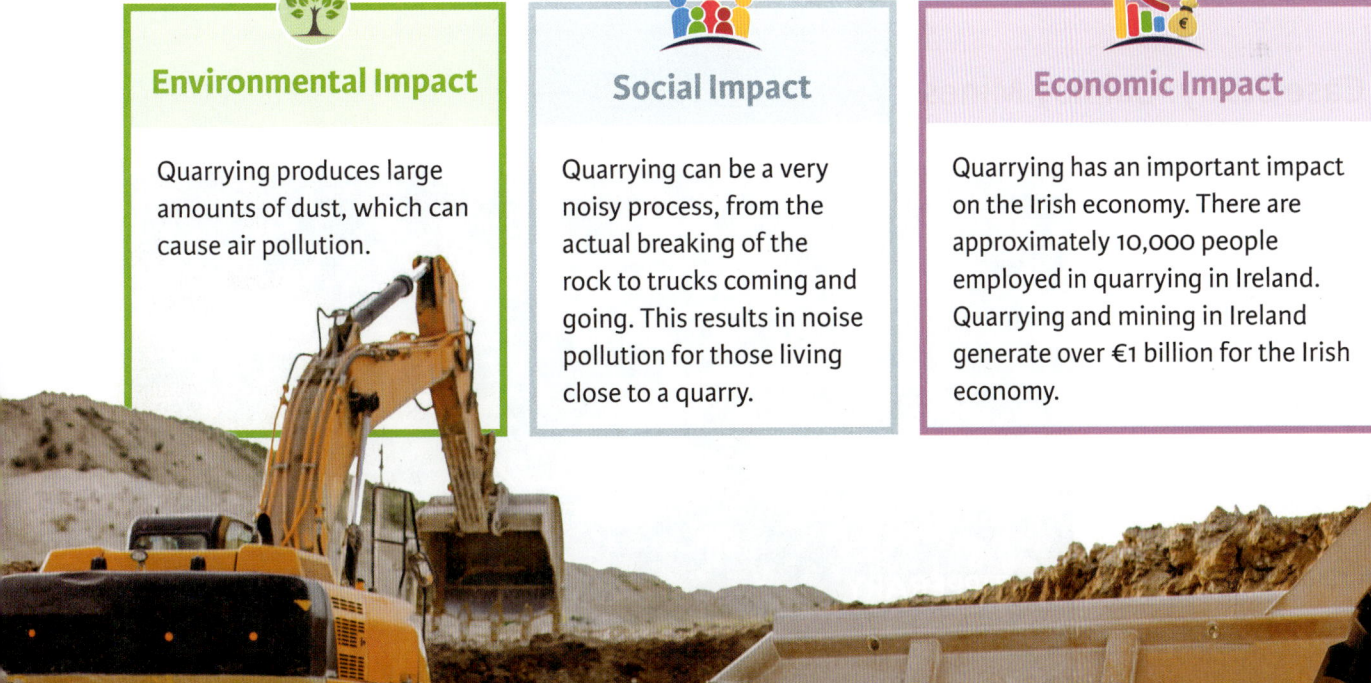

3. Shaft Mining

Shaft mining is the process of drilling vertically into the earth to gain access to minerals beneath the surface. Coal, lead and zinc can be accessed in this way and brought to the surface.

Environmental Impact

Large amounts of dust are created when materials are brought to the surface, which are then carried by the wind. These dust particles can cause many health problems for people who breathe them in.

Social Impact

The landscape above the mines can be vulnerable to sinkholes if the mines below are not correctly managed. This can put people's land or homes at risk.

Economic Impact

The Tara Lead and Zinc Mines in Co. Meath employs 580 people directly and many more indirectly. The company pays €38 million in taxes annually, which provides the government with revenue (money) to pay for services, such as healthcare and education.

Question Time

1. Name the three types of mining.
2. Select one type of mining mentioned in Q1 and describe how it takes place.
3. Explain the environmental impacts of (a) quarrying, (b) drilling and (c) shaft mining.
4. Describe the economic impacts of (a) quarrying, (b) drilling and (c) shaft mining.

Case Study: Galmoy Mines

The Galmoy zinc, lead and silver mines are located in north-west Co. Kilkenny, on the border of counties Laois and Tipperary. This is a rural area. The lands surrounding the mine site are mainly used for livestock grazing.

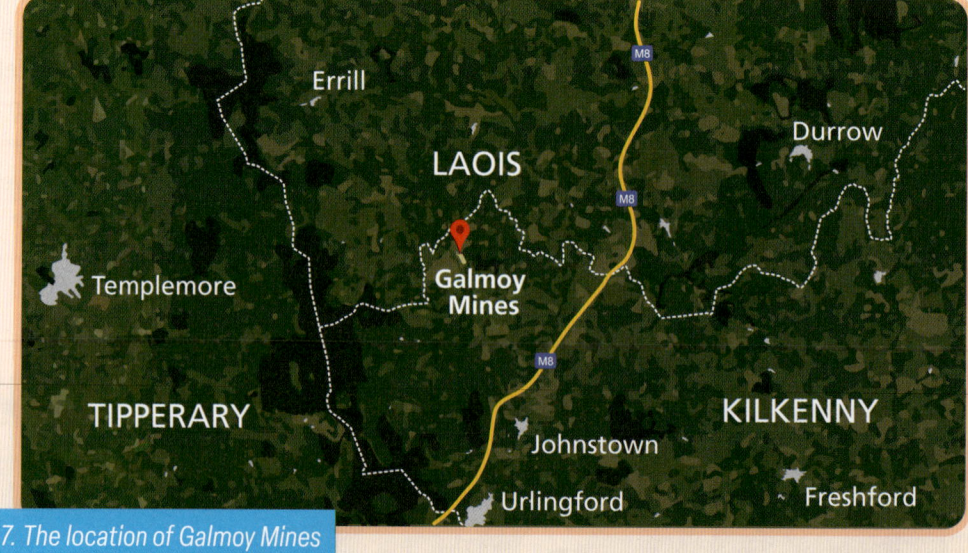

7. The location of Galmoy Mines

5. Rocks

📁 Managing information and thinking

Locate Galmoy Mines in figure 7.

1. Do you think that this is a good location for a mining company to set up when in need of a large number of employees?
2. Do you think this area is well serviced by roads, Wi-Fi, access to services (electricity, water, etc.)? Explain why.

The mine began operating in 1997. The lead, zinc and silver from the mine were transported by truck to New Ross Port, Co. Wexford. It was then shipped to many countries in Europe and around the world. The mine closed in 2014. In 2022, planning permission was granted to reopen the mine.

Environmental Impact

Positives
When the mine was being closed, an artificial wetland was built with four ponds. This caused certain types of birds to migrate to the area and new animal and plant life to grow.

Negatives
There were two reported events of sinkholes appearing on local land. The land on the surface collapsed due to the mining activity below. The land in those areas could not be used for some time.

Social Impact

Positives
The Galmoy Mines were involved in local community sponsorship to support local activities. Many employees at the mine had opportunities to be trained and upskilled in certain areas, which improved their own skill sets.

Negatives
When the plant started operating, the extra traffic to and from the area caused increased noise pollution and vibrations that could be felt by locals.

Economic Impact

Positives
The mine provided over 200 direct jobs for local people. The majority of people working at the mine lived within 30 km of the site. Almost as many jobs were created indirectly by businesses who were local suppliers to the mine.

Negatives
The mine exported its lead and zinc from New Ross Port. When the mine closed, it had a negative impact on the port as it lost income and many people lost their jobs.

The 6-metre sinkhole that appeared on land in Galmoy. In what way would this sinkhole impact on the farmer who owns this land?

Go to YouTube and search for 'Sinkhole appears near Co Kilkenny mine – RTÉ News' to watch the news report on the sinkhole in Galmoy.

Junior Cycle Geography CYCLONE

Q

1. Where is Galmoy Mines located?
2. When did the mines first open?
3. What is a sinkhole?
4. Describe a positive and negative social, environmental and economic impact of Galmoy Mines.

Geography in the News

Galmoy Mines

A total of 190 jobs are set to be created with the planned re-opening of Galmoy zinc and lead mine in Co. Kilkenny and Co. Laois. Shannon Resources sent a planning application to Laois and Kilkenny County Councils to re-start mining activities.

Mining at Galmoy commenced in 1997 and planning consultants for Shannon Resources, Tom Phillips & Associates, said 'significant ore reserves remain underground'.

The planning application states that the re-opening will involve the creation of 100 construction jobs for a one-year period and 90 jobs when operational. The documents state that the mine will have a lifetime of 7 to 10 years once it re-starts operations.

Adapted from the *Irish Examiner*

abc Being literate

Draw around your hands in your copy and fill them up with information.

Step 1: On one hand, list the possible positives of the mine being reopened in Galmoy.

Step 2: On the other hand, list the potential negatives of the mine being reopened.

Step 3: Write an email to Laois and Kilkenny County Councils in response to this news. Your email must contain:
- An opening paragraph that indicates why you are writing
- Your opinion on the reopening of Galmoy Mines
- Three reasons why you feel this way
- A formal closing to sign off.

5. Rocks

EXAM FOCUS

Sample questions

1. (i) **Name** one example of a rock that forms when molten rock cools.

   (ii) Indicate whether the following statement is true or false:
   'Metamorphic rocks were once igneous or sedimentary rocks that were changed by heat and/or pressure.'

2. **Match** the correct rock type named in the box below with the numbered images.

   Shale   Marble   Granite

3. **Describe** the formation of one of the rock types listed below.
   (a) Limestone   (b) Basalt   (c) Marble

Let's get started! Sample starter

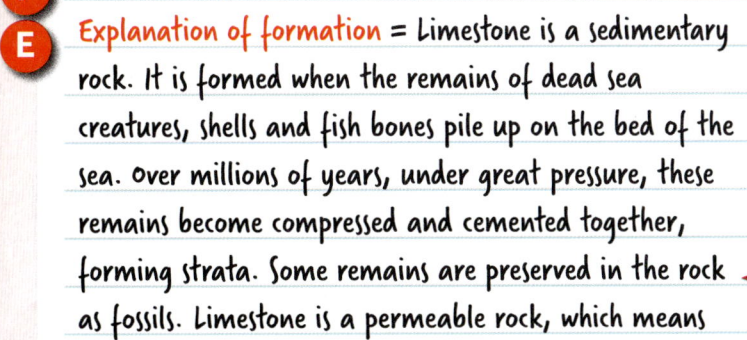

F Limestone

E Explanation of formation = Limestone is a sedimentary rock. It is formed when the remains of dead sea creatures, shells and fish bones pile up on the bed of the sea. Over millions of years, under great pressure, these remains become compressed and cemented together, forming strata. Some remains are preserved in the rock as fossils. Limestone is a permeable rock, which means that water can pass through it. Limestone is also soluble, meaning that it dissolves in water.

E Example =

D Diagram =

... now complete the answer in your copy.

Exam hints!

For this question, we will use **FEED** (**F**eature, **E**xplanation of formation, **E**xample, **D**iagram).
» Start by naming the rock that you have selected.
» Next, give a full explanation of its formation.
» Then state where an example of this rock type can be seen.
» Finally, if space allows, provide a quick labelled diagram.

Complete the sample answer that has been started for you.

59

4. Rock exploitation in Ireland provides many advantages and disadvantages. **Explain** one advantage and one disadvantage.

Let's get started! Sample starter

> One advantage of rock exploitation in Ireland is the creation of jobs.
> One disadvantage of rock exploitation in Ireland is it may cause damage to the local environment.
>
> Creation of jobs
> Jobs are created in the local area where a shaft mine is set up. For example, the Galmoy Mines in Co. Kilkenny provided over 200 jobs for local people. Almost as many jobs were created indirectly by businesses who were local suppliers to the mine.
>
> Damage to the local environment ...

... now complete the answer in your copy.

Exam hints!

For this question, start by naming one advantage and one disadvantage. Then for each one, you must explain why it is a disadvantage or an advantage. You must develop your answer. You could use information learned in your case studies here as evidence and an example. This answer has been started for you. You should use this to complete the question.

Learning Outcomes: 1.9, 2.2

6

Natural Energy Resources

key words
- Natural energy sources
- Renewable
- Non-renewable
- Finite
- Fossil fuels
- Geothermal energy
- Biomass
- Exploitation
- Crude oil
- Hydrocarbons
- Offshore
- Onshore
- Carbon footprint
- Acid rain
- Solar energy
- Solar panels
- Solar farms
- Wind power
- Wind farms

Learning Intentions

You will be able to:
- Explain what natural energy sources are
- Outline the difference between renewable and non-renewable energy sources
- Evaluate the environmental, economic and social consequences of exploiting energy resources
- Understand how we can exploit renewable energy sources in Ireland.

6.1 What Are Energy Resources?

Energy resources are ones that can produce heat, make electricity or move objects depending on what they are used for. **Natural energy resources** are resources created naturally by the earth. They can be **renewable** or **non-renewable**. Our energy resources are used mainly for:

- **Domestic purposes**: Homes require heat and light. Appliances such as televisions, washing machines, phone chargers and computers need power.
- **Transport**: Cars, trucks, planes, trains, etc. need energy sources to run.
- **Industry and business**: Factories and businesses need light, heat and power to run their machines, offices, etc.

61

Non-Renewable and Renewable Energy Sources

Energy sources are either renewable or non-renewable.

Non-Renewable Energy

Can be used only once and will run out eventually. It is also called 'finite' energy.

Examples

- Oil
- Gas
- Coal
- Peat

Fossil fuels are made from the compressed remains of dead organisms. Oil and gas were formed from dead marine material. Coal and peat were formed from dead plant material.

Renewable Energy

Can be used repeatedly and will not run out.

Examples

- Solar power (sun)
- Wind energy
- Tidal energy (waves)
- Geothermal energy (heat energy generated and stored in the earth)
- Biomass energy (burning waste wood or vegetable products)
- Hydroelectric power (water)

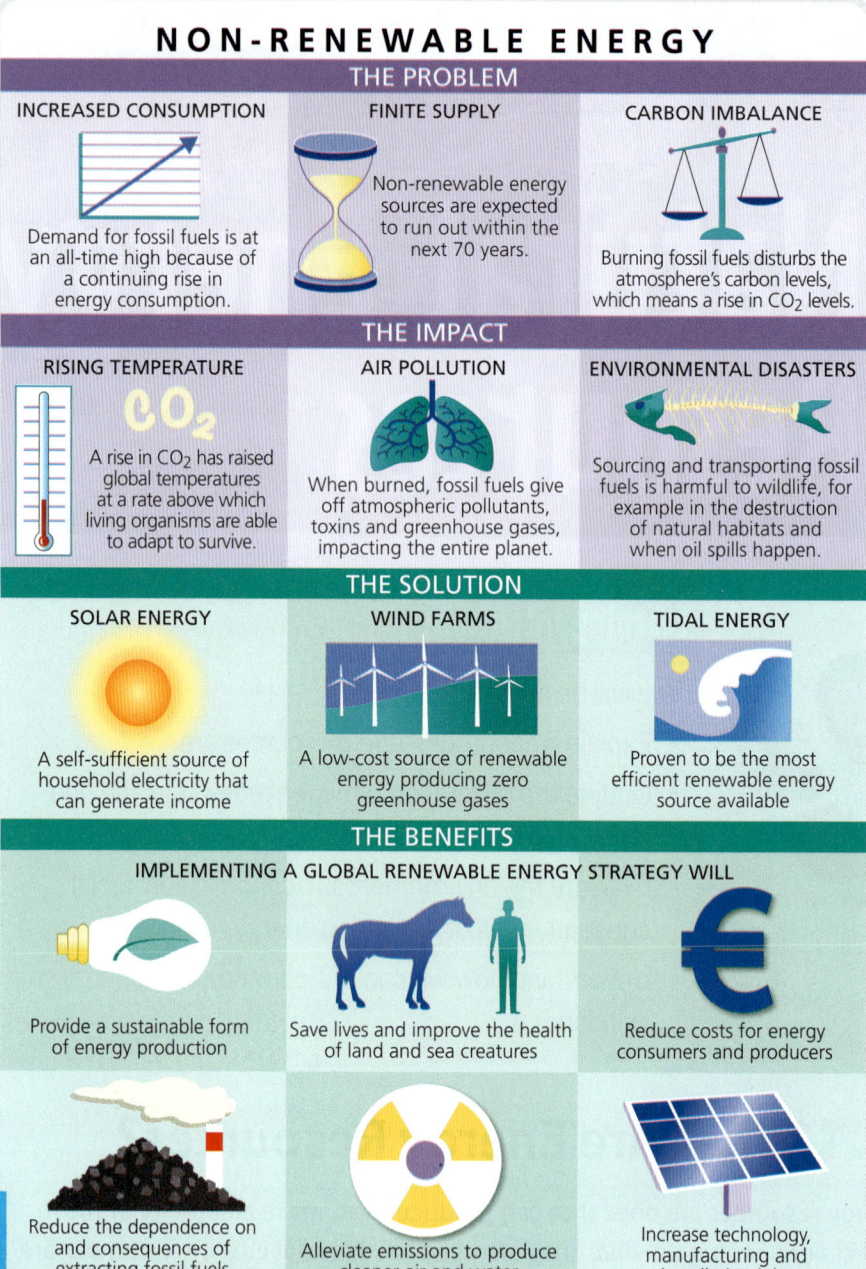

1. The problems with and solutions for non-renewal energy

Managing information and thinking

Study figure 1 and answer these question:

1. When is it estimated that fossil fuels will eventually run out?
2. What are the environmental impacts of using non-renewable energy?
3. Pick two benefits of using renewable energy sources and explain them to the best of your ability.

6. Natural Energy Resources

Question Time

1. Explain what energy resources are.
2. List two uses for energy resources. Select one of these and explain it in detail.
3. Study figure 1 and answer the questions below:
    (a) List one problem with using a non-renewable source of energy.
    (b) Describe an impact that using a non-renewable source of energy can have.
    (c) List one benefit of switching to a renewable source of energy.

6.2 Exploitation of a Non-Renewable Resource: Oil

Exploitation means to make use of and benefit from a resource.

Case Study: Saudi Arabia and Oil Exploitation

Saudi Arabia, a country in the Middle East, was the leading producer of **crude oil** among the OPEC countries in 2020. It was the source of 12% of the world's crude (unrefined) oil.

OPEC (Organization of the Petroleum Exporting Countries) is an intergovernmental organisation (more than one government is in charge of it), founded by Saudi Arabia, Iran, Iraq, Kuwait and Venezuela in 1960. It has 13 current members, and its main function is to control the price of oil.

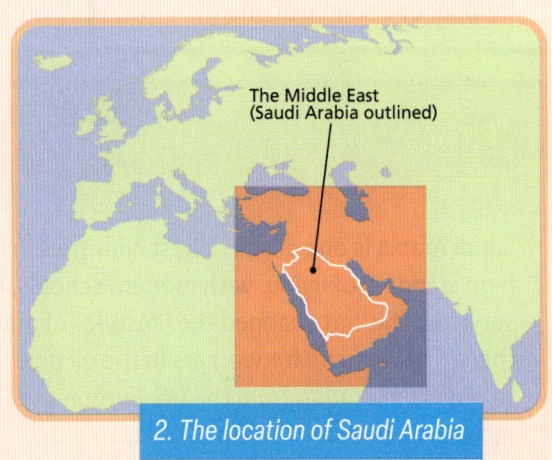

2. The location of Saudi Arabia

In figure 3, we can see that Saudi Arabia had an average daily production of 9.2 million barrels (9,200 × 1,000) per day in 2020. This was a reduction from 12 million barrels per day as demand for oil declined due to the Covid-19 pandemic.

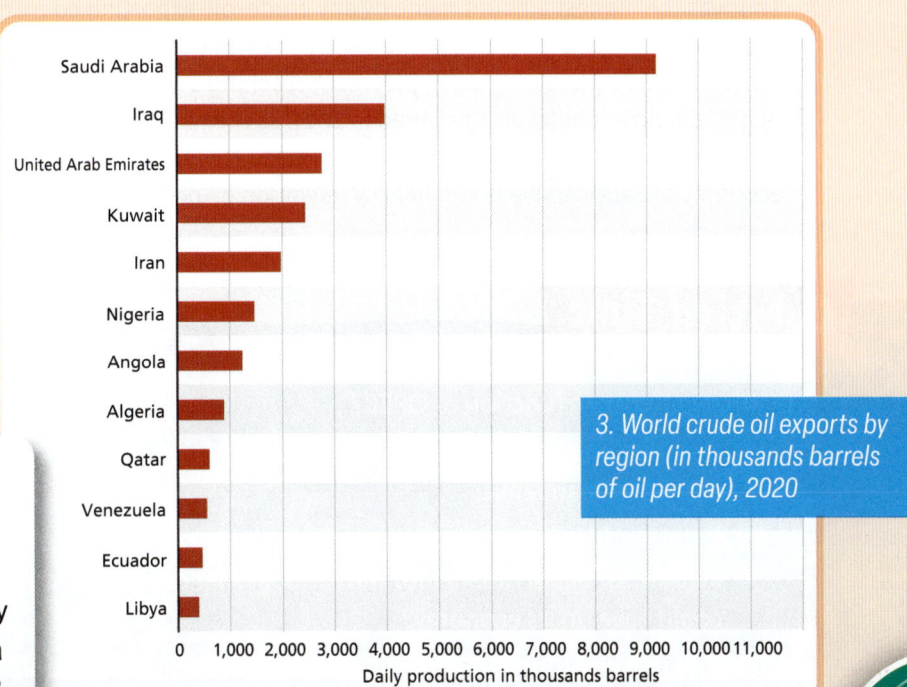

3. World crude oil exports by region (in thousands barrels of oil per day), 2020

Q

1. Which region produces the second highest number of barrels of oil per day?
2. Can you suggest one reason why the Covid-19 pandemic caused a reduction in the demand for oil?

63

To understand the effects of oil production for Saudi Arabia, we can look at environmental, economic and social consequences.

Environmental Impact

Saudi Arabia has some of the highest CO_2 emissions in the world. In 2020, CO_2 emissions for Saudi Arabia were over 588 million tonnes. It has pledged to cut its carbon emissions to zero by 2060. CO_2, or carbon, emissions are caused by the burning of fossil fuels, such as oil. This will prove difficult as oil remains the country's number one source of energy.

Social Impact

The Bedouin people once lived nomadic lifestyles, moving from place to place to source water and grazing for their animals. Once the production oil began, the Bedouins' traditional grazing lands were taken over. Now, many Saudi Bedouins live settled lifestyles, mainly in urban areas.

Bedouins in 1930

Economic Impact

Saudi Arabia is one of the richest countries in the world. It has a very high standard of living, with modern schools, hospitals and other services. This has changed the lifestyles of many of the people living there. The demand for workers in the oil production industry has led to people migrating from the US, Europe and Egypt to work there. This increase in people has helped cities to grow.

Saudi Arabia's crude oil exports in 2020 had a value of US$113.7 billion. This was the highest amount made by any crude-oil-exporting country that year. The global decrease in the demand for oil caused by the Covid-19 pandemic lead to unemployment reaching a record high of 15.4%. Employment levels recovered in 2021, as the number of people employed in oil production rose to over 68,000. This shows how the economy of Saudi Arabia is still heavily reliant on oil production.

Before oil

After oil

Question Time

1. What is the main function of OPEC?
2. (a) Describe one economic impact that Saudi Arabia experienced when the demand for oil decreased during the pandemic.
   (b) If the world begins to move away from the burning of fossil fuels, this will decrease demand and income for Saudi Arabia into the future. What can Saudi Arabia do to plan for the impact this could have on its economy?
3. Identify one environmental impact oil production in Saudi Arabia is having on the world.

6. Natural Energy Resources

6.3 The Search for Hydrocarbons in Irish Waters

There are large deposits of oil and natural gas – also called **hydrocarbons** – under the seabed off the Irish coast. Ireland has the right to claim any resources such as these found off our coast up to a distance of just over 19 km. That equals nearly 90 million hectares of seabed to explore.

Where Are Ireland's Oil and Gas Deposits?

Ireland has **offshore** (at sea) and **onshore** (on land) sites to explore for oil and gas. Figure 4 shows 15 locations, but more than 60 are under investigation for gas discovery around the Irish coast. The Irish government no longer issues new licences to multinational (operating in more than one country) companies to explore Irish land for oil and gas. Ireland is shifting to renewable energy sources.

Hopefully in the near future Ireland will be far less reliant on non-renewable energy. The infographic below and the article on the next page show Ireland's reliance on oil and gas in 2020.

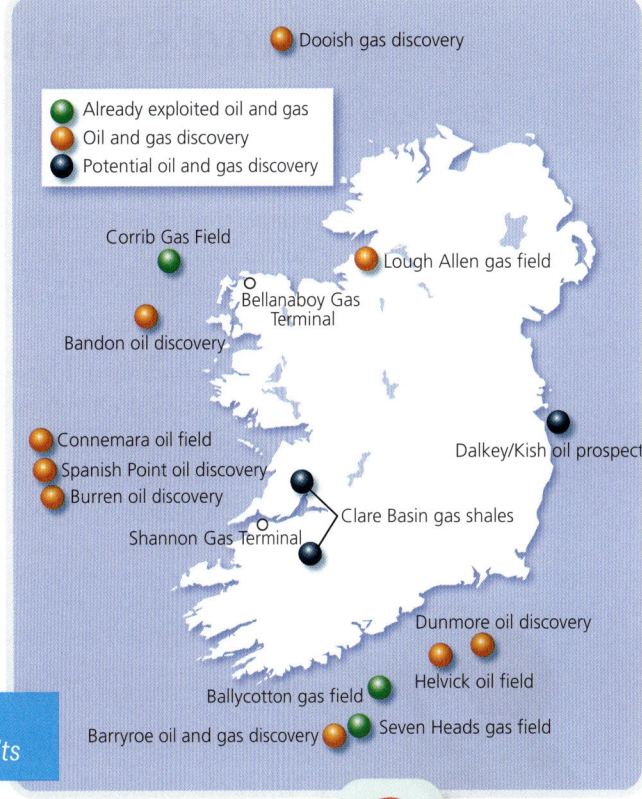

4. The location of some of Ireland's oil and gas deposits

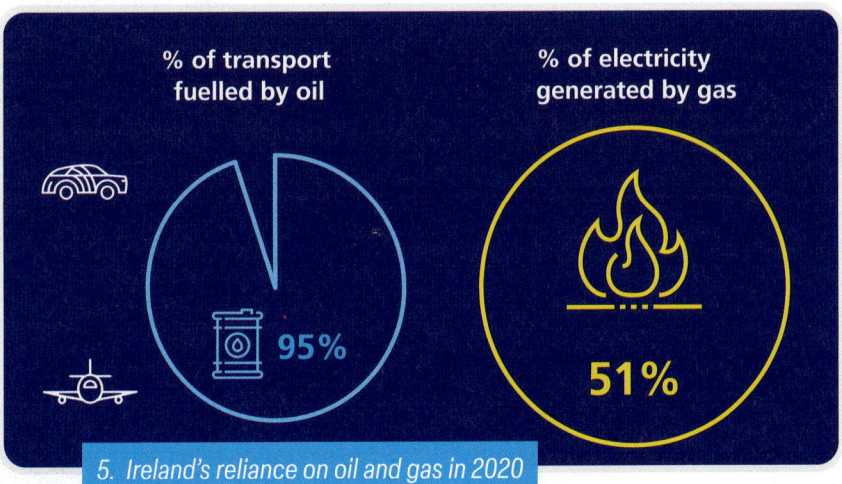

5. Ireland's reliance on oil and gas in 2020

Q

1. Looking at the map in figure 4, locate and name the three sites already being exploited for gas.
2. Where is the majority of oil and gas exploration happening? Why do you think there is no exploration labelled on the map off the northern coast?

Working with others

In pairs answer these questions:
1. What types of transport are reliant on oil as an energy source?
2. Name a company reliant on oil to fuel its transport in Ireland today.

Geography in the News

Ireland's Reliance on Gas and Oil

In 2020, 46% of our country's electricity was generated from natural-gas fuelled power stations. Until 2014, Ireland was importing all its gas using a system of pipelines connecting us to Scotland. This changed when the Corrib Gas Field (discovered in 1996) came online in December of 2015. Corrib accounts for up to 56% of our gas needs.

Ireland's dependency on oil is the fourth highest in the EU. Ireland has one operational oil refinery in Whitegate, in Cork. It refines oil from the North Sea and Africa and provides 40% of Ireland's transport and home heating needs. Oil makes up by far the largest share of our energy imports, accounting for 73% of total energy imports in 2018. The vast majority of Ireland's crude oil imports came from the UK.

With a growing population, increased car ownership, the expansion of Dublin Airport and continued reliance on fossil fuels as our main energy sources, Ireland needs to start looking for its own oil and gas supplies. This, however, is expensive to do. Sourcing non-renewable sources of energy to replace oil and gas is now a top priority of Irish governments looking to decrease Ireland's **carbon footprint**.

1. **What** percentage of Ireland's electricity is generated by natural gas?
2. From **where** did Ireland import all its gas until 2014?
3. **What** big change happened to Ireland's gas supply in December 2015?
4. **What** percentage of Ireland's oil is refined locally? Where is it refined?
5. **What** factors are driving Ireland's continued reliance on oil as an energy source?

Consequences of Oil and Gas Exploitation in Ireland

Environmental Impact

Ireland failed to decrease its 2013–2020 level of CO_2 emissions and paid €150 million in EU fines. In 2020, Ireland's carbon emissions were estimated to be 57.70 million tonnes, a decrease of 3.4% from 2019 due to the Covid-19 pandemic. The country is struggling to meet the 2020–2030 targets dues to continued exploitation of oil and gas.

Social Impact

Tax collected from non-renewable energy use funds the government construction of roads and buildings such as schools and hospitals. The continued use of non-renewable energy may also lead to taxes increasing to cover fines.

Economic Impact

Direct and spin-off jobs are created by the continued exploitation of oil and gas. In 2016, there were an estimated 571 direct and indirect jobs in exploration and extraction of natural gas. Jobs lost as Ireland switches to renewable energy can be balanced with jobs created by new sustainable energy industries.

Who Benefits from Ireland's Oil and Gas Discoveries?

The economic benefits of oil and gas discoveries to Ireland are not as high as might be expected. To attract companies like Shell to investigate potential discoveries in Ireland and its waters, the Irish government agreed to take just 25% of any profits made. It was set at this low level, compared to other European countries, in order to encourage companies to explore the Irish seabed.

As well as that, any money spent on investigating potential oil and gas in Ireland is subtracted from profits made. The benefits of oil and gas exploration and exploitation in Ireland must be weighed against the environmental, economic or social costs to ensure it is not just the multinational companies that profit.

Question Time

1. In your opinion, do the economic benefits of oil and gas exploration outweigh the negatives? Explain your answer, giving reasons.
2. (a) Where is Ireland's only onshore oil refinery located?
   (b) Where does the oil it processes come from?
3. Why does Ireland not fully reap the benefits of any oil or gas discovered off its shores?

6.4 Exploitation of a Non-Renewable Resource: Peat

Peat is a natural resource that was used for centuries as a source of heat and energy in Ireland.

Exploitation of Ireland's Bogs

Bord na Móna was responsible for the sustainable exploitation of peat in Ireland. There were once 10 active peat power plants in Ireland burning peat to generate energy. All 10 have now stopped operating, with the last harvesting of peat taking place in 2018. Bord na Móna will now move to producing biomass (burning waste wood or vegetable products) and wind energy. It will use its land to grow sources for biomass energy, e.g. willow trees, and to install wind turbines.

Environmental Impact

Bord na Móna has now stopped harvesting peat for energy. Trees will be replanted to restore the bogland and wildlife habitats. Bord na Móna will continue to produce energy, but it will be through biomass and wind energy.

Economic Impact

Bord na Móna employs a workforce of 1,500 people. Its employees spend money locally, sustaining the economy of the area.

Junior Cycle Geography CYCLONE

Q

Examining figure 6:
1. What percentage of Ireland's electricity comes from non-renewable resources?
2. What percentage of Ireland's electricity comes from renewable resources?

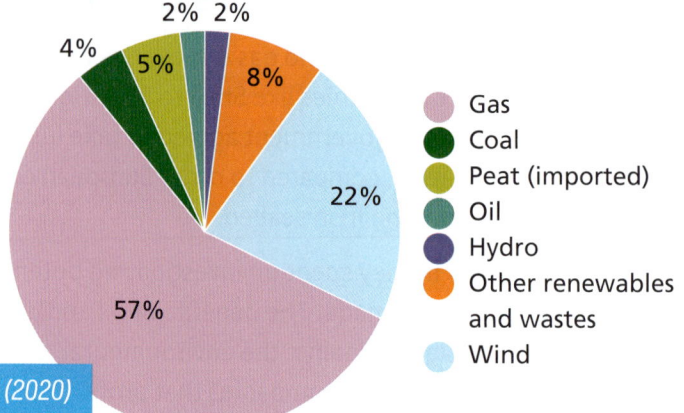

6. Electricity creation by fuel source in Ireland (2020)

Geographical skills: Map Work

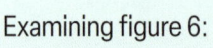

1. Oil reserves have been discovered off the Galway coast at M 31 22. Explain one economic, one environmental and one social consequence of this for Ireland. (See p. 84 in Chapter 7: Geographical Skills: Ordnance Survey Maps for information on how to read map coordinates.)
2. Gas reserves have been discovered off the coast of Galway. A proposed refinery is to be located at M 296 243. Suggest one reason why this might not be a suitable choice of location.
3. (a) Give a six-figure grid reference for a more suitable location.
   (b) Explain why you think your chosen location is more suitable.

Exam hint!
When answering Q2 and Q3, think of the potential dangers of having unrefined gas made safer on land. Think of what is close to the proposed refinery mentioned in Q2.

6.5 Environmental Consequences of Non-Renewable Energies: Acid Rain

Acid rain is a consequence of the use of non-renewable energy sources. Acid rain is no longer regarded as a major threat to the environment in Europe, mainly because EU member states have reduced their emissions of the gases that cause acid rain. However, as long as we continue to burn fossil fuels, there will be some acid rain.

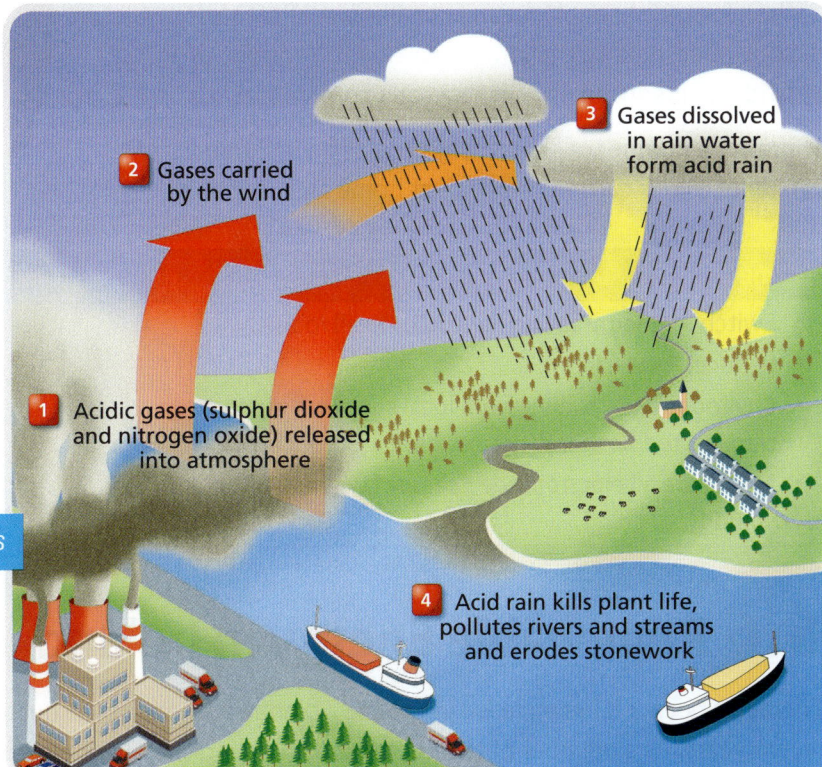

7. How acid rain forms

Damage to Farmlands and Forests

The acidic nature of acid rain leaches nutrients available to plants from the soil and so affects agricultural productivity.

Acid rain also affects the quality of agricultural products. It can damage the leaves of vegetables such as spinach and destroy tomatoes, as seen here.

Acid rain weakens trees, damages leaves, limits the levels of nutrients available to the trees or poisons them slowly with toxic substances released by the soil. Here we see spruce trees in a Polish forest destroyed by acid rain.

Acid rain damages buildings and structures because it dissolves the exposed stone or rusts the metal.

Question Time

1. Draw and label a diagram showing how acid rain forms.
2. Explain two negative effects of acid rain.
3. In Europe, why is acid rain no longer as great a threat as it once was?

Reducing the Effects of Non-Renewable Energy at Home and at School

You may have heard of the term 'carbon footprint'. Carbon footprint simply means the amount of carbon dioxide released into the atmosphere by the actions of an individual person, an organisation or a community.

There are a number of simple ways you as an individual can reduce your carbon footprint at home and at school.

FUN FACT! Some electronic devices use electricity when they are on standby mode. These are called vampire devices! So make sure all devices are switched off completely.

Home	School
• Switch to energy saving lightbulbs in all rooms. • Reduce, Reuse, Recycle. • Buy foods and produce from local suppliers as foods transported further distances have a larger carbon footprint. • Make sure all electronic devices are switched off and not left on standby.	• Walk, cycle or use public transport to get to and from school. • Switch to motion-sensor lightbulbs in rooms and corridors. • Make sure all electronic devices are switched off and not left on standby.

6.6 Renewable Energy in Ireland

There has been a huge increase in the use of renewable energy resources in Ireland in recent years. As a member of the EU, Ireland launched the Climate Action Plan 2021 to put Ireland on a more sustainable path. To do this, Ireland has committed to reducing harmful gas emissions by 51% by 2030.

Solar Energy

Solar energy is the heat and light from the sun that is harnessed and converted directly into other forms of energy, such as electricity.

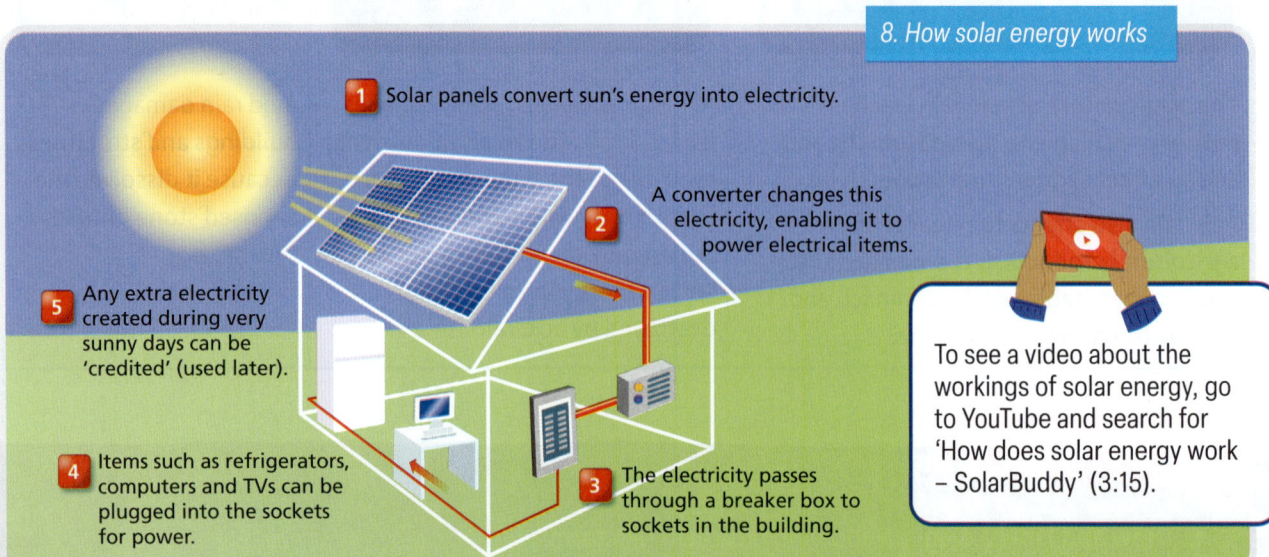

8. How solar energy works

1. Solar panels convert sun's energy into electricity.
2. A converter changes this electricity, enabling it to power electrical items.
3. The electricity passes through a breaker box to sockets in the building.
4. Items such as refrigerators, computers and TVs can be plugged into the sockets for power.
5. Any extra electricity created during very sunny days can be 'credited' (used later).

To see a video about the workings of solar energy, go to YouTube and search for 'How does solar energy work – SolarBuddy' (3:15).

6. Natural Energy Resources

Advantages	Disadvantages
• Renewable energy source • Clean, with no waste product, no atmospheric emissions • Solar panels are inexpensive to repair and maintain • Technology is now more affordable to install as grants are available	• Not possible at night (though advances in solar power technology mean that solar energy can be stored in batteries for later use) • Weather dependent; cloudy weather lessens the amount of solar energy created

Question Time

1. According to figure 8, what happens to any extra electricity created but not used by the home or business?
2. (a) Name one advantage and one disadvantage of solar energy.
   (b) Do you think the advantages of solar energy outweigh the disadvantages, or do you think other sources of renewable energy might be better suited to Ireland?

Solar Energy in Ireland

Many hot countries, such as Saudi Arabia, have solar farms (large areas of land covered with solar panels). Ireland currently does not have these. However, BNRG Renewables, an international company with its headquarters in Dublin, has made an agreement with the Irish government to build almost two dozen solar farms at a cost of more than €220 million.

Managing information and thinking

Examine figure 9, which shows rates of average hours of sunshine in Ireland per year. Where would you locate solar farms in Ireland based on the information in this map. Why?

9. Average sunshine in Ireland per year

KEY
1,100
1,150
1,200
1,250
1,300
1,350
1,400
1,450
Annual sunshine (hours)

Installing solar panels on the roofs of Irish homes is becoming increasingly popular.

Environmental Impact

Homes with solar panels do not need to use an oil/gas boiler, solid fuel stove or electric immersion heater to heat water for much of the year. This reduces CO_2 emissions.

Economic Impact

• Solar panels can reduce electricity and gas/oil bills.
• They also create employment: several companies have been set up in Ireland for the purpose of installing solar panels, e.g. Solar Home Ireland.

Junior Cycle Geography CYCLONE

1. What are the advantages of installing solar panels?
2. What are the disadvantages of relying on solar panels for electricity/heat?

Wind Power

Wind power is the use of airflow through wind turbines to power generators to create electricity.

Wind farms consist of many individual wind turbines that are connected to an electric grid. These farms can be located offshore or onshore.

FUN FACT! Windmills have been in use since 2000 BCE. They were developed in Persia and China, where farmers used wind power to pump water and grind mill grains.

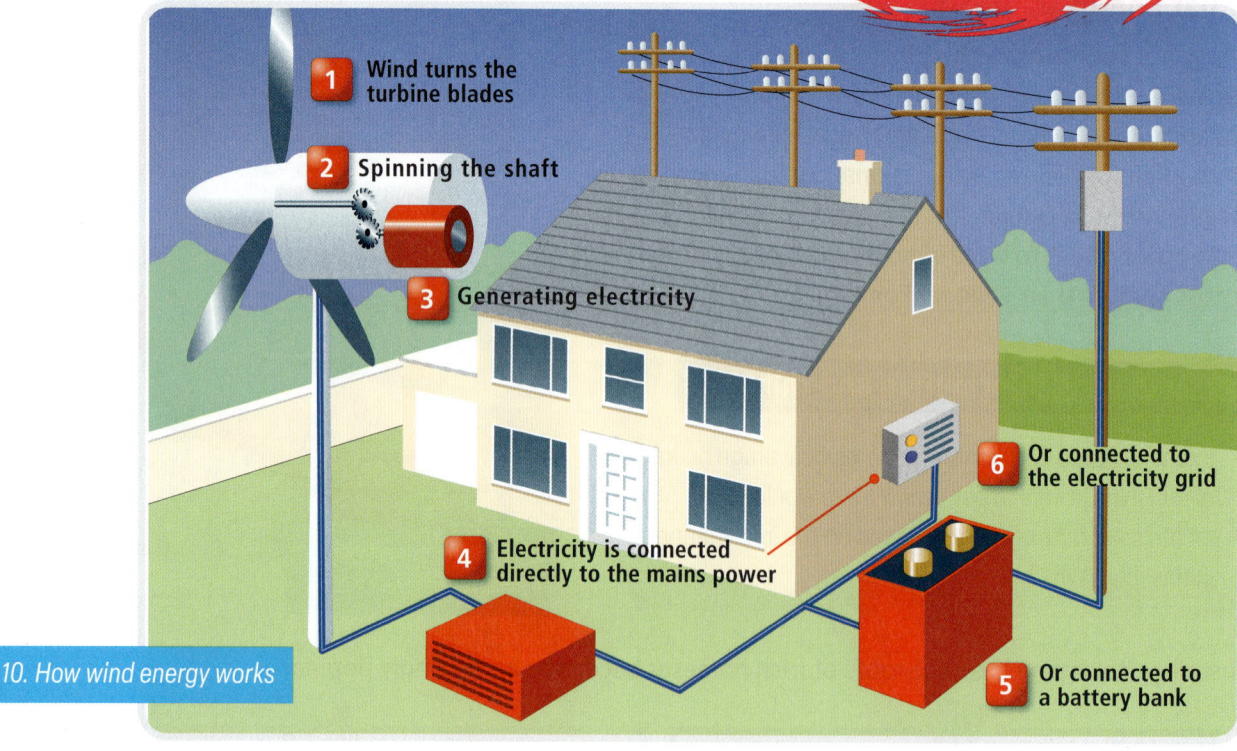

10. How wind energy works

Advantages	Disadvantages
• Renewable energy source • No waste product, no atmospheric emissions • Can be built almost anywhere	• Noise pollution: the air passing through the blades can make a whining sound • Can ruin the scenic view of areas • Concern for avian life (birds and bats) that may fly into the rotors of wind turbines

6. Natural Energy Resources

Wind Power in Ireland

There are over 400 wind farms on the island of Ireland. Ireland's first wind farm was opened in 1992 at Bellacorick, Co. Mayo.

Ireland also has offshore wind farms located in the Irish Sea, around 10 km off the coast of Arklow. These supply energy to the county of Wicklow.

Environmental Benefit

Wind energy is a renewable and clean source of energy. In 2021, 86% of all Ireland's electricity from renewable sources came from wind energy. This meant a reduction in CO_2 levels. Also, offshore wind farms do not disturb the peace of towns or villages.

Economic Benefit

- Harnessing wind energy has the potential to create 7,250 jobs for the construction industry as more wind farms are planned for Ireland. There are approximately 6,000 people employed in wind energy companies in Ireland.
- As well as creating employment, the companies building these wind farms are mostly Irish. This means that the money spent on building these farms stays within our economy.

Managing information and thinking

With the person sitting next to you, examine this photo of a wind farm located in Co. Donegal. Why do you think this was a suitable location in which to build a wind farm?

Question Time

1. (a) Name two advantages and two disadvantages of wind-generated electricity.
   (b) Do you think the advantages outweigh the disadvantages? Explain why.
2. Explain the environmental and economic benefits of wind energy for Ireland.
3. Imagine you live near the site for a proposed wind farm. Write an email to your local council either in favour of or against the wind farm's location. Provide three reasons to support your view.

Tidal Energy

Tidal energy is a renewable energy source produced by the natural rise and fall of tides. Studies carried out by the Sustainable Energy Authority of Ireland (SEAI) have shown that Ireland has the potential to generate up to 75% of its electricity from tidal energy. Ringaskiddy in Cork and Galway Bay have been chosen as the locations where tidal energy can be tested and developed.

In 2021, Galway-based company ÉireComposites was awarded an EU grant of €3 million to develop a new underwater turbine that generates clean energy from tides.

Advantages
- Tides easily predictable
- Reliable and renewable source of energy
- Produces no atmospheric emissions or other waste

Disadvantages
- Construction of power plants expensive
- Negative influence on marine life forms
- Location limits
- Strength of sea waves unreliable

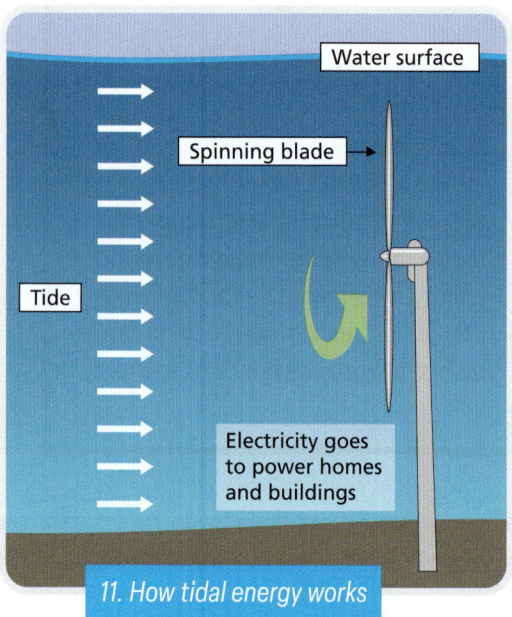

11. How tidal energy works

Environmental Benefits
- Tidal energy has many environmental benefits. The tides are easy to predict, which allows for the maximum amount of energy to be created.
- It is not expensive to maintain.
- It produces no atmospheric emissions or other waste.

Economic Benefits
- If tidal energy generators were installed around the coast, the government would spend less on imported energy sources.
- Jobs would be created in areas such as research, construction, operation and maintenance of these turbines.

 Managing information and thinking

Why would a sheltered bay be a poor location for a turbine?

Question Time

1. Briefly explain the environmental and economic benefits of tidal energy for Ireland.
2. Can you think of any disadvantages of tidal energy? (Consider the maintenance involved in tidal energy systems because of their location.)

6. Natural Energy Resources

EXAM FOCUS

Sample questions

1. Study the pie chart which shows the different sources of energy Ireland used in 2020 and answer each of the following questions.

   (a) **What** was the most used non-renewable resource in Ireland in 2020?

   (b) **What** percentage of Ireland's energy was created by renewable energy sources in 2020?

   (c) **Name** one type of non-renewable energy used in Ireland.

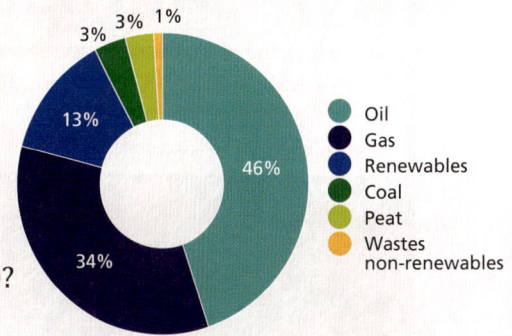

% share primary energy by fuel 2020
- Oil 46%
- Gas 34%
- Renewables 13%
- Coal 3%
- Peat 3%
- Wastes non-renewables 1%

2. (a) **Name** three renewable sources of energy.

   (b) **Explain** two environmental benefits of using one of the renewable sources you named in part (a) above.

3. (a) **Describe** three ways an individual could reduce their carbon footprint.

   (b) **Explain** two possible disadvantages of using a renewable energy resource.

4. (a) Choose a non-renewable energy source you have studied. **Explain** an economic, social and environmental impact of its exploitation.

   or

   (b) Choose a renewable energy source you have studied. **Explain** an economic, social and environmental impact of its exploitation.

Exam hints!

When attempting a longer question like 2b, first note the action verb and what it is asking. The question is asking for two benefits so make sure each benefit you choose is clearly different. Use any facts and figures you can recall from your studies to support your answer.

Questions like these are a good opportunity to use a case study you have examined to improve your answer. You can use information from a case study to help support your answer.

Lets get started! Sample starter

Non-Renewable Energy Source: Oil

Case Study: Exploitation of Oil: Saudi Arabia

Saudi Arabia produces 12% of the world's crude oil. The country was the leading producer of crude oil among the OPEC countries in 2020. The exploitation of this non-renewable resource has a number of economic, social and environmental impacts.

Economic: (mention the value of oil exploitation to Saudi Arabia in terms of jobs and value)

Social: (think standard of living)

Environmental: (think CO_2 emissions) ...   ... now complete the answer in your copy.

5. Read the article and answer the questions.

Electrical Supply Is Brighter Than Ever with Our First Solar Farm

Neoen's solar farm in Wicklow was the first large-scale ground-mounted solar farm to supply the Irish national grid.

Ireland's first ever solar farm opened in Wicklow in April 2022. The first large-scale ground mounted solar farm to supply electricity has been opened near Ashford in Co. Wicklow.

Millvale solar farm incorporates 33,600 solar panels and covers a quarter of a kilometre square. It was developed by French company, Neoen, a leading producer of renewable energy.

The farm will provide enough power for approximately 3,600 homes every year.

It is estimated that the electricity it generates will prevent 4,800 tonnes of greenhouse emissions, helping Ireland to meet its 2030 climate change agreements.

(i) **Where** did Ireland's first ever solar farm open?

(ii) **How** many solar panels are located on this farm?

(iii) **List** one social, one economic and one environmental impact this solar farm will have on Ireland.

Exam hints!

When presented with a news article in the exam, read it carefully and highlight any key words. You can use direct quotes from the article to back up your answer. The information needed to answer the questions may not be presented in the same order as the questions asked. For the social impact in (iii), think about installation and maintenance.

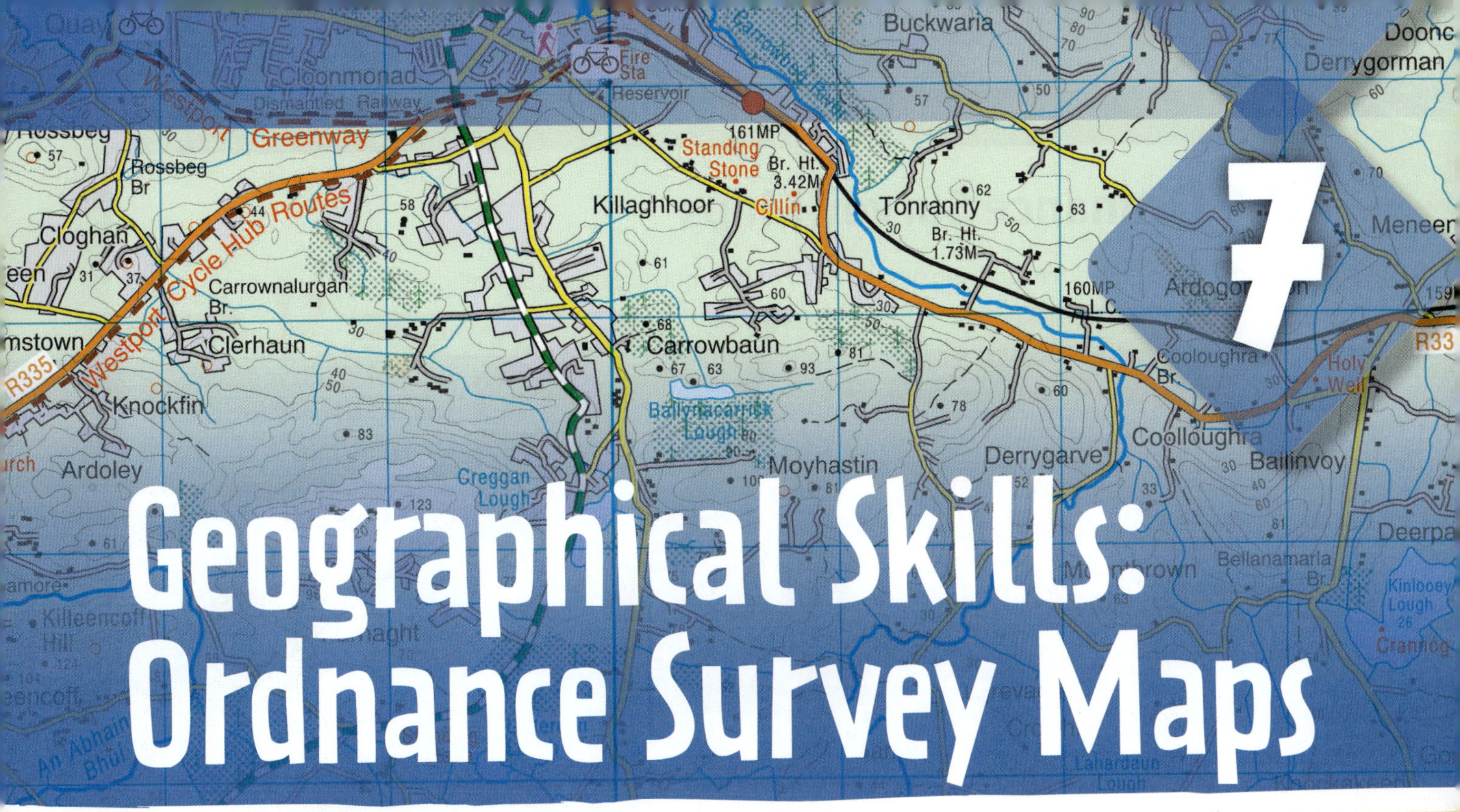

Geographical Skills: Ordnance Survey Maps

7

Element: Geographical Skills

You will be able to:

- Develop your graphicacy through reading and interpreting visual stimuli and data sets
- Investigate geographical information, by suitably interpreting, analysing and presenting data.

Learning Intentions

You will be able to:

- Use and interpret information on different scale Ordnance Survey maps.

- Map
- Scale
- Ordnance Survey (OS) map
- Straight-line distance
- Curved-line distance
- Grid squares
- National Grid
- Sub-zones
- Eastings
- Northings
- Grid reference
- Four-figure grid reference
- Six-figure grid reference
- Altitude
- Height
- Contours
- Triangulation stations
- Spot heights
- Slope
- Cross-sections
- Relief
- Drainage
- Settlement
- Density
- Dispersed
- Linear
- Nucleated

7.1 Ordnance Survey Maps

A **map** is a scaled-down drawing or plan of all or part of the earth's surface.

Q What skills do you think you would need to become a cartographer?

FUN FACT! Cartography is the study of maps and map-making. Someone who makes maps is called a cartographer.

What Is Scale?

Scale is the relationship between a distance on a map and its corresponding distance on the ground. For example, if the scale on a map is 1 cm = 0.5 km (1 cm:0.5 km), on the map, 10 cm would equal a distance of 5 km (10 × 0.5 = 5 km).

On an **Ordnance Survey (OS) map**, the same scale can be shown in three different ways (see figure 1).

Scale is given as a **ratio**. The ratio **1:50 000** means that any one unit of measurement on the map corresponds to 50 000 similar units on the ground.

Linear scale is a ruled line divided into kilometres and parts of a kilometre (or into miles and parts of a mile). It allows us to measure distance directly from the map.

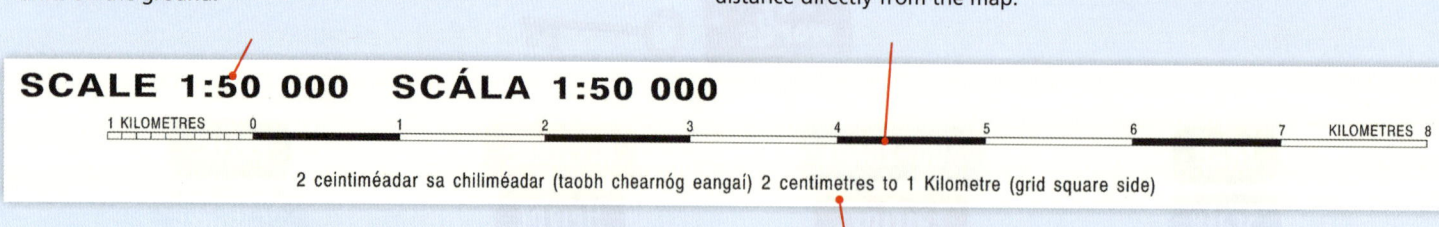

*This scale will be included in your exam papers.

Statement of scale simply states what the scale is (in this case, it is 2 cm to 1 km).

1. Scale*

Maps can be divided into two distinct types:
1. Small-scale maps show large areas in little detail.
2. Large-scale maps show small areas in greater detail.

7. Geographical Skills: Ordnance Survey Maps

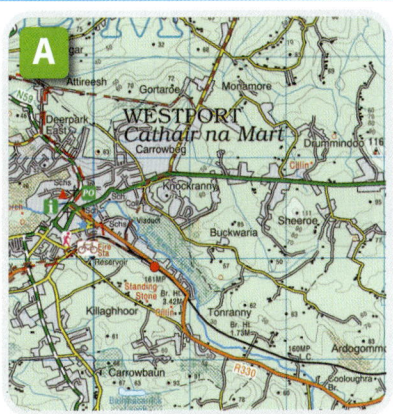

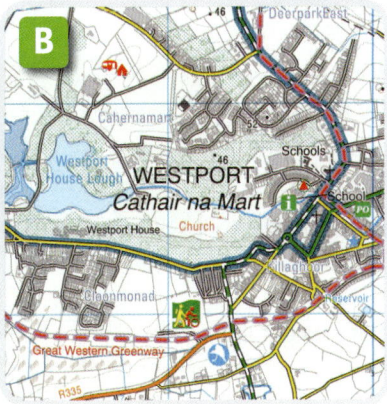

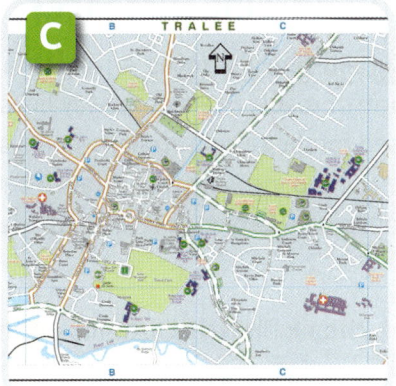

1:50 000 scale map

Amount of detail: You can use a 1:50 000 map for directions when walking but it will only show larger features such as roads and large paths.

1:25 000 scale map

Amount of detail: A 1:25 000 map will give you enough detail for directions to go walking, running and off-road cycling.

1:10 000 scale map

Amount of detail: These maps are used for town planning. Car navigation systems also use 1:10 000 scale maps.

 Working with others

Examine the three images A–C and answer these questions:

Which map would be useful for:

1. Someone wishing to locate a town or village in Ireland?
2. Someone wishing to locate a particular landmark in a town or county?
3. Someone wishing to plot a route around a city or town?

7.2 How to Measure Distances on a Map

Straight-Line Distances

- The shortest distance between any two points is the **straight-line distance**, or 'as the crow flies'.

To calculate straight-line distance, use this method:

- **Step 1:** Place a straight edge of paper on the map so that it passes through the two points you want to measure.
- **Step 2:** Mark the edge of the paper where the first point and last point touch it.
- **Step 3:** Place the edge of the paper on the map's linear scale and measure precisely the distance between the two marks.

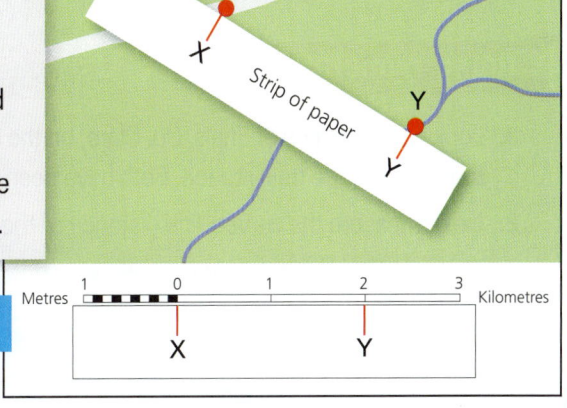

2. Measuring straight-line distance

79

Curved-Line Distances

Curved-line distances on an OS map take the form of roads, rivers or railway lines. The method used for measuring straight-line distances cannot be used to measure the distance between two points on a curved line. Instead, use this method:

- **Step 1:** Divide the road into a number of straight line segments.
- **Step 2:** Using a straight edge of a piece of paper, lay it on the starting point of the distance you wish to measure. Mark it on the paper. Mark each straight line segment along your piece of paper. You now have a number of straight-line distances, which are easier to measure.
- **Step 3:** Mark point A on the linear scale and measure the distance of the curved road. Remember to place point A on the 0 of the linear scale.

3. Measuring curved-line distance

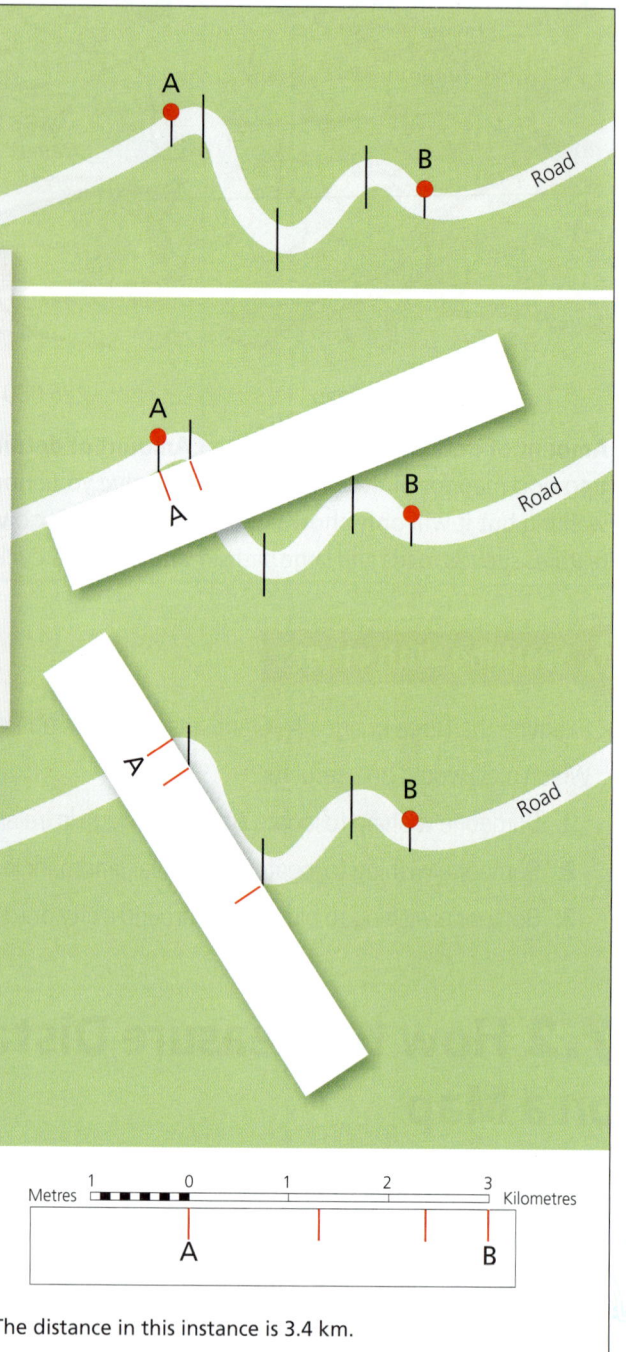

The distance in this instance is 3.4 km.

FUN FACT!

Maps became much more accurate in the eighteenth, nineteenth and twentieth centuries. As the European powers took over territories in the Americas, Africa and Asia, maps became essential for managing the new empires.

Being numerate

1. Looking at the map of Ennis, Co. Clare, on the next page, measure the straight-line distance between the crannóg and the train station. Both have been marked with a blue X.
2. Next, measure the curved-line distance of the railway line between the two red Xs marked on the map.

7. Geographical Skills: Ordnance Survey Maps

Calculating Area on Maps

Ordnance Survey maps with a scale of 1 cm:50 000 cm (which means 1 cm on the map is 50 000 cm on the ground or 1 cm is 500 m) have boxes called **grid squares**. The side of a full grid square measures 1 km.

Calculating a Regular-Shaped Area

Use this method to measure any regular-shaped (rectangular) area:

- **Step 1:** Count the number of grid squares across the base of the area you want to measure.
- **Step 2:** Count the number of squares going up the side of the area.
- **Step 3:** Multiply both numbers. This gives you your answer in square kilometres (sq km/km²).

Calculating an Irregular-Shaped Area

The total area of each grid square on a 1:50 000 map is 1 km². Use this method to calculate the approximate area of any irregular-shaped feature, e.g. a lake, a bay, a forest:

- Tick off and count all the squares that are at least half-filled by the feature. This number will represent the approximate area of the feature in km².

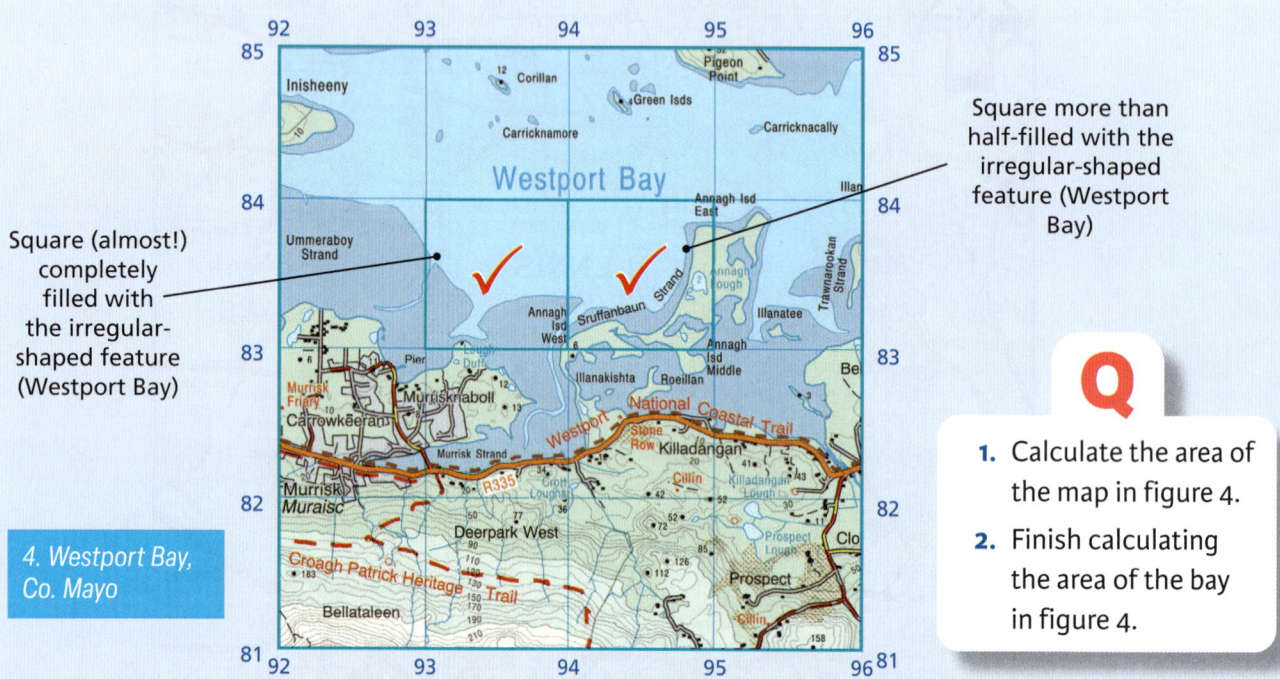

Square (almost!) completely filled with the irregular-shaped feature (Westport Bay)

Square more than half-filled with the irregular-shaped feature (Westport Bay)

4. Westport Bay, Co. Mayo

Q
1. Calculate the area of the map in figure 4.
2. Finish calculating the area of the bay in figure 4.

7.3 Directions on an OS Map

Directions on OS maps are given in the form of compass points.

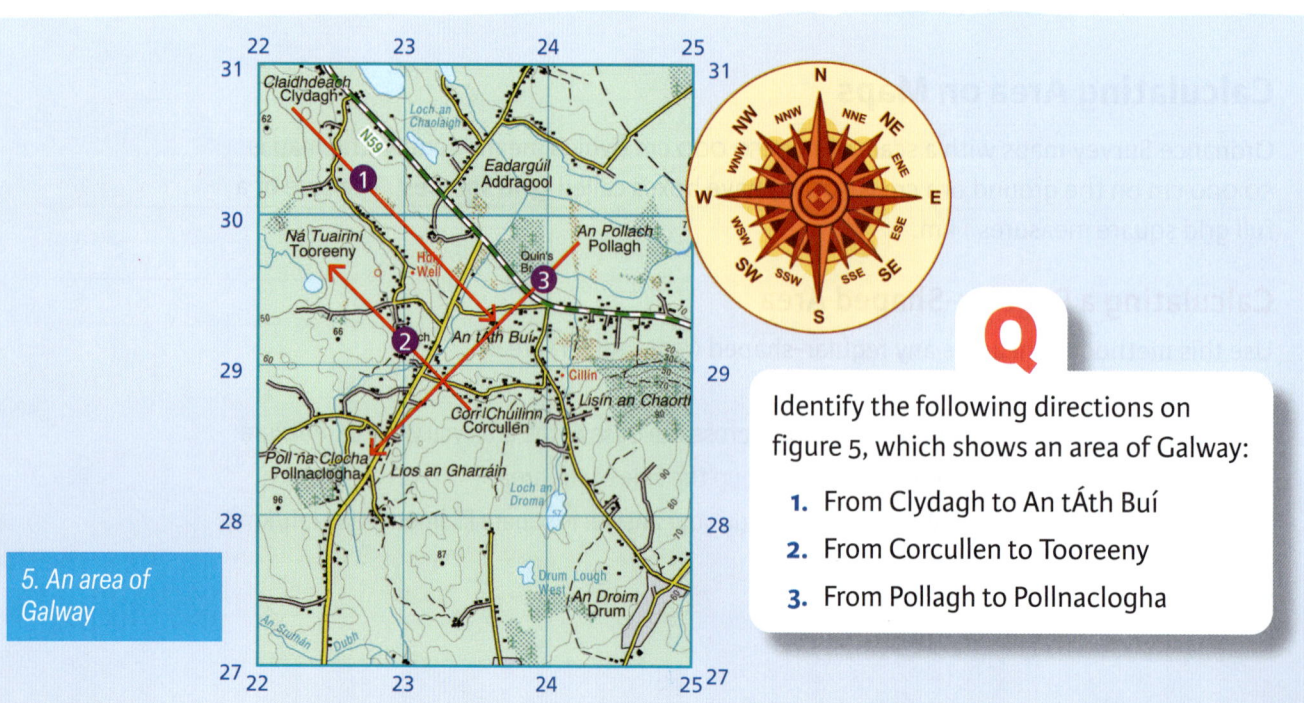

5. An area of Galway

Q

Identify the following directions on figure 5, which shows an area of Galway:

1. From Clydagh to An tÁth Buí
2. From Corcullen to Tooreeny
3. From Pollagh to Pollnaclogha

7. Geographical Skills: Ordnance Survey Maps

7.4 Locating Places: The National Grid

To be useful, maps have to provide a way to identify what areas they are showing. The Irish **National Grid** reference system is used to locate areas on Ordnance Survey maps.

- On the back of an OS map you will find a small map of Ireland. The map is divided into 25 lettered squares called **sub-zones**. (The letter I is left out so as not to confuse it with the number 1.)

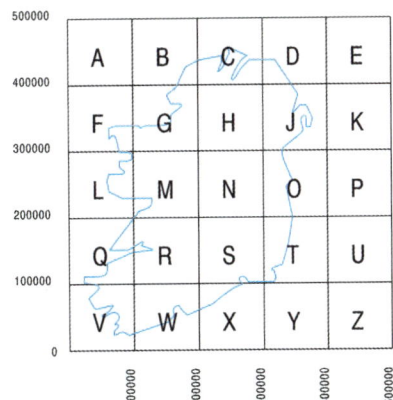

Q Examine the grid (figure 6). Identify the location of your school and record what sub-zone/s it is found in.

6. The Irish National Grid

- Each OS map has a large letter outlined in blue printed on it. This indicates the sub-zone from which the map is taken. If an OS map covers more than one sub-zone, the letters of each sub-zone will be included.

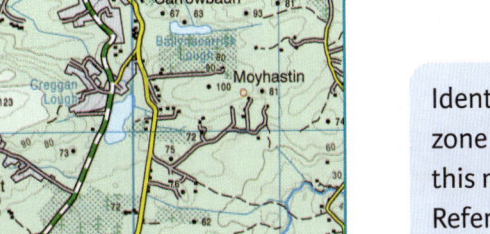

Q Identify the sub-zone letters on this map fragment. Referring to figure 6, which part of Ireland do they indicate?

In order to successfully identify features on an OS map, each sub-zone is divided into smaller squares by a grid of blue lines.

The vertical lines (up and down) are **Eastings**. Their values increase towards the east. We read them across the map in an eastwards direction (so from left to right).

The horizontal lines (across) are **Northings**. Their values increase towards the north. We read them in a northwards direction (so from bottom to top).

- Eastings and Northings are numbered from 00 minimum to 99 maximum.
- We read Eastings first, then Northings.

Grid References on an OS Map

Grid references are used to locate a specific location on an OS map. There are two types of grid references: **four-figure grid references** and **six-figure grid references**.

Four-Figure Grid References

Four-figure grid references are used to give the location of a single square on the OS map. They give the general location of features within that square.

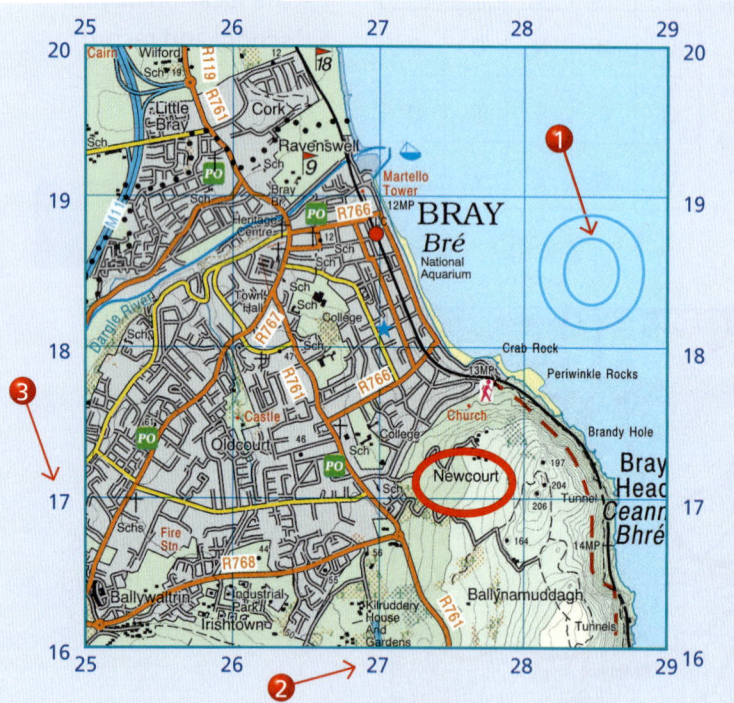

How to calculate a four-figure grid reference

To get the four-figure grid reference number for Newcourt, Co. Wicklow, in this OS map fragment, use LEN (sub-zone **L**etter, **E**asting, **N**orthing):

- **Step 1:** Start by giving the sub-zone letter located on the map.
- **Step 2:** Next, give the two-digit number of the Easting on the bottom side of the square.
- **Step 3:** Then give the two-digit number of the Northing on the left side of the square.

Therefore, the four-figure grid reference for Newcourt is: O 27 17.

Q What is the four-figure grid reference for the Martello Tower?

Six-Figure Grid References

Six-figure grid references are used to give the exact location of a feature within a grid square.

To get a six-figure grid reference, we must imagine that the distance between each pair of Eastings and between each pair of Northings is divided by 10, as in decimal measurement. This will enable us to give a third figure for both the Easting and Northing readings.

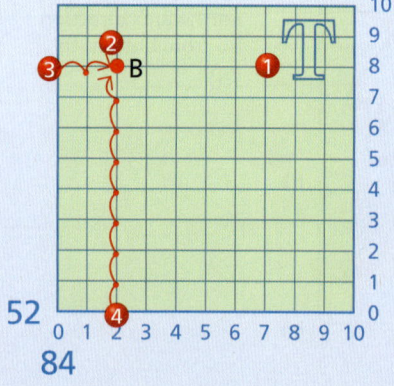

How to calculate a six-figure grid reference

To get the six-figure grid reference number for point B in this enlarged square taken from an OS map:

- **Step 1:** First identify the sub-zone the OS map is taken from (the letter). In this case the sub-zone is 'T'.
- **Step 2:** Identify which square point B is in using four-figure grid referencing: T 84 52.
- **Step 3:** Now count the number of small squares east of 84 to find point B. In this example, it is 2 lines east of 84.
- **Step 4:** Now count the number of lines north of 52 to find point B. In this example, it is 8 lines north of 52.

If we add these new figures to our four-figure grid reference, we get: T 842 528.

7. Geographical Skills: Ordnance Survey Maps

7.5 Identifying Symbols on an OS Map

Symbols are used on an OS map to show features. The symbols can be found on the back of OS maps in the form of a legend (a key to the map).

 Working with others

With the person sitting next to you, group as many of the symbols in the OS legend as you can into the following headings: 1. Transport, 2. Tourist attractions, 3. Services (e.g. school, hospitals).

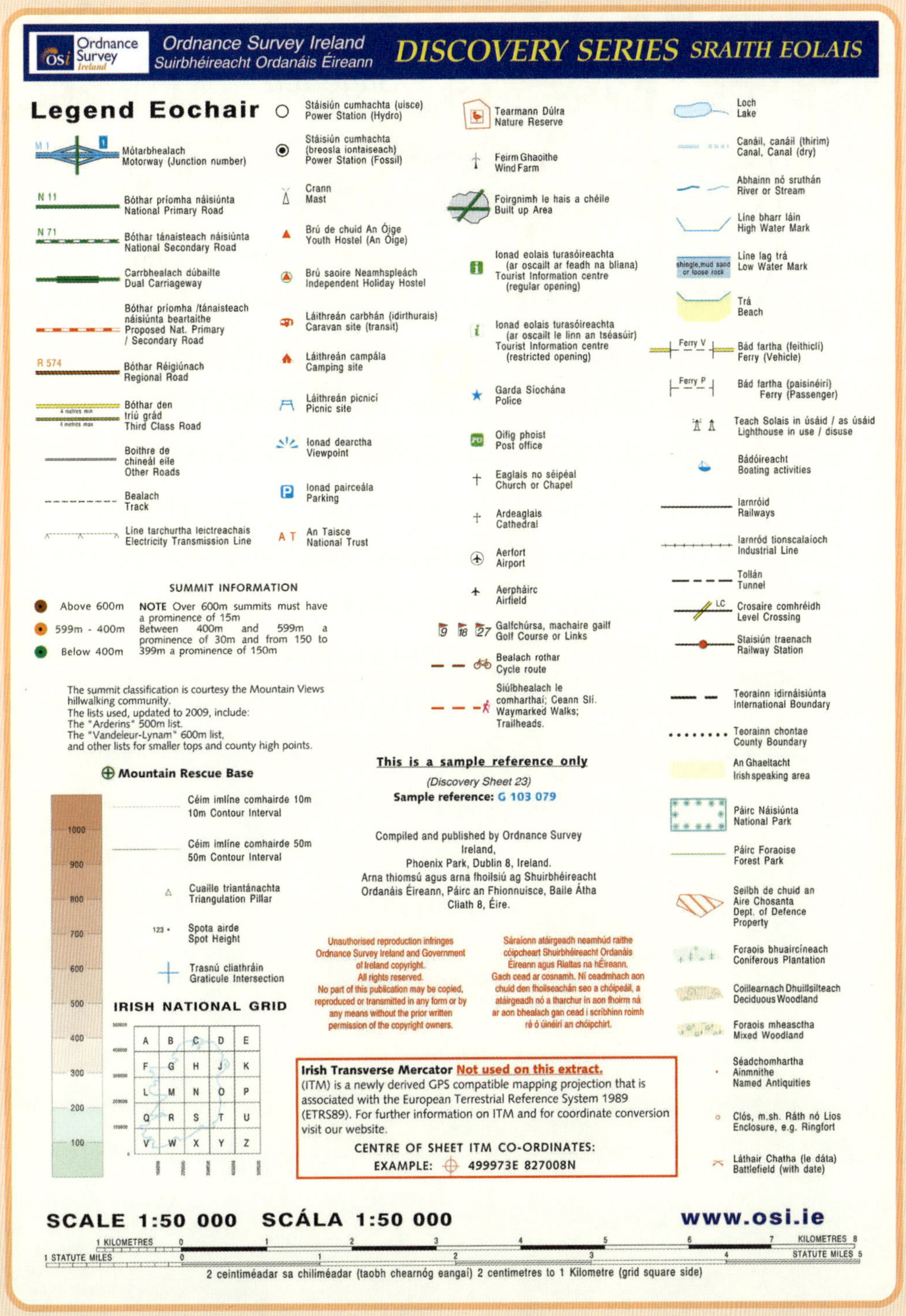

Junior Cycle Geography CYCLONE

Sligo Town Area, Co. Sligo

Q

Give six-figure grid references for each of the following features on the OS map of Sligo above:

1. The train station
2. A spot height under 40 m
3. A named antiquity (Crannóg)
4. A youth hostel.

7. Geographical Skills: Ordnance Survey Maps

Mullingar, Co. Westmeath

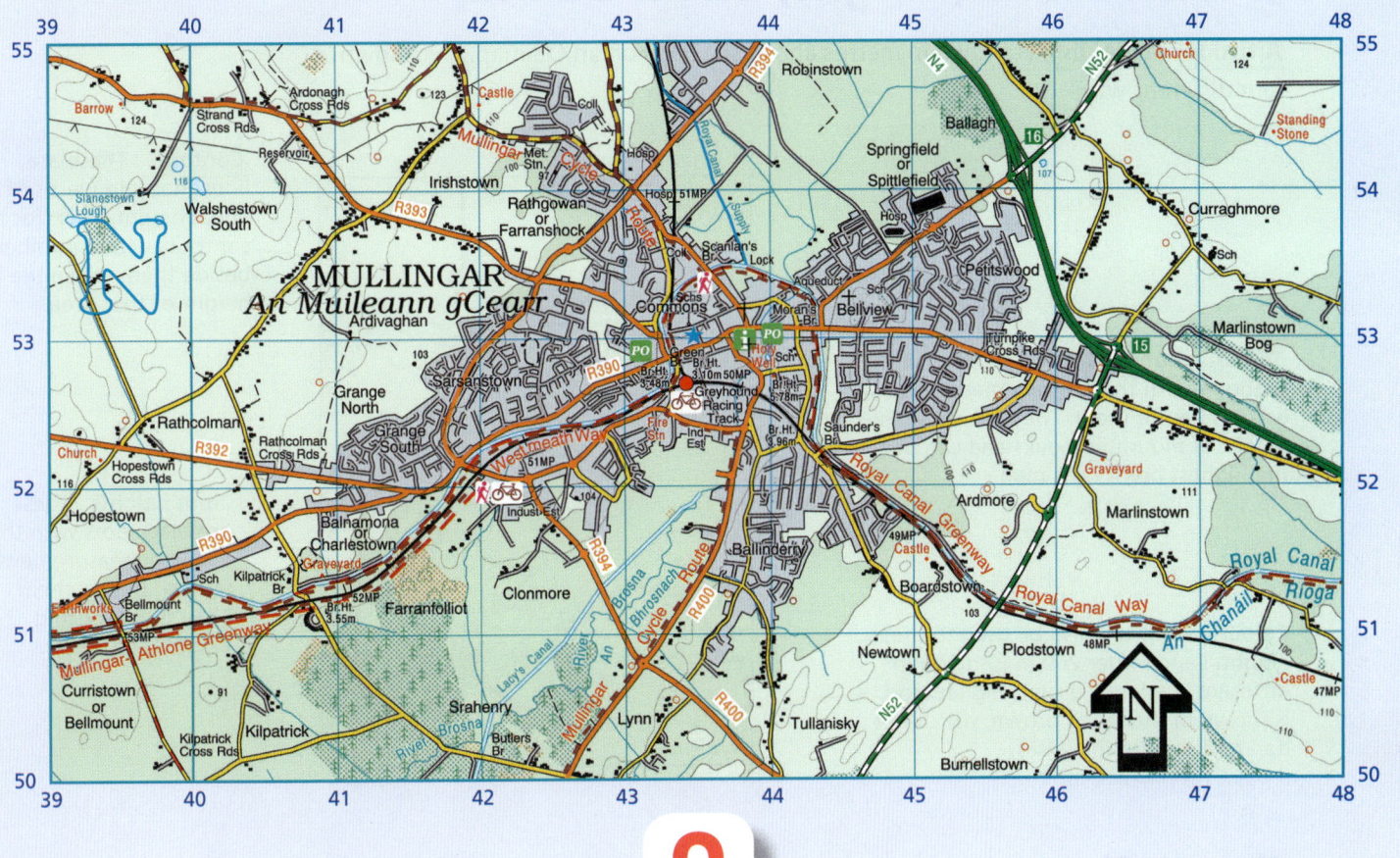

Q

Examine the OS map of Mullingar and answer the following questions:

1. In what sub-zone is Mullingar located?
2. In what direction would a person be travelling if they left Mullingar via the R393 heading left? (Be as detailed in your answer as possible.)
3. Calculate the area of the OS map extract.
4. Calculate the area of the built-up area of Mullingar town.
5. Give the four-figure grid reference for the forested area to the southwest of Mullingar.
6. Which type of forest is this: deciduous or coniferous?
7. Give the six-figure grid reference for the castle located in the very southwest of the map.
8. Give a six-figure grid reference for a spot hight under 100 m located on the map.
9. What tourist attraction is found at N 436 534?
10. Apart from the R393 road what other human-made feature runs through N 39 54 and N 40 54?

87

7.6 Altitude (Height) on an OS Map

Altitude, or **height**, is shown in metres above sea level on an OS map. It is shown in four different ways.

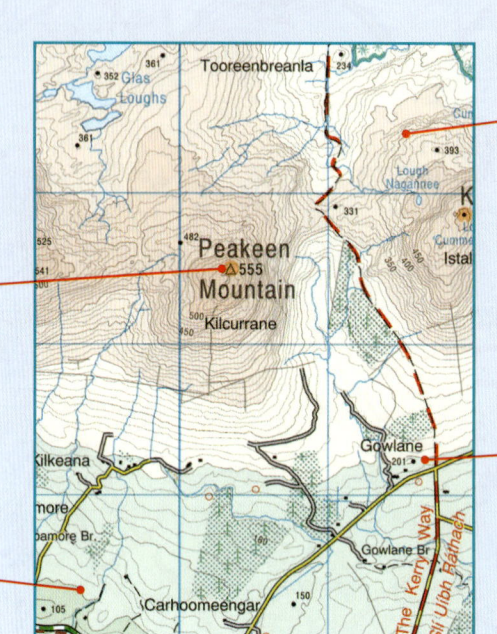

Contours: These are lines that join places of equal height. Look for the contours with a number beside them to see the height of that area.

Triangulation stations: These are black triangles with the altitude measurement written beside them. They can show the height of mountain peaks and can be the highest point on a map.

Spot heights: These are black dots with the altitude measurement written beside them. They show specific heights for places.

Colour coding can also be used to show height. Land under 200 m is shown in different shades of green. Higher land is shown in shades of brown, which get darker as height increases.

Showing Slope

Slope describes the steepness of the land. The closer the contour lines are on a map, the steeper the slope of the land.

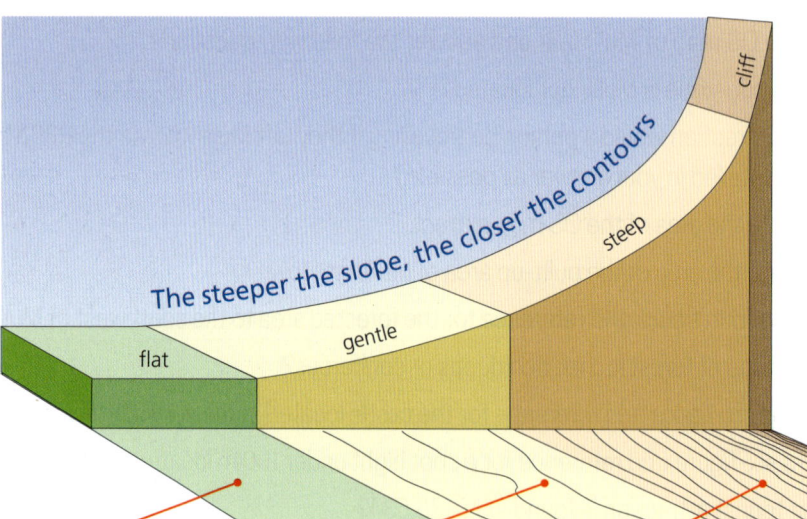

7. Showing slope

Flat Land	Gentle Slopes	Steep Slopes	Cliffs
On an OS map, this will be shown by an absence of contour lines.	On an OS map, gentle slopes will be indicated by contour lines that are widely spread out.	On an OS map, contour lines closely packed together indicate a steep slope.	On an OS map, the presence of cliffs is indicated by contour lines almost on top of each other.

7. Geographical Skills: Ordnance Survey Maps

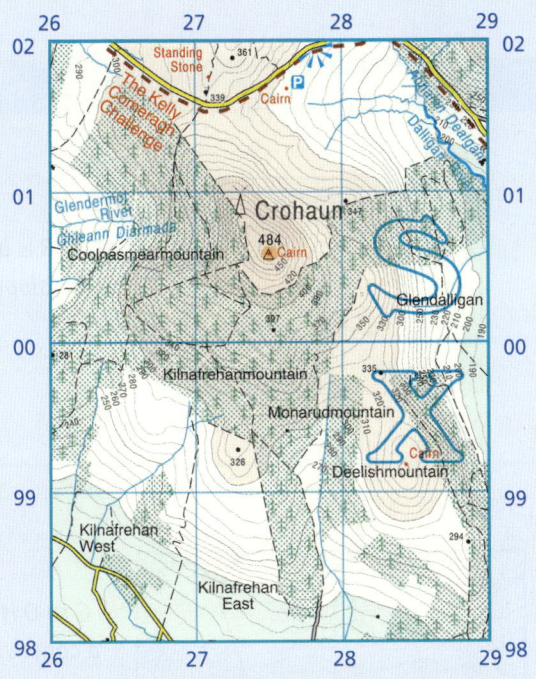

Q

1. Identify the height of the triangulation station you can see on this map segment.
2. What is the second-highest point on this map?
3. Give the four-figure grid references for areas showing: (a) flat land, (b) gently sloping land, (c) steep land, (d) 'cliff' land. (Note that one square may show evidence of different heights or types of land.)

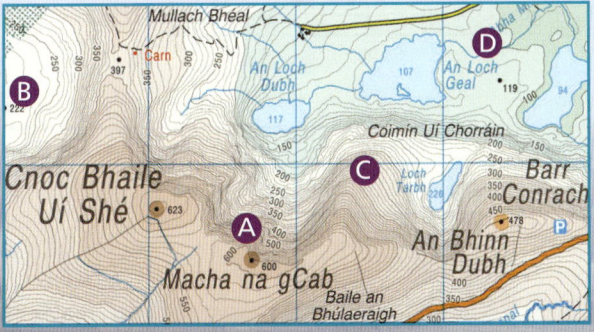

Q

Copy and fill in this table by matching the letter on the map with the appropriate description.

Flat land		Steep slope	
Gently sloping		Cliffs	

Types of Slope on an OS Map

Slope	Side view	On OS map	
Even Slope The contours are evenly spread. The increase in steepness is gradual.			A to B is an even slope.
Concave Slope The contours begin by being widely spaced, then become closer and closer. This means the slope is gentle at first and then becomes quite steep.			C to D is a concave slope.
Convex Slope The contours are close together at the beginning and gradually become more widely spaced at the top. The slope is very steep at its base and evens out at its peak.			E to F is a convex slope.
Compound Slope This slope is a combination of concave and convex slopes.			G to H is a compound slope.

8. Types of slope

7. Geographical Skills: Ordnance Survey Maps

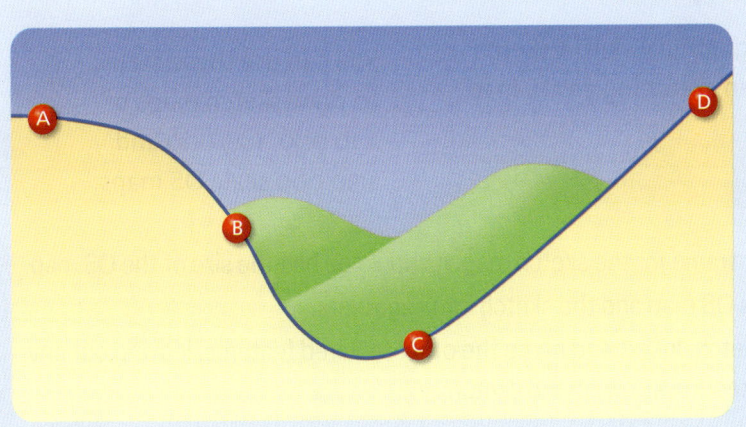

Q Identify the types of slope between each of the following places:
1. A and B
2. B and C
3. A and C
4. C and D

Cross-Sections

Cross-sections, sometimes called profiles, give us a side view of the landscape. This allows us to see the overall height of the landscape and whether it increases or decreases gradually or suddenly.

Cross-sections can be drawn from the contour lines on an OS map.

Look up maps.scoilnet.ie for other cross-section samples.

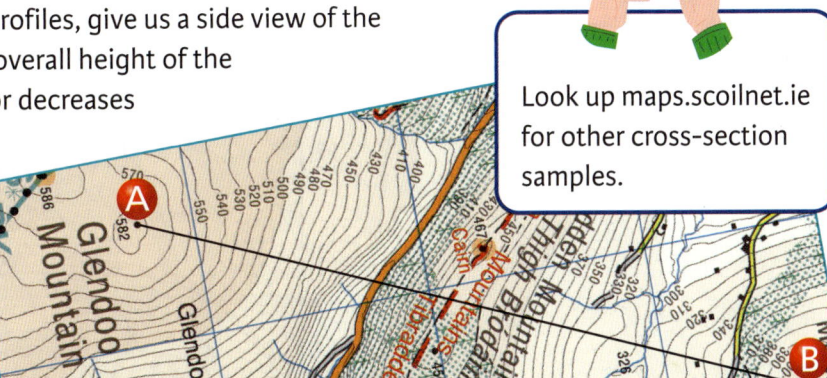

Q
1. Name the mountain labelled A on the map.
2. What is the spot height of this mountain?
3. What type of slope is labelled 1 on the cross-section of this map?
4. What human-made feature, labelled 2, is located on the cross-section?
5. What natural feature occurs at the place labelled 3 on the cross-section?
6. What type of vegetation, labelled C on the cross-section, is evident on the map?

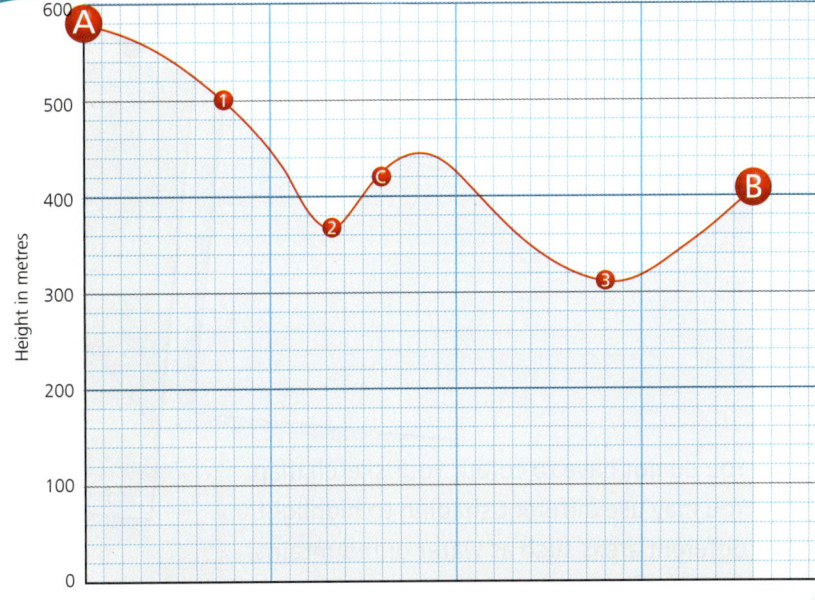

Elevation Profile

7.7 Drawing Sketch Maps

Sketching OS maps is another important geographical skill. A sketch map is a hand-drawn copy of an OS map. When sketching an OS map, there are a number of steps to follow.

You can use these steps to draw sketch maps of 1:10 000, 1:25 000 and 1:50 000 scale OS maps.

How to draw a sketch map
- **Step 1:** Measure the map. Draw a rectangular frame for the sketch map. It should be half the size of the OS map.
- **Step 2:** Using a pencil, lightly divide both the OS map and the sketch into segments.
- **Step 3:** Insert the coastline (if any) on the sketch. Insert and name the other required features, leaving out any unnecessary details. Indicate north with a direction arrow on the sketch.
- **Step 4:** Give the sketch a title.
- **Step 5:** Beside or below your sketch, provide a key/legend that details the key items as drawn on your sketch.

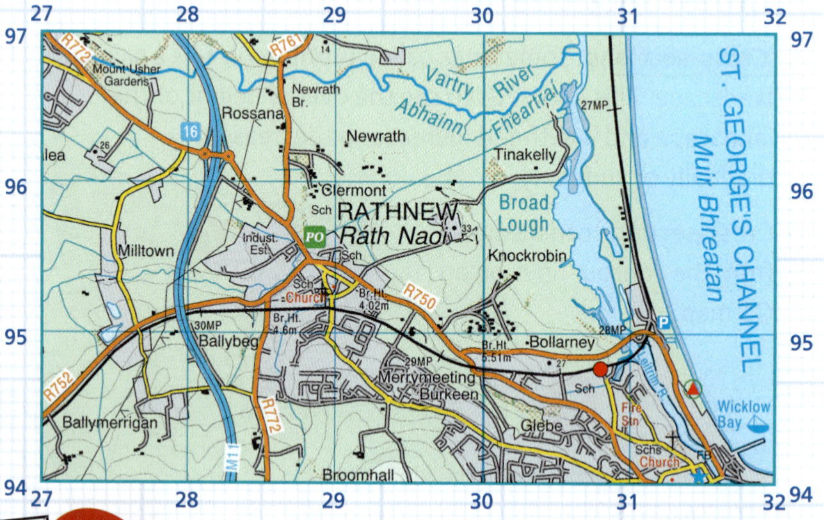

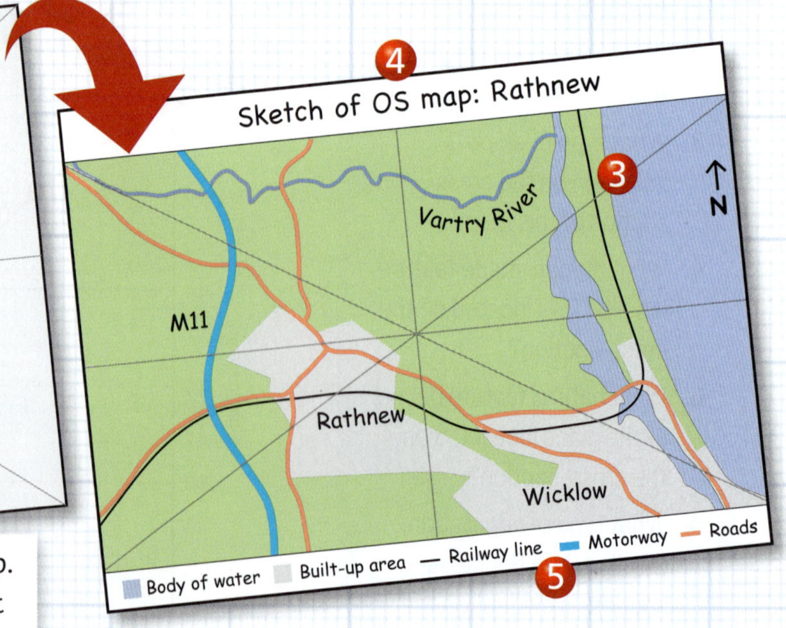

You do not have to colour your sketch map. You may use different shading to highlight different areas such as water or mountains.

7. Geographical Skills: Ordnance Survey Maps

Swords Area, Co. Dublin

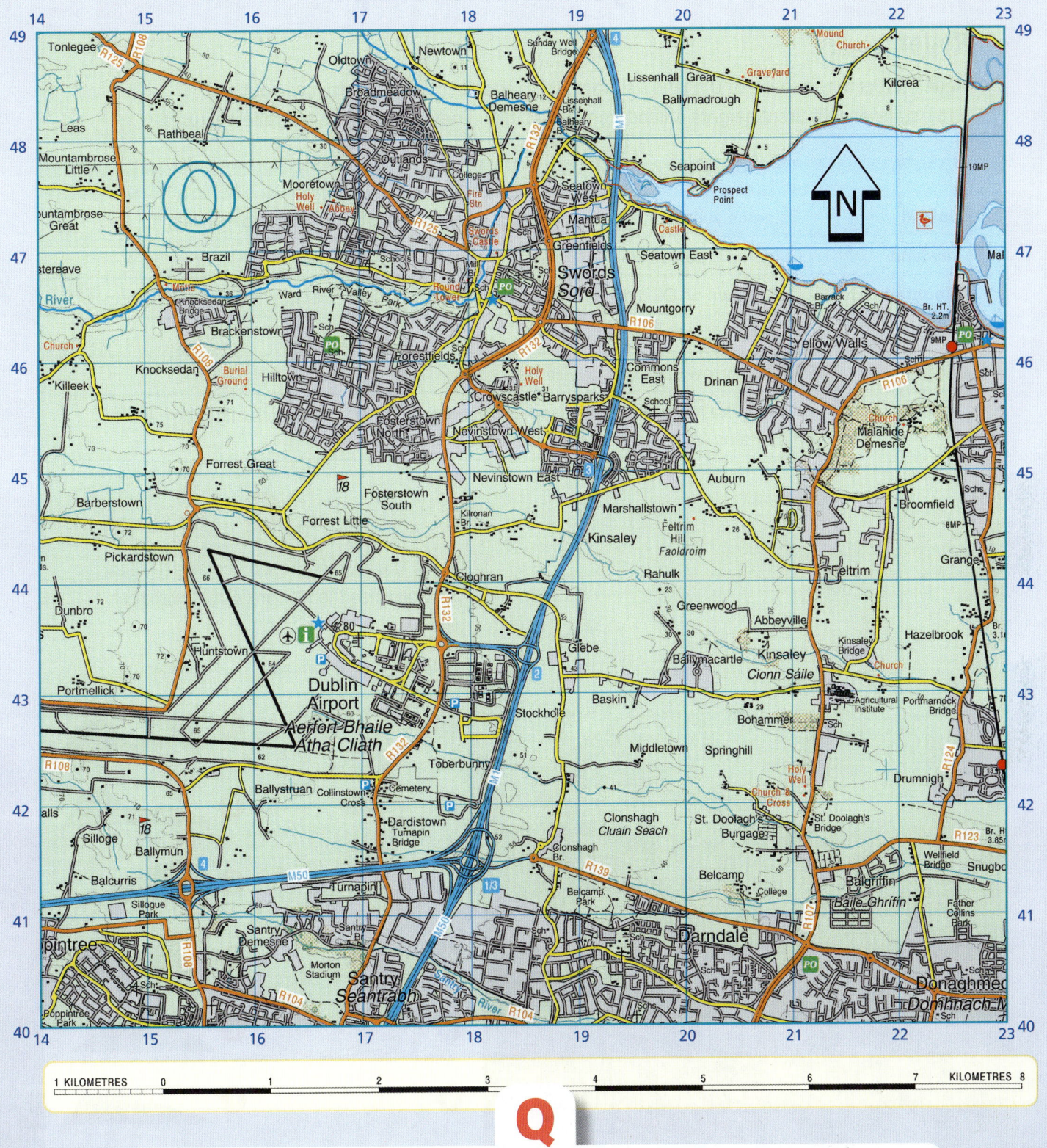

Q

Draw a sketch map of the Swords area as shown above. On your sketch map show the following:

- The bay
- The built-up area of Swords
- Dublin Airport
- The M50/M1 motorway
- The R132 road
- A spot height of your choice.

7.8 Reading the Physical Landscape

Relief

Relief refers to the shape of the land's surface. It is important to be able to make out and understand the different heights of land from an OS map. There are six different relief features.

Relief Feature	Definition
Upland	Coloured yellow or brown on OS map – a large area of land over 200 m in height
Lowland	Coloured green on OS map – a large area of land less than 200 m in height
Mountain	Dark brown on OS map – a steep-sided landform that is over 400 m in height
Hill	A raised area of land under 400 m
Ridge	A long narrow upland area with steep sides
Valley	A valley is a lower part in the land, between two higher parts which might be hills or mountains. Valleys can be created by rivers or glaciers.

Q

Match the relief of these mountains (1–6) as you would see them on an OS map with their side view (A–F).

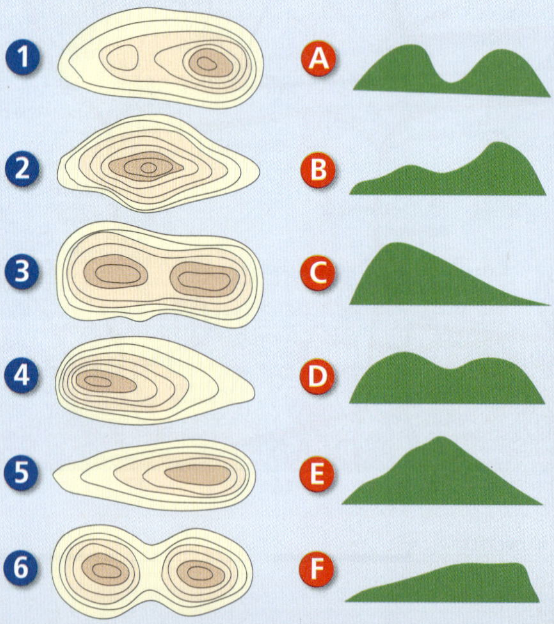

Relief on OS Map Contours   **Side View**

7. Geographical Skills: Ordnance Survey Maps

Galtymore Mountain, Limerick–Tipperary Border

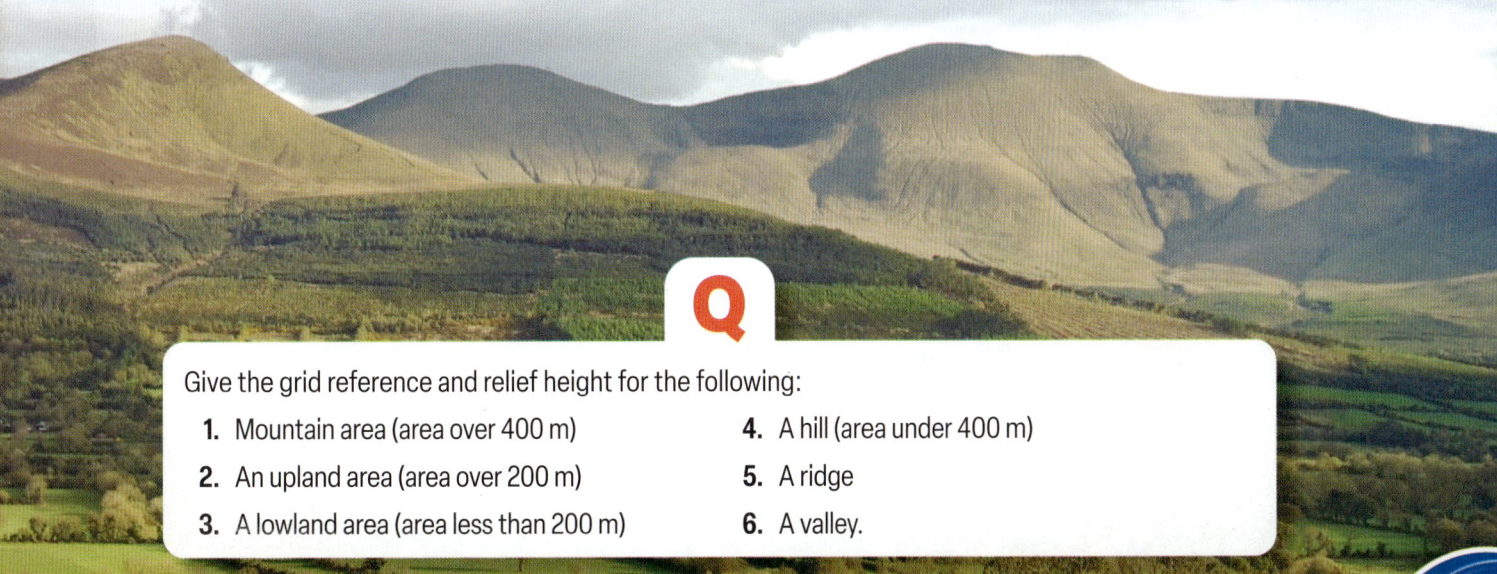

Q

Give the grid reference and relief height for the following:

1. Mountain area (area over 400 m)
2. An upland area (area over 200 m)
3. A lowland area (area less than 200 m)
4. A hill (area under 400 m)
5. A ridge
6. A valley.

Water Features

These are the symbols used to identify water features on a 1:50 000 OS map.

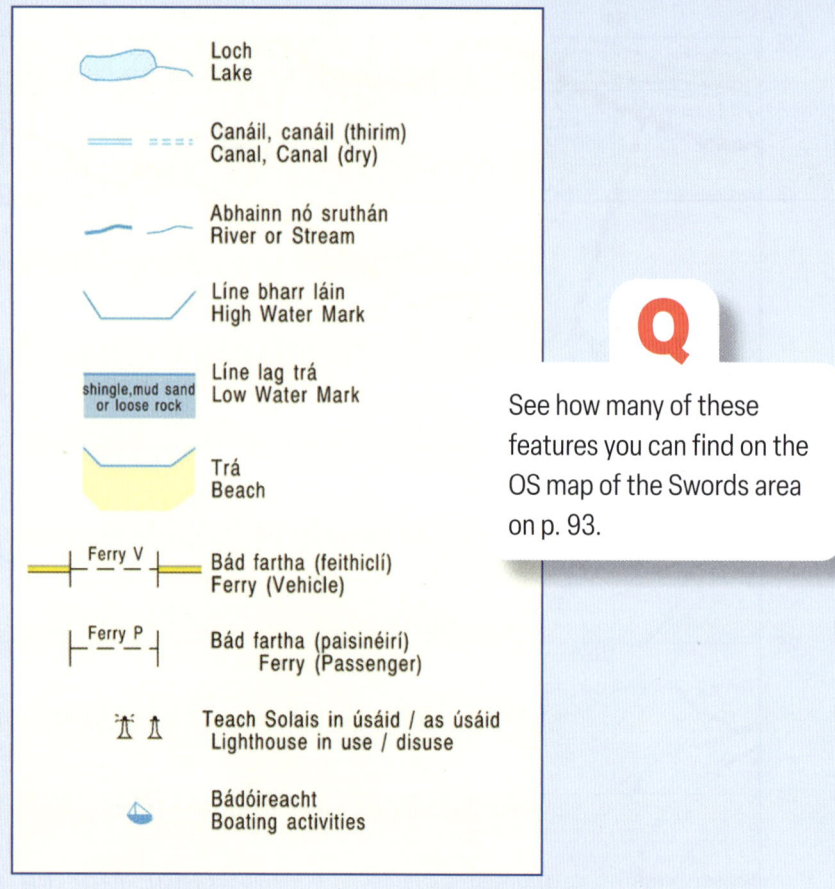

> **Q** See how many of these features you can find on the OS map of the Swords area on p. 93.

Drainage

Drainage refers to the way that water flows on the land's surface. It relates to rivers, lakes and marshes.

OS maps sometimes allow us to identify places that are well-drained and places that are poorly drained or flooded/likely to flood. There are a number of ways to identify each type of area.

A well-drained area will have:

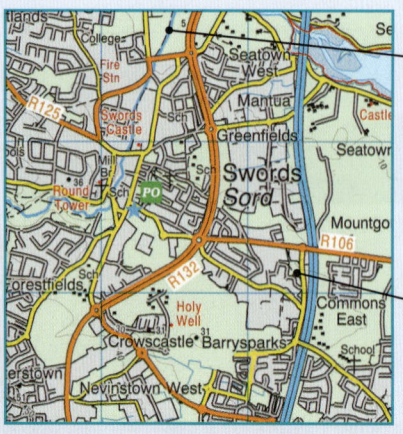

- Very few areas of surface water. A river is the only example in this map segment.
- Examples of settlement and roads, despite the river. We can take this to mean that the river is unlikely to flood and will probably not disrupt the area.

7. Geographical Skills: Ordnance Survey Maps

A poorly drained area may have:

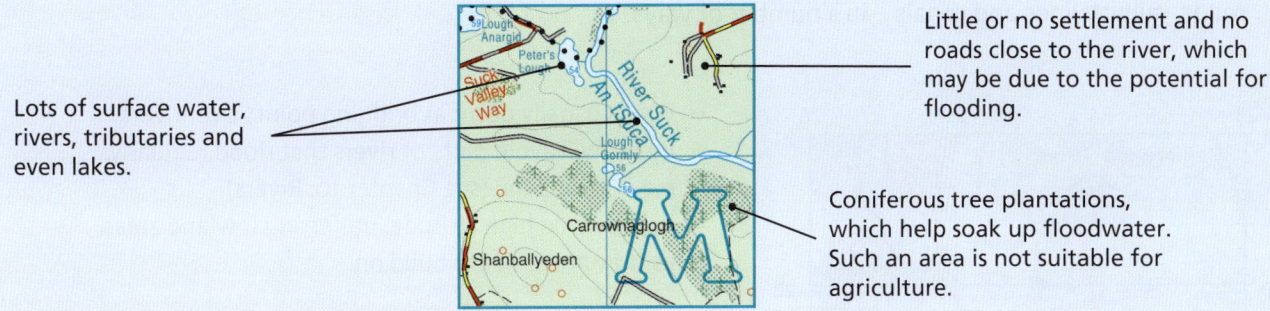

Lots of surface water, rivers, tributaries and even lakes.

Little or no settlement and no roads close to the river, which may be due to the potential for flooding.

Coniferous tree plantations, which help soak up floodwater. Such an area is not suitable for agriculture.

7.9 Reading the Human Landscape

Transport

In Ireland, most transport is carried out on roads. Transport also takes place by rail, air, sea, river and canal.

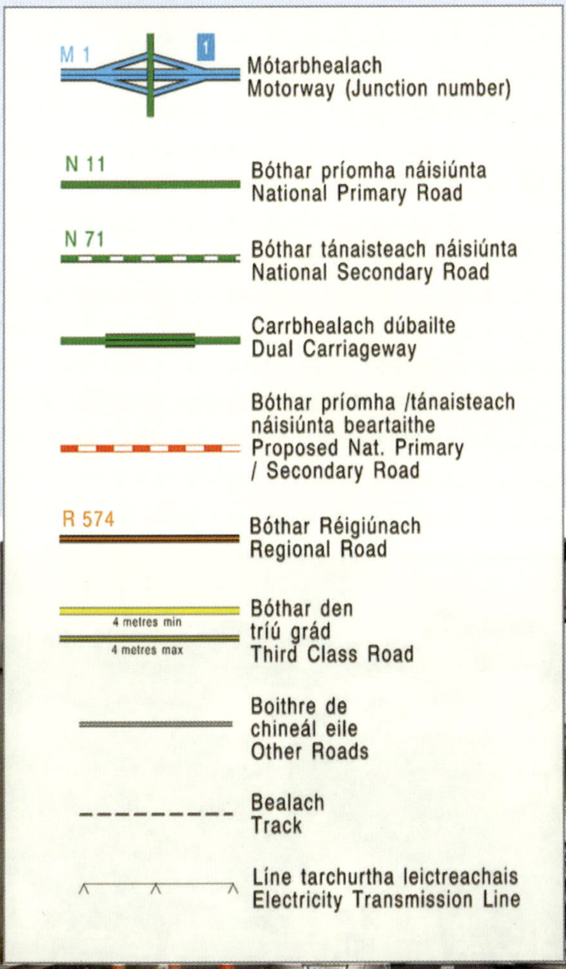

M 1 Mótarbhealach / Motorway (Junction number)

N 11 Bóthar príomha náisiúnta / National Primary Road

N 71 Bóthar tánaisteach náisiúnta / National Secondary Road

Carrbhealach dúbailte / Dual Carriageway

Bóthar príomha /tánaisteach náisiúnta beartaithe / Proposed Nat. Primary / Secondary Road

R 574 Bóthar Réigiúnach / Regional Road

Bóthar den tríú grád / Third Class Road (4 metres min / 4 metres max)

Bóithre de chineál eile / Other Roads

Bealach / Track

Líne tarchurtha leictreachais / Electricity Transmission Line

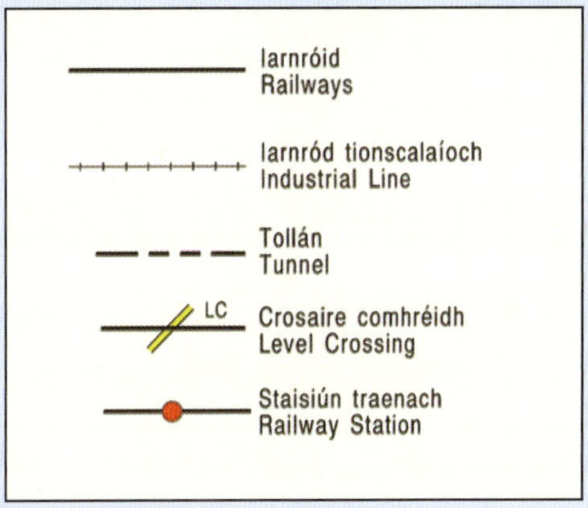

Iarnród / Railways

Iarnród tionscalaíoch / Industrial Line

Tollán / Tunnel

LC Crosaire comhréidh / Level Crossing

Staisiún traenach / Railway Station

Q Which types of road do you think are the most used for transportation? Why?

Junior Cycle Geography CYCLONE

How Landscape Can Influence Route Ways

Relief and drainage can influence route ways – displayed on OS maps in the form of roads, railway lines and canals – in a number of ways.

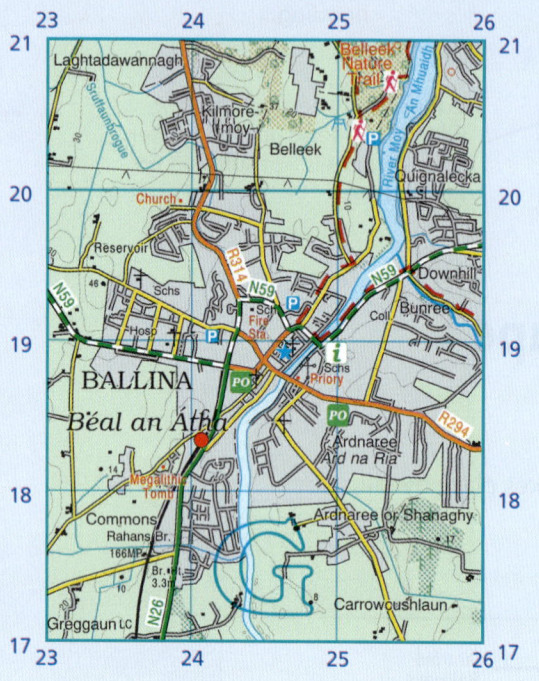

- Roads usually occur at bridging points over rivers.
- Roads will avoid parts of rivers that flood regularly, e.g. a floodplain (see Chapter 10: Rivers).
- Roads avoid high altitude, preferring lowland areas, as they are easier to build on.

Working with others

1. Working with the person next to you, examine the OS map section showing Ballina and identify:
   - A bridging point
   - A road that avoids bodies of water
   - An alternate route way (not a road – think historically)
2. Identify the height of the land in this OS map section and comment on how it has influenced the location of Ballina.

7.10 Settlement on an OS Map

A **settlement** is a place where people live. There are a number of different types of settlement that can be seen on an OS map.

Former/Historic settlement	Ancient or historical sites of where humans lived
Rural settlement	Small villages and isolated houses
Urban settlement	Cities and towns
Place names	Common place names

98

7. Geographical Skills: Ordnance Survey Maps

Former/Historic Settlement

On an OS map, former/historic settlements are called 'antiquities' – an object or building, for example, from the ancient past. They are shown and named in red print and have a red dot or red circle beside them. Two crossed swords indicate a battlefield.

- • Séadchomhartha / Ainmnithe / Named Antiquities
- ○ Clós, m.sh. Ráth nó Lios / Enclosure, e.g. Ringfort
- ✕ Láthair Chatha (le dáta) / Battlefield (with date)

Ballyshannon, Co. Donegal

Q Ballyshannon claims to be the oldest continually settled place in Ireland. Use six-figure grid referencing to identify four examples of antiquities, which may be evidence of former/historic settlement in the Ballyshannon area.

Rural Settlement

There are a number of factors that affect the **density** (the amount of settlement) and location of rural settlement.

Altitude	Most human settlement is found on land below 200 m (coloured green on OS maps). Land above that is often too cold, wet and windy to live on.
Slope	Most building is done on flat or gently sloping land. Farming is also easier on this type of land.
Aspect	This refers to the direction in which the slope faces. South-facing slopes receive more sun and therefore are preferable for settlement.
Drainage	Well-drained land is better for settlement. Marshes or rivers that are prone to flooding are avoided where possible.

99

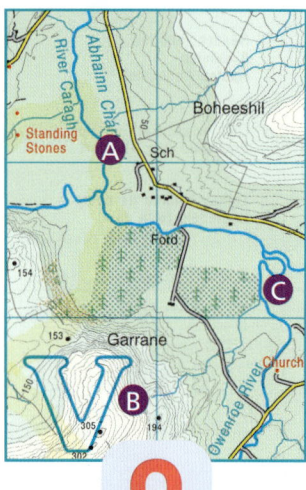

Rural Settlement Patterns

Settlement patterns mean the general way or shape that houses are distributed (spread out) on an OS map. Rural settlement generally has three patterns: dispersed, linear and nucleated. Definitions of these types of settlement can be found in Chapter 21: Rural and Urban Settlement in Ireland.

Houses are represented by little black squares on an OS map.

Dispersed settlement patterns

Q Explain why people would live near area A on this map segment, but not in the areas labelled B and C.

Linear settlement patterns

Nucleated settlement patterns

7. Geographical Skills: Ordnance Survey Maps

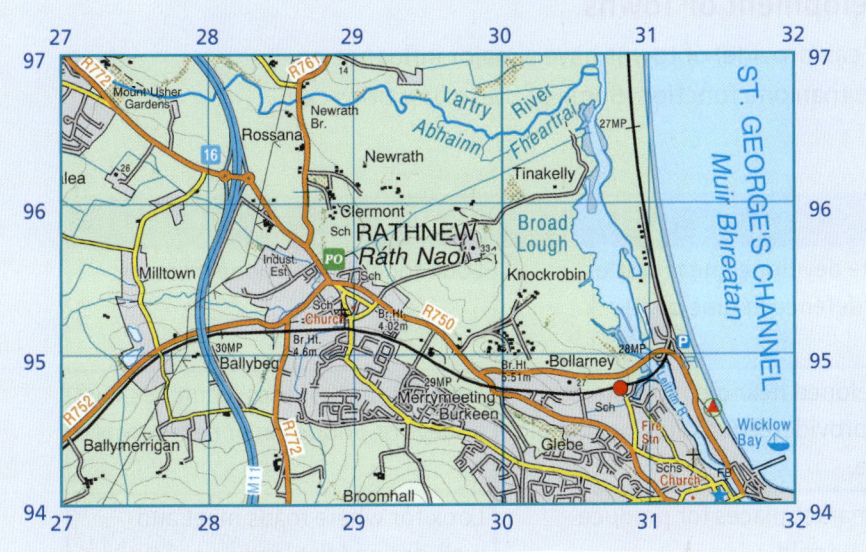

Q Looking at the OS map segment of Rathnew, Co. Wicklow, use six-figure grid referencing to identify an example of:

1. Dispersed settlement
2. Linear settlement
3. Nucleated settlement.

Urban Settlement

Urban settlement refers to settlement in towns and cities. There are a number of factors that influence the location of urban settlement.

Flat or gently sloping land	Flat or gently sloping land is favourable for the construction of buildings and roads.
Where transport routes (road, rail, other) meet	Villages, towns and cities can develop at the point where transport routes meet. These are called nodal points or route focus points. These types of place may encourage the development of trade.
Bridging points on rivers	There has to be a bridge where rivers and roads meet. In the past, bridges were very expensive to construct so roads would converge where one existed. This made areas surrounding bridges attractive locations to begin settlement.
Coastal locations	Ports developed at settlement points along the coast to allow trading. This can provide jobs and encourage further settlement.

Services/Functions and the Development of Towns

The services or functions (things a town can provide) of towns have a major influence on their development. Towns can have more than one function. Some of these functions can be identified using an OS map.

Function	Explanation	OS Map
Defence (function of the past)	Towns would have developed near castles. Castles provided defence in time of attack.	Look for castles in or near towns.
Church (function of the past)	Some towns developed near old monasteries or churches, which provided education, health and spiritual guidance.	Look for monasteries, churches, abbeys or priories in or near towns.
Market (function of past and present)	Towns provided market places for produce farmed nearby to be sold.	Look for where roads meet and well-drained flat land.
Port	Ports can be either large cargo ports, like Cork or Dublin, or smaller fishing ports like Killybegs in Donegal.	Look for piers, docks, lighthouses, beacons or sheltered bays.
Tourist/recreational	Some towns developed because of the recreational or leisure activities they provide.	Scenic: Mountains, lakes, forests Coastal: Beaches, cliffs Cultural: Antiquities Facilities: Camp sites, caravan parks youth hostels, golf courses, etc.
Manufacturing	Most towns include some type of manufacturing or factory. These are not shown by a symbol on an OS map. Some towns feature industrial estates, which contain several factories in one area.	On an OS map, look for the abbreviation Ind. Est. They may be found in or near the town.
Transport	Some towns can develop because of the transport facilities they provide.	Look for settlements near national roads or motorways. Also, look for airports and canals.
Other services	Almost all towns provide a variety of commercial services, shops, offices, etc. Other recreational services, such as restaurants and bars, can also help an area to develop.	Some of these services, e.g. schools, colleges, Garda stations, hospitals, will be indicated on an OS map.

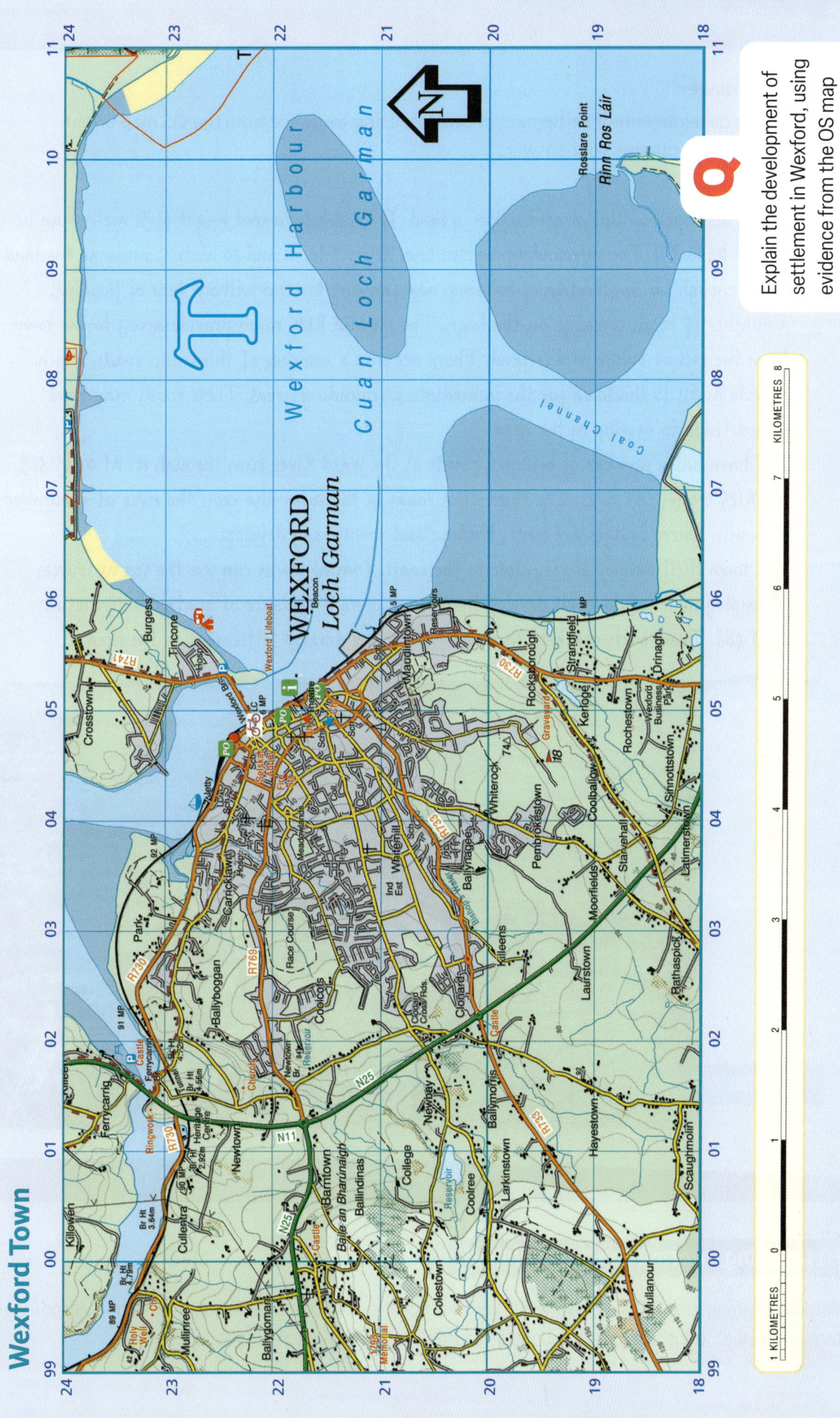

Q Explain the development of settlement in Wexford, using evidence from the OS map to support your answer.

103

Sample Answer

Explain the development of settlement in Swords, using evidence from the OS map on the opposite page to support your answer.

> Swords developed on flat or gently sloping land. For example, a spot height of 31 metres can be seen at O 187 457. The surrounding contour lines descend to 30 and 20 metres, meaning the land is flat enough for construction of buildings and farming to occur without fear of flooding.
>
> A number of roads converge on the town. The M1 and R125 roads provide access to the town from the rest of Dublin and Ireland. There are also a number of third class roads, which provide access to the town for the immediate surrounding areas. These roads may have helped trade to develop in the area.
>
> The town has a number of bridging points as the Ward River runs through it. At O 183 469, the R125 crosses this river. The third-class roads in the town also cross the river at a number of points. Where bridges are built, trading and towns may develop.
>
> The town itself developed very close to the coast. Coastal towns can use the sea as sources of employment should ports develop there, and there is evidence of boating activities at O 211 468. These ports can help trading and further develop settlement in the area.

Aerial photo of Swords

📁 Managing information and thinking

Without referring to an OS map, how would you describe the development of your own town or village? Use the terms 'flat land', 'transport routes', 'bridging ' and/or 'coastal settlement' in your answer.

Swords Area

Junior Cycle Geography CYCLONE

7.11 Large Scale Maps

1:25 000 Scale Maps

A 1:25000 map gives enough detail for directions to go walking, running and off-road cycling.

On a 1:25 000 map, 1 cm on the map represents 25,000 cm or 250 m on the ground.

Grid Referencing on Large Scale Maps

Grid referencing on 1:25 000 maps is the same as for 1:50 000. It uses sub-zone letters and Easting first followed by Northing (LEN).

Q

1. Locate the Great Western Greenway walk on the Westport map.
2. Follow its course as it enters the map at L 97 83 and leaves the map at L 98 85.
3. Identify the following features or activities that the walk comes close to using four-figure and six-figure grid references:
   - A forested area (four figure)
   - A reservoir (four figure)
   - The point at which the walk crosses the Carrowbeg River (six figure)
   - A caravan park (six figure).
4. To increase tourism in the area, the local council would like you to plan another walkway. Create a new walkway using six-figure grid references to locate the start and end point of your walk in and around Westport. Use the directions such as 'Head northwest' or 'Head southeast'. Give four-figure and six-figure grid references of any features/activities your new walkway will come close to.

7. Geographical Skills: Ordnance Survey Maps

1:10 000 Scale Maps

On a 1:10 000 scale map, 1 cm on the map represents 10 000 cm or 100 m on the ground. This scale is used for street maps.

Grid Referencing

Grid referencing on 1:10 000 maps is similar to small-scale maps: Easting first followed by Northing (EN), e.g. A1 or D3.

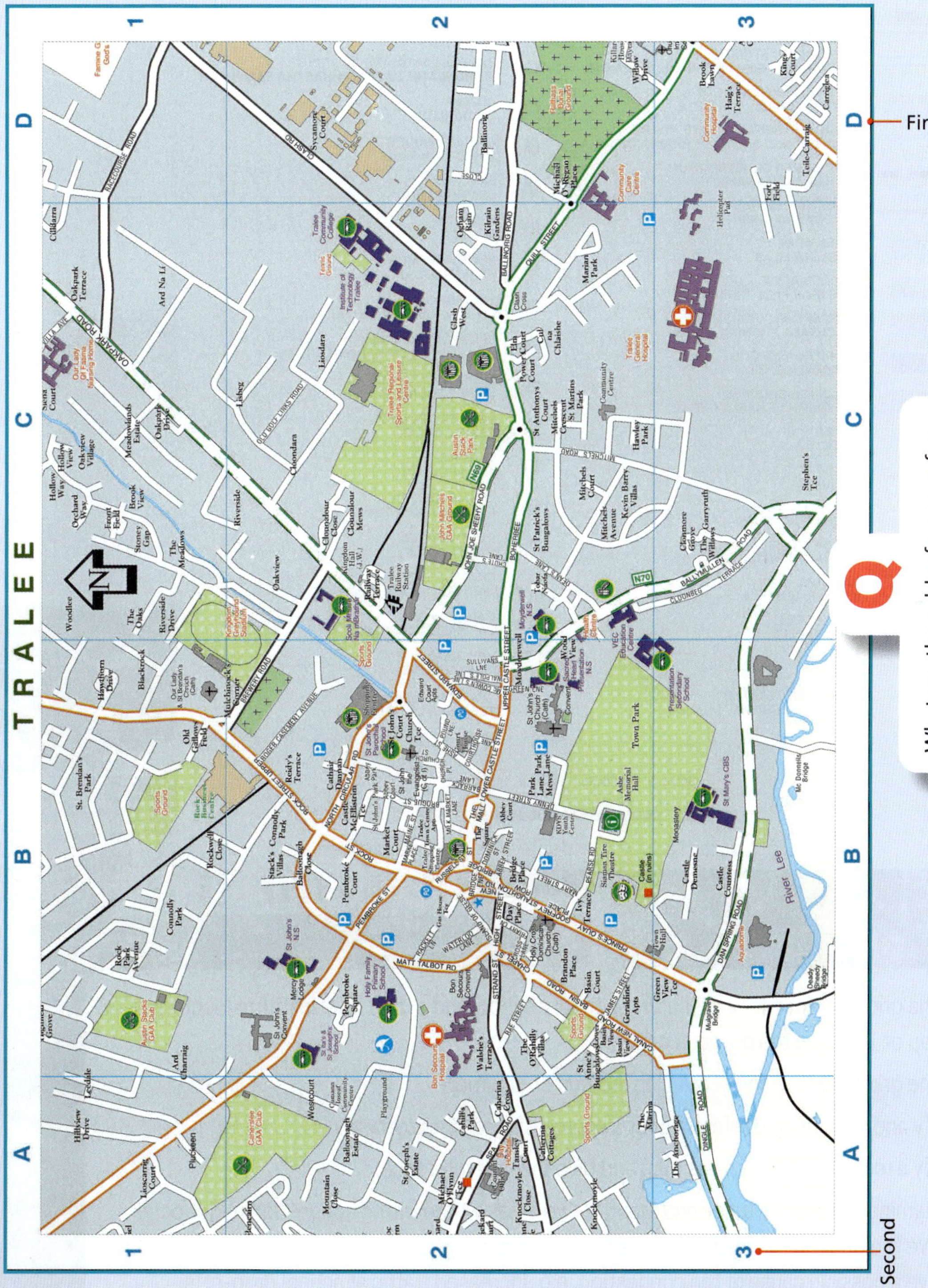

Q What are the grid references for:
(i) Two hospitals
(ii) The sports and leisure centre
(iii) Austin Stacks GAA club?

Junior Cycle Geography CYCLONE

The legend for 1:10 000 maps is different from those for 1:50 000 and 1:25 000 maps. Different symbols are needed as more detail can be seen on 1:10 000 maps.

Legend for 1:10 000 Maps

Q

1. Give the grid reference for a primary/secondary school and a third level institute located on the map of Tralee on the previous page.
2. Give the grid reference for a GAA club located on the map.
3. Identify and give the grid reference for a source of local employment.
4. Identify a residential area on the map and give the grid reference.
5. (a) Examine the map of Tralee and identify two areas where traffic congestion might occur.
   (b) Give the grid reference of the two areas you chose.
   (c) Give a possible cause of the congestion, e.g. increased traffic at school drop-offs.

7. Geographical Skills: Ordnance Survey Maps

Recreation and Tourism on OS Maps

Many tourist or recreational attractions and services can be found on OS maps.

PHYSICAL FEATURES
HUMAN FEATURES

- **Sea** (swimming, fishing, sailing, wind surfing)
- **Car park**
- **Boating activities**
- **Railway station** (access)
- **Offshore island** (scenic)
- **Village** (probable accommodation and other services)
- **Caravan park and camping site** (accommodation)
- **Golf course**
- **Sandy beach** (sunbathing, picnicking, children playing)
- **Cliff-like, rocky coast** (scenic)
- **National roadway** (good access)
- **River/lake** (scenic, angling)
- **Antiquity** (historical interest)
- **Viewpoint** (scenic spot)
- **Mountain** (scenic, walking, climbing)
- **Trekking path** (long-distance walks)
- **Youth hostels**
- **Boating activities**
- **Picnic site with car park**

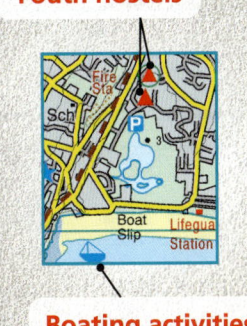

109

Junior Cycle Geography CYCLONE

Sample questions

OS Map 1: Bundoran, Co. Donegal

7. Geographical Skills: Ordnance Survey Maps

Q

1. Draw a sketch map of the area shown on the map. On your sketch map, label the following:
   - The coastline
   - The N15 road
   - The built-up area of Bundoran
   - The Drowes River and any tributary
   - Evidence of historical settlement
   - A feature of coastal deposition.
2. Suggest one reason for the low population density at G 79 56.
3. Give a four-figure grid reference for an area of forestry located on this map.
4. The town of Bundoran provides a variety of services. With reference to the OS map, locate and describe three of these services. (See Recreation and Tourism on OS Maps on p. 109.)
5. Calculate in square kilometres the area of this map.
6. Calculate in square kilometres the amount of area covered by water in this map.
7. State the general direction from:
   (a) The tourist information centre in Bundoran to Rosfriar Point
   (b) Bundoran to the highest point on the map.

Junior Cycle Geography CYCLONE

OS Map 2: Rosslare, Co. Wexford

7. Geographical Skills: Ordnance Survey Maps

Examine the Ordnance Survey map of Rosslare. Answer each of the following questions:

In your copy, complete the sketch map of the Rosslare area by showing and labelling each of the following:

- The N25 national primary road
- The railway line
- One example of tourist accommodation
- One named tourist attraction.

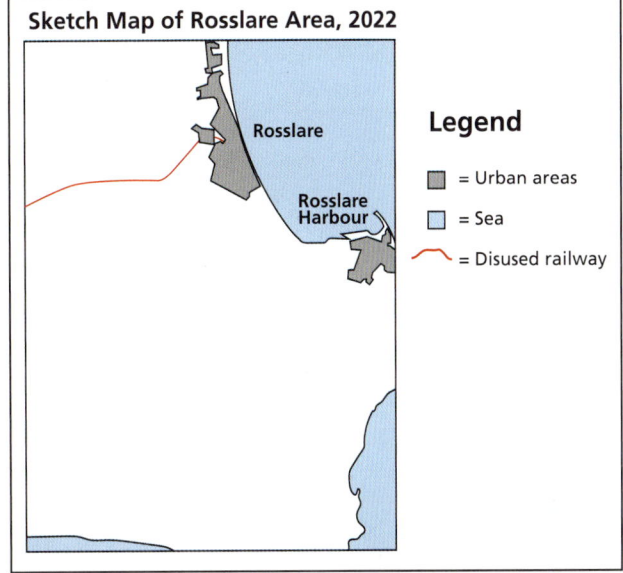

OS Map 3: Lusk, Co. Dublin

Examine the 1:10 000 Ordnance Survey map of Lusk on the next page and answer each of the following questions.

1. (a) The table below lists evidence of four functions in the town of Lusk. Use the Ordnance Survey map of Lusk to complete the table in your copy by answering each of the following:

    (i) Name the function most associated with each piece of evidence.

    (ii) State the grid square where each piece of evidence can be found on the map. One has been completed for you.

Function	Evidence of function	Grid square on map
Residential	The Forge housing estate	B3
	Lusk Community College	
	St MacCullins Church	
	Lusk Town Centre	

    (b) Measure the length of Rathmore Road, in kilometres, from the Raheny Roundabout to the Dublin Road Roundabout.

    (c) In your copy, complete the sketch map of Lusk on the next page by showing and naming each of the following:

    - The full route of Rathmore Road
    - A post office
    - Lusk Heritage Centre
    - One named recreational area.

113

Junior Cycle Geography CYCLONE

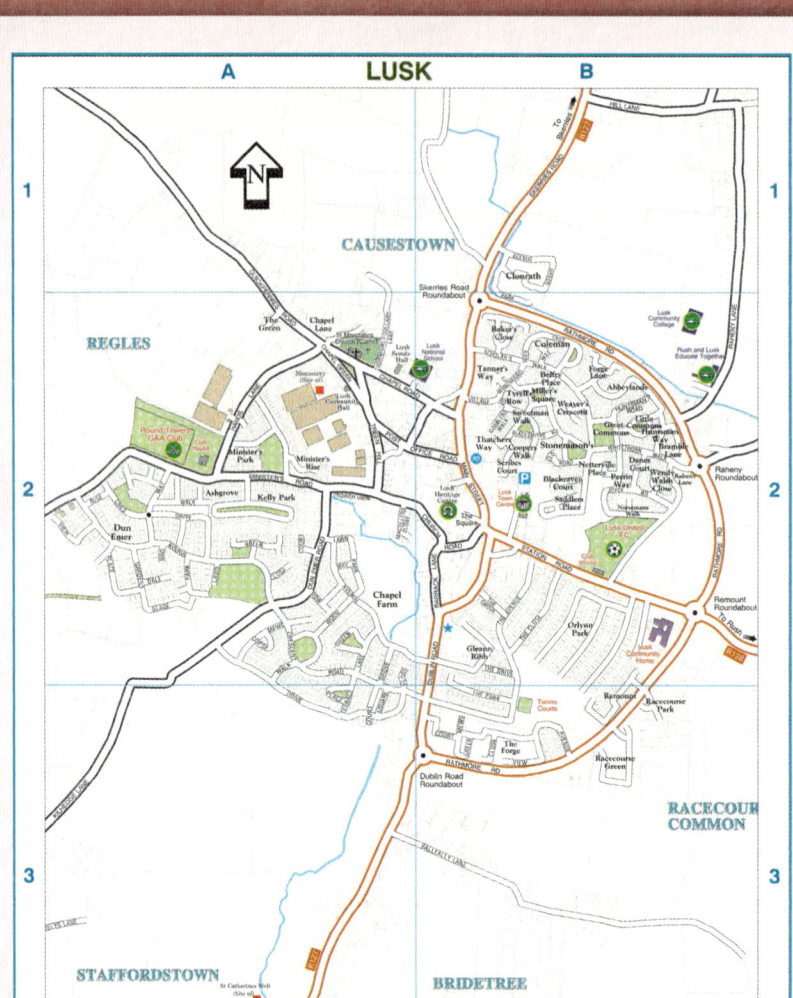

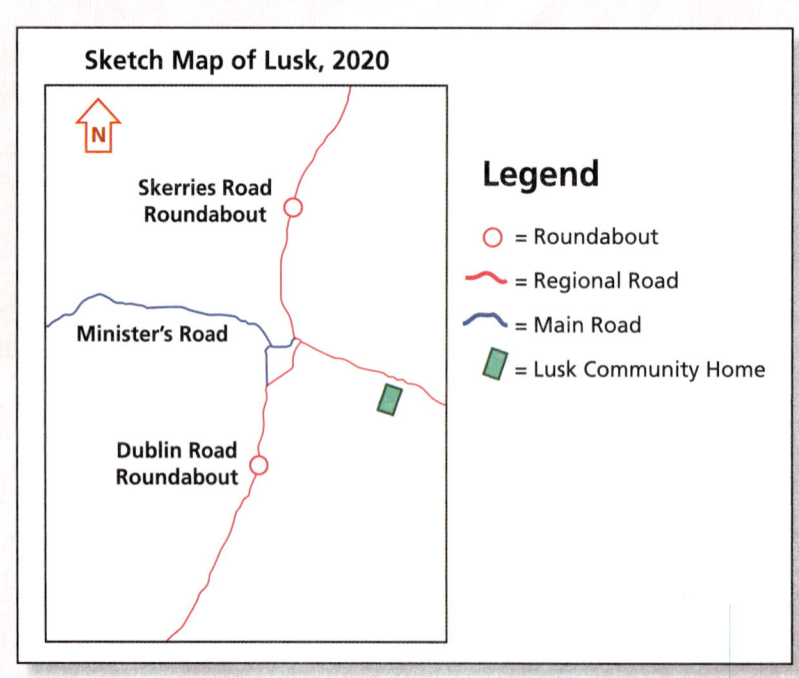

Learning Outcomes: 1.3, 2.9

Weathering

8

key words

Learning Intentions

You will be able to:
- Discuss the process and the effects of mechanical, chemical and biological weathering on the landscape
- Describe how the physical landscape of the Burren has attracted tourism to this region.

Denudation
Weathering
Erosion
Mechanical
Chemical
Biological
Freeze-thaw
Joint
Scree
Carbonation
Permeable
Bedding plane
Karst
Clint
Grike
Swallow hole
Passages
Caves
Stalactites
Stalagmites
Pillar

8.1 The Landscape

Our landscape did not always look as it does today. We know that millions of years ago our mountains were much higher than they are now. Over time, these mountains were worn down. Our landscape is dynamic. This means that it is always changing.

Imagine laying a new pathway – think of how smooth and flat it would be. Then think forward 10 years … how does your pathway look now?

 Being creative

Think for two minutes about the differences you see in these two photos. Then, in pairs, discuss how the road has changed. What may have caused these changes? Prepare to share your thoughts with the class.

A newly paved road

… and 10 years later

8.2 Denudation

Rocks on the earth's surface are constantly being worn down by a number of processes known as **denudation**. Denudation is caused by **weathering** and **erosion**.

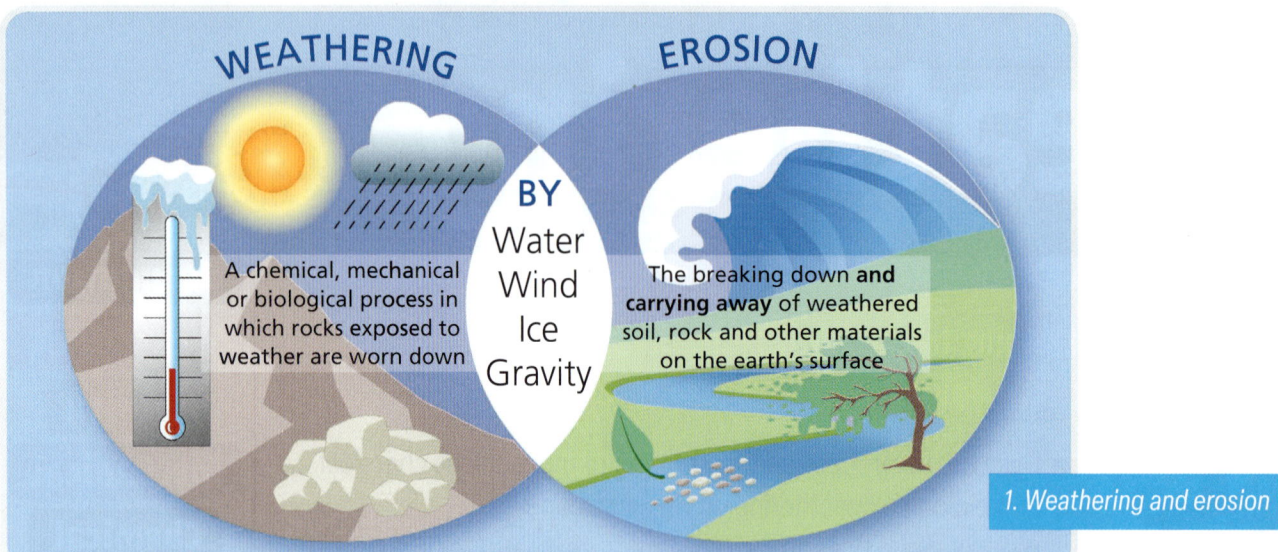

1. Weathering and erosion

Weathering is the process that physically changes solid rock into smaller pieces or sediments. When weathering takes place, the broken sediments remain in place. The three types of weathering are **mechanical** weathering, **chemical** weathering and **biological** weathering.

Erosion is the process that breaks down and moves the sediments away from their original positions. Erosion is caused by water, wind, moving ice (glaciers) and gravity. Once broken, the material will be transported elsewhere and deposited.

FUN FACT! Plants and microbes also eat away at rocks. All living things need elements such as potassium and iron, and these minerals are found within rocks.

Mechanical Weathering

Mechanical weathering is the physical breaking down of rocks into smaller pieces.

Freeze-Thaw Action

The most common type of mechanical weathering is **freeze-thaw** action, also known as frost shattering. Freeze-thaw action can be very active in upland areas such as mountains.

1. Rain water gathers in the cracks (**joints**) in rocks.
2. During winter or when the temperature drops below 0 °C, the water freezes to become ice. This puts pressure on the rock as the ice expands by up to 10%. As the temperature rises, the water thaws and melts again.
3. Repeated freezing and thawing result in the joints expanding and the rock becoming weak and breaking into smaller pieces. These pieces then roll down the slope and gather, where they are known as **scree**.

8. Weathering

By day — ① By day water collects in cracks in the rock.
By night — ② At night the water freezes and expands.
Over time — ③ Repeated freeze-thaw action causes the rock to shatter, and it falls to the bottom of the slope as scree.

2. Freeze-thaw in action

Being creative
Partially fill a plastic bottle with water and mark the height of the water in the bottle with a permanent marker. Place the plastic bottle into the freezer overnight to allow the water to freeze. Note how much the water has expanded when you remove the bottle the next day.

To see the process of weathering and freeze-thaw in action, go to YouTube and search for 'Weathering and erosion – freeze-thaw weathering (Lammas Science)' (1:20).

Q
1. What is the name of the broken material you see here in the photo?
2. Where would you expect to find it as a result of freeze-thaw action: at the top or bottom of mountains?

Question Time

1. What is denudation?
2. Describe the difference between weathering and erosion.
3. Name the three types of weathering.
4. What is mechanical weathering?
5. Using a labelled diagram/s, explain freeze-thaw action.

Biological Weathering

Biological weathering takes place when rocks are worn away by animals, plants and living organisms (creatures). For example, trees and plants can grow within the cracks in rocks. As the roots grow bigger, they push open the cracks, making them wider and deeper. Over time, the growing tree or plant eventually breaks the rock apart.

Weeds pushing up through the cracks in the pavement

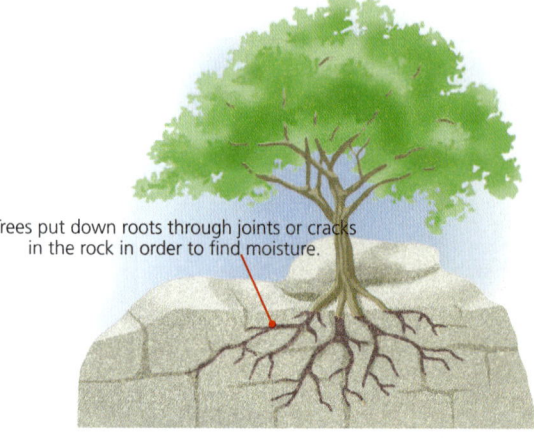

Trees put down roots through joints or cracks in the rock in order to find moisture.

3. Biological weathering in action

Managing information and thinking

1. How do you think biological weathering caused the cracks you see in figure 3 to form?
2. Take a walk around your school or local area and take photographs of any biological weathering that you can identify. Then display your images in class.

Chemical Weathering

Chemical weathering is caused by chemical changes, when rock is dissolved or decays.

Carbonation

Carbonation is an example of chemical weathering.

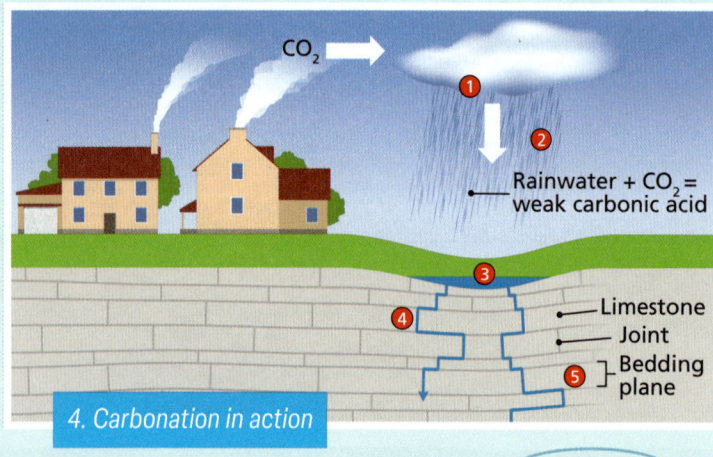

4. Carbonation in action

5. The calcium carbonate in limestone dissolves in the same way as when you drop soluble tablets into water!

❶ As rainwater falls and passes through the air, it dissolves carbon dioxide.

❷ Carbon dioxide mixes with the rainwater to form a weak carbonic acid.

❸ Limestone in the ground contains calcium carbonate. When the weak carbonic acid in the rainwater falls onto the limestone, a chemical reaction takes place.

❹ The calcium carbonate in the limestone begins to dissolve. This process is known as carbonation.

❺ Limestone is a **permeable** rock, which means that rainwater can easily pass through it. It has vertical cracks, which are known as joints, and horizontal **bedding planes**.

8. Weathering

Karst Landscapes

Karst landscapes are areas of land made up of limestone. The effects of carbonation are best seen in karst areas. The overlying soil and plants have been removed, and the bare limestone rock is exposed (uncovered). An example of a karst region in Ireland is the Burren in Co. Clare.

Managing information and thinking

Looking at this map, can you identify three countries that appear to have large karst regions? Use an atlas or Google Maps to help you.

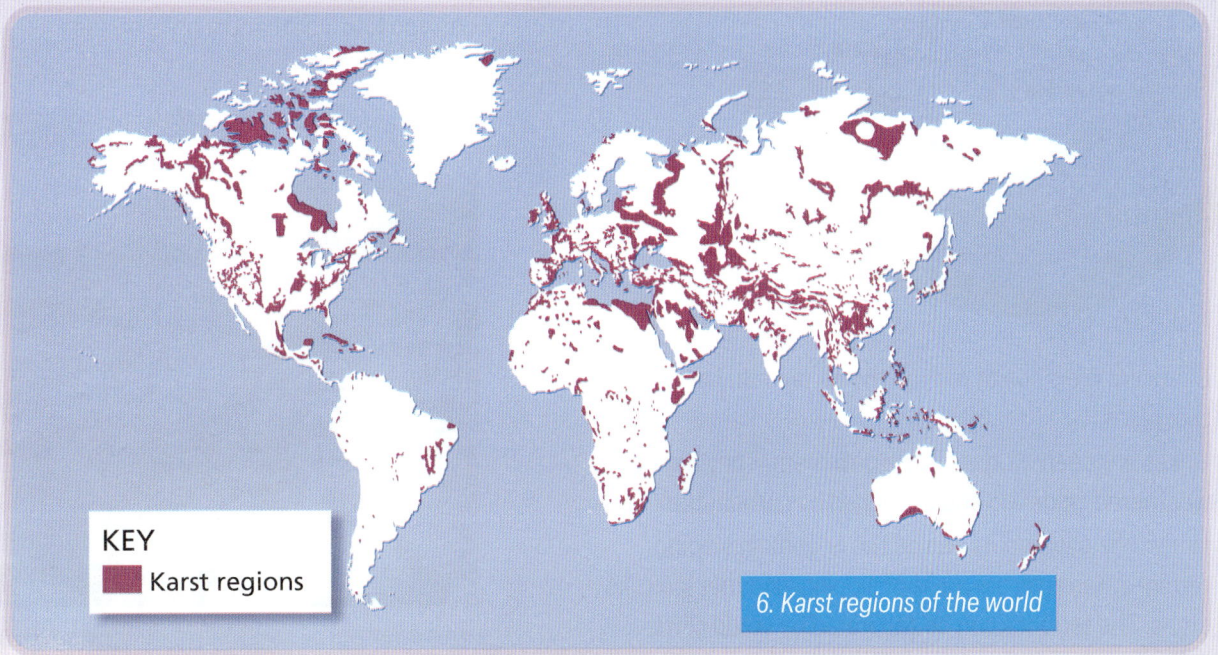

KEY
■ Karst regions

6. Karst regions of the world

The Burren, Co. Clare

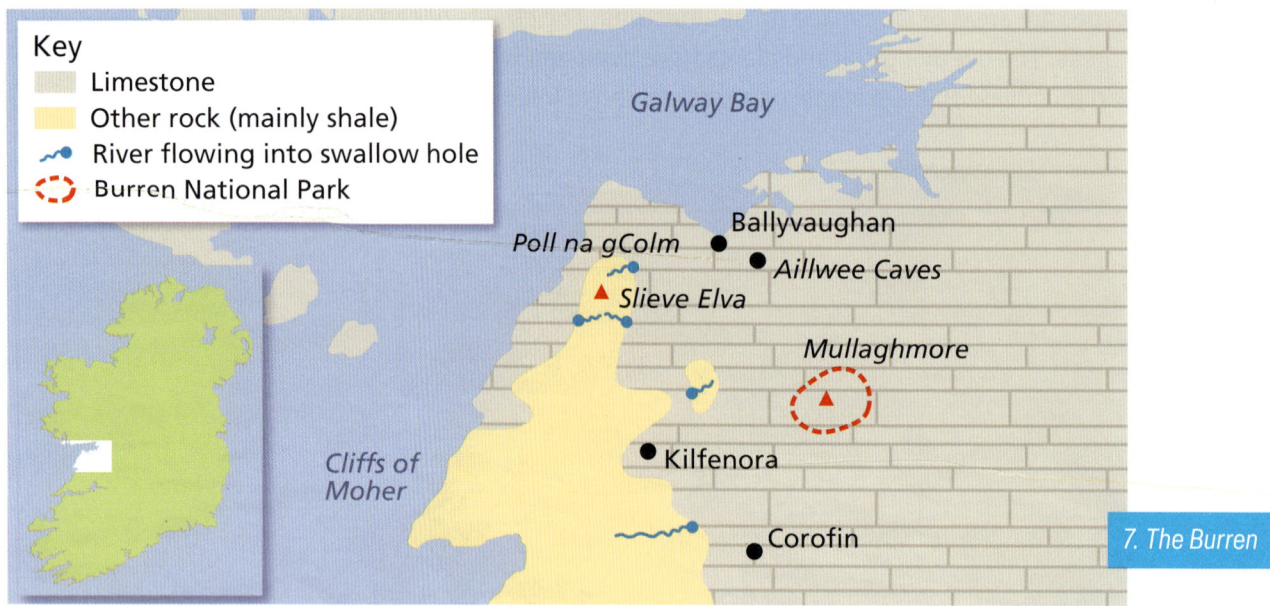

7. The Burren

Weathering by carbonation has made a unique and interesting landscape in the Burren. Both on the surface and below the ground, there are features formed by carbonation.

Surface Features

Limestone Pavement

The bare, rocky karst surface is called a limestone pavement. The pavement is made up of **clints** and **grikes**.

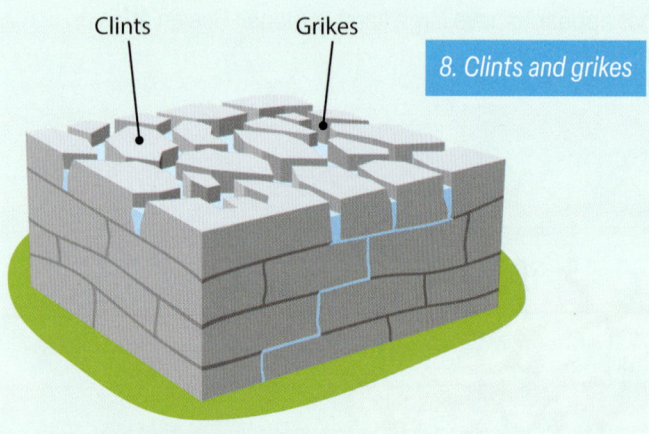

8. Clints and grikes

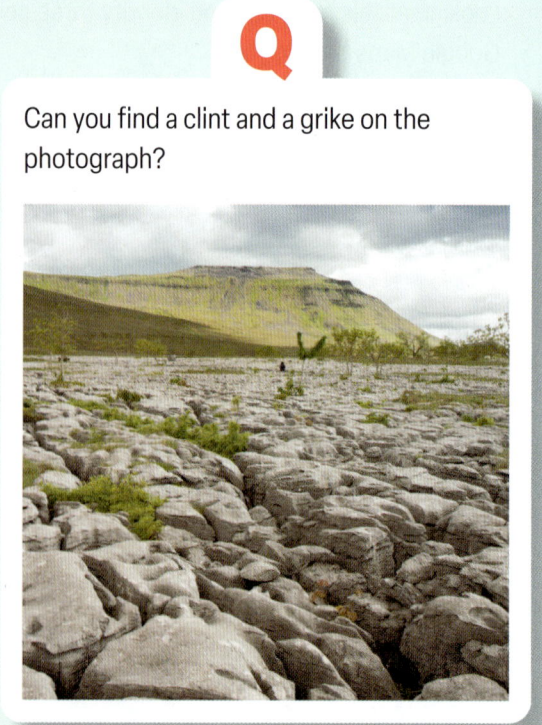

Q Can you find a clint and a grike on the photograph?

- Clints are the slabs of rock on the surface that separate the grikes.
- Grikes are deep grooves in the pavement. They become weathered, widened and deepened by carbonation.
- Limestone pavements can be flat or contain sloping terraces. Terraces are areas of sloping land that look like steps. The terraces can be separated by steep cliffs.

8. Weathering

Swallow Hole

When a river enters a karst landscape, it can disappear underground through grikes. The grikes become enlarged, forming big vertical holes known as **swallow holes**. Swallow holes vary in size depending on how large or small the river flowing into them is. Poll na gColm is an example of a swallow hole in the Burren.

Rivers can disappear underground through a swallow hole and re-emerge at some point further downstream.

Working with others

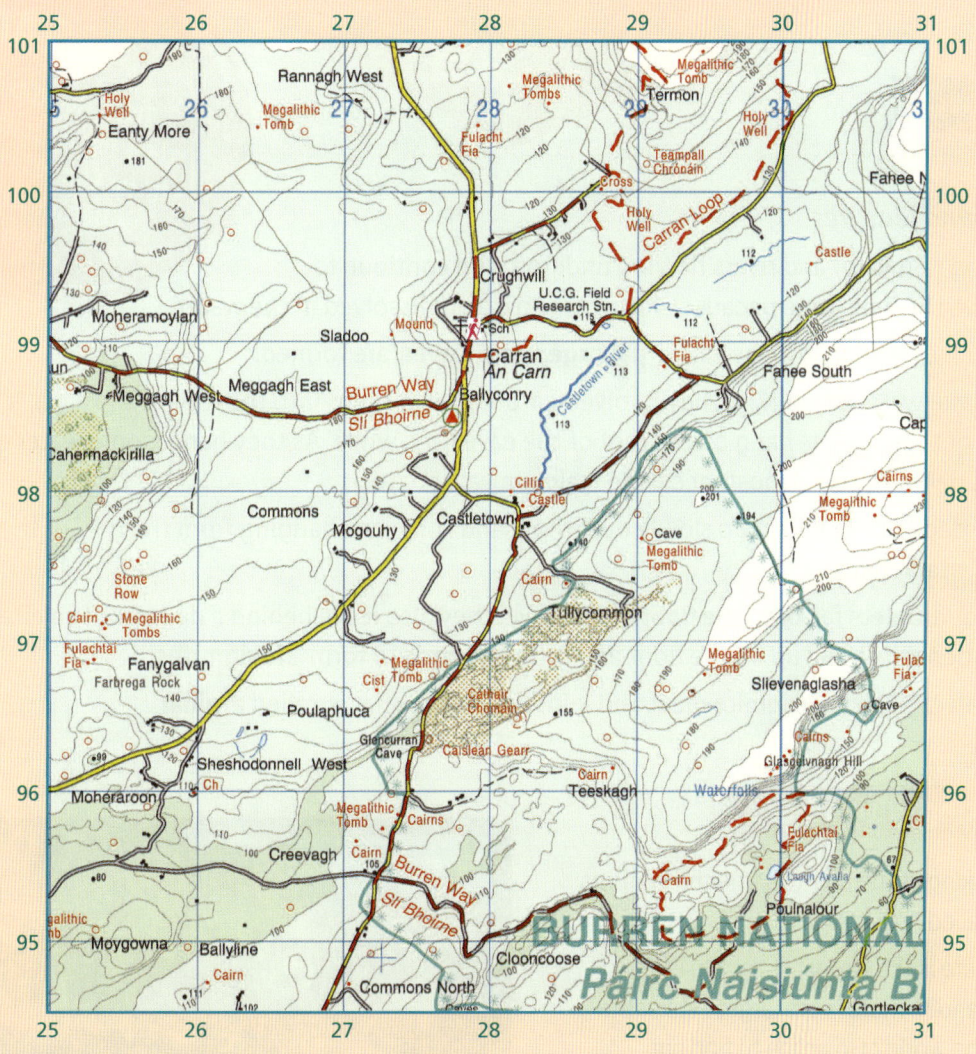

In pairs, look at this OS map of the Burren. Can you identify places on the map where swallow holes might exist? Are there any rivers that seem to disappear?

121

Limestone Cave

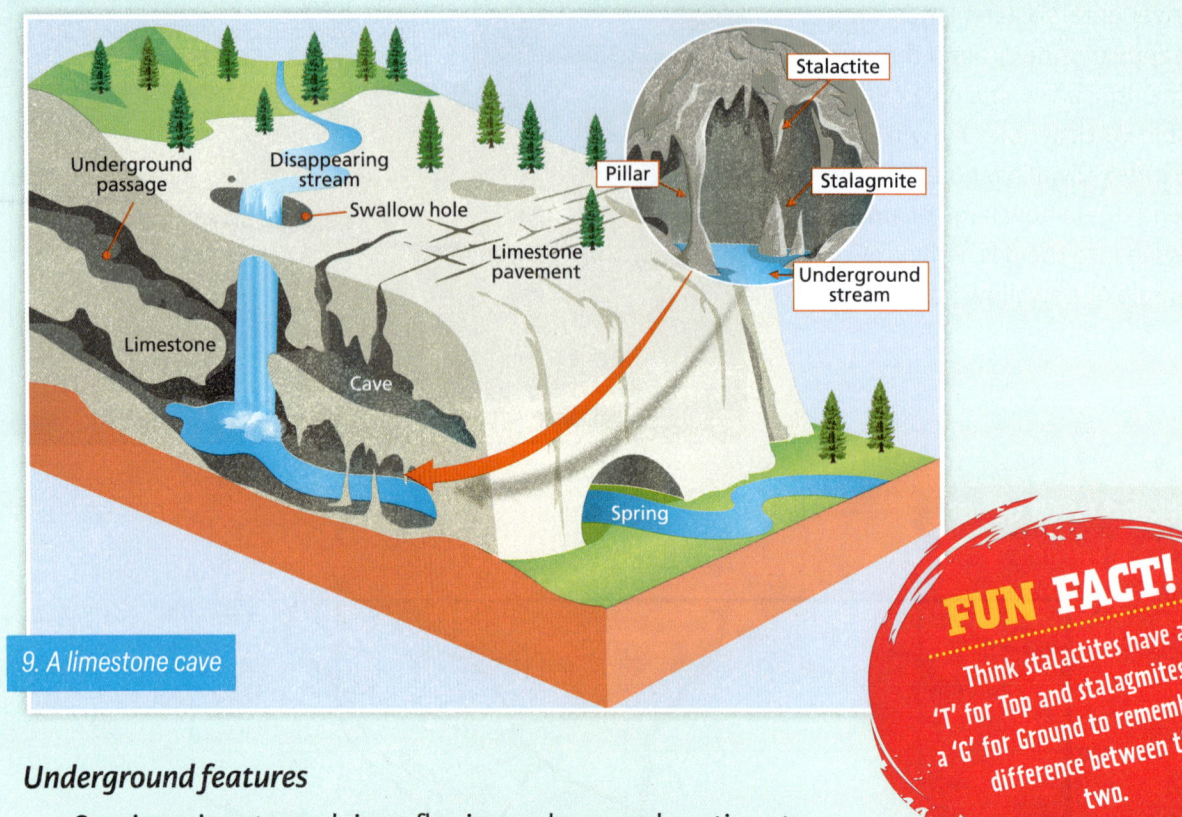

9. A limestone cave

FUN FACT!
Think stalactites have a 'T' for Top and stalagmites have a 'G' for Ground to remember the difference between the two.

Underground features

1. Seeping rainwater and rivers flowing underground continue to dissolve the limestone because of the carbonic acid dissolved in the water.
2. Gradually, large open areas called **passages** and **caves** are formed.
3. When the water containing the dissolved carbonic acid seeps into the roof of the cave, it begins to drip. Some of the droplets hang on the roof of the cave for a while. As they hang, some water evaporates and leaves behind a tiny deposit of calcite (calcium carbonate).
4. These deposits can build up slowly over thousands of years, hanging from the roof of the cave. They are called **stalactites**.
5. If water droplets fall onto the floor of the cave, they also leave behind a deposit of calcite. Over time these deposits build up on the floor and grow upwards to form **stalagmites**.
6. If a stalactite and stalagmite meet and join, they form what is called a **pillar**.

Working with others

1. Looking at the photo, name the features labelled A, B and C.
2. Where is the largest stalactite in Ireland? Where is the largest stalactite in the world? Go online to find out.

A limestone cave

8. Weathering

Question Time

1. What is chemical weathering?
2. Explain the term 'carbonation'.
3. What is a karst landscape? Give an example of a karst landscape in Ireland.
4. Name one surface and one underground landform found in the Burren.
5. With the aid of a labelled diagram, describe how the following were formed: swallow hole, limestone cave, stalactite, stalagmite, pillar.

8.3 Sustainability and Tourism in the Burren

The weathering of limestone in the Burren has created a unique landscape that has attracted tourists from all over the world. Tourism can give an area a great economic boost, but it is also important that we protect our natural landscape. We must make sure to use it in a sustainable way so that it can be enjoyed by everyone into the future.

Search online for 'RTÉ video A Rocky Place 1992' (3:04). The video explains the natural beauty of the Burren and outlines the concerns locals had back in the 1990s to a proposed visitor centre.

Car Park Controversy in the Burren

A PROPOSED car park in the Burren National Park has sparked another planning controversy. The Burren Action Group (BAG) has condemned a new planning application by the National Parks and Wildlife Service (NPWS) for a 27-space car park, and have lodged an objection with the local planning authority.

The NPWS insists that their intention is to improve road safety, traffic congestion and access problems for the local communities. It claims there will be 'no significant impact on the structure of the local environment'. There will be no bins, picnic or toilet facilities, no water supply and no lighting. The car park will operate from dawn to dusk on weekdays and at weekends.

The BAG believes that there are alternative sites that can be used for parking. According to the group, 'There is the potential to develop a management plan that respects and promotes the needs of the local people, local environment and those wishing to engage in sustainable low impact tourism in and around the national park.'

(Adapted from *The Clare Champion*)

1. **What** has been proposed to be built in the Burren National Park?
2. **Why** do the National Parks and Wildlife Service want to build this new development?
3. (a) **How** do the Burren Action Group feel about the new development?
   (b) **Outline** two reasons why they may feel this way.
4. In your opinion, **why** is sustainable tourism important in areas of natural beauty?

Junior Cycle Geography CYCLONE

Sample questions

1. **Examine** the 1:50 000 Ordnance Survey map extract below and answer each of the following questions. The landscape shown on this map extract has been shaped by different physical processes including weathering.

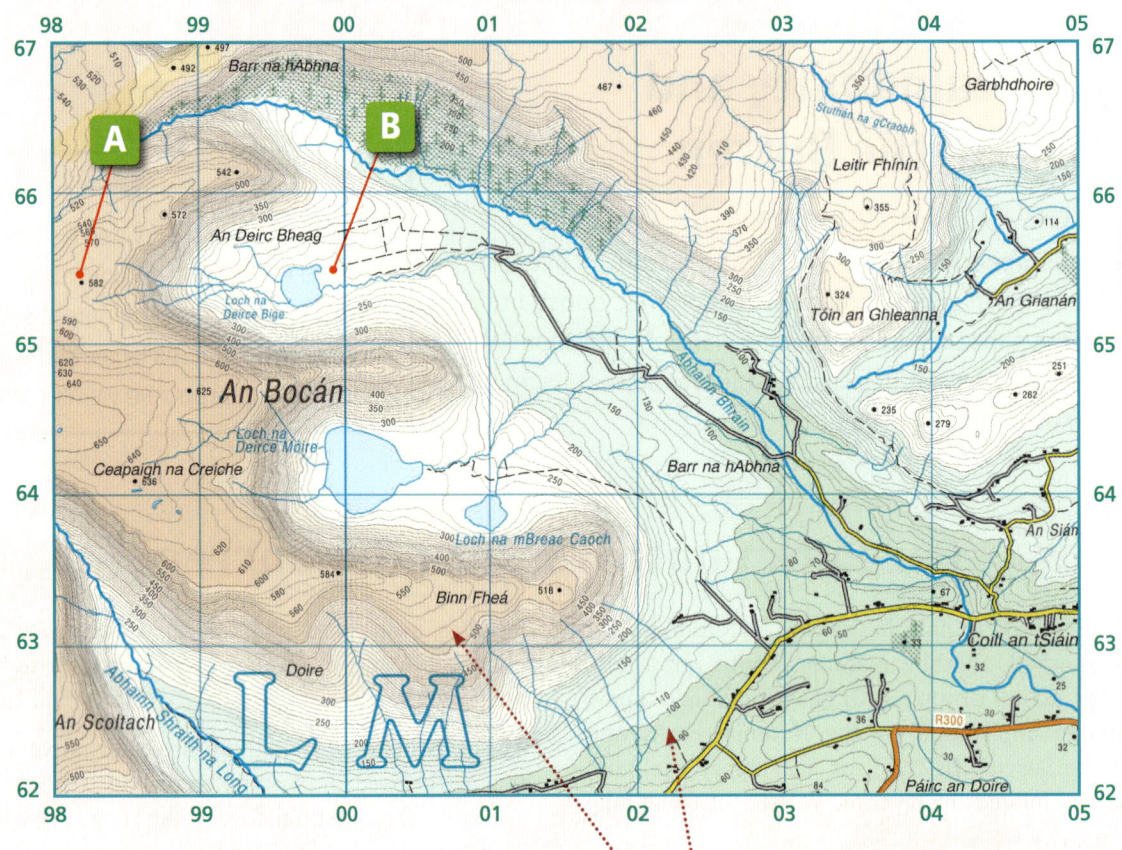

(i) At which site on the map extract above, A or B, is freeze-thaw action most likely to occur?

(ii) **State** one reason why freeze-thaw action is most likely to occur at the site you chose.

(SEC sample paper, Q6 (i) & (ii))

2. **Name** and **explain** one process of mechanical weathering that you have studied. Use a labelled diagram to illustrate your answer.

Exam hints!

Remember: green shaded areas on the map show lowland areas and the light brown areas are upland.

The action verb here is 'State'. This means you must provide a concise statement with little or no supporting argument. In other words, your answer should be brief.

This answer has been started for you on the next page. Use this to help you complete the question. You can use FEED to provide structure and give your answer a clear layout.

8. Weathering

Let's get started! Sample starter

F Mechanical weathering: Freeze-thaw action

E Explanation: This is a type of mechanical weathering. Rain water gathers in the joints in rocks. In winter or when the temperature drops below 0 °C, the water freezes to become ice. This puts pressure on the rock as the ice expands by up to 10%. As the temperature rises ...

E Example of where this can be seen:

D Diagram/s:

... now complete the answer in your copy.

Exam hints!

For this question, we will use **FEED** (**F**eature, **E**xplanation of formation, **E**xample, **D**iagram).

Learning Outcomes: 1.3, 2.8

9 Mass Movement

Learning Intentions

You will be able to:
- Explain what mass movement is and how it happens
- Describe both slow and fast types of mass movements using examples of each
- Investigate the impact mass movement can have on humans.

Key words

- Gravity
- Regolith
- Gradient
- Deforestation
- Unstable
- Soil creep
- Mudflow
- Saturated
- Bog burst
- Landslide

9.1 Mass Movement

Mass movement is the movement of loose material down a slope under the influence of **gravity**. The loose material is made up of soil, rocks and mud and is known as **regolith**.

👍 Staying well

Look at this photo. We can see the loose regolith moving down the hillside.

1. If you lived in this area, what consequences could mass movement have for you and your family? Write down two consequences.
2. What do you think could be done to prevent mass movement like this from impacting on the people who live here? Write down one thing that you think could help.

126

9. Mass Movement

Factors That Influence Mass Movement

Slope or Gradient
The steeper the **gradient** (another word for slope), the faster the loose material can move down the slope. A gentler slope will have a slower movement of regolith.

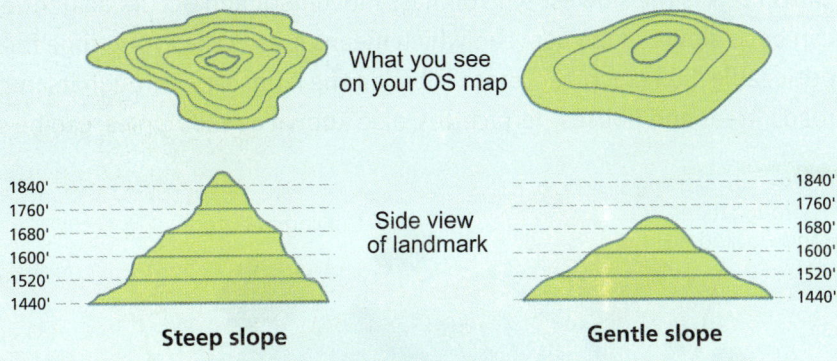

1. How slope is shown on an OS map

Vegetation
Roots of plants and trees keep soil together and make it more stable. This can help to prevent mass movement. When **deforestation** (the removal of trees) occurs, the soil on a hillside can move more freely down the slope.

Water
Water makes the regolith heavier, causing it to move faster down the slope.

Human Activity
Humans can speed up mass movement by making upland areas **unstable** if they cut into hillsides to build roads. Cutting into the base of a slope can cause the hill above it to collapse.

Plant roots binding soil

Soil saturated by water is heavier and flows down the hillside

Hill collapsing under a roadway

9.2 Types of Mass Movement

The main types of mass movement are shown here in figure 2. The movements occur at different speeds depending on the steepness of the slopes on which they occur.

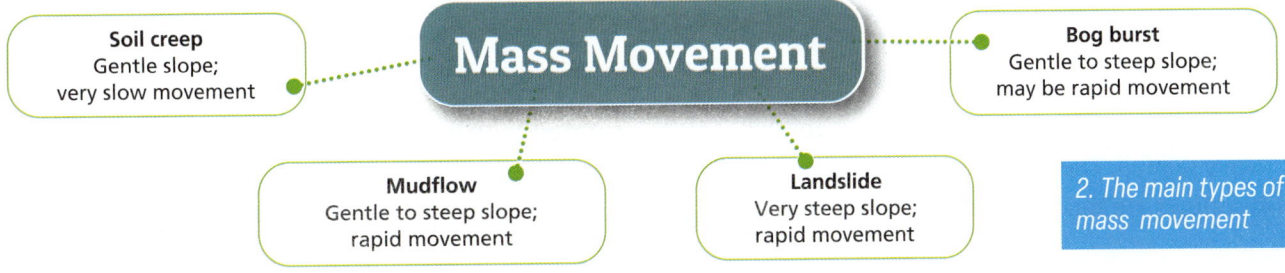

2. The main types of mass movement

Soil Creep

Soil creep is the slowest type of mass movement and occurs on gentle slopes. Movement can be as slow as 1 cm per year, which means it can take a long time for the effect to be noticeable. Soil creep becomes visible on the landscape through its impact on walls, roads, trees and houses. Terracettes, also known as soil ripples, can be seen on the land.

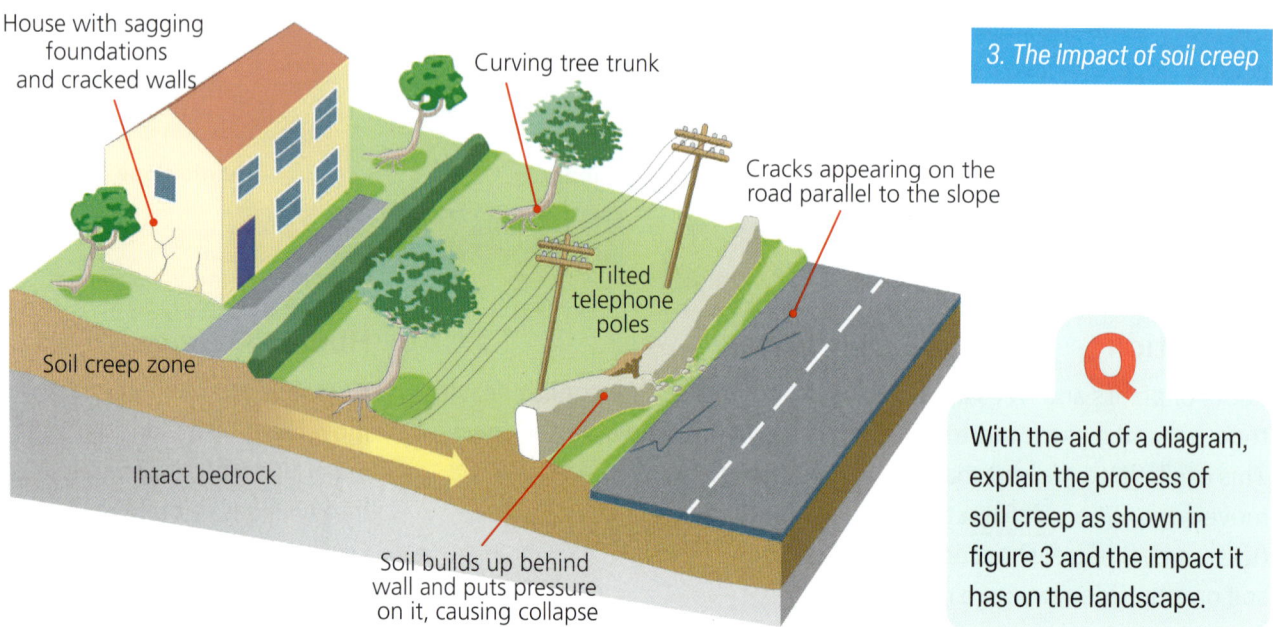

3. The impact of soil creep

Q With the aid of a diagram, explain the process of soil creep as shown in figure 3 and the impact it has on the landscape.

Working with others

Study the photographs shown here. Then discuss with the person next to you what evidence you can see that soil creep is happening in these areas. Share your thoughts with the class.

Mudflow

Mudflows occur when the soil and regolith become **saturated** with water, usually after heavy rainfall. They mix together to form mud, which then moves rapidly downslope. Mudflows are the fastest type of mass movement, with speeds of up to 100 km per hour.

9. Mass Movement

Geography in the News

Mudflow in India's Assam region, May 2022

At least 11 people have died in floods and mudflows triggered by heavy rain in India's northeast region of Assam. Nearly 200,000 people in the region were cut off from the rest of the state, as roads and bridges were either blocked by mudflows or washed away.

The state's disaster management agency said nearly 700 villages were underwater. The army sent helicopters to help with rescue efforts.

People had limited drinking water and sparse food supplies. Communications were also cut off and roads and railways were closed. The mudflow had washed away houses and fields of crops.

Mudflows and floods are common in India's northern region. Scientists say they are becoming more frequent as global warming plays a role in the melting of glaciers there.

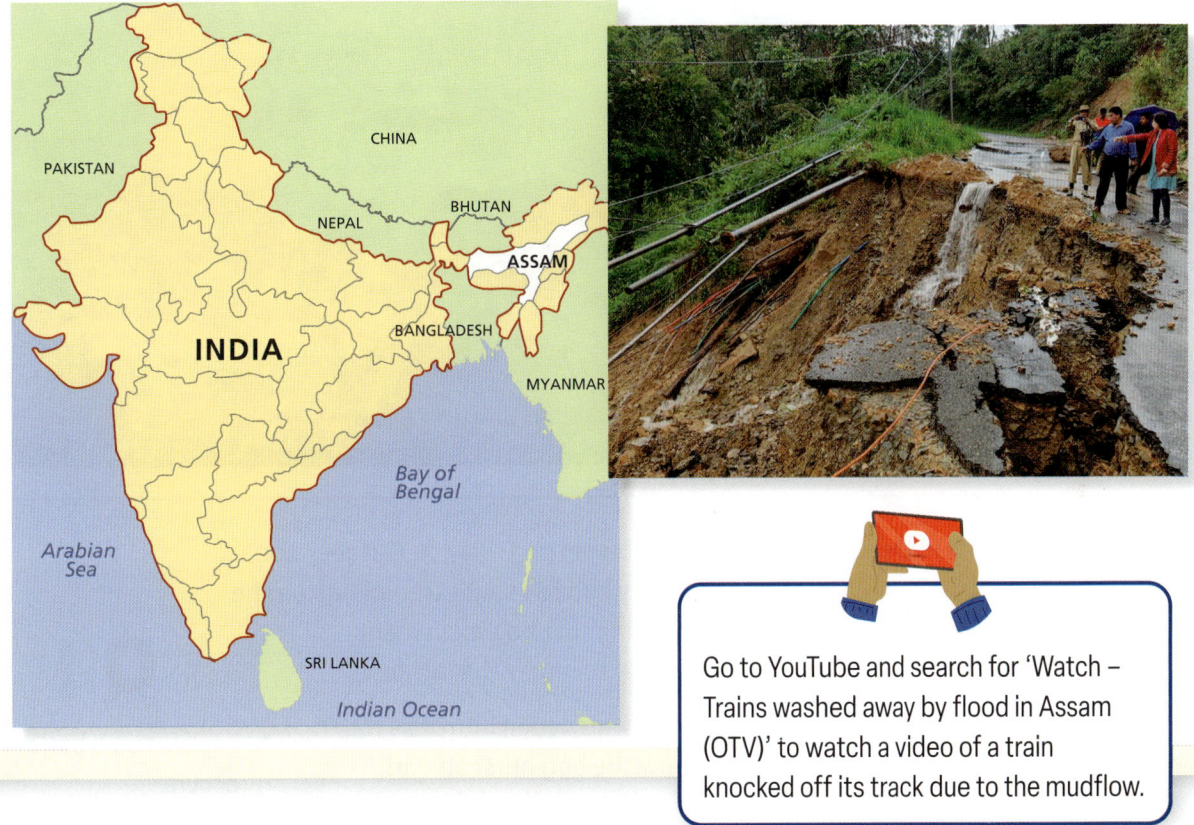

Go to YouTube and search for 'Watch – Trains washed away by flood in Assam (OTV)' to watch a video of a train knocked off its track due to the mudflow.

1. **Describe** how a mudflow occurs.
2. **What** triggered the mudflow in Assam, India?
3. **How** did the army respond to this natural disaster?
4. **Describe** the impact of the mudflow in Assam. In your answer, mention one social and one environmental consequence.
5. **How** is climate change contributing to increased mudflows in this region?

Bog Burst

In some parts of Ireland, we find areas covered in thick layers of peat or turf. **Bog bursts** can take place in these areas after long periods of heavy rainfall. If the rain water cannot drain or soak away, the peat can become saturated with water. Peat can then flow downslope and cause great damage over a large area.

Case Study: Derrybrien Bog Burst, 2003

Managing information and thinking

Look carefully at this image, which shows the Derrybrien bog burst.

1. Describe the damage shown.
2. How do you think this incident affected the local economy?

The frightening events in Derrybrien, Co. Galway, began on Friday, 31 October 2003. Some locals working in the area later said that they had felt slight tremblings on the bog. It was not until 10.45 a.m. the following morning that bog movement was more visible, and a local man phoned the gardaí to report the incident.

As Saturday afternoon and evening passed, locals watched in horror as the bog burst began to speed up. Locals recalled how the bog burst spread out close to 50 metres in width, with one man explaining how he first noticed a ripple effect on the bog before seeing hundreds of tonnes of peat flowing down the hillside. The bog burst came to a halt on Saturday night about half a kilometre downhill when a specially built dam finally stopped the flow.

1. Explain how a bog burst occurs.
2. Name one economic, one social and one environmental consequence of a local bogburst.

Impact

Homeowners were left stranded, and private bog plots and even livelihoods were destroyed. Boil water notices were put in place and the Shannon Regional Fisheries Board warned of an environmental disaster.

9. Mass Movement

Landslide

A **landslide** is the very rapid movement of regolith down a very steep slope. A landslide occurs when the slope becomes unstable, which can be caused by road-building, quarrying, earthquakes or sea erosion.

Geography in the News

Landslide in Ecuador, February 2022

A landslide in Ecuador's capital city Quito has caused major damage. Rain had weakened the hillside, sending loose regolith downslope to collide with homes and a sports field.

The security department reported that at least 24 people were killed and another 47 were injured. Eight houses collapsed, with many more damaged.

Rescue workers and local neighbours began looking through the ruins to locate survivors. Soldiers were sent to help with the search and rescue efforts. Parts of Quito were left without power after electrical poles were brought down. The city council said that residents impacted by the landslide had been moved to emergency shelters.

Scientists have said that climate change is increasing the risk of heavy rains around the world with the warmer atmosphere holding more water.

1. **Describe** what a landslide is and how it occurs.
2. **Where** did the landslide in Ecuador take place?
3. **How many** people lost their lives or were injured by the landslide?
4. **Describe** the response to this natural disaster.
5. **How** has climate change increased the risk of events such as this?
6. The two images below show the effects of mass movement on a coastline. **Identify** what type of mass movement this is and give two reasons for your answer.

Mass movement affecting a coastline

Sample questions

1. Read the article below and answer each of the following questions.

Landslide in Indian village of Malin leaves 17 dead and 200 missing

At least 17 people were killed and 200 people were trapped as 44 houses in Malin village in India were flattened by a landslide that hit after heavy rain. Environmentalists said large-scale deforestation had made the place vulnerable. An expert on landslides said 'the cause of the landslide appears to be the clearing of land on the hill for farming and the removal of trees'.

(i) In what country did this landslide take place?

(ii) **Name** one effect of the landslide mentioned in the article.

(iii) The article states that one cause of this landslide was the 'removal of trees'. Briefly **explain** how the removal of trees can lead to a landslide.

(iv) If this land was reforested (replanted with trees), it would help to protect against future landslides. **Describe** another benefit of planting trees.

(SEC sample paper, Q1 (a) (i) –iv))

Exam hints!

Here you must think back on the influence that vegetation can have on mass movement events.

For this question, you can use your knowledge on sustainability and climate change. How might planting trees be positive for the environment?

9. Mass Movement

2. (i) Is the mass movement shown in the photograph above occurring quickly or slowly? Use evidence from the photograph to support your answer.
   (ii) **Name** and **outline** one influence on the rate of mass movement.
   (iii) Select one form of mass movement mentioned below, and **explain** how it occurs.

   Mudflow   Landslide   Bog burst

Exam hints!

Question (ii) has two parts: you must first name the influence and then outline its influence. You must make sure to answer both to achieve the full marks.

In part (iii), make sure to refer to how gradient, human activity, water and vegetation all have an influence.

10 Rivers

Learning Outcomes: 1.5, 1.10, 2.7, 3.4

Learning Intentions

You will be able to:
- Outline the key characteristics and landforms found along the three stages of a river
- Describe, with the aid of diagrams, how the processes of erosion, deposition and transportation shape a river
- Discuss how people interact with and manage surface processes in rivers
- Describe how rivers impact on the location and origin of settlement in Ireland.

Course · Attrition · V-shaped valley · Levee
Erosion · Solution · Interlocking spurs · Ox-bow lake
Transportation · Traction · Waterfall · Delta
Deposition · Saltation · Mature · Distributaries
Hydraulic action · Suspension · Flood plain
Load · Meander · Old
Abrasion · Youthful · Alluvium

10. Rivers

A river is fresh water flowing across the surface of the land, usually to the sea. Rivers are found on every continent and on nearly every kind of land.

10.1 The Journey of a River

Figure 1 shows the journey a river takes from its source to the sea.

FUN FACT! Most of the world's major cities are located near the banks of rivers.

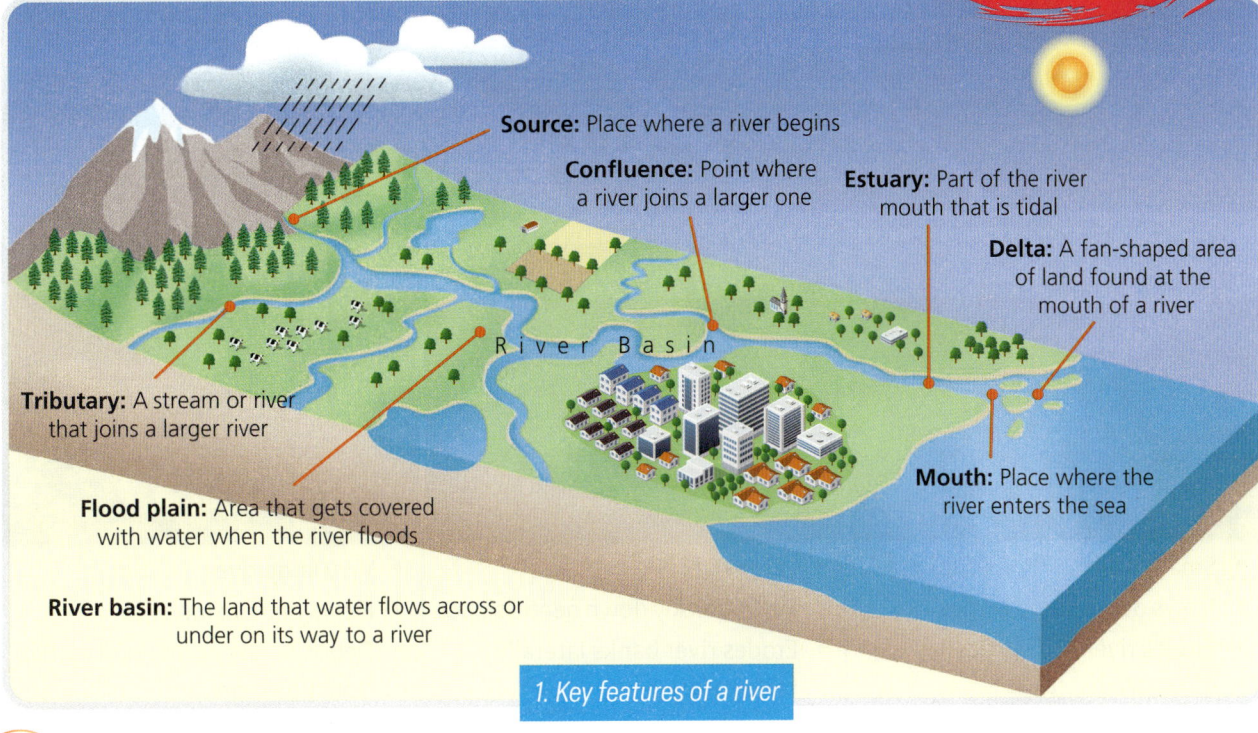

Source: Place where a river begins

Confluence: Point where a river joins a larger one

Estuary: Part of the river mouth that is tidal

Delta: A fan-shaped area of land found at the mouth of a river

Tributary: A stream or river that joins a larger river

Flood plain: Area that gets covered with water when the river floods

River basin: The land that water flows across or under on its way to a river

Mouth: Place where the river enters the sea

1. Key features of a river

Working with others

Using an atlas or Google Maps, and working with the person next to you, identify two cities in Ireland located on rivers and two foreign cities located on rivers.

Question Time

Can you match these three features of a river with the correct photograph?

Flood plain   Tributary   Mouth

A

B

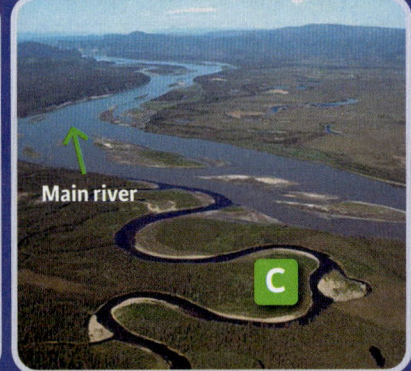
Main river
C

Junior Cycle Geography CYCLONE

10.2 The Stages of a River

The journey a river takes from its source to its mouth is called the **course** of the river. This course is broken up into three different stages.

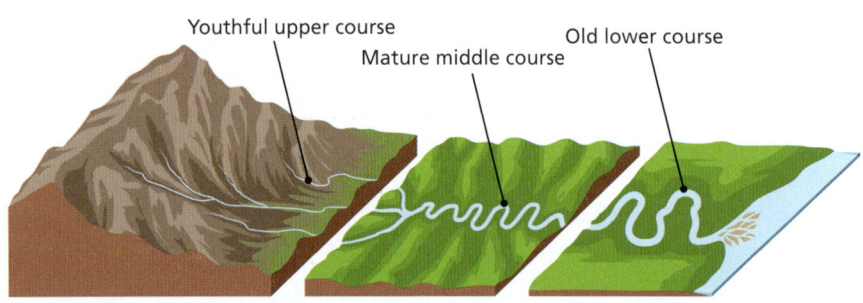

2. The three stages of a river

Upper Course	Middle Course	Lower Course
• Small river • Flows down steep slopes • Erodes river bed vertically (downwards)	• Large river • Flows quickly down gentle slopes • Erodes river banks laterally (sidewards)	• Very large river • Flows on flat slopes • Deposits lots of material carried from upstream
• Narrow V-shaped valley • Steep valley sides • Load = boulders and rocks	• Wide U-shaped valley • Gentle valley sides • Load = smoother stones and pebbles	• Wide valley • Flat • Load = smooth sand and silt
Features	**Features**	**Features**
• V-shaped valley • Interlocking spurs • Waterfalls	• Meanders • Wide valleys	• Ox-bow lakes • Deltas • Flood plains • Levees

Question Time

1. What is a flood plain?
2. What is the name for the path that the river takes to get to the sea?
3. Draw and label a diagram to show the three stages of a river.
4. Name one feature found in each stage of a river's course.

10.3 The Processes of a River

We will examine the three processes of a river which change the landscape.

3. The processes of a river

1. **Erosion**: The river wears away the landscape.
2. **Transportation**: The river carries the eroded material away.
3. **Deposition**: The river drops the material it is carrying.

River Erosion

River erosion is also known as fluvial erosion. The river erodes the landscape using many processes.

1. The fast-moving water breaks and wears away material from the bed and banks of the river. This process is known as **hydraulic action**.

2. The broken material that is now carried along by the river is called its **load**. The load is thrown against the bed and banks of the river and wears them down. This process is known as **abrasion**.

The motion of water hitting against the river bed/banks

Scraping or wearing away

4. The water can dissolve some of the minerals, e.g. limestone, on the river bed and banks. This process is known as **solution**.

3. The load itself is broken down further as particles bounce off one another and become smooth and rounded. This process is known as **attrition**.

Minerals are dissolved in the water and carried along

The impact of the rock grains hitting off one another

4. River erosion

River Transportation

As a river moves from source to the sea, it carries, or transports, its load. The load is transported in a number of ways by the river. River transportation is also known as fluvial transportation.

1. Large pebbles and stones are rolled and dragged along the river bed by the force of the water. This method is known as **traction**.
2. Small pebbles are bounced along the river bed. This method is known as **saltation**.
3. Tiny particles are held in **suspension** in the water – the movement of water is fast enough to keep them from dropping to the river bed.
4. Dissolved minerals like limestone are carried in solution so cannot be seen.

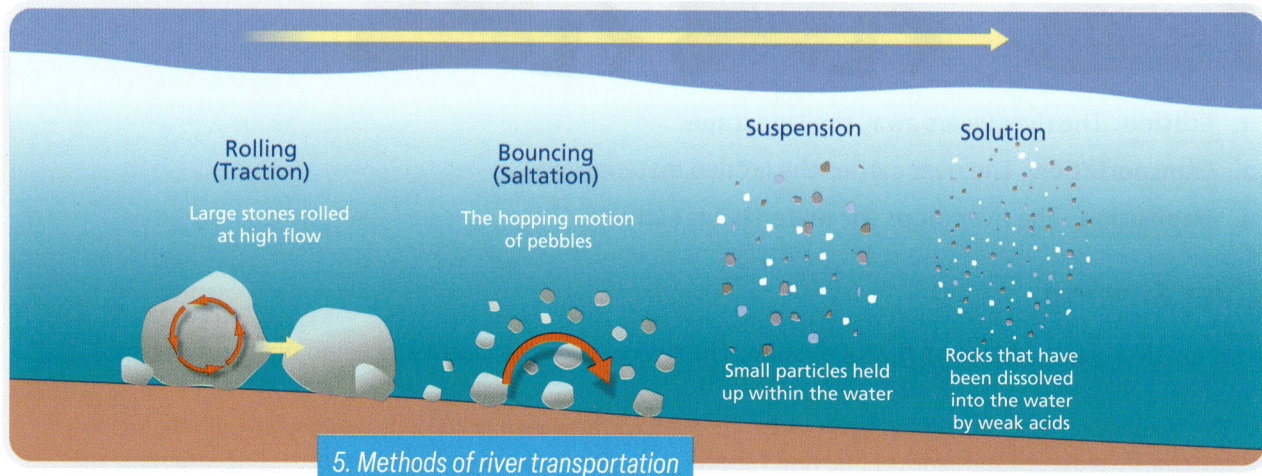

5. Methods of river transportation

River Deposition

River deposition is also known as fluvial deposition. Rivers deposit, or drop, their load when they begin to slow down and lose energy. The heaviest material such as rocks and pebbles is deposited first. This happens:

1. When a river reaches flat ground
2. When a river enters a lake or the sea
3. At the inside bend of a **meander** (which we will look at in more detail later in this chapter).

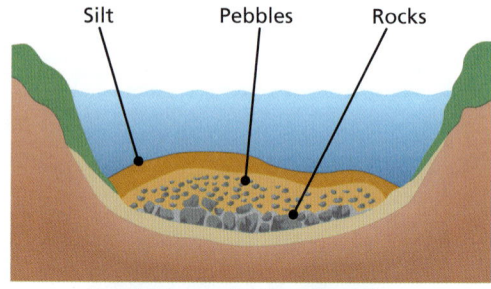

Heaviest material deposited first

6. River deposition

> **Managing information and thinking**
>
> From smallest to largest particle size, put in order the ways a river transports its load and explain each.
> 1. Solution, 2. Bouncing, 3. Rolling, 4. Suspension

Question Time

1. What three processes occur in a river to change the landscape?
2. Explain hydraulic action.
3. Write a definition of the following terms: (a) traction, (b) solution, (c) abrasion, (d) attrition
4. Give two reasons why a river may deposit its load.

10.4 Landforms of a Youthful River

Feature: V-Shaped Valley

When a river is in its **youthful** stage, it erodes vertically (downwards). This forms a valley with steep sides and a narrow floor. The is called a **V-shaped valley** as it looks like the letter V. V-shaped valleys are a feature of erosion in a river.

Example: A V-shaped valley can be seen in the youthful stage of the River Liffey.

How a V-Shaped Valley Forms

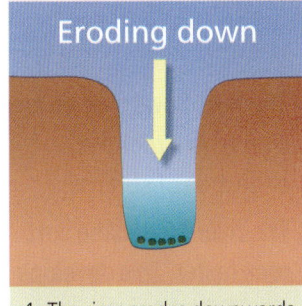 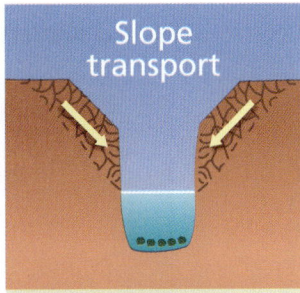 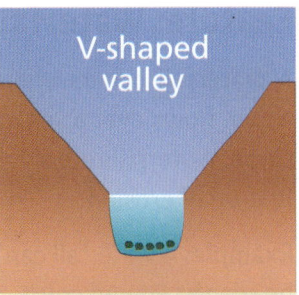

1. The river erodes downwards by hydraulic action. The valley is deepened by vertical erosion.

2. As the river cuts down, the steep sides are attacked by weathering. This breaks up and loosens the soil and rock.

3. The loosened material slowly creeps down the slope because of gravity or is washed into the river by rainwater. The river carries it away.

4. The end result is a steep-sided valley that has the shape of the letter 'V'.

7. The formation of a V-shaped valley

> **Managing information and thinking**
>
> In your copy, sketch or trace the stages of formation of a V-shaped valley. Make sure to include labels.

Feature: Interlocking Spurs

Interlocking spurs are areas of high ground that jut out on either side of the river and appear to lock together.

Example: Interlocking spurs can be seen in the upper course of the River Slaney.

Interlocking spurs

Junior Cycle Geography CYCLONE

Working with others

With a partner, point to where you think the interlocking spurs are on this photo.

Feature: Waterfalls

A **waterfall** occurs where a river flows over a vertical slope. Waterfalls are a feature of erosion.

Example: A waterfall can be seen at Powerscourt, Co. Wicklow.

How a Waterfall Forms

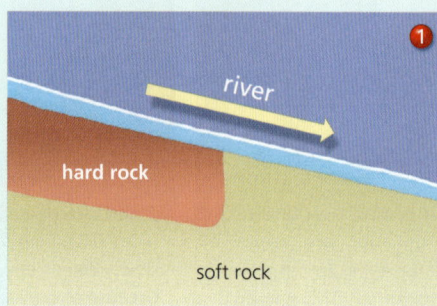

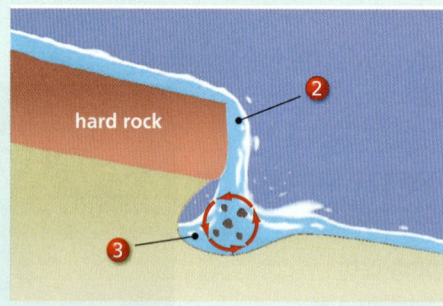

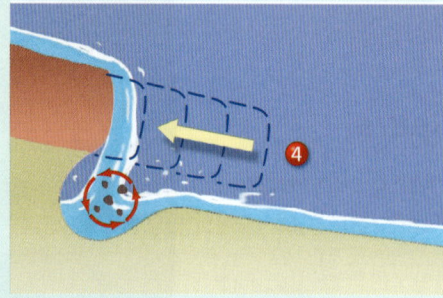

8. The formation of a waterfall

1. The river flows over an area where hard rock lies on top of an area of soft rock.

2. The river erodes the soft rock faster. This develops a vertical drop where the hard rock and soft rock meet. This drop is called a waterfall.

3. At the base, a plunge pool develops due to the force of the falling water (hydraulic action) along with its load swirling around at the base which wears away the base by abrasion.

4. The rock above the plunge pool is undercut and left hanging. It eventually collapses into the plunge pool below. As this process continues, the waterfall slowly erodes its way upstream.

Powerscourt Waterfall, Co. Wicklow

Question Time

1. Describe the formation of a V-shaped valley, using a labelled diagram to support your answer.
2. What are interlocking spurs?
3. Explain, with the aid of labelled diagrams, how waterfalls form. Where in Ireland might you find a waterfall?

10.5 Landforms of a Mature River

Feature: Wide Valleys

In the **mature** stage of a river, erosion occurs laterally (sideways) rather than vertically (downwards). This creates a wide river valley that is not very steep. Interlocking spurs have been removed by lateral erosion, which widens the valley floor. Weathering and mass movement have also made the valley sides less steep.

Feature: Flood Plains

A **flood plain** is an area of flat land on either side of the river channel found in the mature and **old** stages of a river. Flood plains are covered in fertile soil, known as **alluvium**. This rich and fertile soil is full of minerals, making it excellent for farming.

Example: An example of a flood plain can be found in the lower course of the Shannon.

Working with others

Look at the photograph of a wide valley for a couple of minutes. Then, in pairs, discuss these questions and write down your answers.

1. State two reasons why this settlement may have been built at this location.
2. Why might you not wish to build a factory too close to this river? Give two reasons for your answer.

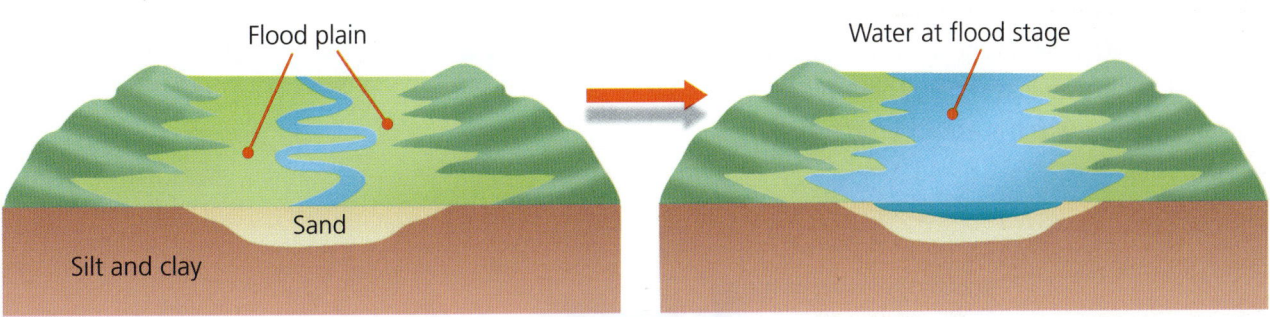

9. A flood plain

Junior Cycle Geography CYCLONE

📁 Managing information and thinking

Look at these photographs of Bandon, Co. Cork, before and after flooding (showing the view up and down the same street). Then answer these questions:

1. Do you think that this area has experienced flooding in the past? Explain your answer.
2. In your opinion, will there be a large or small financial cost to the local economy? Explain your answer.
3. What measures could the local government put in place to prevent this from happening again?

Feature: Meanders

A meander is a winding curve or bend in a river.

Example: Meanders can be seen in the middle and lower course of the Rivers Shannon and Boyne.

How a Meander Forms

Meanders are formed by a combination of erosion and deposition in the mature and old age stages of a river.

1. Water flows more quickly on the outside of the river bend. The river erodes the bend by a combination of hydraulic action and abrasion. As the current is faster, it undercuts the bank.
2. Water flows more slowly on the inside bend of the river. Material is deposited on the inside bend as the river loses energy here. A point bar can form, which is a build-up of river deposits.

10. How a meander forms

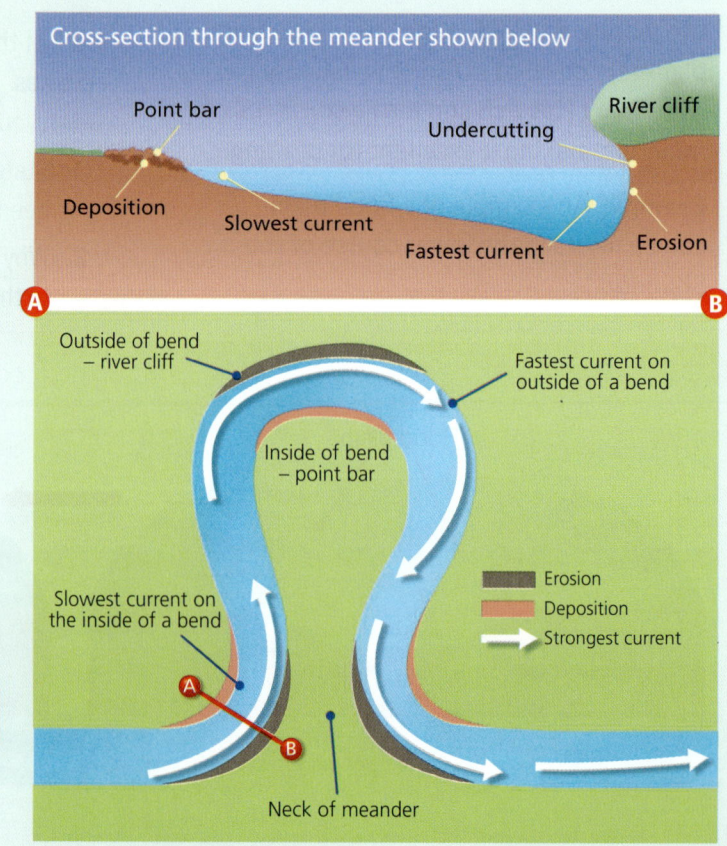

10. Rivers

Question Time

1. Explain why wide valleys are found in the mature stage of a river.
2. Why is the soil on a flood plain good for growing crops?
3. Explain why, in your opinion, it would be important to know if your land is on a flood plain before building a home or business.
4. Describe with the aid of a labelled diagram how a meander is formed.
5. Look at this photo of a meander. For each of the points labelled A and B, state whether erosion or deposition is occurring. What evidence can you see to support your answer?

10.6 Landforms of an Old River

Feature: Levees

A **levee** is a build-up of alluvium on the banks of a river and is a feature of the old stage river. Levees are caused by floods and are a feature of deposition.

Example: Levees can be seen in the lower course of the River Liffey and the River Moy, Co. Mayo.

How a Levee Forms

1. During times of low flow in a river, the load that is being carried by the river is dropped onto the river bed. This raises the height of the bed.
2. During times of flood, the water flows out more easily over the top of the channel and onto the surrounding land. As it does this, it loses energy and deposits its load.
3. The heavy coarse material is deposited on the river banks. The finer silt is deposited further away onto the flood plain.
4. After many floods, the river builds up a bank on either side.

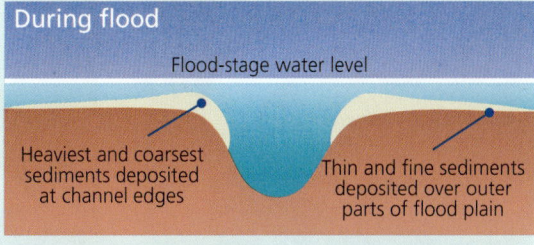

11. How a levee forms

143

This levee provides a natural walkway. Levees can also be called dikes or embankments.

Feature: Ox-Bow Lakes

Ox-bow lakes are horseshoe-shaped lakes found on the flood plains of old rivers. They are formed by river erosion and deposition.

Example: An example can be seen on the River Moy.

How an Ox-Bow Lake Forms

1. Erosion by hydraulic action and abrasion takes place on the outside banks of the meander. The neck of land between the meander begins to narrow.
2. During times of flood, the water has increased energy. It takes the shortest course by breaking through the neck and begins to flow in a new, more direct course.
3. Deposits of alluvium now build up and seal off both ends of the meander. The old section of the meander is now cut off from the main river and becomes known as an ox-bow lake. Over time, the lake may dry up.

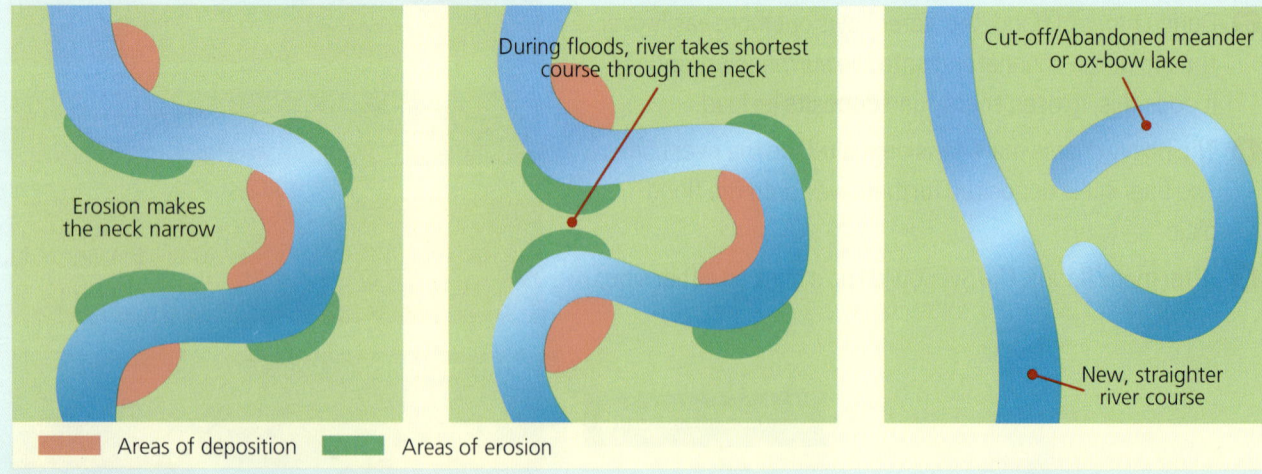

12. How an ox-bow lake forms

10. Rivers

 Communicating

Look at the photo and pinpoint the ox-bow lake. On a piece of paper, trace from the river to the ox-bow lake and try to discover what the old course of the meander may have looked like. After you have done this, turn to the person next to you and compare your work.

Feature: Delta

A **delta** is a fan-shaped area of land found at the mouth of a river. It is a feature of river deposition.

Example: An example can be seen at the Shannon estuary, Co. Limerick.

How a Delta Forms

1. When a river reaches a lake or the sea, it slows down and loses the power to carry sediment.
2. The sediment is then dropped at the mouth of the river. Some rivers drop so much sediment that waves and tides cannot carry it all away. This forms the delta.
3. The large deposits build up over time at the mouth of the river and block the river's entry to the sea. The river then has to break through the delta in many small channels called **distributaries**.

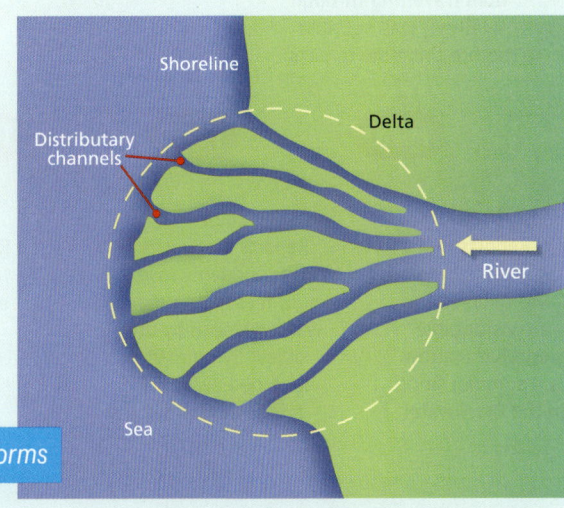

13. How a delta forms

 Working with others

Take one minute to examine this photo. Then with your partner, answer this question: What material is being deposited here to make the delta?

Question Time

1. What is a levee?
2. (a) With the aid of a labelled diagram, explain how an ox-bow lake is formed.
   (b) Name the location of an ox-bow lake in Ireland.
3. Explain how a delta is formed.

Junior Cycle Geography CYCLONE

10.7 Rivers and Human Interaction

Case Study: Flooding in Mallow

The town of Mallow in Cork has a long history of flooding from the River Blackwater. Records as far back as 1628 detail damage to structures in the town from flooding. It has been a common occurrence in the town ever since.

Between 2006 and 2009, a flood relief scheme for Mallow was constructed. The aim of the relief scheme was to try to prevent serious damage to local homes and businesses from flooding.

1600 Reports that 'this day past the Blackwater, by reason of the greatness and height thereof, he [a man travelling through Mallow] was not able before this time to pass'

1853 Reports that extreme flooding had caused 'the bridge [to be] almost totally swept away by a flood event'

1916 Reports of disastrous floods causing extensive damage

1980 Extreme flood event, with serious flooding in Mallow

2004 Significant flooding

1578 Reports of someone who 'was delayed for a considerable period at Mallow by a summer flood in the Blackwater'

1628 Reports that '... the river suddenly rose eight feet higher than was even seen by the oldest living man. It carried away the bridge at Mallow.'

1875 Reports of disastrous floods causing extensive damage to houses, household goods, shops, cattle and crops

1969 Significant flooding, with irreparable damage caused

1988 Significant flooding

2006–2009 Construction of the relief scheme

Being numerate

1. How many large flood events were recorded from 1916 to 2004?
2. Which two flood events had the shortest time between them?

Managing myself

If you lived in Mallow, how would you feel if the flooding affected your home or school?

Flooding in Mallow, 2009

10. Rivers

Achievements of the Flood Relief Scheme

- Provided protection against a one-in-a-hundred-years flood event
- Developed a flood warning system for the people of Mallow
- Built flood defences:
  - flood defence walls
  - storm drains
  - pumping stations
- Designed and constructed all defence walls with future climate change in mind

Economic and Social Impact

Led to a reduction in:
- Damage to homes and businesses
- Stress and anxiety experienced by local people from the threat of future flooding
- Financial loss to local businesses
- Disruption to the local community

Managing information and thinking

Describe the flood defence measures in Mallow as seen in photos A–D.

Council car park

Bridge Street – Before

Bridge Street – After (1)

Bridge Street – After (2)

Question Time

1. Why was a flood relief scheme needed in Mallow?
2. Name two things that the relief scheme aimed to achieve.
3. Explain two benefits of the flood relief scheme for Mallow.

Junior Cycle Geography CYCLONE

Geography in the News

Floods hit Bangladesh and India, May 2022

Heavy rain caused widespread flooding in parts of Bangladesh and India, leaving millions stranded and at least 57 dead during May 2022.

In Bangladesh, about 2 million people have been stranded by the worst floods in the country's northeast for nearly 20 years. The rains have washed away villages, inundated roads, damaged crops and cut off access to drinking water and electricity for some people.

Many parts of Bangladesh and neighbouring regions in India are prone to flooding, and experts say that climate change is increasing the likelihood of extreme weather events around the world. Every extra degree of global warming increases the amount of water in the atmosphere by about 7% which impacts on rainfall levels.

In Assam region, nearly 90,000 people have been moved to state-run relief shelters as water levels in rivers run high and large areas of land remain under water.

1. **When** did this flood event occur?
2. **Where** did the flood event impact?
3. **Who** was impacted by this flood event?
4. **What** was the damage caused?
5. **What** might be the economic and environmental consequences of this event for locals?
6. **What** impact is climate change said to have had?

Case Study: Rivers and Settlement

Water affects where people choose to settle. Historically, people settled near rivers, so that they had a supply of water for cooking, washing and drinking, for transport and for agriculture.

Norman Settlement in Ireland

The Normans came to Ireland in 1169, crossing over from England and Wales. They landed in the southeast of the country, at Bannow Bay, Co. Wexford. At this time, Ireland had very few roads and much of the country was covered with forest. Many of the Normans' early settlements were located in the south and east of Ireland. They usually settled on land that was flat and fertile so that they could grow food and keep animals.

Norman settlements were usually located near rivers for a number of reasons:

1. Rivers gave them a supply of drinking water.
2. Without a good road network, rivers provided the Normans with suitable transport routes.
3. Rivers could offer protection, acting as defensive moats around their castles.

14. The location of Bannow Bay

15. Norman settlements on Irish rivers

Q

1. List and explain three reasons why the Normans chose to settle close to rivers.
2. Find out the name of another castle in Ireland that is beside a river and was built by the Normans. Are there any close to you?

The Norman settlement of Trim Castle, built on the River Boyne

Sample questions

1. **Examine** figure 16, which shows a landscape that has been shaped by different physical processes including erosion, transportation and deposition by rivers and the sea.

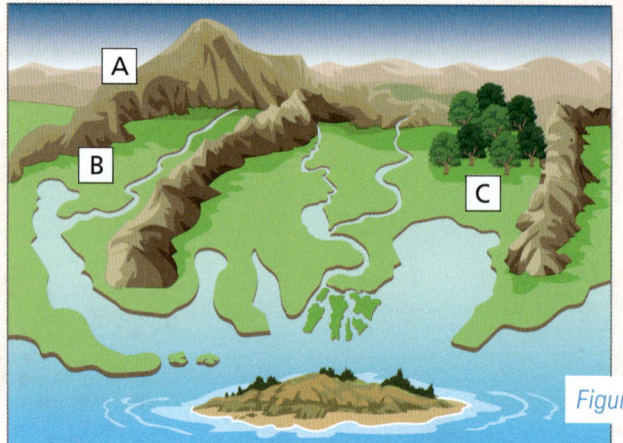

Figure 16

(i) Figure 16 shows an area along the coast similar to the area that Ireland's first settlers arrived at long ago. If you were one of the first settlers in the area shown in figure 16, would you choose to settle at site A or B or C?

(ii) Explain in detail why you would choose to settle at this site.

(SEC sample paper, Q4 (b))

Exam hints!

Look closely at the image and identify items that may be useful to you as an early settler. Is it easier to build your home on flat or hilly land? What materials will you use to construct your home? Where might you find or grow food and source drinking water?

2. **Identify** the following statements as true or false:
   (a) A river source is generally found in high mountainous areas.
   (b) A narrow, fast-flowing river would generally mean it is in a youthful stage.
   (c) It is uncommon to find a settlement along a river.
   (d) Hydraulic action is a form of river deposition.
   (e) Deposition mainly occurs on the inside bend of a meander.
   (f) Freeze-thaw action is a type of weathering.
   (g) Abrasion is a type of deposition.

These are quick, short-answer questions that you may be faced with in the exam. You simply state True or False next to each.

3. **Select** one feature or landform in the photographs below and explain how the processes of fluvial erosion or deposition formed the feature. **Support** your answer using diagrams.

A

B

C

Exam hints!

Look at images A, B and C. Start by naming the feature or landform that you feel most confident with. You can then use FEED to explain how the processes of erosion or deposition have shaped the landscape to form this feature. Make sure to name and explain the processes. You must include a diagram or diagrams as the question has asked for this.

For this question, we will use **FEED** (**F**eature, **E**xplanation of formation, **E**xample, **D**iagram).

11 The Sea

Learning Outcomes: 1.5, 1.10, 2.7, 2.9, 3.4

Learning Intentions

You will be able to:
- Describe, with the aid of diagrams, how the processes of erosion, deposition and transportation shape the coastline
- Discuss how people interact with and manage surface processes at the coast
- Discuss the impact that coastal processes can have on coastal areas.

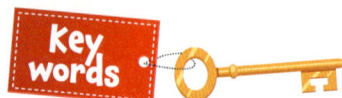

Fetch	Load	Sea arch	Tombolo
Swash	Longshore drift	Sea stack	Groynes
Backwash	Bay	Sea stump	Sea walls
Hydraulic action	Headland	Blowhole	Gabions
Abrasion	Differential erosion	Beach	Rock armour
Compressed air	Sea cliff	Sand dunes	
Attrition	Wave-cut platform	Sand spit	
Solution	Sea cave	Sand bar	

11. The Sea

11.1 Ireland's Coastline

You may have spent time on a sandy beach in Co. Wexford, visited the steep Cliffs of Moher in Co. Clare, or explored the Maghera sea caves in Co. Donegal. Ireland has a varied coastline. However, a basic difference can be seen between western shores exposed to the full force of the Atlantic Ocean and the more sheltered east coast.

Communicating

Examine the map in figure 1, paying close attention to the western and eastern coastlines.

1. Describe how the shape of the western coastline is different to that of the eastern coastline.
2. Can you think of any reasons why the coastlines may be different?

1. Ireland's contrasting coastline

11.2 Coastal Processes

The sea is constantly shaping our coastline. Waves are eroding, depositing and transporting material along the coast.

Erosion by the Sea

Waves

Waves are formed from wind moving across the surface of the sea. The strength of the wind and the length of sea that it passes over (called the **fetch**) affect the size of the waves. When waves reach the shallow water at the coast, they break. The water that moves up the shore is known as the **swash**, and the water that moves back down the shore is the **backwash**.

Types of Wave

Constructive Waves	Destructive Waves
These build up a beach. Material is carried in the strong swash and is deposited up the beach. The backwash is weak.	These can destroy a beach. They are very high waves. The backwash is strong and transports beach material out to sea. The swash is weak.

Processes of Coastal Erosion

1. **Hydraulic action**: This is the power of water as it crashes against the coast.
2. **Abrasion**: Waves pick up stones and throw them against the coast. These stones abrade the coast.
3. **Compressed air**: Air in rocks becomes trapped by the incoming waves. The trapped air puts pressure on the rocks. When the water retreats (moves out), the air expands and the pressure drops. This repeated compression and release causes the rocks to shatter.
4. **Attrition**: Stones carried by the waves hit off each other. Over time they are worn down and smoothed.
5. **Solution**: Certain rocks, such as limestone, are dissolved by the water.

> **Being creative**
>
> Using the first letter of each of the processes of erosion (H, A, C, A, S), create a fun rhyme or mnemonic to help you remember them.

Transportation by the Sea

Material carried by the sea (sand, silt, mud, pebbles) is known as its load. The load is moved up the shore by the force of the incoming waves (the swash), and back down when it is pulled by the retreating waves (backwash). This is known as longshore drift.

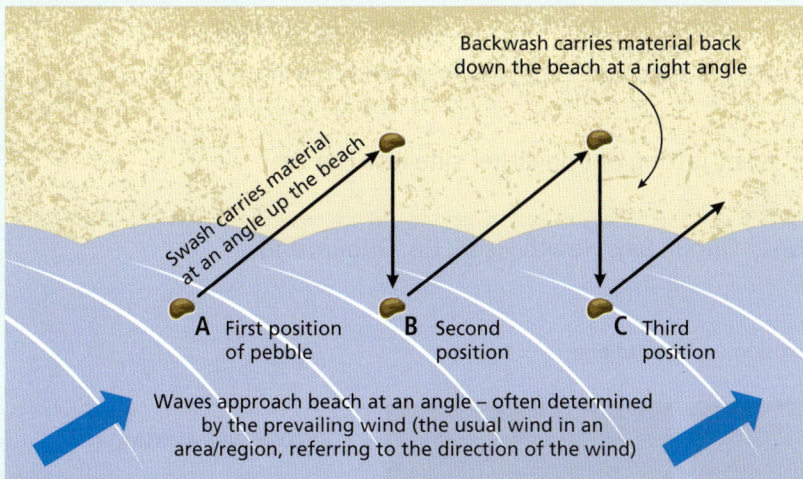

Process

1. Waves approach the shore from the side.
2. The swash pushes the material up the beach at an angle.
3. The backwash drags the material back down the beach at a right angle.
4. As the processes repeat, the material is moved in a zig-zag pattern along the shore.

2. Longshore drift

Deposition by the Sea

When the power of waves is reduced, the sea will drop its load because it loses energy. This can happen in:

- Sheltered bays
- Gently sloping coastal areas
- Shallow water.

Question Time

1. (a) Name the two types of wave.
   (b) Outline the difference between the two types.
2. Explain the terms swash and backwash.
3. Name the five processes of erosion at the coast.
4. Explain how compressed air has an impact on the coastline.
5. What is the load carried by the sea made up of?
6. Using a labelled diagram, describe the process of longshore drift.
7. Where does the sea drop its load?

11.3 Landforms of Sea Erosion

Feature: Headlands and Bays

How a Headland and Bay Forms

1. **Bays** and **headlands** are formed on coastlines where an area of soft rock is eroded more quickly than the hard rock in the headland. This is known as **differential erosion**.
2. Soft rock, such as sandstone, is more easily eroded and begins to form a hollow in the coast known as a bay.
3. The area of hard rock, such as granite, is left jutting out on either side into the sea. This is called the headland.

Examples: Headlands and bays can be seen in Dublin Bay, Howth Head, Co. Dublin, and Liscannor Bay, Co. Clare.

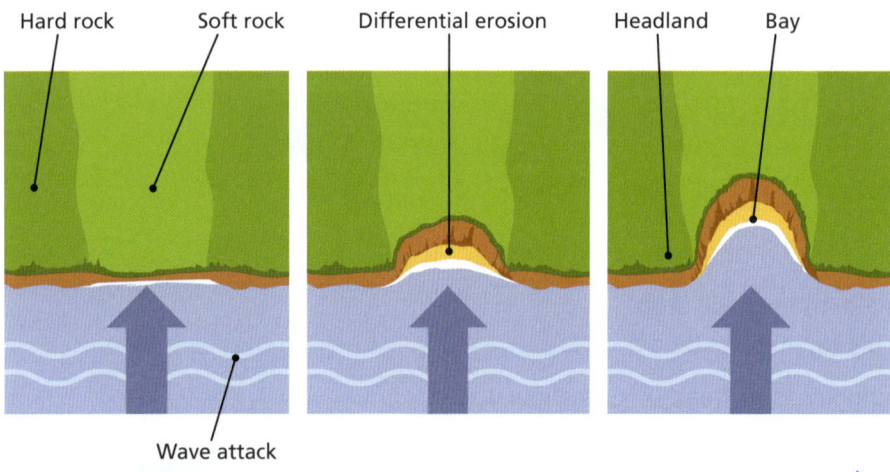

Processes at work
- Hydraulic action
- Abrasion
- Attrition
- Compressed air

3. How a headland and bay is formed

Communicating

With the person next to you, take two minutes to examine this photo of Liscannor Bay. Discuss where most erosion has taken place. Point out the bay and headland.

11. The Sea

📁 Managing information and thinking

1. Pick out each of the following on this OS map: the main headland, a bay with a sandy beach, a bay without a sandy beach.

2. Several 'middens' are shown on the map. Go online to find out what a midden is. What does the presence of middens tell us about the area shown on the map?

Feature: Sea Cliff

How a Sea Cliff Forms

1. Waves erode a notch at the base of the cliff by hydraulic action, abrasion and compressed air.
2. When the notch becomes large enough, the overhanging rock collapses under its own weight.
3. A **sea cliff** has then been formed.
4. The notch continues to erode, so the cliff keeps slowly retreating into the coast. As the cliff retreats, an almost flat, rocky area is created on the shore. This is called a **wave-cut platform**. The wave-cut platform is usually visible at low tide.

Examples: Cliffs can be seen at the Cliffs of Moher, Co. Clare, and at Howth Head, Co. Dublin.

FUN FACT! The Cliffs of Moher stretch for over 8 km and are up to 214 metres high at their highest point. They attract over 1 million tourists each year.

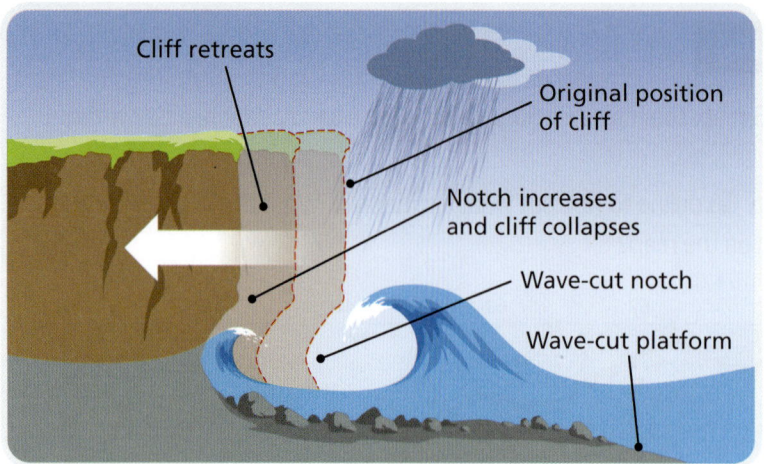

4. How a cliff is formed

Processes at work
- Hydraulic action
- Abrasion
- Compressed air

Feature: Sea Cave, Arch, Stack and Stump

How Sea Caves, Arches, Stacks and Stumps Form

1. Strong waves attack cracks and weaknesses in the base of a cliff. Through hydraulic action, abrasion and compressed air, they erode a large hole at the foot of the cliff.
2. This continues over thousands of years. The hole is known as a **sea cave**.
3. If a sea cave erodes all the way through the headland (or if two caves erode back to back), a **sea arch** will form. This is an archway straight through the headland.
4. If the roof of the arch collapses, a pillar of rock will remain standing alone in the sea. This pillar is known as a **sea stack**.
5. If over time the sea stack collapses into the sea, all that remains is a small stump of rock. This is known as a **sea stump**.

Examples: These can be seen at the Old Head of Kinsale, Co. Cork (sea caves and arch), and Hook Head, Co. Wexford (sea stack).

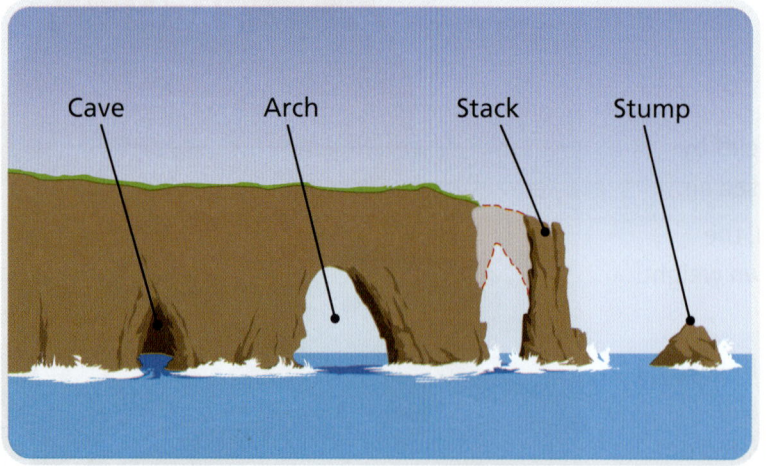

5. A sea cave, arch, stack and stump

Processes at work
- Hydraulic action
- Abrasion
- Compressed air

11. The Sea

Feature: Blowhole

How a Blowhole Forms

A **blowhole** is a passage that goes through the roof of a cave to the land surface above. It is formed when powerful waves compress air that breaks a hole through the roof of the cave.

Example: An example can be found at Downpatrick Head, Co. Mayo.

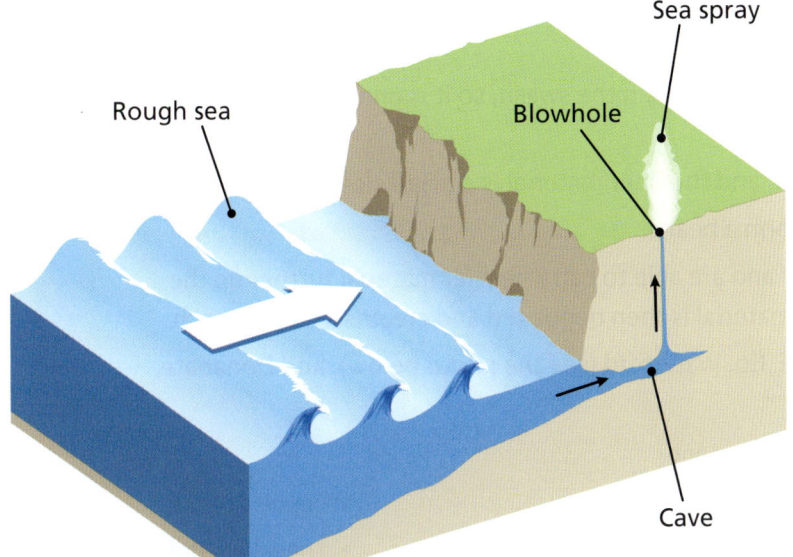

Processes at work
- Hydraulic action
- Abrasion
- Attrition
- Compressed air

6. A blowhole

Managing information and thinking

Name the features labelled A–E.

Question Time

1. Describe, with the aid of a labelled diagram, the formation of a bay and a headland.
2. Explain how a sea cliff is formed.
3. What is a sea cave?
4. Describe how the processes of erosion have helped in the formation of a sea arch. You may use a diagram/s to support your answer.
5. What is a blowhole and where in Ireland can one be seen?

11.4 Landforms of Sea Deposition

Feature: Beach

A **beach** is a gently sloping area of sand, shingle (pebbles) or stones that is found between the high and low tide levels. It is formed by the swash and backwash movements of the waves.

How a Beach Forms

1. When waves break, they lose their energy and drop the load they were carrying. The swash carries this material up the shore and deposits it.
2. In constructive waves, the backwash is weaker than the swash, so it is unable to carry all of the deposited material back out.
3. The heaviest material is deposited first, and the finer material is carried closer to the shoreline.
4. Over time, this material builds up to form a beach.
5. During storms, the waves are stronger and are able to carry heavier material further up the shoreline to the high-tide mark. This material is then deposited and creates a storm beach.

Examples: Beaches can be seen at Tramore, Co. Waterford (see the photo below), and Donabate, Co. Dublin.

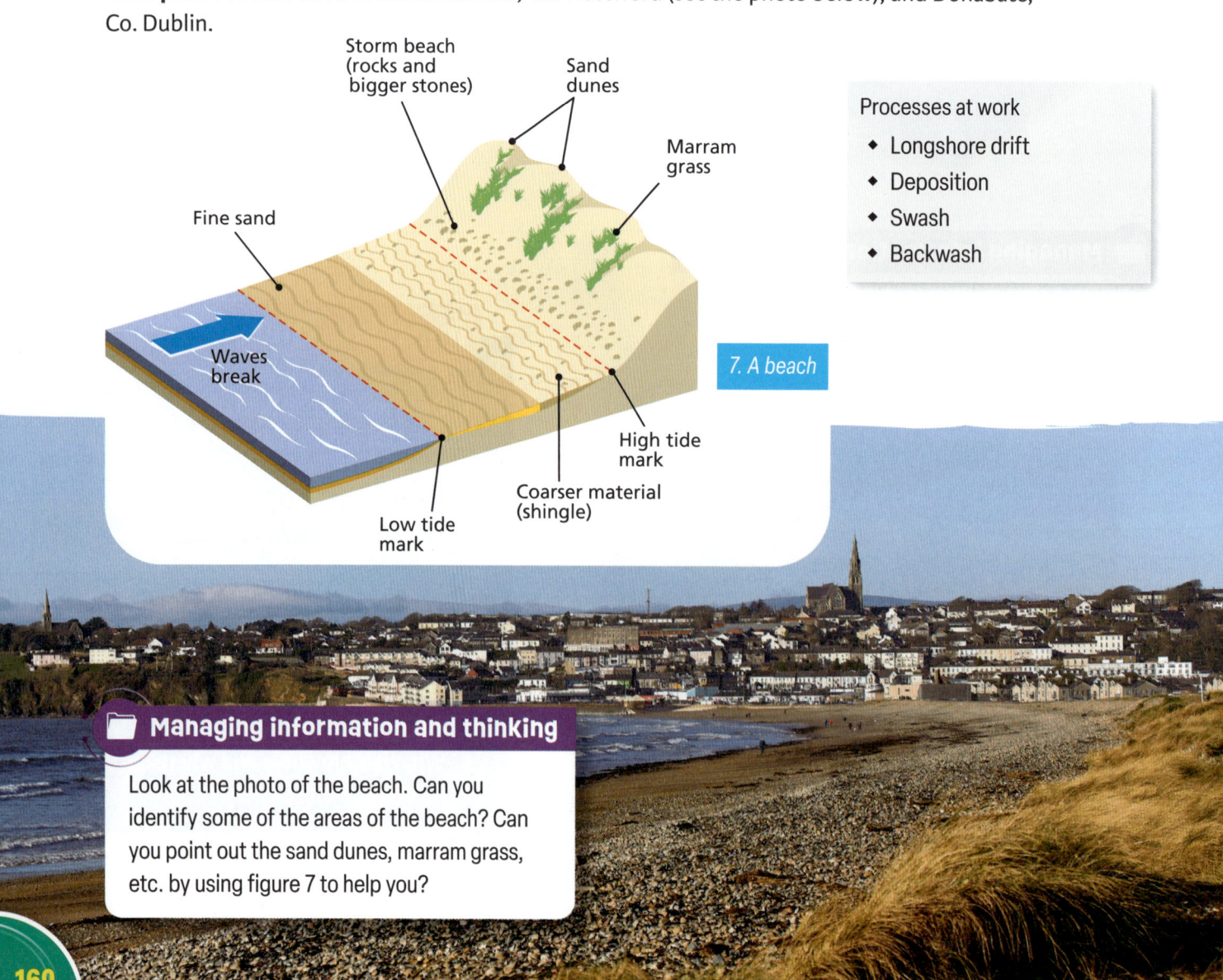

7. A beach

Processes at work
- Longshore drift
- Deposition
- Swash
- Backwash

Managing information and thinking

Look at the photo of the beach. Can you identify some of the areas of the beach? Can you point out the sand dunes, marram grass, etc. by using figure 7 to help you?

Feature: Sand Dunes

Sand dunes are mounds of sand that build up on the shore beyond the high-tide mark at the back of the beach.

How Sand Dunes Form

1. Sand on the beach is dried by the wind.
2. Dry sand is lighter and the wind can blow it inland.
3. The sand becomes trapped by vegetation such as marram grass or other barriers.
4. It then builds up to form sand dunes.
5. Marram grass is sometimes planted on sand dunes to stop the sand blowing further inland. Marram grass has deep roots and is a thick, coarse grass that is resistant to salt.

Example: Sand dunes and marram grass can be found at Enniscrone Strand, Co. Sligo.

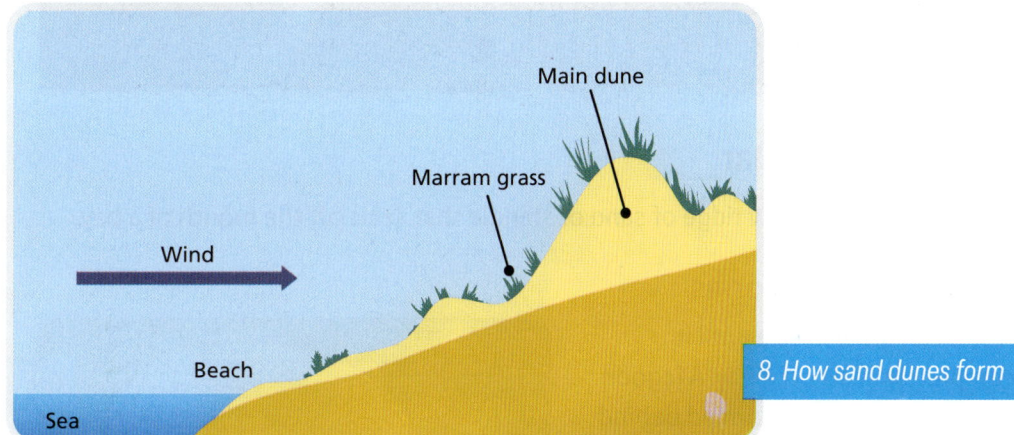

8. How sand dunes form

Feature: Sand Spit

A **sand spit** is a narrow ridge of sand or shingle that is connected to the land on one end and juts out to sea at the other.

How a Sand Spit Forms

1. Longshore drift loses energy when it is interrupted by a sheltered bay.
2. The material carried by longshore drift is then deposited.
3. These deposits build up over time to gradually form a spit.
4. The spit extends across the bay. As it increases in size, vegetation grows on it.

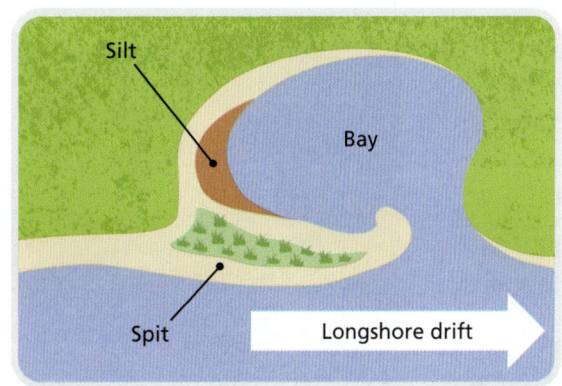

Processes at work
- Deposition
- Longshore drift
- Wind

9. How a sand spit forms

Examples: Sand spits can be seen at Portmarnock, Co. Dublin, and Bannow Bay, Co. Wexford.

La Manga spit in Murcia, Spain, was transformed in the 1960s when the spit was developed for tourism.

Feature: Sand Bar

A **sand bar** is a narrow ridge of sand or shingle that seals off the mouth of a bay.

How a Sand Bar Forms

A spit can grow in size until it completely seals off a bay. A lagoon is a small lake that is formed behind the sand bar. It was originally part of the bay, but it has now been sealed off by the bar.

Examples: Sand bars can be seen at Loch Muirí, Co. Clare, and Lady's Island Lake, Co. Wexford.

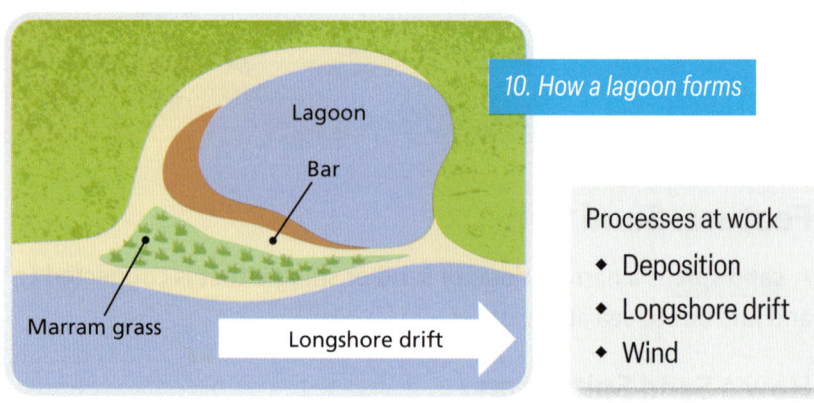

10. How a lagoon forms

Processes at work
- Deposition
- Longshore drift
- Wind

Feature: Tombolo

A **tombolo** is a narrow ridge of sand or shingle that joins an offshore island to the mainland.

How a Tombolo Forms

A spit can grow in length until it reaches a nearby island. The spit is then known as a tombolo.

Example: A tombolo can be found near Roundstone, on the Galway coastline.

11. How a tombolo forms

Processes at work
- Deposition
- Longshore drift

11. The Sea

Iorras Beag peninsula and tombolo, near Roundstone, Co. Galway

📁 Managing information and thinking

Examine this OS map. Draw a sketch map of the area, ensuring to include a key. On your sketch, identify the following:

- The coastline
- A feature of coastal deposition
- A regional road
- A nature reserve.

Junior Cycle Geography CYCLONE

Question Time

1. Draw a beach profile into your copy. On it show and label:
   - Sand dunes
   - Marram grass
   - Storm beach
   - High and low tide marks.
2. What impact can longshore drift have on coastal erosion?
3. Describe the formation of a sand dune.
4. Match each of the following features with the correct term:

A  B  C  D

Sand spit   Tombolo   Lagoon   Sand dune with marram grass

5. (a) Draw a labelled diagram of one feature of coastal deposition.
   (b) Explain how that feature was formed by the processes of coastal deposition.
   (c) Name a location where this feature can be seen on the Irish landscape.

11.5 People and the Sea: Managing Surface Processes

Humans have put many measures in place to defend their coasts from the powerful processes of sea erosion. The following are methods used to defend our coasts against the sea.

Groynes

Groynes are concrete or wooden walls or fences that are built out into the sea at right angles to the coast. They work to reduce longshore drift by trapping the sediment carried by waves. The trapped sand builds up the level of the beach.

Example: An example can be seen at Rosslare, Co. Wexford.

Q How are groynes helping to build up this beach?

11. The Sea

Sea Walls

Sea walls are built to break the power of the incoming waves. They are curved at the top to push the waves back out to sea.

Example: An example can be seen at Bray, Co. Wicklow.

Q Why is the top of a sea wall curved, such as you can see here?

Gabions

Gabions are steel wire cages filled with stones. They are then stacked on top of each other where they act like a sea wall. They are used to slow down or prevent erosion by breaking the power of the waves.

Example: An example can be seen at Lahinch, Co. Clare.

Q Why do you think the stones are put in cages rather than being built as a stone wall?

Rock Armour

Large boulders known as **rock armour** are placed at the base of a cliff, sand dune or sea wall. When a wave breaks, it will hit the rock armour and lose its energy. This reduces erosion of the coastline.

Example: An example can be seen at Strandhill, Co. Sligo.

Junior Cycle Geography CYCLONE

Question Time

1. Name four coastal defence measures.
2. Describe two ways that humans can protect the coastline from the destructive power of the sea. Use diagrams to support your answer.
3. The aerial photographs below show coastal erosion occurring over a 16-year period along a stretch of the coastline. Look closely at the images and answer the questions below.
   (a) What evidence of coastal erosion can you identify between 1996 and 2012?
   (b) Name any coastal defence mechanisms that are visible in the aerial photos?
   (c) Suggest one further coastal defence measure that you would recommend to use in this area. Explain why you think this would be effective?

1996

2006

2012

11.6 Coastal Management

Case Study: North Bull Island, Co. Dublin

North Bull Island is located in Dublin Bay. It exists as a result of decisions made over 200 years ago.

In the early nineteenth century, longshore drift was carrying large amounts of sand and silt into Dublin Bay. The deposits were partially blocking the mouth of the harbour, making it difficult for ships to enter.

11. The Sea

A survey was carried out and it was decided that walls would be built on both sides (north and south) of Dublin Harbour.

North Bull Wall was completed in 1821. Longshore drift now deposited material behind the North Bull Wall. This helped to build a large sand spit, which developed into North Bull Island.

The island reached its current length in 1902. It is approximately 5 km long and 1 km wide at its widest point.

Managing information and thinking

Locate both the North and South Bull Walls on this OS map.

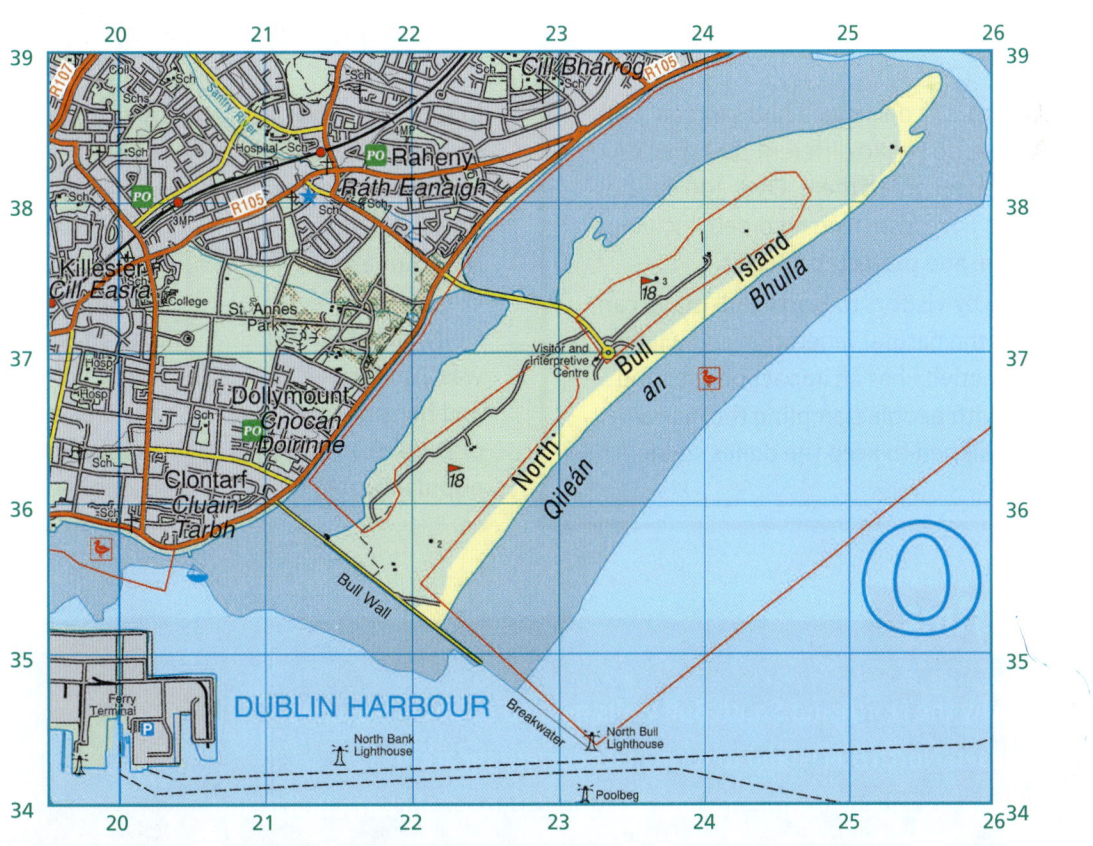

167

Junior Cycle Geography CYCLONE

Q Name all the features of coastal deposition that you can see in this aerial photograph of Dublin Bay.

Environmental Impact

- The island was declared a bird sanctuary in the 1930s. Up to 180 different bird species have been recorded there. Over 300 species of plants have also been recorded, including some rare and protected species.
- Littering by visitors to North Bull Island can damage the natural vegetation and wildlife. Human activity has an impact on the sand dunes, with people trampling the marram grass designed to keep the dunes stable.

Social Impact

In 1885, the Royal Dublin Golf Club was established on the island, followed by St Anne's Golf Club in 1921. Dollymount Strand was also formed and became one of Dublin's most popular beaches.

Economic Impact

A visitor and interpretative centre was built in 1986. The centre has a lecture theatre, a natural history museum and an aquarium. Large numbers visit the area and spend money in the local economy.

Question Time

1. Explain why the North and South Bull Walls needed to be constructed.
2. What caused North Bull Island to form in its current location?
3. Discuss one social, environmental and economic impact of North Bull Island.

11. The Sea

EXAM FOCUS

Sample questions

1. **Examine** the photograph of a coastal area shown below and answer each of the following questions.

(i) In your copy, match each of the labels 1 to 5 on the photograph with the feature that best matches it in the table below. One has been completed for you.

Feature	Number
Headland	2
Sea stack	
Beach	
Cliff	
Bay	

(ii) The house labelled X on the photograph is a holiday home for tourists. **Describe** one reason why the area on the photograph would attract tourists.

(iii) **Explain** how longshore drift transports material along the coast. You may also draw a diagram if you wish.

(2022 SEC exam paper, Q3 (a) (i)–(iii))

2. The photograph shows an area at risk of damage due to coastal erosion.

Junior Cycle Geography CYCLONE

You are concerned about the future of this area and that there may be a collapse. **Write** a letter to your local councillor suggesting how the coast could be protected from further erosion. Your letter must include each of the following:

(i) An example of a coastal defence measure that would be suitable to protect the area from future erosion.

(ii) An explanation of how the coastal defence measure you named will protect the area from the processes of coastal erosion. You may also draw diagrams if you wish.

(SEC sample paper, Q3 (b) (i) & (ii))

Exam hints!

The question is asking you to write a letter. In the letter, you should write a brief opening paragraph and then get straight into your answer. It has been started for you below, so you should copy and complete this in your copy.

Lets get started! Sample starter

Dear Councillor Moran,

I am writing to you about a very serious concern of coastal erosion in my local area. I am concerned that there may be a collapse. I have a suggestion on how this situation can be solved …

The coastal defence measure that would be suitable to protect this area from the processes of coastal erosion is:

I will explain how this measure can be used …

… now complete the answer in your copy.

11. The Sea

CBA 1 Geography in the News

Dooagh Beach Reappears – 33 Years After Being Washed Away

A beach that disappeared more than 30 years ago, returned to an island off the County Mayo coast in 2017.

The sand at Dooagh, Achill Island, was washed away by storms in 1984, leaving only rocks and rock pools. However, following a freak tide, hundreds of tonnes of sand were deposited, recreating the old 300-metre stretch of golden beach.

Before

After

According to Sean Molloy, manager at Achill Tourism: 'Before it disappeared, the beach had been there for as long as living memory, almost continuously, until 1984–85. … then when we had a cold snap over Easter, the wind was coming in from the north. It was very constant and steady and it must have transported eroded material in from elsewhere.'

He said the bulk of the sand was deposited in just over a week, leaving locals delighted.

Excitement was short lived however, as in January 2019, the beach disappeared once again during a period of bad weather.

2022

To read a news report on the reappearance and disappearance of this beach, search online for 'Coast to ghost: Irish beach vanishes after brief reappearance – Guardian'.

171

1. Where is Dooagh beach located?
2. When did the beach reappear?
3. Describe one positive impact that the appearance of the beach may have had on the local community?
4. As a local hotel owner, write a piece that will be included in an Irish tourist magazine about the beach in Dooagh.
   (a) Tell them a little about the history of the beach.
   (b) Explain why people should come to visit the area and stay in your hotel.
5. Investigate one of the three 'Geography in the News' coastal events listed. Use the success criteria below to complete this activity.

Search online for 'Portrane dad fears family will be homeless by Christmas as coastal erosion ravages seaside house'.

Search online for 'How Irish Rail is fighting an "alarming increase in erosion" near vulnerable railway lines'.

Search online for 'Hundreds of family homes in Galway are at serious risk from coastal erosion'.

SUCCESS CRITERIA

I Must
- Outline where the coastal protection issue is taking place
- State when it is/was occurring
- Describe who is involved/ who might be impacted
- Outline what is happening in the area
- Outline why it is occurring.

I Should
- Describe in my opinion how the situation has or should be managed
- Suggest suitable measures that can be taken to protect the coastal area involved.

I Could
- Draw a sketch of the coastline involved, using Google or scoilnet maps of the areas
- Draw diagrams to demonstrate the measures I feel are required to protect the coastal region selected.

Peer Assessment
Swap your work with a classmate. You must read their work and tell them two things that they have done very well, and suggest one thing that they could improve on based on the success criteria above.

Redrafting
Taking on board your classmate's comments on your work, make any changes that you think will improve your work, reviewing the success criteria again if needs be. When you are ready, show your work to your teacher.

Learning Outcomes: 1.5, 1.10, 2.7, 2.9

12

Glaciation: The Work of Moving Ice

Learning Intentions

You will be able to:

- Describe, with the aid of diagrams, how the processes of erosion, deposition and transportation shape the glacial landscape
- Explain how glaciation has impacted on Co. Wicklow.

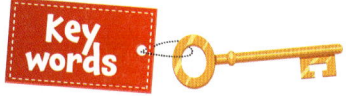

key words

Climates	Friction	Tarn	Englacial
Lowland areas	Meltwater	Pyramidal peak	Subglacial
Glacier	Pluck	Arête	Moraines
Ice sheets	Embedded	Paternoster lakes	Drumlins
Ice Age	Abrade	Truncated spur	Erratics
Interglacial periods	Glaciated valley	Hanging valley	Esker
Erodes	Cirque (corrie)	Supraglacial	Outwash plain

173

12.1 Glaciation

About 2 million years ago, the climates (weather conditions in an area measured over 30 years) of countries such as Ireland became much colder. Year after year, snow fell. It began to gather in upland areas, gradually turning to solid ice. The ice began to move slowly to lowland areas under the influence of gravity as great rivers of ice called glaciers. Some glaciers melted, while others joined together to form ice sheets. The ice sheets (figure 1) covered huge areas of the earth's surface. An ice age had arrived.

Fox Glacier in New Zealand

Glacier

Glacial meltwater flowing from the snout (front)

During the ice age, there were warmer periods known as interglacial periods. The average length of an interglacial period is 100,000 years, so it is possible that we are now living in an interglacial period and that the ice sheets will return!

During the last ice age, 14,000 years ago, almost one-third of the earth's surface was covered in ice. Ice sheets still cover Antarctica and Greenland today.

Managing information and thinking

1. Locate Ireland on the map shown in figure 1. Was it affected by an ice sheet?
2. Using an atlas or Google Maps if necessary, name 10 countries affected by an ice sheet.

1. Ice sheets of the last ice age

12. Glaciation: The Work of Moving Ice

12.2 Processes of Glacial Erosion

As a glacier moves, it **erodes** the landscape by two main processes.

1. Plucking

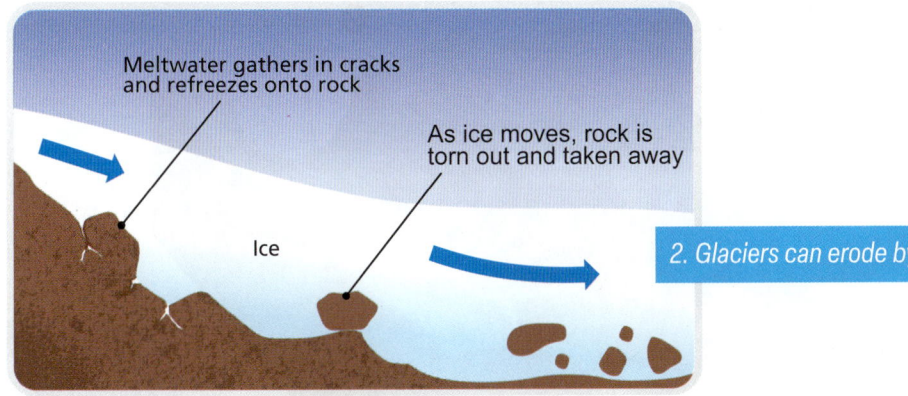

2. Glaciers can erode by plucking

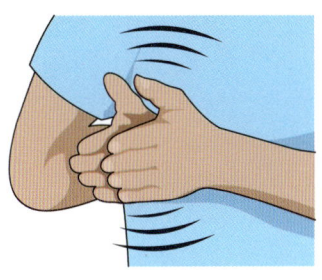

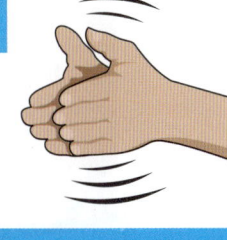

Rub your hands together. Can you feel the friction created between them?

As ice moves, it causes **friction** between the ice and the ground below. This friction creates heat, which causes the ice at the base to melt. The **meltwater** flows into the cracks in the rock. The water then refreezes around the rocks and sticks to the glacier. When the ice moves forward again, it can pull or **pluck** chunks of the rock out of the ground.

2. Abrasion

Plucked rocks become **embedded** in the base of the glacier. As the glacier moves, the rocks **abrade** (scrape and smooth) the surface over which they pass.

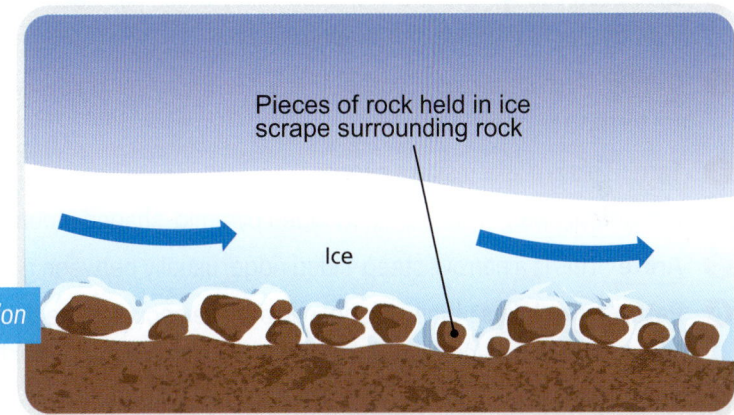

3. Glaciers can also erode by abrasion

FUN FACT! Seventy-five per cent of the world's total supply of fresh water is frozen in glaciers.

175

Features of Glacial Erosion

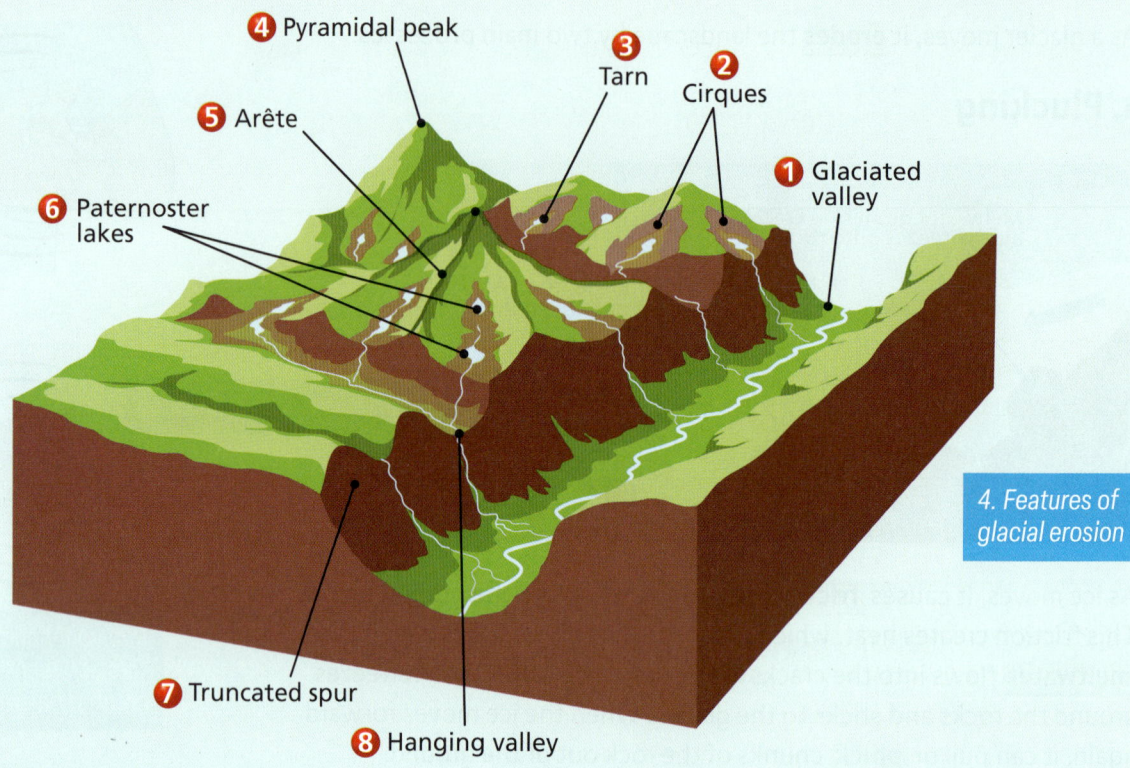

4. Features of glacial erosion

1. **Glaciated valley**: This is a V-shaped valley which has been straightened and flattened by the movement of a large glacier to become a U-shaped valley.
2. **Cirque**: Also known as a 'corrie', this is a large hollow on the side of a mountain. It has three steep sides and is the birthplace of a glacier.
3. **Tarn**: This is a lake inside a cirque.
4. **Pyramidal peak**: This is a steep-sided pyramid-shaped mountain that was eroded on all sides by many cirques.
5. **Arête**: This is a narrow, steep-sided ridge, usually between two cirques.
6. **Paternoster lakes**: When a long, narrow lake occupies the floor of a glaciated valley, it is called a ribbon lake. When a few ribbon lakes are linked, they are called paternoster lakes.
7. **Truncated spur**: This was originally an interlocking spur that was eroded, having its 'head' cut off as the glacier moved through the valley.
8. **Hanging valley**: This is a small tributary valley that hangs above the main glaciated valley.

Question Time

1. Where in the world can we still find ice sheets today?
2. Ice wears away the landscape by erosion.
   (a) Name the two processes of glacial erosion.
   (b) Explain how each process erodes the landscape.
3. Explain the following terms: (a) truncated spur, (b) hanging valley, (c) pyramidal peak, (d) paternoster lake.

12. Glaciation: The Work of Moving Ice

Feature: U-Shaped Glaciated Valley

A U-shaped glaciated valley was originally a V-shaped river valley. It became straightened and deepened by a glacier filling it, which changed it to look more U-shaped.

How a U-Shaped Glaciated Valley Forms

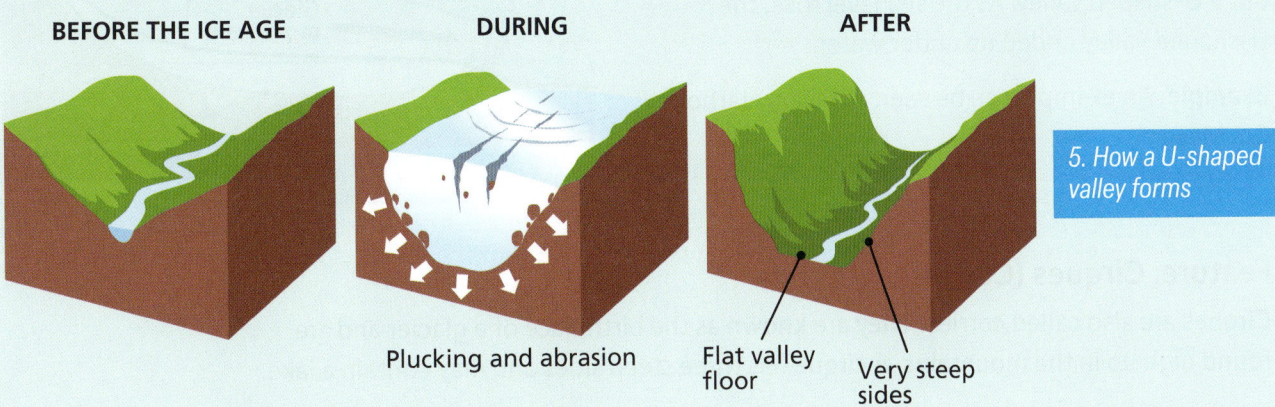

5. How a U-shaped valley forms

1. During an ice age a glacier fills a river valley.
2. The base of the glacier melts into the underlying rocks due to the heat caused by friction as the glacier moves. Then the meltwater refreezes, and as the glacier moves on, it plucks chunks of rock out of the ground.
3. Plucked rocks become embedded in the base of the glacier. As the glacier moves, the rocks abrade the rock surface they pass over.
4. The river's once V-shaped valley has now been reshaped into a U-shaped valley.

Examples: U-shaped valleys can be seen in Glendalough, Co. Wicklow, and Doolough Valley, Co. Mayo.

Q What evidence of glaciation can you identify in this image?

U-shaped valley, Glendalough

Feature: Fjords

How a Fjord Forms

A fjord is a U-shaped valley found under the sea. They were formed when the glacier retreated after carving out a U-shaped valley. As the sea level rose, the U-shaped valley ended up under water.

Example: An example can be seen at Killary Harbour, Co. Mayo.

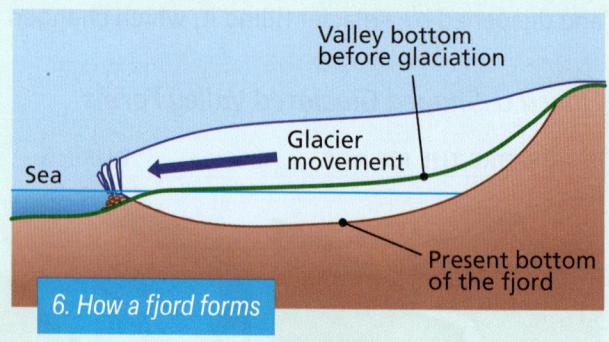
6. How a fjord forms

Feature: Cirques (Corries)

Cirques are also called corries. They are known as the birthplace of a glacier and are found high up in the mountains. A cirque has three steep sides and may contain a lake.

How a Cirque Forms

❶ Snow collects in a mountain hollow high up in the mountain. With repeated snowfalls, the snow compacts to form ice. A glacier is born.

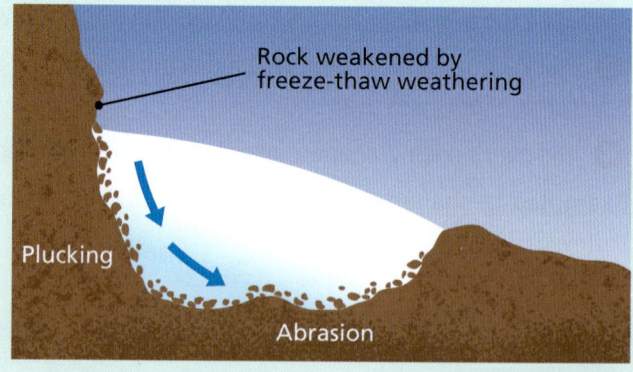

❷ Through plucking and abrasion, the hollow grows deeper. Freeze-thaw action is also at work here. (Look back at Chapter 8 to remind yourself what freeze-thaw action is.)

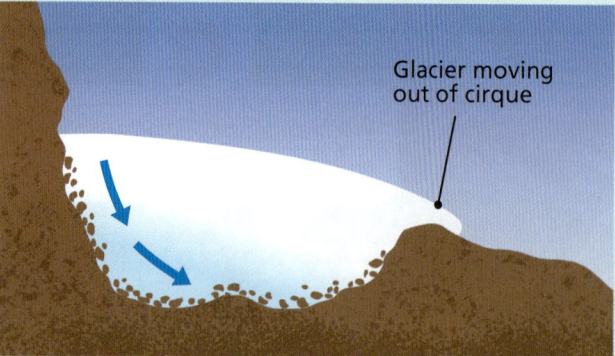

❸ When the glacier is big enough, it begins to flow over the edge of the cirque. It then begins its journey.

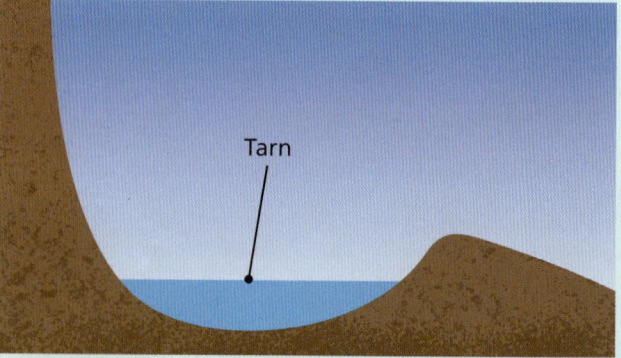

❹ When the glacier melts, a lake called a tarn may be trapped in the cirque hollow.

12. Glaciation: The Work of Moving Ice

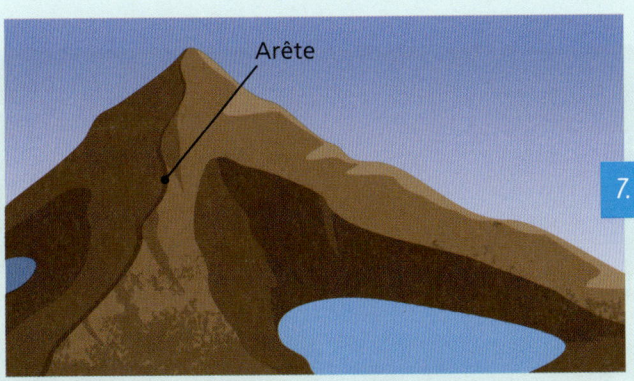

5. When two cirques form back to back, the ridge between them is known as an arête.

7. How a cirque forms

Examples: Examples include the Devil's Punch Bowl, Co. Kerry, and Coumshingaun Lough in the Comeragh Mountains, Co. Waterford.

Managing information and thinking

Coumshingaun Lough

Examine the photograph of the Comeragh Mountains and the OS map. Then answer the following questions in your copy.

1. Identify the feature labelled A in the photograph and describe how it was formed.
2. At what time of year do you think this photograph was taken? Explain your answer.
3. On the OS map, locate the lake shown in the photograph. Give the name of the lake and a six-figure grid reference for its location.
4. Looking at the OS map, how high above sea level is this lake?

Question Time

1. What would a U-shaped valley have looked like prior to glaciation?
2. What is a fjord?
3. What is a tarn?
4. What name is given to the ridge between two cirques?
5. Select one feature of glacial erosion.
   (a) Explain how that feature was formed.
   (b) Name one example of that feature.
   (c) Draw a labelled diagram of that feature.

12.3 Transport and Deposition by Moving Ice

As a glacier moves, it transports eroded material. These materials are carried on the surface of the ice (**supraglacial** transport), within the ice (**englacial** transport) and underneath the glacier (**subglacial** transport).

1. **Supraglacial**: These are materials found on the surface of the glacier, along the top or sides. These materials are carried along as the glacier moves. This material has usually fallen down the mountain and landed on the sides or top of the glacier.
2. **Englacial**: This is any regolith (loose material) trapped within the ice, including material that has fallen down cracks in the ice.
3. **Subglacial**: This is loose regolith trapped underneath the glacier and dragged along in the bottom of the ice.

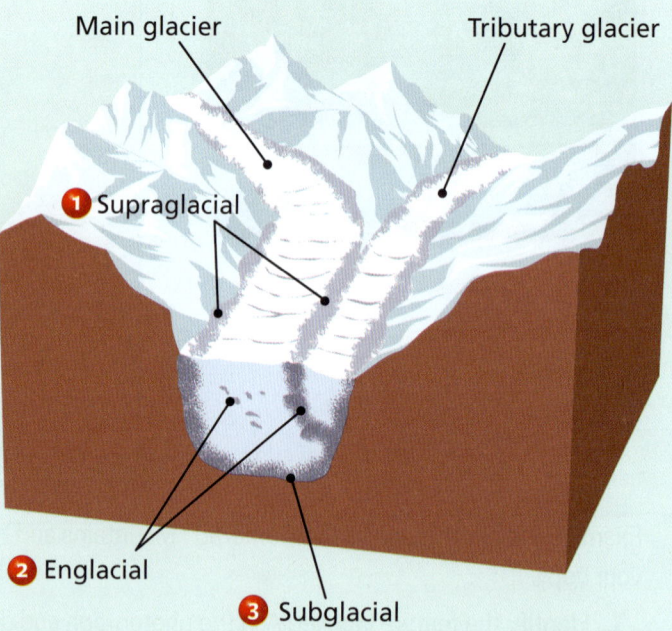

8. How material is transported by a glacier

Question Time

1. Name and describe the three ways that glaciers transport materials.
2. Draw and label a diagram showing how glaciers transport eroded material.

12. Glaciation: The Work of Moving Ice

Features of Glacial Deposition and Transportation

Now let's look at some of the distinctive landscape features that were created by glacial transportation and deposition. The features of glacial deposition are mainly seen in lowland areas. As a glacier moves slowly down the valley to the lowland, it carries eroded material with it. When it melts, it begins to deposit its material, creating features.

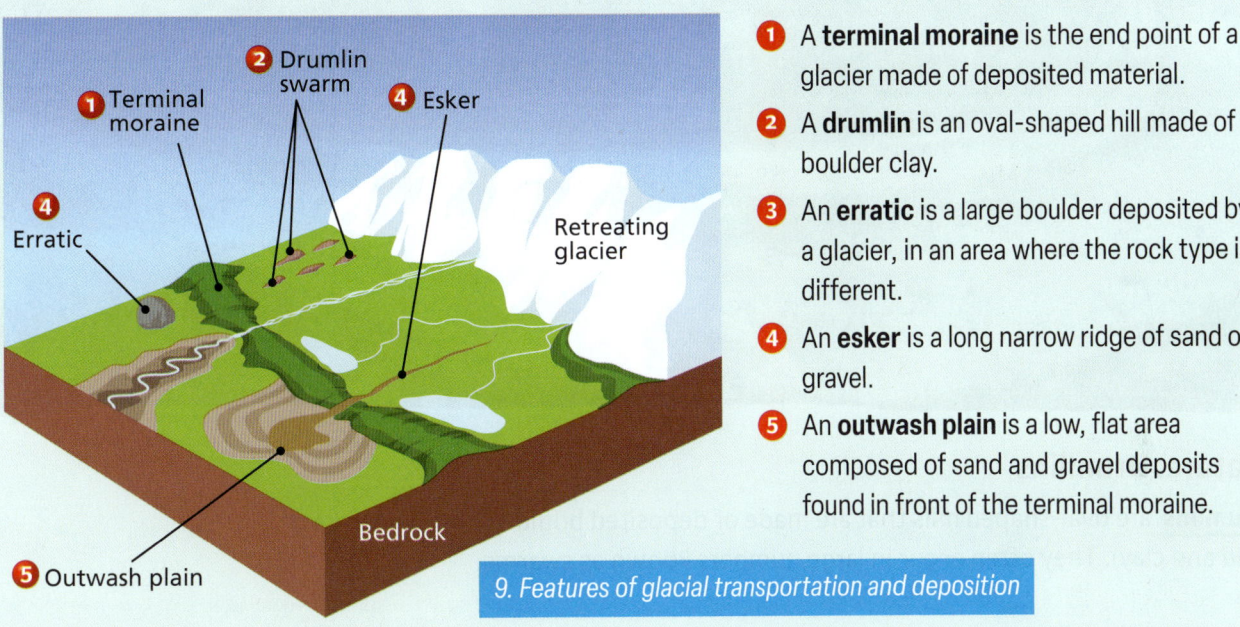

① A **terminal moraine** is the end point of a glacier made of deposited material.

② A **drumlin** is an oval-shaped hill made of boulder clay.

③ An **erratic** is a large boulder deposited by a glacier, in an area where the rock type is different.

④ An **esker** is a long narrow ridge of sand or gravel.

⑤ An **outwash plain** is a low, flat area composed of sand and gravel deposits found in front of the terminal moraine.

9. Features of glacial transportation and deposition

Feature: Moraines

A **moraine** is a mass of rocks and other material carried down and deposited by a glacier.

① A **recessional moraine** is a moraine marking a temporary stop in the retreat (moving back) of a glacier.

② A **lateral moraine** is found at the side of the glacier. It is made of broken rock and soil that fell down the mountain and gathered at the side. This is supraglacial material.

③ A **terminal moraine** is found at the front of the glacier. It is made up of material that is pushed ahead of the glacier. A terminal moraine marks the end point of the glacier.

④ A **medial moraine** is found in the middle of a glacier when two glaciers combine. Their two lateral moraines join together. This is englacial material.

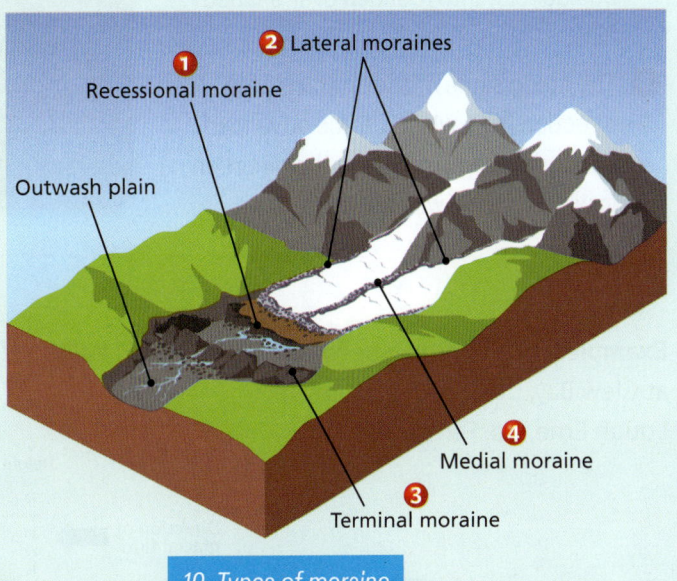

10. Types of moraine

181

Junior Cycle Geography CYCLONE

Q What types of moraine are shown at A and B? State whether each moraine is subglacial, englacial or supraglacial.

Feature: Drumlins

Drumlins are oval-shaped hills that are made of deposited boulder clay (a mixture of sand and clay). They often occur in large numbers known as swarms.

How Drumlins Form

1. Glacial ice deposits boulder clay in irregular heaps.
2. The ice retreats and then advances again. This time it shapes and smooths the boulder clay into rounded oval-shaped hills.
3. The steep slope of the drumlin is the direction from which the ice advanced. The gentle slope points to the direction in which the ice was travelling.

Examples: Drumlins can be found at Clew Bay, Co. Mayo, and Lower Lough Erne, Co. Fermanagh.

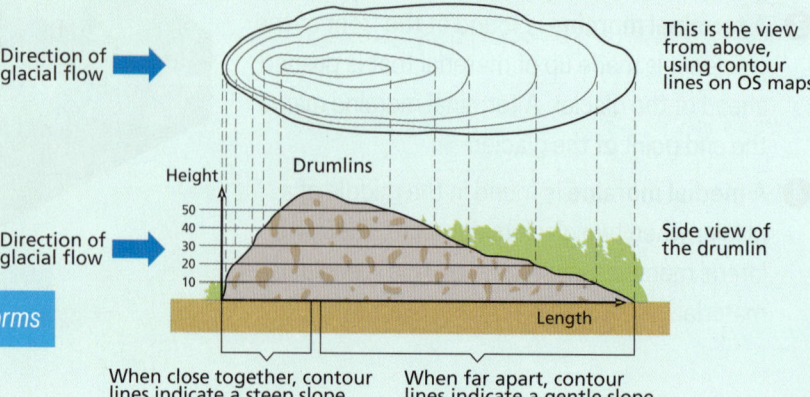

11. How a drumlin forms

12. Glaciation: The Work of Moving Ice

Working with others

The photograph and OS map show a drumlin swarm in Clew Bay, Co. Mayo. These drumlins were partially submerged (drowned) when the sea rose at the end of the Ice Age.

1. How do you know by the shape of these islands that they are drumlins?
2. The place labelled A on the photograph is shown as Rosnakilly on the OS map. Using the map to help you, name the points labelled B and C and the island labelled D on the photograph.
3. Is there any evidence on the photograph or map that the drumlins shown may have been inhabited in ancient times?

Feature: Erratics

How an Erratic Forms

Erratics are large boulders that were deposited by ice in an area where the rock type of the area is quite different to the rock type of the boulders. Because of this, the boulders look out of place in the landscape in which they now stand. Erratics can tell us how far and in what direction ice has travelled.

Example: Erratics of granite can be found in the limestone area of the Burren, Co. Clare, as seen in the photograph.

A granite erratic in the Burren

183

Question Time

1. Name three types of moraine.
2. What is a drumlin?
3. What is an erratic?
4. Name one feature of glacial deposition.
   (a) Explain how the feature was formed.
   (b) Name one example of that feature.
   (c) Draw a labelled diagram of the feature.

Features of Glacial Meltwater

At the end of an ice age, when global temperatures began to rise and glaciers started to melt, the meltwater flowed out from the melting ice sheet and deposited material across the land surface. The heaviest material, such as stones and gravel, was dropped first, then the lighter material, such as sand and silt.

Feature: Esker

An **esker** is a long, winding ridge of sand or gravel.

How an Esker Forms

1. As the ice melts, large tunnels of meltwater flow in tunnels beneath the ice.
2. When a river leaves an ice tunnel, it slows down immediately and deposits material at the mouth of the tunnel.
3. As the ice slowly melts back, the deposited material is dropped in the form of a long, narrow ridge of sand and gravel.

Example: An example can be seen at the Trim esker system, Co. Meath.

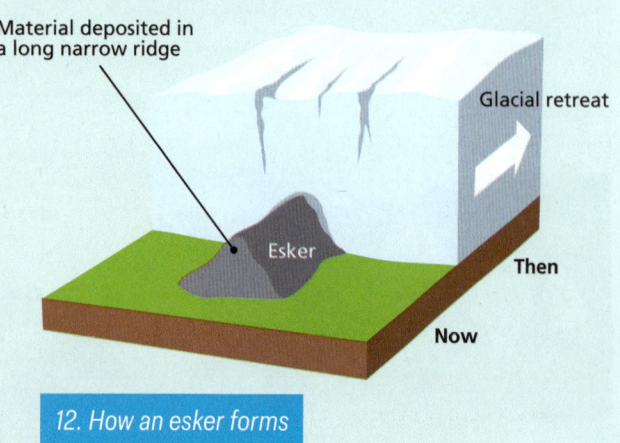

12. How an esker forms

Working with others

An esker

Take 30 seconds to examine the photograph of an esker. Then turn to the person next to you and together identify the esker in the image. Compare your answer with those given by your classmates.

12. Glaciation: The Work of Moving Ice

Feature: Outwash Plain

An **outwash plain** is a flat area of sand and gravel, found in front of a terminal moraine.

How an Outwash Plain Forms

1. As the ice sheet melts, huge amounts of meltwater flow out of it.
2. The water flushes large quantities of sand and gravel onto the lowland beyond the front of the ice.
3. As the meltwater loses energy, it deposits the heaviest material first, followed by the lighter material.

Example: The Curragh, Co. Kildare, seen in the photograph below, is an outwash plain.

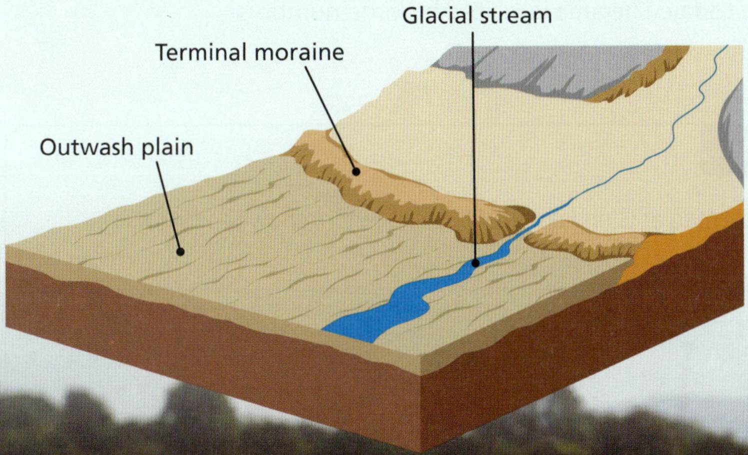

13. How an outwash plain forms

The Curragh

Question Time

1. At the end of the last ice age, describe what happened to the meltwater as glaciers began to melt?
2. What is an esker?
3. Outline what an outwash plain is.
4. Name one feature of meltwater deposition.
   (a) Explain how that feature was formed.
   (b) Name one example of that feature.
   (c) Draw a labelled diagram of that feature.

185

12.4 Glaciation and People

The landscape of Co. Wicklow provides us with a good example of how glacial activity has influenced how people interact with their environment.

Case Study: How People Interact with and Use the Wicklow Landscape

Glendalough is a glacial valley in Co. Wicklow. During the last ice age, the valley was carved out by the ice into the U-shape we see today. The area is visited by large numbers of tourists throughout the year.

Managing information and thinking

Draw a sketch map of the area shown on this OS map of Glendalough. On your sketch, show the following:

- The Glenealo River and the Upper Lake in Glendalough
- An area of forestry
- An upland area of more than 400 m elevation (height)
- Two areas that can be used for recreational activities
- A landform of glacial erosion or deposition.

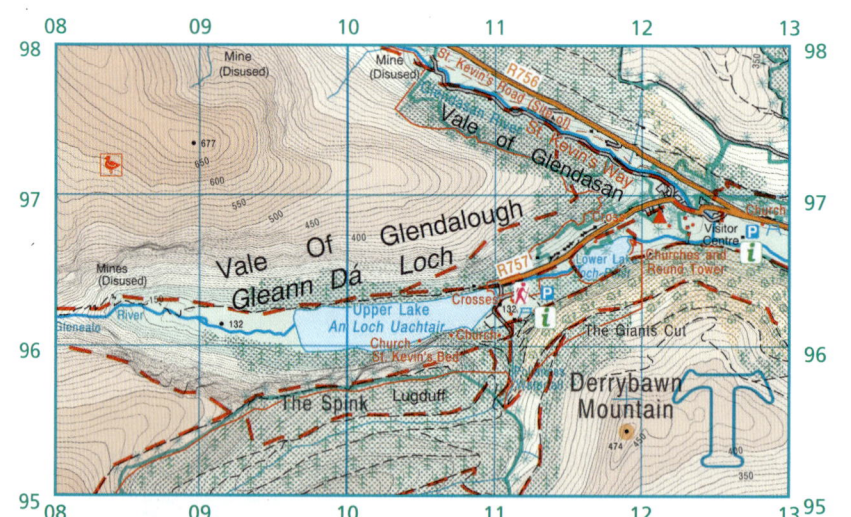

Walking trails in Glendalough

12. Glaciation: The Work of Moving Ice

Impact of Glaciation in Co. Wicklow

Environmental Impact

- Glacial lakes have provided natural reservoirs used for generating hydroelectric power in Co. Wicklow. Hydroelectric energy is a form of renewable energy created by the power of moving water.
- Turlough Hill is the central control point for all hydroelectricity generation throughout Ireland. During the highest electricity demand periods, water is released from an upper reservoir and flows through four turbines into lower Lough Nahanagan – a tarn lake.

Social Impact

- The Wicklow Mountains National Park, of which Glendalough is a part, covers an area of 20,000 hectares, making up much of upland Wicklow. It attracts over 1 million visitors every year.
- There are many walking trails of varying distance and difficulty around Glendalough. Within the valley itself, there are nine colour-coded walking trails.
- Lakes found in the Glendalough Valley – created as a result of ice thaw after the last ice age – attract anglers. The valley is also home to a variety of flora (plants) and fauna (animals).
- Local businesses reap the rewards, as visitors stay in hotels and B&Bs throughout the county and spend money in the local economy.

Economic Impact

- County Wicklow is known as the Garden of Ireland. The steep topography (landscape) created by glaciation can make certain types of farming difficult. Glaciers also removed much of the rich soil cover in upland areas, leaving them unsuitable for highly productive agriculture.
- Almost 22% of its land area is devoted to forestry, making it one of the most important land uses in the county. A government grant scheme encourages farmers to plant trees, especially on poorer quality land. Forestry also provides employment, both directly and indirectly, for many local people.

Search online for 'Presentation welcome to ESB's Turlough Hill Power Station' to watch a presentation on how electricity is generated at Turlough Hill.

1. Describe how glaciation has had an impact socially, economically and environmentally in Co. Wicklow.
2. What is hydroelectric power?

Turlough Hill

Junior Cycle Geography CYCLONE

Sample questions

1. **Examine** the 1:50 000 Ordnance Survey map extract found on p. 124 and answer each of the following questions. The landscape shown on this map extract has been shaped by different physical processes including river erosion, glacial erosion and weathering.

   (a) **Circle** the two processes of glacial erosion listed in the box below.

   Plucking   Solution   Longshore drift   Saltation   Abrasion   Compression

   (b) On a hike through the area shown on the map extract on p. 124, your friend asks you how this landscape was formed. **Write** the answer you would give to **explain** how the processes of glacial erosion have shaped the landscape shown on the map extract. **Refer** to the map extract in your answer.

   (SEC sample paper, Q6 (b) & (c))

2. Below is a diagram showing features of glacial deposition. Select one feature and **describe** how the processes of glacial deposition have resulted in its formation.

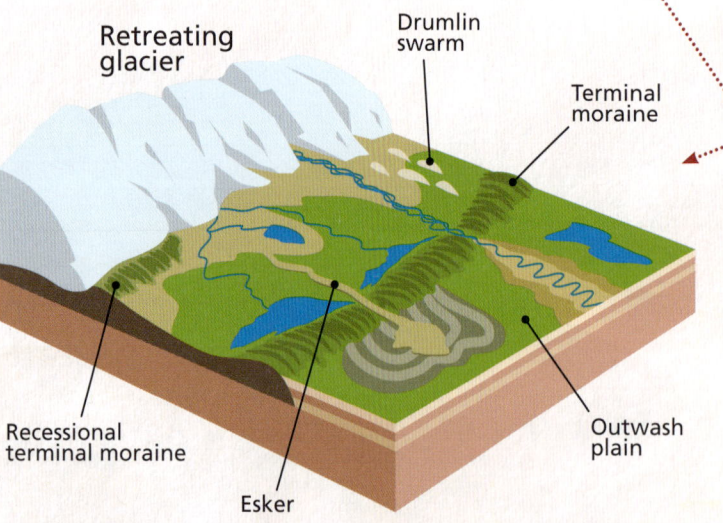

Exam hints!

You should answer this question by following these steps:
» State your chosen processes of glacial erosion.
» Next, give the grid reference for features of glacial erosion found on the OS map extract.
» Finally, explain how the processes of glacial erosion have created these features on the landscape.

Spend time looking over the features. Select the feature that you are most confident with. Use **FEED** here to give the answer a formal structure. This will ensure that you leave nothing out.

For this question, we will use **FEED**: **F**eature, **E**xplanation of formation, **E**xample, **D**iagram.

Learning Outcomes: 1.8

13
Measuring and Forecasting Weather

Learning Intentions

You will be able to:

- Describe what weather is
- Understand the importance of weather forecasting
- Explain how the atmosphere impacts the weather
- Name the different conditions recorded to produce a weather forecast
- Name and describe the different instruments used to measure these conditions
- Make a selection of instruments and use them to measure and record the weather and produce a weather forecast.

key words

- Weather forecast
- Weather
- Met Éireann
- Atmosphere
- Weather station
- Weather satellites
- Atmospheric pressure
- Millibars (mb)
- Barometer
- Barograph
- Isobars
- High atmospheric pressure
- Low atmospheric pressure
- Degrees Celsius/Centigrade (°C)
- Thermometer
- Maximum and minimum thermometer
- Average (mean) temperature
- Rain gauge
- Precipitation
- Millimetres
- Surface run-off
- Climate graphs
- Cirrus
- Cumulus
- Stratus
- Okta scale
- Wind vane
- Anemometer
- Kilometres per hour
- Beaufort scale
- Water vapour
- Hygrometer
- Relative humidity
- Saturated
- Wet and dry bulb hygrometer
- Campbell-Stokes recorder

Junior Cycle Geography CYCLONE

13.1 Weather Forecasting

Have you ever watched the **weather forecast** on television? If you have, you will be familiar with some of the images presented below. Each image shows a different type of **weather**. Every minute of every day **Met Éireann** (Ireland's leading provider of weather information) records and observes numerous weather types throughout the country to create weather forecasts like you see below.

1. Forecasting different types of weather

What Is Weather?

Weather is the state of the **atmosphere** (the gases surrounding the earth) at any given time. Weather can be hot or cold, wet or dry, calm or stormy, clear or cloudy.

The atmosphere is made up of different layers of gases which surround the earth. In this chapter, we will learn how the atmosphere can affect the many types of weather.

With the person sitting next to you, examine the weather maps. Write down the type of weather conditions you believe each map is showing. After completing the chapter, check back to see how many you got correct.

2. Layers of the earth's atmosphere

13. Measuring and Forecasting Weather

> **Managing information and thinking**
>
> 1. Which layer of the atmosphere is the highest, according to figure 2?
> 2. Between which layers is the ozone layer located?
> 3. In your opinion, why might satellites work best in the outermost layer, the exosphere? (Hint: think about what they are used for.)

Why Is Forecasting the Weather Important?

- Many people rely on weather forecasts for the jobs they do. For example, farmers need to know if it is going to rain before they fertilise land, because rain can wash fertilisers off, which then flow into nearby rivers and pollute them.
- Measuring and recording the weather in a region over a long period of time identifies the climate of the region.

> **Q** Can you think of any other reasons why forecasting the weather might be important? (Hint: think about the social and economic benefits.)

Weather Stations and Gathering Weather Data

Weather Stations

A **weather station** is a facility that uses equipment and instruments to measure and observe the weather. Irish weather stations all around the country use technology to gather, measure and record weather conditions on a daily basis.

3. Weather stations around Ireland

> **Managing information and thinking**
>
> Locate your nearest weather station. Go to www.met.ie and search for 'Weather observing stations' to investigate when the station was set up and what it measures.

How Weather Data Is Gathered and Broadcast

1. **Weather satellites** hosting various instruments orbit the earth, scanning the earth's atmosphere, gathering data and forming images.

2. The data gathered by satellites, along with the data collected by weather stations, is measured and 'read' by experts working at Met Éireann, Ireland's national meteorological service. They use this information to create weather maps and predict the weather.

3. The weather forecast is then broadcast by meteorologists (weather forecasters) from Met Éireann live on television and radio, published in newspapers, and made available in apps and online via sites such as Twitter.

Met Éireann, based in Dublin

What Data Is Gathered?

Weather stations gather data on:

1. Atmospheric pressure
2. Temperature
3. Precipitation (e.g. rain, hail, sleet, snow)
4. Wind direction and speed
5. Humidity
6. Sunshine

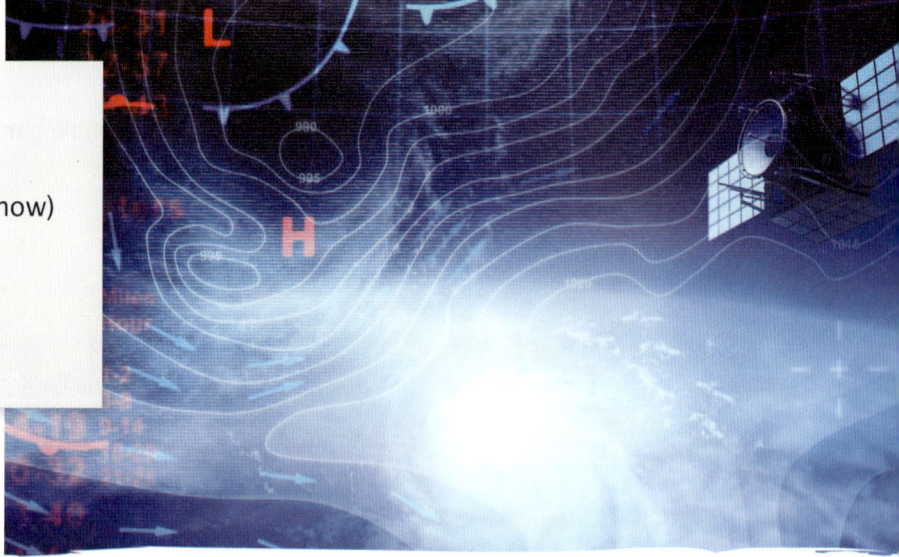

We will look at how gathered data for each of these conditions is measured and read to forecast the weather.

Question Time

1. Name the five layers of the atmosphere.
2. Explain why weather forecasting is important.
3. Describe each stage involved in gathering data to create a weather forecast.

13. Measuring and Forecasting Weather

13.2 Forecasting the Weather

Once the data has been gathered, it needs to be measured and read in order to forecast the weather.

1. Pressure

Atmospheric pressure is the weight of the atmosphere pushing down on the earth. It is measured in units called **millibars (mb)** by an instrument called a **barometer** (figure 4).

How a Barometer Works

- A very sensitive vacuum chamber inside the barometer responds to atmospheric pressure.
- The main needle on the face moves clockwise as pressure increases, and anticlockwise as pressure decreases.
- The second needle is used as a marker. It can be moved by rotating the stud in the centre of the face. Aligning the marker with the main needle 'marks' the existing pressure. This can be useful for observing pressure changes later.

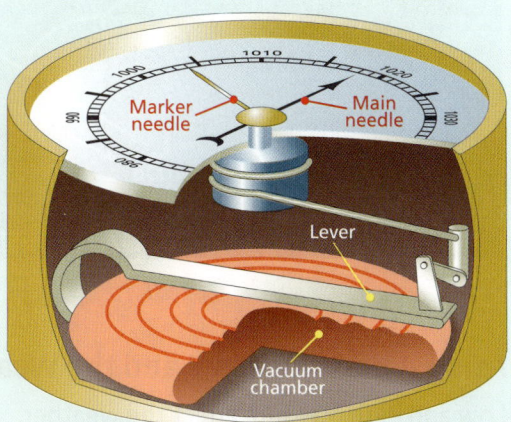

4. A barometer

In a weather station, a **barograph** (a barometer that records its readings on a moving graph; figure 5) measures and records atmospheric pressure.

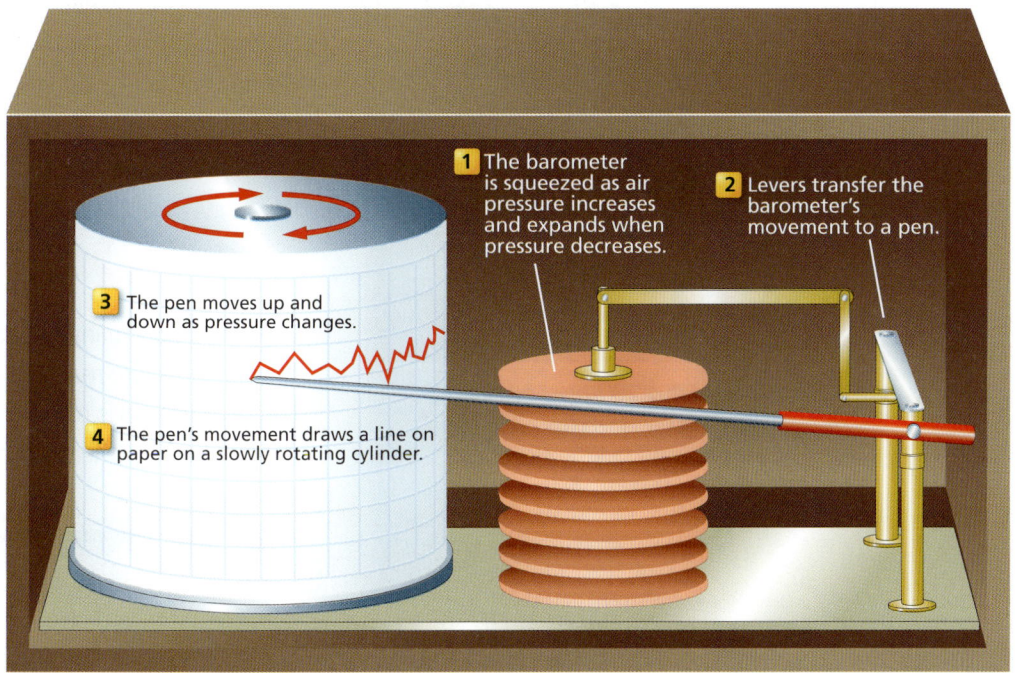

5. How a barograph works

Junior Cycle Geography CYCLONE

Atmospheric Weather on a Weather Map

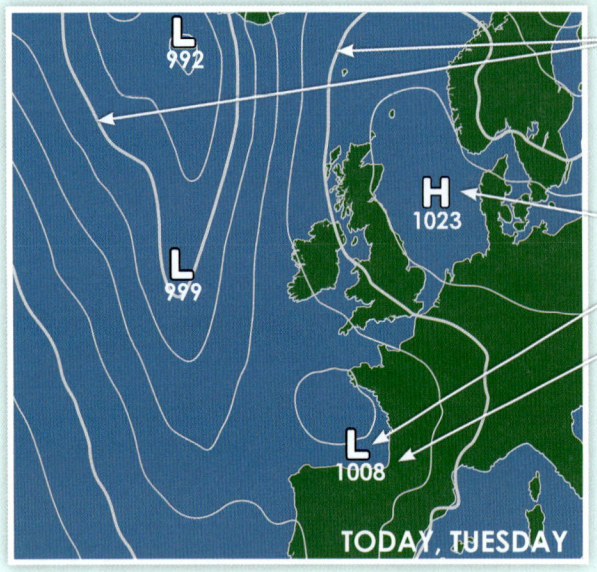

The lines used on a weather map to display atmospheric pressure are called **isobars**.
These connecting lines show areas experiencing similar atmospheric pressure.

'H' is used to show areas of **high atmospheric pressure**.

'L' is used to show areas of **low atmospheric pressure**.

The number indicates the pressure in millibars.
Anything greater than 1013 mb is considered high pressure.
Anything below 1013 mb is considered low pressure.

6. Atmospheric pressure on a weather map

Atmospheric Pressure and the Weather

Low and high pressure affect the weather.

Low Pressure and the Weather

- **Wind**: Quite strong. Indicated by isobars that are close together.
- **Cloud**: Significant cloud cover.
- **Precipitation** (any type of rain, hail, sleet or snow): High levels of precipitation, e.g. rain.

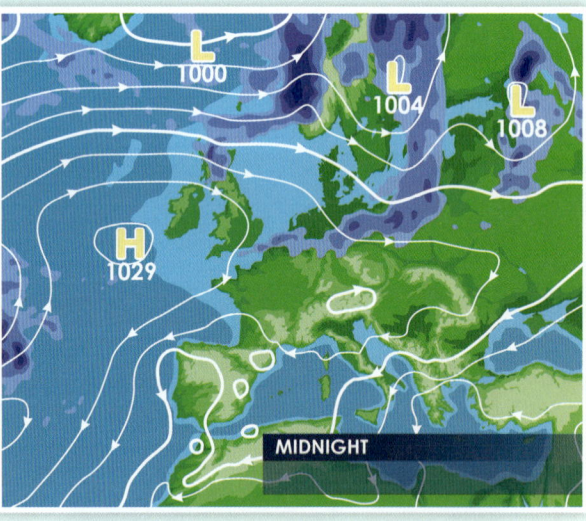

7. High and low pressure on a weather map

High Pressure and the Weather

- **Wind**: Light. Indicated by widely spaced isobars.
- **Cloud**: Little or no cloud cover.
- **Precipitation**: Little or no precipitation. Weather is mainly dry.

Q

In figure 7, how many millibars is the high pressure reading? Do you think Ireland is experiencing good or bad weather at this time? Why?

13. Measuring and Forecasting Weather

2. Temperature

Temperature is measured in **degrees Celsius/Centigrade (°C)** using an instrument called a **thermometer**. A special type of thermometer called a **maximum and minimum thermometer** (figure 8) is used in weather stations.

Centigrade is the old name for 'Celsius'.

How a Maximum and Minimum Thermometer Works

A maximum and minimum thermometer is a U-shaped glass tube on which temperature readings are marked. The tube contains alcohol on its 'minimum' side, a vacuum (space) on its 'maximum' side and mercury in between. It also contains two metal markers inside the tubing on either side of the mercury. The metal markers show the maximum and minimum temperatures reached.

When the temperature decreases, the alcohol contracts and causes the metal marker on the minimum side of the tube to 'retreat' so that it marks the lowest temperature reached.

When the temperature increases, the alcohol expands and causes the metal marker on the maximum side of the tube to be pushed up so that it marks the highest temperature reached.

8. A maximum and minimum thermometer

 Being numerate

What are the maximum and minimum readings shown on the maximum and minimum thermometer in figure 8? (Readings are taken from the bottom of the metal markers.)

Temperature on a Weather Map

On a weather map, the **average (mean) temperature** for an area is displayed in the form of a number.

9. Average temperature

Q According to figure 9, what is the highest average temperature/s being experienced by Ireland on this particular afternoon, and where is it recorded? (Refer back to the map of weather stations on p. 191 to help you.) Which counties are experiencing the lowest temperatures? Why might this be the case?

3. Precipitation

A **rain gauge** (figure 10) is used to measure **precipitation**, in liquid (mainly rainfall) form, in **millimetres**.

The rain gauge consists of an outer cylinder which contains a funnel that directs water into a measuring cylinder. It is partially buried in the ground in an open space away from buildings or trees.

The measuring cylinder is graduated (has markings on the side) to show precipitation levels in millimetres.

Q Why do you think the rain gauge is buried away from buildings and trees?

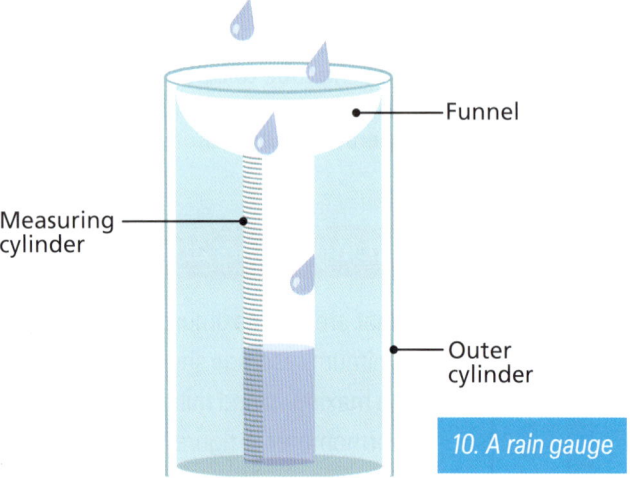

10. A rain gauge

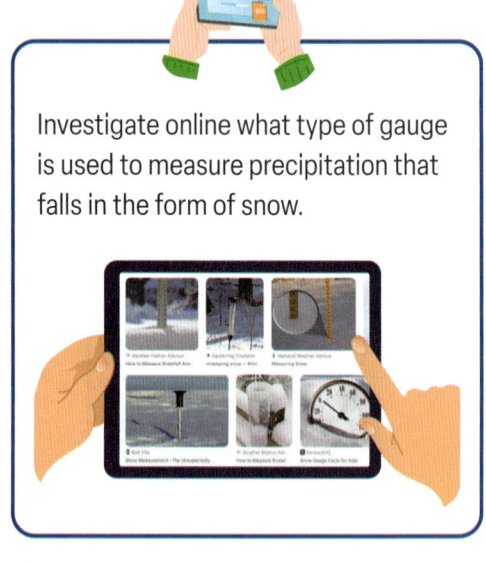

Investigate online what type of gauge is used to measure precipitation that falls in the form of snow.

The Water Cycle

The water cycle shows the continuous movement of water (in different states such as liquid or gas) within the earth and its atmosphere. Without this continuous cycle, there would be no precipitation.

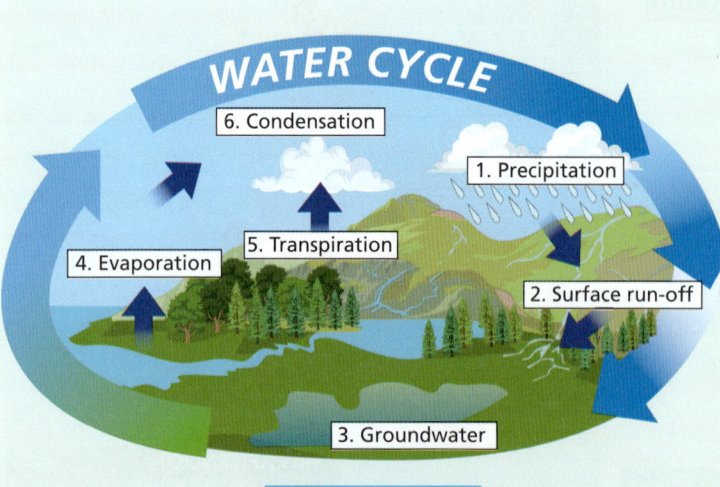

11. The water cycle

1. **Precipitation**: rain, snow, sleet or hail that falls to the ground
2. **Surface run-off**: water which runs above ground instead of filtering through it
3. **Groundwater**: water that exists underground
4. **Evaporation**: heat turns the water from the seas, oceans and lakes into gas (vapour)
5. **Transpiration**: the transfer of water vapour from vegetation to the atmosphere
6. **Condensation**: as the water vapour rises, the temperatures drop and the vapour turns back into liquid and falls to the ground starting the cycle again

13. Measuring and Forecasting Weather

Types of Rainfall

Rainfall is one type of precipitation that the water cycle can create. There are three types of rainfall.

1. Relief Rain

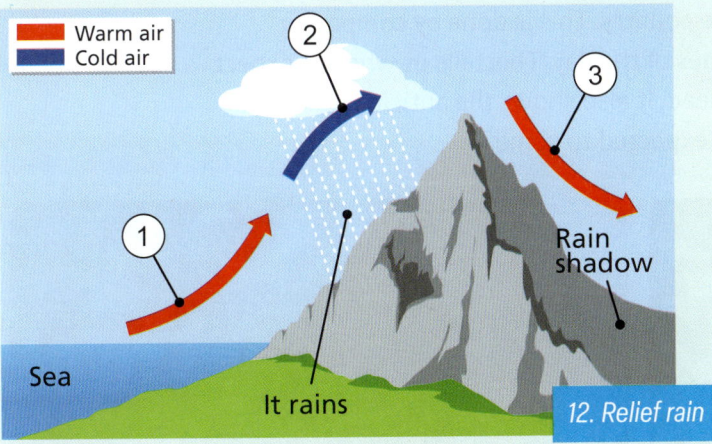
12. Relief rain

- ❶ Warm moist air travels over a large body of water like an ocean.
- ❷ It reaches a mountain, rises and begins to cool and condense. This forms clouds, bringing rain.
- ❸ Once it passes over the mountain, the air begins to warm up again and forms a rain shadow, which is a dry area.

2. Convectional Rain

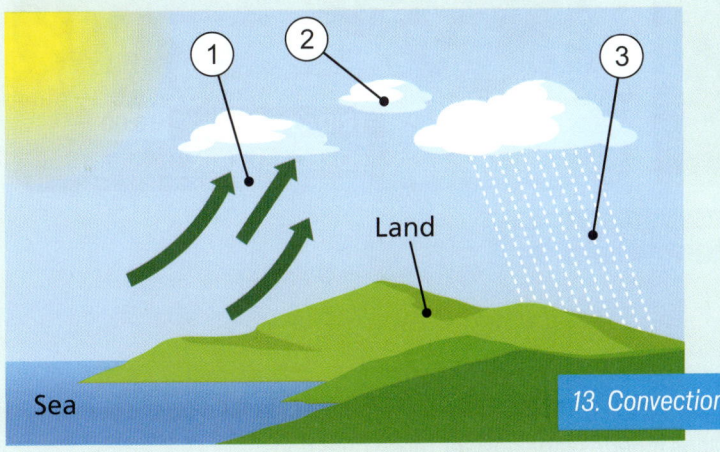
13. Convectional rain

- ❶ This type of rainfall occurs during warm weather when the land is heated.
- ❷ This creates warm air which rises and is known as a convection current.
- ❸ As the warm air rises, it cools and condenses forming large clouds which can produce thunderstorms.

3. Frontal Rain

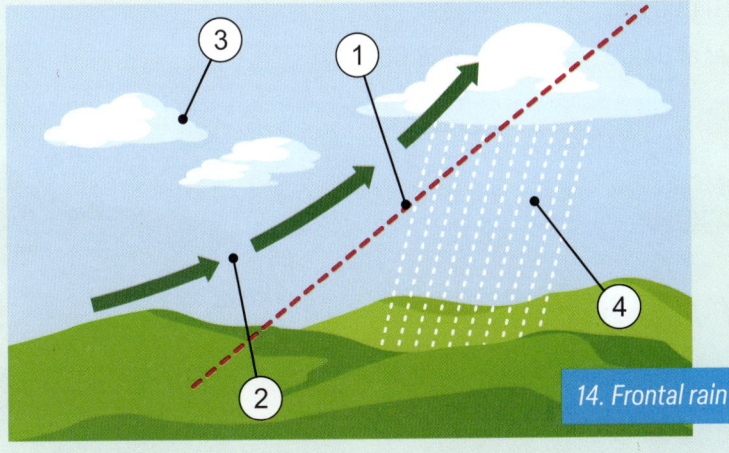
14. Frontal rain

- ❶ This rainfall occurs when a warm front of air meets a cold front. A front is a boundary between two different types of air.
- ❷ The heavier cold air will sink to the ground and the warm air will rise above it.
- ❸ When the warm air rises, it cools and condenses.
- ❹ Clouds form, bringing rain.

Junior Cycle Geography CYCLONE

Rainfall on a Weather Map

- Rain on a weather map is displayed as different shades of blue.
- The lighter the shade of blue, the less rainfall a particular area will experience. The darker the shade of blue, the more rainfall expected.
- The rain can be tracked as it moves across the country. This is done by comparing rainfall maps of the same area at different times of the day. This tells us when to expect rainfall later that day and during the week ahead. It also shows the anticipated rainfall patterns and from which direction the rain is expected to come.

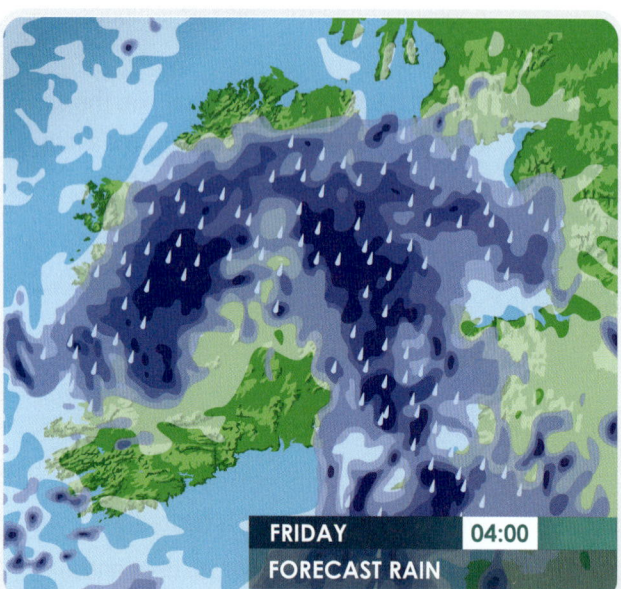

15. Rainfall weather maps

 Working with others

With the person sitting next to you, examine the two weather maps. They show rainfall in Ireland on the same day at different times. Using an atlas or Google Maps if necessary, answer these questions:

1. Name a county experiencing no rainfall on this day at 18:00 hrs.
2. Name a county experiencing low rainfall at 18:00 hrs.
3. Name a county experiencing high rainfall at 04:00 hrs.
4. Name a county that experiences dry weather at 04:00 hrs and rain at 18:00 hrs.

13. Measuring and Forecasting Weather

Climate Graphs

Climate graphs are used to illustrate average temperature and average precipitation of an area over the 12 months of the year.

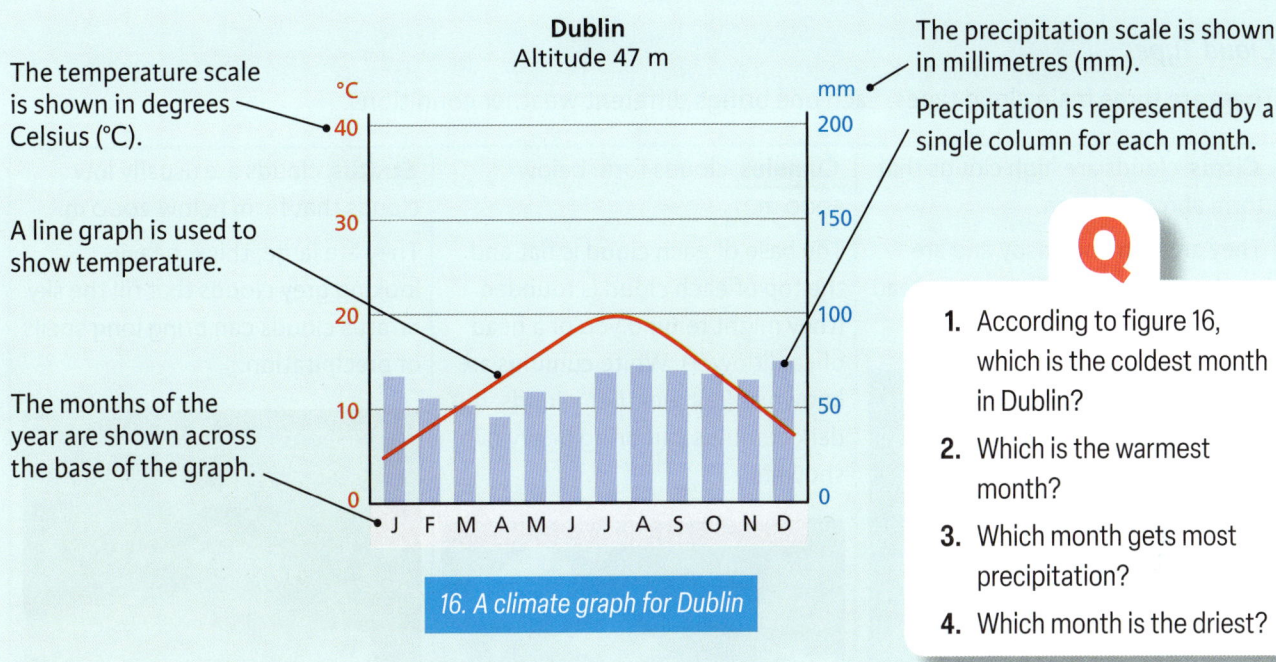

16. A climate graph for Dublin

The temperature scale is shown in degrees Celsius (°C).

A line graph is used to show temperature.

The months of the year are shown across the base of the graph.

The precipitation scale is shown in millimetres (mm).

Precipitation is represented by a single column for each month.

Q

1. According to figure 16, which is the coldest month in Dublin?
2. Which is the warmest month?
3. Which month gets most precipitation?
4. Which month is the driest?

Being numerate

Looking at figure 16, calculate the average precipitation for Dublin over the months of June, July and August. Calculate the average temperature for the months of October, November and December.

Question Time

1. Name the instruments that measure the following:
   (a) Atmospheric pressure   (b) Temperature   (c) Precipitation
2. What units of measurement are the following measured in?
   (a) Atmospheric pressure   (b) Temperature   (c) Precipitation
3. What symbols indicate low and high pressure on a weather map? Briefly describe the weather associated with low and high pressure.
4. List the three types of rainfall.
5. Explain the key stages of the water cycle.
6. Choose two types of rainfall and compare and contrast how they form.

Exam hint!

When a question asks you to 'compare and contrast', you must describe what is similar and different about two or more things.

Clouds

What Are Clouds?

Clouds are groups of tiny water droplets or ice crystals held in the atmosphere. The droplets are so small and light that they can float in the air.

Cloud Types

There are three main cloud types. Each one brings different weather conditions.

Cirrus clouds are high clouds that form above 8000 m. They are thin and wispy and are made up of tiny ice crystals instead of water droplets.	**Cumulus** clouds form below 5000 m. The base of each cloud is flat and the top of each cloud is rounded (they might remind you of a head of cauliflower). White cumulus are known as 'fair weather' clouds, but dark cumulus can bring heavy rain showers.	**Stratus** clouds are usually low clouds that form below 2000 m. They are large, thick, heavy-looking grey clouds that fill the sky. Stratus clouds can bring long spells of precipitation.

Q Which type/s of clouds can you see today? Are these clouds likely to bring rain?

Being creative

Using the information in the table showing cloud types, create a one-page poster displaying the height at which each type of cloud is formed, a colour scale showing dark stratus clouds and white cumulus and cirrus clouds, and one other key piece of information about each cloud type.

How to Measure Cloud Cover

Cloud cover is measured and recorded using the **okta scale**. Cloud cover is estimated in terms of how many eighths of the sky are covered in cloud:

from 0 oktas (completely clear sky) …

to 8 oktas (completely overcast).

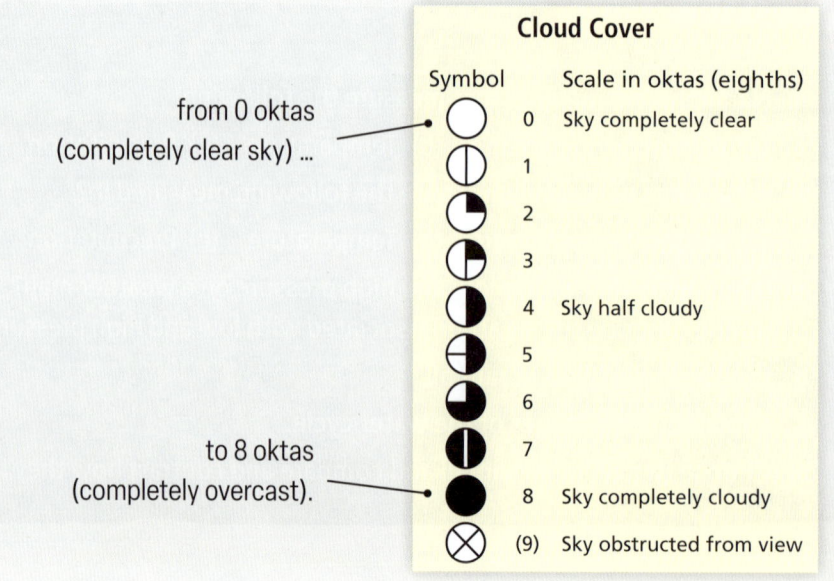

13. Measuring and Forecasting Weather

Cloud Cover on a Weather Map

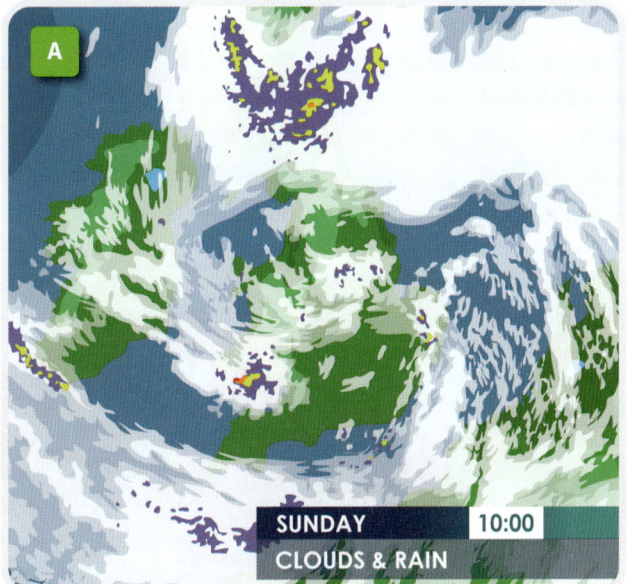

17. Weather maps showing cloud cover

Q Examine these two weather maps showing cloud cover over Ireland and the United Kingdom.
1. Which map shows more cloud cover Ireland, A or B?
2. Which map shows the greater likelihood of precipitation for Ireland? Explain your answer.

4. Wind Direction and Speed

A **wind vane** (see the photograph on the next page) is used to indicate the direction of wind (the moving air in the atmosphere).

The speed or strength of wind is measured using an **anemometer** (see the photograph on the next page) in units of **kilometres per hour** (km/h).

How Wind Is Created

Wind is created when the sun warms air in the atmosphere. This heated air begins to rise and cold air rushes in to take its place. This constant moving of warm and cold air creates wind.

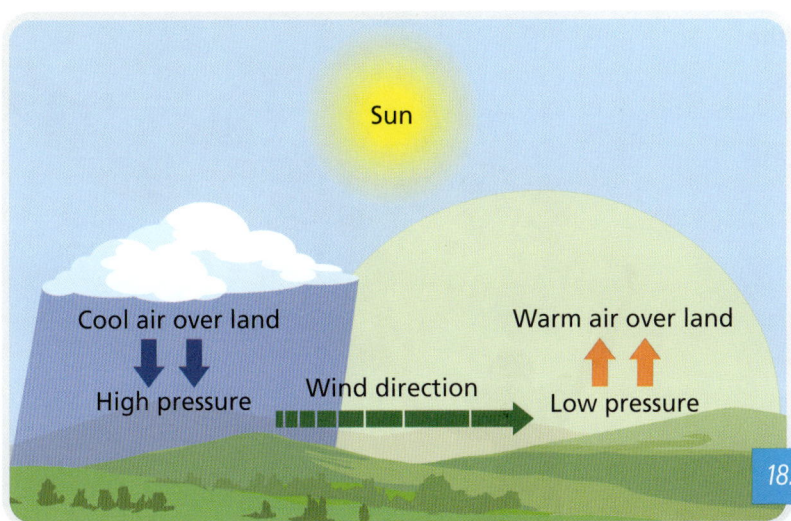

18. How wind is created

Being creative

To feel the movement of air or wind in action, have a classmate stand by the closed door of your classroom, and then open it quickly. Ask your classmate what they experience/feel.

201

Junior Cycle Geography CYCLONE

How a Wind Vane Works

A wind vane

The wind vane is a free-moving arrow on a high mast.

The tail of the arrow is wide and blown forward by the wind.

This allows the head of the arrow to point in the direction from which the wind is blowing.

Q According to the arrowhead, from which direction is the wind blowing?

FUN FACT!
The rooster you see on top of the wind vane is just for decoration! They came into fashion in the ninth century when Pope Nicholas ordered that all weather vanes feature them. Can you think of what they might symbolise? Think of what happened at Easter.

How an Anemometer Works

An anemometer

Three cups rotate when the wind blows. The stronger the wind, the faster they move.

The number of times the cups rotate is used to calculate wind speed. It is recorded and displayed on the meter.

The Beaufort Scale

In 1810, an Irish-born admiral named Francis Beaufort invented a scale for measuring the 'force' (strength) of wind by observing its effects on trees, water and other familiar objects. Modern shipping weather forecasts still refer to wind 'force' as used in the **Beaufort scale**.

13. Measuring and Forecasting Weather

Force 0: Calm, wind less than 1km/h, water like a mirror

Force 1: Light air, wind 1–5 km/h, sea has scaly ripples

Force 2: Light breeze, winds 6–11 km/h, water rippled into wavelets

Force 3–4: Gentle to moderate breeze, winds 12–28 km/h, flags blow about

Force 5: Fresh breeze, winds 29–38 km/h, moderate waves, many whitecaps

Force 6: Strong breeze, winds 39–49 km/h, large branches shake, difficulty with umbrellas

Force 7: Moderate gale, winds 50–61 km/h, sea heaps up, white foam blown about

Force 8: Gale, winds 62–74 km/h, hard to walk against the wind, sea has high waves with breaking crests and spray

Force 9: Strong gale, winds 75–88 km/h, slight damage to trees and buildings

Force 10: Whole gale, winds 89–102 km/h, severe damage to trees and buildings

Force 11: Storm winds, 103–117 km/h, widespread damage

Force 12: Hurricane, winds over 117 km/h, devastation

Q Observe today's weather. Going by the Beaufort scale, what is the force of the wind? Compare your answers as a class.

Wind on a Weather Map

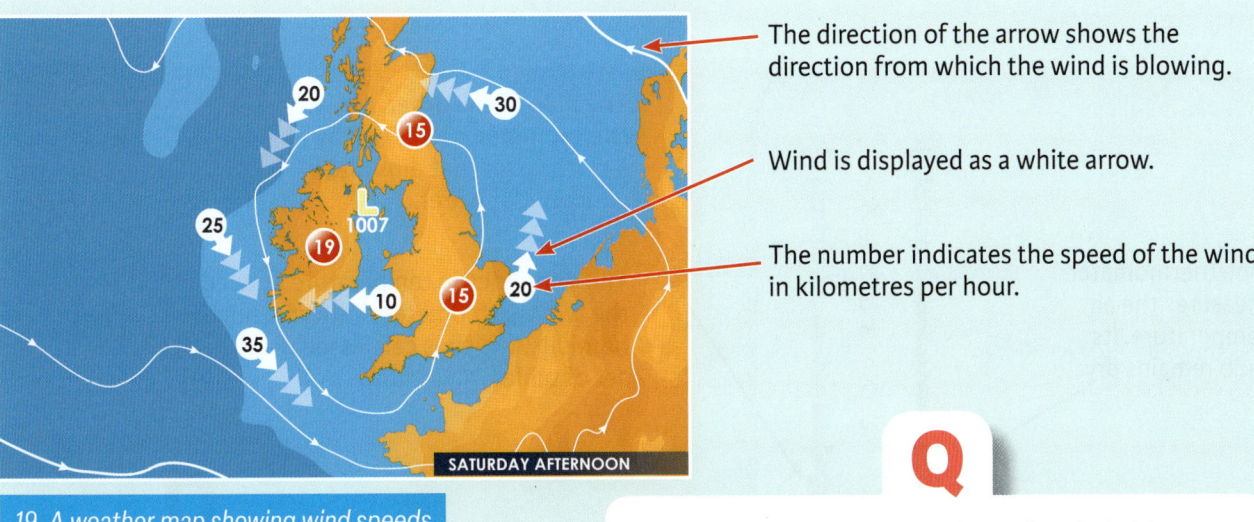

The direction of the arrow shows the direction from which the wind is blowing.

Wind is displayed as a white arrow.

The number indicates the speed of the wind in kilometres per hour.

19. A weather map showing wind speeds

Q Examine figure 19. What speeds are the winds (a) over Co. Wexford, (b) off the west coast of Kerry, (c) over the east coast of Scotland, (d) off the east coast of England? You may use Google Maps to help you locate each place.

203

5. Humidity

Humidity is the amount of water vapour in the air. A hygrometer (see photograph and figure 20) is used to measure relative humidity, which is expressed as a percentage.

Humidity and the Weather

When weather forecasters speak about humidity, they use the term 'relative humidity'.

The term 'humidity' refers to water vapour in the air. Warm air can hold more water vapour than cool air can. When the air can hold no more water vapour, it is said to be saturated. Any extra water vapour then condenses into tiny droplets to form clouds.

FUN FACT!
On a hot, humid day it often feels hotter than it actually is. This is because there is so much water vapour in the air that our sweat cannot evaporate to help cool us.

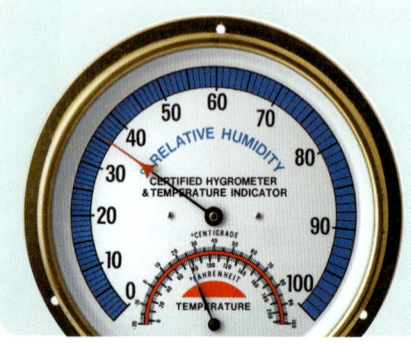

The photograph shows a hygrometer. It measures temperature and humidity. Inside it, a very thin piece of paper is attached to a coil. When it absorbs air moisture, the paper expands, which causes the coil to rotate, which moves the humidity needle. What is the relative humidity shown by the hygrometer here?

A hygrometer

How a Wet and Dry Bulb Hygrometer Works

Let's look at how a wet and dry bulb hygrometer works.

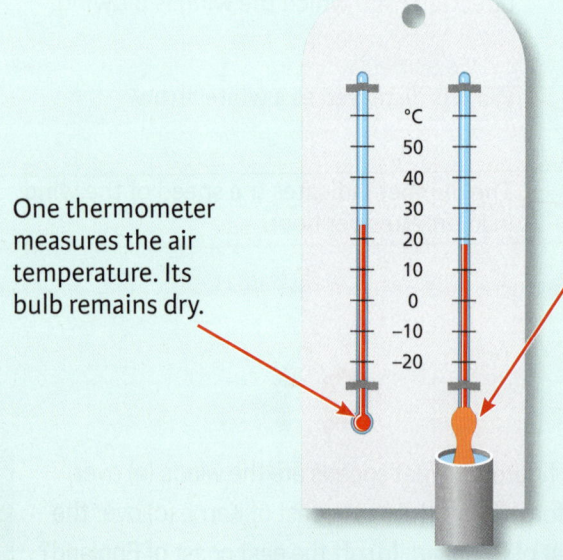

One thermometer measures the air temperature. Its bulb remains dry.

The bulb of the other thermometer is kept wet by a cloth cover that dips into a container of water.

The difference in temperature between the two thermometers indicates relative humidity, which is calculated using a special set of tables.

20. A wet and dry bulb hygrometer

What is the temperature of the dry bulb? What is the temperature of the wet bulb?

13. Measuring and Forecasting Weather

Storage of Instruments

Some weather forecasting instruments, such as the barograph, the maximum and minimum thermometer and the hygrometer, are stored in a wooden box called a Stevenson screen.

The Stevenson screen is painted white to reflect the sun.

Gaps in the sides allow air but not direct sunlight to enter the box.

The Stevenson screen allows for more accurate measurements to be recorded.

6. Sunshine

A **Campbell–Stokes recorder** measures the hours of sunshine in a day.

A Campbell–Stokes recorder

A solid glass ball concentrates the sun's rays onto a removable strip of card, which is placed behind the ball.

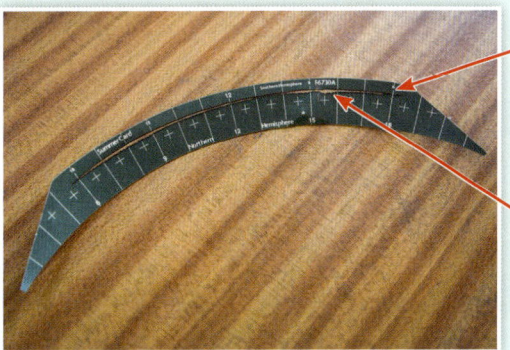

The card is marked at hourly intervals.

As the sun shines through the ball, it scorches a mark on the paper.

Question Time

1. Name the three main types of clouds.
2. Which cloud type can bring heavy rain showers?
3. Choose one type of cloud and describe its appearance.
4. What unit of measurement is used to measure wind speed?
5. Explain how a wind vane or an anemometer works.
6. (a) What is relative humidity?
   (b) Which instrument is used to measure it?
7. (a) What does a Campbell–Stokes recorder measure?
   (b) How does it work?

Junior Cycle Geography **CYCLONE**

CBA 2 My Geography Moment

MAKE YOUR OWN WEATHER INSTRUMENTS

Make Your Own Barometer

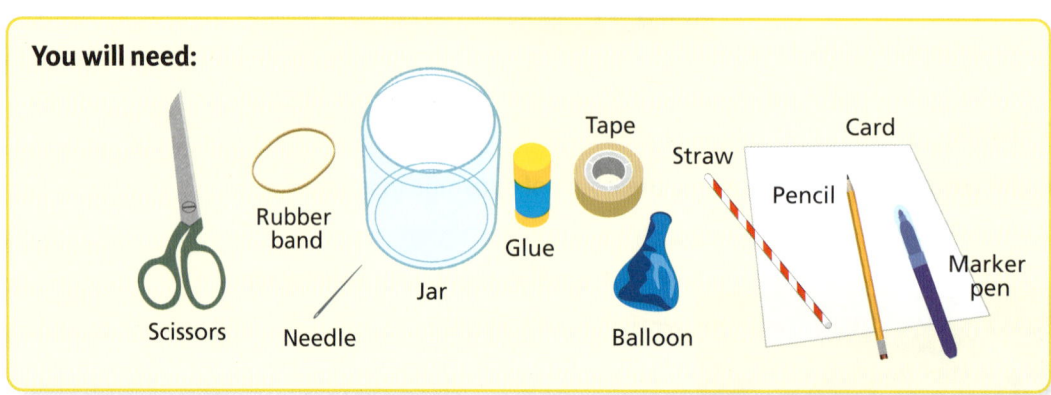

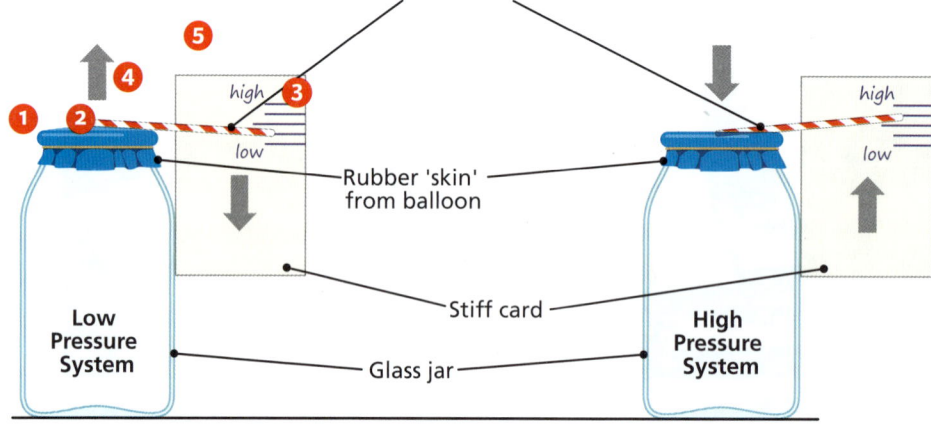

1. Cut the neck off the balloon and stretch the main piece (the 'balloon' part) over the opening of the glass jar – the tighter the better. (Make sure the balloon has not been blown up at any point.)

2. Stretch the rubber band around the balloon and opening of the jar to hold the balloon in place and to seal it, so no air can get in or out.

3. Glue a needle to one end of the straight straw (do not use a bendy straw, or cut the bendy piece off the straw and use the remaining straight piece). This will act as your pointer.

4. Glue the other end of the straw to the centre of the balloon covering and gently tape in place until the glue has dried and holds the straw in place itself.

5. Place your barometer beside a wall and tape a piece of card to the wall at the point that the needle hits. On the card, write 'High' above and 'Low' under the needle's point. Use a pencil or pen to record the atmospheric pressure three times a day: in the morning, at lunchtime and at the end of the school day.

Make Your Own Wind Vane

You will need: Cinema drink cup and lid, Straw, Card, Glue, Small rocks, Scissors, Push pin, Pencil with eraser, Marker

1. Make two 1 cm-long cuts into the ends of a straight straw. The cuts should be in line with each other.
2. Cut a triangle shape and a square shape from the card.
3. Place the cardboard triangle in one cut of the straw, and the cardboard square in the other cut, adding a little glue to hold them in place. It should now look like an arrow.
4. Fill your drinks container with the small rocks, and glue the lid of the container in place.
5. Turn the container upside down and stick a pared pencil (lead-side in) through its centre. Glue the pencil in place if necessary.
6. Push the pin through the straw, exactly mid-way down its length, and then into the rubber head of the pencil.
7. Use a compass to find north, south, east and west and write them onto the sides of the cup.

Make Your Own Rain Gauge

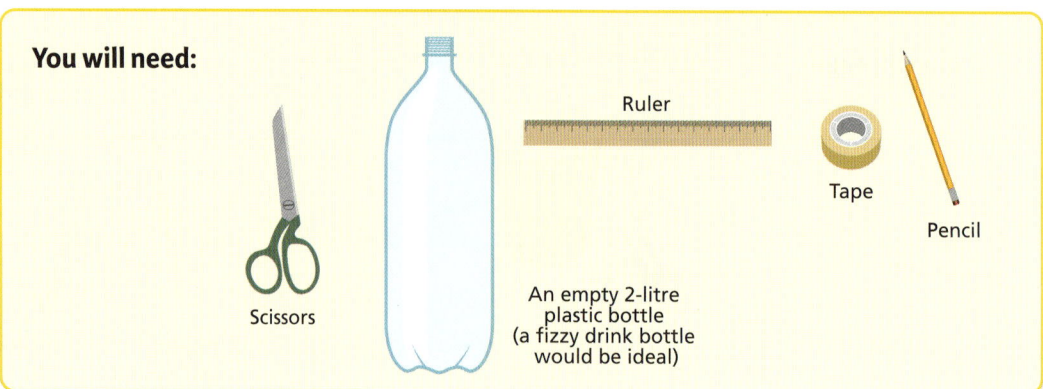

1. Carefully cut around the plastic bottle, about two-thirds of the way up.
2. Turn the top part of the bottle upside down and place it inside the bottom part, fixing it in place using the tape.
3. Using a ruler, write a scale in centimetres on a piece of tape and fix it to the side of your bottle.
4. Find an open space away from trees in which to place your rain gauge.
5. Dig a hole and bury your rain gauge so that the top is sticking about 5 cm out of the ground. This will stop the rain gauge from blowing down on windy days. You will need to remove your rain gauge from the ground each time you want to take a reading.

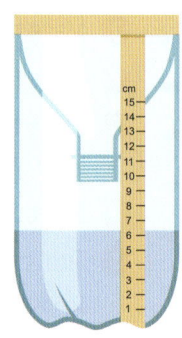

Make Your Own Okta Scale

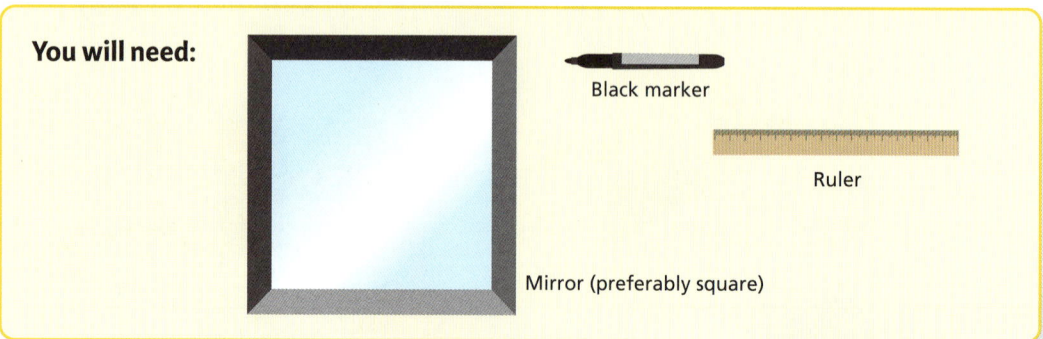

1. Use the marker and ruler to divide your square mirror into eight equal parts.
2. Place your mirror flat on the ground away from any trees or buildings that might obstruct its ability to reflect the sky.
3. Use the okta scale to count how many of the squares in the mirror are at least half-filled with reflected cloud cover. This indicates the okta scale for the section of sky reflected.

13. Measuring and Forecasting Weather

EXAM FOCUS

Sample questions

1. In your copy, match each of the images below, labelled A to D, to the weather instrument that matches it by writing the correct letter for each instrument in the table below.

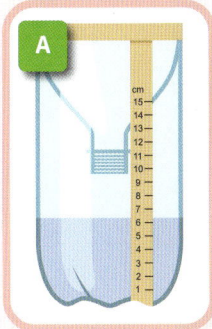

Weather Instrument	Letter
Wind/weather vane	
Campbell–Stokes recorder	
Rain gauge	
Anemometer	

2. (a) Copy and complete this diagram of the water cycle, filling in the key words correctly.

Condensation

Groundwater

Evaporation

Surface run-off

Transpiration

(b) **Explain** what is occurring at each stage of the water cycle using the key words from part (a).

209

3. (a) **Name** one person who depends on accurate weather forecasts for their job.
   (b) **Explain** one reason why the weather forecast is important to their job.
4. This map shows Ireland's mountains. Explain the type of rainfall you might expect to occur at A labelled on the map.

13. Measuring and Forecasting Weather

CBA 2 My Geography Moment

BECOME A METEOROLOGIST

Create your own weather forecast and become a weatherperson (meteorologist) for your school. This is a group task for three–four students.

Your teacher will assign each group a specific instrument to make from the list given.

Each group will be responsible for making their instrument and placing it in the best spot for it to work properly. Instruments can be set up at school or at home. Each group will monitor their instrument for one week.

You must record daily readings in your copy. At the end of the week, you must present your findings as a group.

Instrument	Instruction
Barometer	Place it in a shaded area indoors (this will act like a Stevenson screen). Decide as a group where the safest place to keep your instrument will be. Take readings twice a day: in the morning and at lunchtime.
Wind vane	Place outside in an open area, e.g. garden or field, away from obstructions, buildings or trees. Decide as a group where the safest place to keep your instrument will be. Take readings twice a day: in the morning and at lunchtime.
Rain gauge	Dig a hole and bury your rain gauge so that the top is about 5 cm above the ground. This will stop the rain gauge from blowing down on windy days. Record once a day, in the morning.
Okta scale	Place outside in an open area, e.g. garden or field, away from obstruction. Unlike the other instruments, it does not have to remain outdoors for the entire day. Take readings twice a day: in the morning and at lunchtime.

Instruments such as the maximum and minimum thermometer, hygrometer and Campbell-Stokes recorder are very difficult to make, so for these you can follow this digital skill instead:

Download the Met Éireann/RTÉ Weather app/s to your phone or tablet or check their websites daily and record the particular weather conditions measured by these instruments.

Junior Cycle Geography CYCLONE

Success Criteria

We Must

- Discuss as a group our roles: who will make the instrument, who will monitor the instrument on each day and who will present the findings.
- Follow the 'How to make' instructions carefully for our assigned instrument, making sure to complete each step. Instruments can be made in class or at home.
- Check the readings from the instrument, as outlined in the instructions on the previous page.
- Share our individual weather instrument readings with the other groups to allow them to successfully complete their own weather reports.
- Write a brief weather report using all of the information gathered over the last week.
- Prepare and practise our weather report, which will be two minutes in length.

We Could

- Tweet our daily findings to the Met Éireann or RTE Weather Twitter profiles using the hashtag #localweather.
- Using our phone, tablet or camera, pre-record our weather forecast to show as a video clip to the rest of the class.

We Should

- Use climate graphs where possible to present our instrument's readings.
- Compare and contrast our daily findings with those of the Met Éireann/RTE Weather websites or Twitter profiles for our local area to see how accurate our findings are.
- Do a trial run of our weather forecast before we present it to the class.

Peer Assessment

In your group, carefully review the success criteria and check that you have followed each step and fulfilled what was asked of you.

Redrafting

Taking into account your self-assessment notes, make any changes that you think are now needed. When you are happy, you can prepare to present your weather report to the class.

Learning Outcomes: 1.7, 2.8

14

Severe Weather

Learning Intentions

You will be able to:
- Describe what severe weather is
- Explain how tropical storms form and how they are categorised
- Describe the impacts of tropical storms on an area.

key words

Tropical storms

Tropics

Hurricanes

Cyclones

Typhoons

Saffir–Simpson hurricane wind scale

14.1 What Is Severe Weather?

Severe weather is a weather event that is significantly different from the usual weather an area experiences. It also puts people and animals in danger or causes considerable damage to buildings. This may take place over one day or a period of time.

A

B

C

D

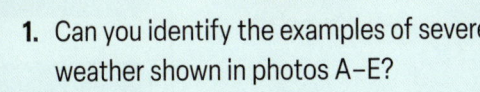

Q

1. Can you identify the examples of severe weather shown in photos A–E?
2. Suggest what changes to normal weather conditions may have occurred to create these severe events.

E

213

Junior Cycle Geography CYCLONE

14.2 Tropical Storms

Tropical storms are huge storms that develop in the **Tropics** (a region of the earth north and south of the equator). In the US and the Caribbean, they are called **hurricanes**, in Australia and southeast Asia they are known as **cyclones**, and in Japan they are called **typhoons**.

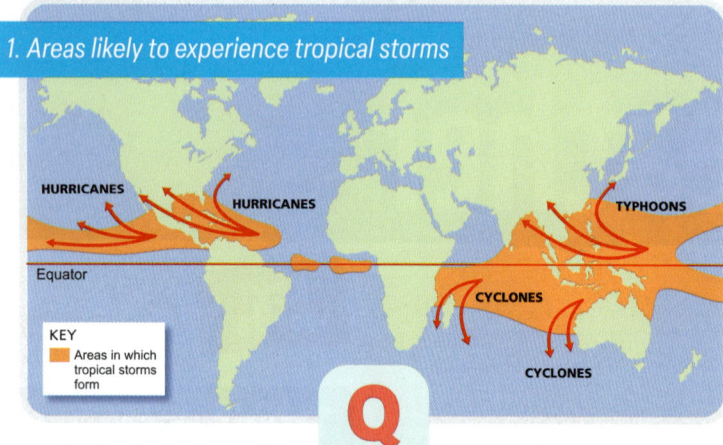

1. Areas likely to experience tropical storms

Q Using an atlas or Google Maps, name four countries likely to experience a tropical storm according to figure 1.

How Do Tropical Storms Form?

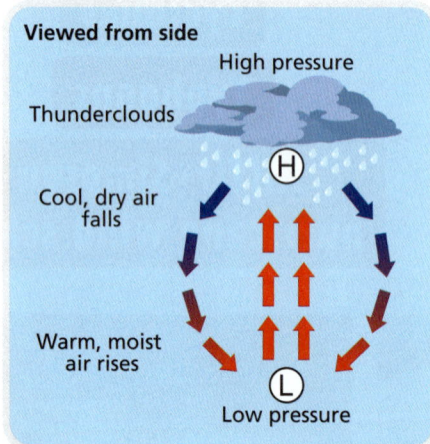

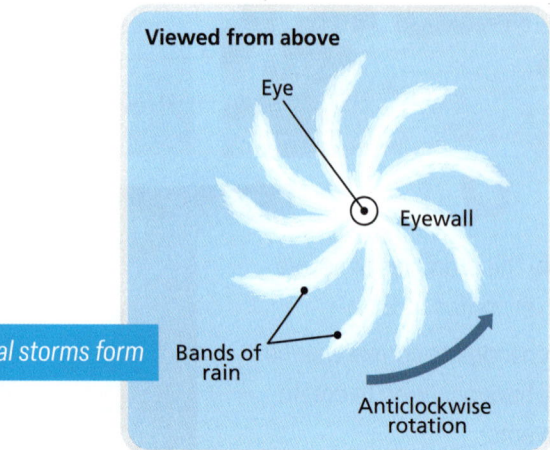

2. How tropical storms form

1. Tropical storms start to form over tropical oceans, usually during late summer and early autumn. They need two things to get them started:
   1. heat
   2. moist air.

2. The warm, moist air rises. As it rises, it cools and condenses (gas changes to a liquid).

3. This creates an area of low pressure.

4. As the earth rotates, it creates winds, which blow inwards towards the centre of the low pressure area. In the northern hemisphere, the air moves anticlockwise.

5. Water is drawn upwards from the ocean.

6. When wind speed reaches 120 km per hour, the storm is officially labelled a tropical storm.

7. The storm develops an 'eye' in its middle (an area of calm weather).

8. The storm is carried across the ocean by winds, getting stronger as it goes.

9. When it reaches land, friction with the land's surface slows it down and weakens it.

214

14. Severe Weather

Hurricane Categories

Hurricanes are measured using a scale called the **Saffir-Simpson hurricane wind scale**, a 1–5 rating based on a hurricane's sustained wind speed. The scale estimates potential damage.

Saffir-Simpson hurricane wind scale

Category	Sustained Winds	Damage
1	119–153 km/h	Very dangerous winds; will produce some damage
2	154–177 km/h	Extremely dangerous winds; will cause extensive damage
3 (major)	178–208 km/h	Devastating damage will occur
4 (major)	209–251 km/h	Catastrophic damage will occur
5 (major)	252 km/h or higher	Catastrophic damage will occur

FUN FACT!
The diameter of a hurricane is measured from one side to the other. Hurricanes can have a diameter of over 900 km!

Being numerate

Draw the Saffir-Simpson scale as a bar chart in your copy. Label the x-axis 'Categories 1–5'. Label the y-axis 'Sustained Winds'. (You can choose either the lowest or highest recorded wind speeds.)

FUN FACT!
The World Meteorological Organization (WMO) is responsible for naming hurricanes created in the North Atlantic Ocean. The WMO keeps six lists of 21 male and female names that are used in rotation, and recycled every six years.

Question Time

1. Explain what severe weather is.
2. What are tropical storms called in the following places?
   (a) The US
   (b) The Caribbean
   (c) Australia/Southeast Asia
   (d) Japan
3. Draw and label a diagram explaining how a hurricane forms.

Junior Cycle Geography CYCLONE

Case Study: Hurricane Dorian 2019

Hurricane Dorian is the most powerful Category 5 Hurricane to hit the Bahamas since records began. On 23 August 2019, a low-pressure system formed in the Atlantic Ocean. By 24 August, it had become a tropical storm. Four days later, it was classified as a hurricane as it had reached wind speeds of 350 km/hr and higher.

The islands of Abaco and Grand Bahama were the worst affected. At least 65 people were confirmed to have died on just these two islands, with many more missing. It was estimated that 13,000 homes were severely damaged or destroyed.

3. The pathway of Hurricane Dorian

Q

1. On what date did Hurricane Dorian make land, according to image A above?
2. In what direction did Storm Dorian move in image B, as it travelled from:
    (a) The Atlantic to the Bahamas
    (b) The Bahamas to North America

These images are taken from a satellite used to track tropical storm movement. They show the movement of Hurricane Dorian from 2 to 6 September 2019.

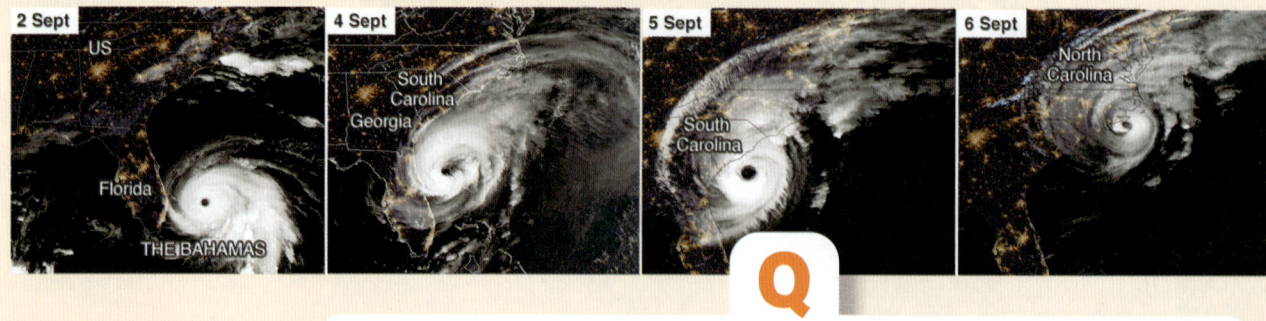

Q

1. Why do you think satellites are used to track hurricanes?
2. Based on the satellite images, on which date do you think the storm was at its strongest?

Economic and Social Impacts of Hurricane Dorian

Economic Impact

- Dorian caused about US$3.4 billion in damages, which is equal to one-quarter of the gross national product (total value of all finished goods and services produced by a country's citizens in a given financial year) of the Bahamas.
- The main airport in Grand Bahama was destroyed. Roads were littered with debris and fallen trees, and fuel supplies were limited.
- Hurricane Dorian affected about 2,500 small- and medium-sized businesses in the Bahamas. All businesses on the hardest-hit east side of the island were forced to close.

Social Impact

- The death toll stands at 74 people, although 245 people were still missing a year later.
- Some 29,500 people were left homeless and/or jobless.
- The added crisis of Covid-19 and another hurricane season in 2020 severely hindered recovery efforts.

Emergency Response to Hurricane Dorian

Governments and aid agencies, such as Mercy Corps and World Vision, provided short- and long-term help in the countries affected by the hurricane.

Examples of short-term aid	Examples of long-term aid
Emergency shelter provided for those who lost their homes	Seed distribution to grow crops and agricultural support for farmers who lost their livelihoods
Food and medical supplies provided to thousands of families across the affected countries	Rebuilding of roads and buildings such as hospitals and schools
Water purifiers provided to ensure clean water	Provision of counselling to help people cope with their losses

A Mercy Corps volunteer delivers potable (drinkable) water to communities on Grand Bahama Island, January 2020

Junior Cycle Geography CYCLONE

Preparing for Hurricane Season in the Bahamas

The government of the Bahamas has created simple steps to prepare its people for hurricane season in the hopes of reducing the devastating social and economic impacts.

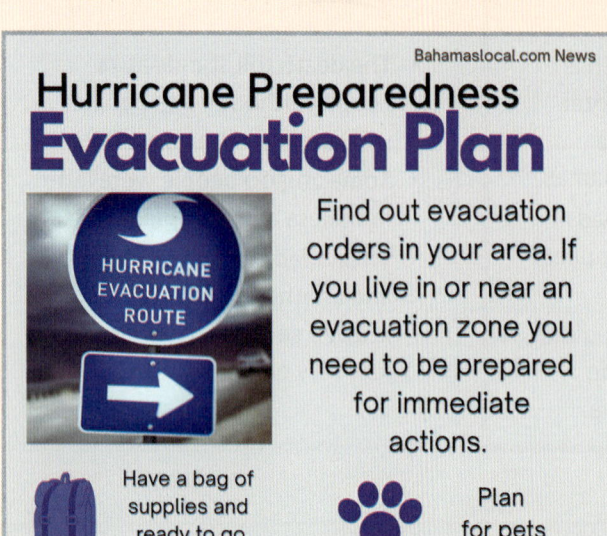

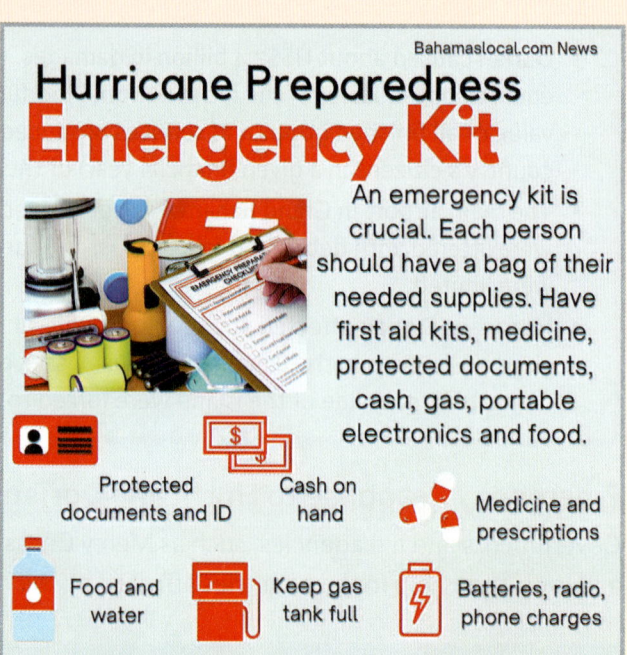

Q

1. Study the Hurricane Preparedness Plan for the Bahamas. List one suggestion to reduce:
   (a) An economic impact
   (b) A social impact.
2. How might the Covid-19 pandemic have impacted the recovery effort of aid agencies in the Bahamas after Hurricane Dorian hit?

14. Severe Weather

Geography in the News

STORM OPHELIA – IRELAND'S FIRST EVER 'SEVERE WEATHER' ALERT FOR THE ENTIRE COUNTRY

For the first time in its meteorological history, Met Éireann issued a 'Status Red' warning, its highest storm category, for Ireland, with the expected approach of Storm Ophelia on 16 October 2017. Although weather warnings are usually issued only 24 hours in advance of a storm's expected landfall, this warning was issued 48 hours in advance.

The southwest was the first region to experience the force of the storm, which spread countrywide as the hours passed. Gusts of up to 191 km/h hit the Cork coast, the highest wind speeds ever recorded in Ireland.

Five people were killed as a result of the storm. The ESB confirmed that more than 450,000 customers were left without electricity for a time, and it was estimated that Ophelia caused an economic loss of €1 billion, owing to the shutdown of businesses and industry while the storm passed. Schools and colleges were also ordered to close for two days while the storm followed its path over the country. One school that felt the full force of Storm Ophelia was Douglas Community School in Cork, which had the roof of its gym ripped off.

1. **Why** did the closure of businesses result in an economic loss of €1 billion?
2. Do you think the government was right to order the closure of schools across the country for two days? **Think** about this from your perspective, from your teacher's perspective and from your parents'/guardians' perspective.

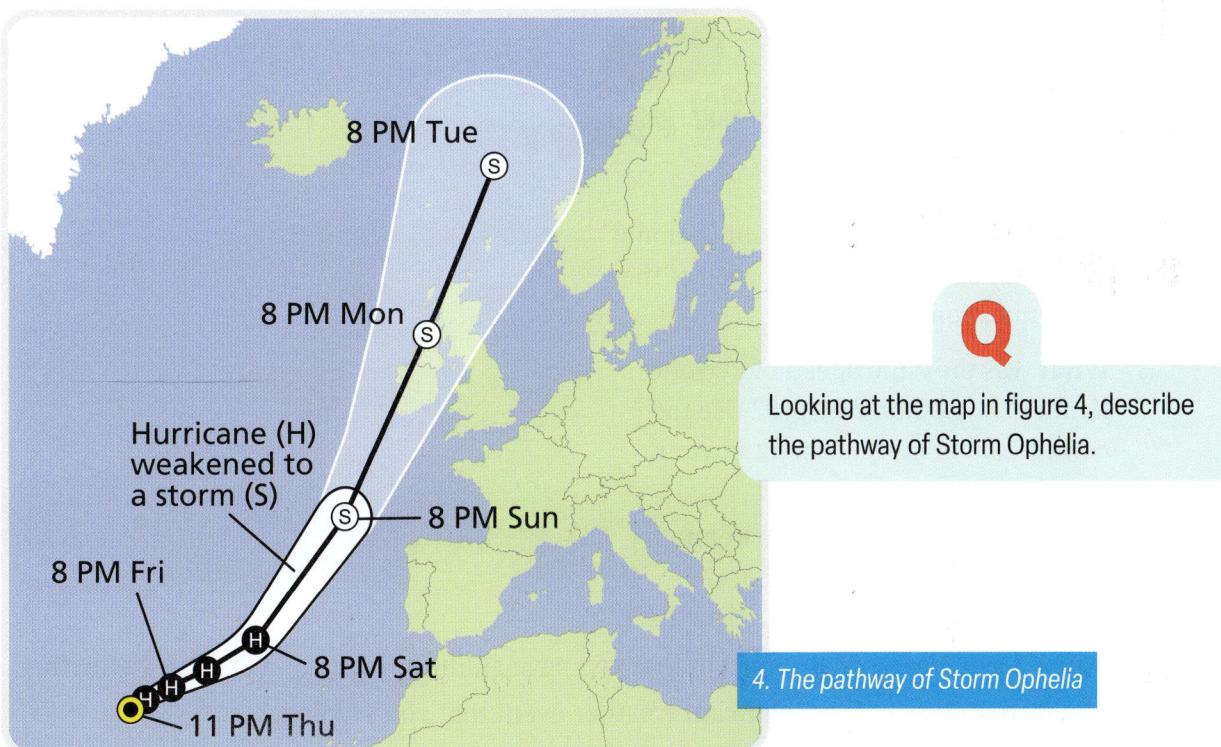

Q Looking at the map in figure 4, describe the pathway of Storm Ophelia.

4. The pathway of Storm Ophelia

Junior Cycle Geography CYCLONE

Being creative

Take two minutes to examine these images of destruction caused by Storm Ophelia. Now write a diary entry describing how Storm Ophelia affected your local area. (Be as creative as you wish.)

EXAM FOCUS

Sample questions

1. Study the map showing the path of Cyclone Kenneth from 2019 and answer each of the following questions.

   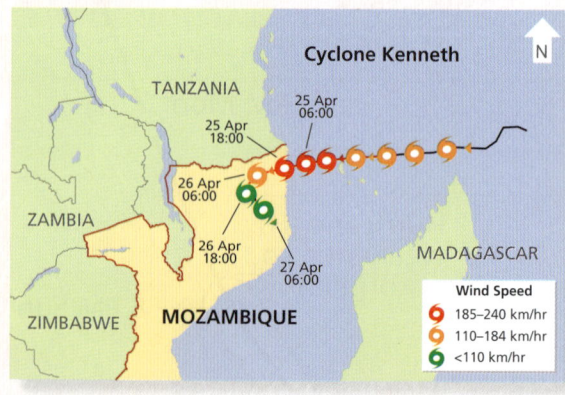

   (i) **What** was the wind speed of Cyclone Kenneth at 6:00 on 25 April? Use the correct unit of measurement in your answer.

   (ii) **What** direction was Cyclone Kenneth travelling before it came ashore?

       Westerly   Easterly   Northerly

   (iii) **Name** a significant weather event that you have studied, other than Cyclone Kenneth.

   (iv) **List** three effects of the significant weather event that you studied.

14. Severe Weather

2. News headline: 'Climate experts warn that Ireland could face hurricanes/cyclones due to climate change.'

    (i) **Explain** one potential social and one economic impact that could occur.

    (ii) **Describe** the steps Irish people could take to reduce the effects of a cyclone/hurricane.

3. (i) Other than an earthquake, **name** one example of a natural disaster you have studied and state where it happened.

    (ii) **Explain** what caused the natural disaster.

    (iii) **Explain** two economic and two social effects of the natural disaster on the region.

Let's get started! Sample starter

> (i) Storm Dorian, the Bahamas 2019
>
> (ii) Hurricanes start to form over tropical oceans, usually during late summer and early autumn.
> They need two things to get them started: heat and moist air.
> The warm moist air rises. As it rises, it cools and condenses (gas changes to a liquid).
>
> *[Complete this part of the answer.]*
>
> (iii) Economic Impacts: 1. Dorian caused about US$3.4 billion in damages, which is equal to one-quarter of the Bahamas' gross national product. ...
> Social Impacts ...

... now complete the answer in your copy.

Exam hints!

The action verb here is 'Describe'. This means you must give a detailed account in words. (Think practically – refer back to the the Bahamas' preparedness plan on p. 218.)

Clearly state where your chosen natural disaster occurred: was it in Ireland or abroad? Give as much detail as you can. Continue to explain How/What Caused it?, including as much detail as you can recall. One economic impact has been completed for you. Write one more as well as two social impacts.

Learning Outcome: 1.6

15 Global Climates

key words
- Climate
- Weather
- Characteristics
- Hot
- Temperate
- Cold
- Equatorial
- Savanna
- Hot desert
- Warm temperate oceanic
- Cool temperate oceanic
- Tundra
- Boreal
- North Atlantic Current (NAC)
- Latitude
- Prevailing wind/air mass
- Local climates
- Altitude
- Aspect
- Lapse rate
- Relief rainfall
- Leeward

Learning Intentions

You will be able to:
- Explain what a climate is
- Classify global climates into hot, temperate (warm/cool) and cold climates
- Outline the factors that affect Ireland's climate
- Discuss the factors that affect the climate of a particular area/locality.

15.1 What Is a Climate?

Climate is the average temperature, wind, humidity, snow and rain in a place over the course of years.

To classify the climate of an area, certain **weather** conditions, such as level of rainfall or hours of sunshine, must be measured. These conditions are measured for 30 years or more before an area's climate is determined.

15.2 Classifying Global Climate Types

There are many different types of climate found in different parts of the world. Each has its own **characteristics**, e.g. the region where it is found, its average temperature, hours of sunshine and levels of precipitation.

There are three broad categories of climate: **hot**, **temperate** and **cold**. Each of these includes a number of climatic types:

Hot Climates	Temperate Climates	Cold Climates
Equatorial	Warm temperate oceanic (Mediterranean)	Tundra
Savanna	Cool temperate oceanic	Boreal
Hot desert		

Latitude: measurement of location north or south from the equator

To help tell the difference between different climate types, we will examine certain characteristics that are unique to each one:
- Location (where they are found)
- Temperature
- Levels of precipitation (e.g. rain, hail, sleet or snow).

Working with others

Study the table above. With the person sitting next to you, take turns picking a climate type and stating which category it belongs to. For example, 'Boreal is a cold climate'.

Junior Cycle Geography CYCLONE

Hot Climates

Figure 1 shows the locations of the world's hot climate types.

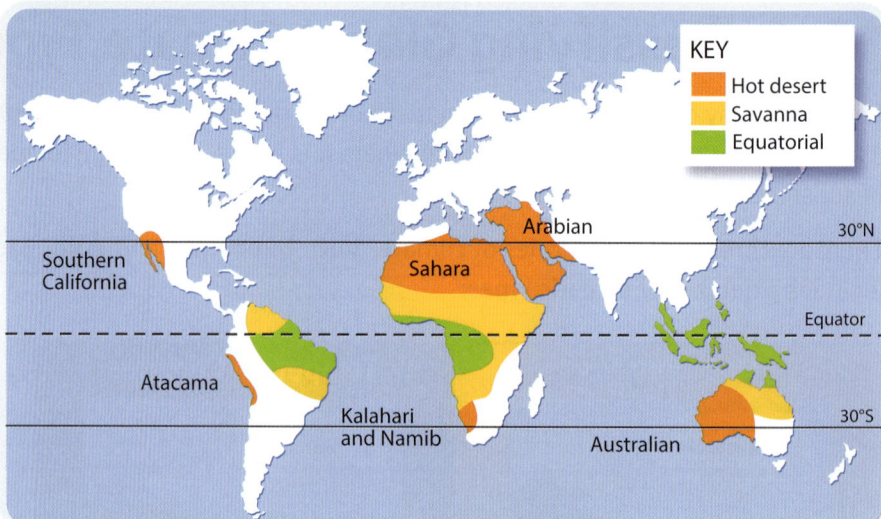

1. Hot climates around the world

Characteristics	Equatorial	Savanna	Hot Desert
Location	Found on or near the equator, e.g. North Brazil	Found between 5–15 degrees north and south of the equator, e.g. parts of Brazil and Nigeria	• Found between 15–30 degrees north and south of the equator, e.g. the Sahara • Usually found on the western side of countries and continents
Temperature	Hot all year round, average temperature 27 °C	Warm all year round. • **Summers**: Very wet. Temperatures can reach 30 °C • **Winters**: Long dry season. Temperature is a little cooler at 20 °C	**Daytime**: Temperatures can reach between 40 and 50 °C **Night-time**: With no cloud cover to trap the heat, temperatures can drop to 0 °C.
Precipitation	**Type**: Rainfall. Equatorial climates have twice the average rainfall of Ireland	**Dry season**: Savanna climates have very little rainfall, 20 mm at most. No rain between December and February **Wet season**: Can be as much as 1,000 mm of rain	Deserts have less than 250 mm of precipitation per year
Vegetation	Rainforest, e.g. the Amazon, Brazil	Grassland, e.g. plains of East Africa	Cacti

Question Time

1. Outline the characteristics of a hot climate you have studied under the following headings: Location, Temperature, Precipitation and Vegetation.
2. Study the table above and arrange equatorial, savanna and hot desert in order from highest to lowest in terms of rainfall and temperature.
3. Give one reason why you think Ireland does not have a hot climate.

15. Global Climates

Temperate Climates

Figure 2 shows the locations of the world's warm and cool temperate oceanic climate types.

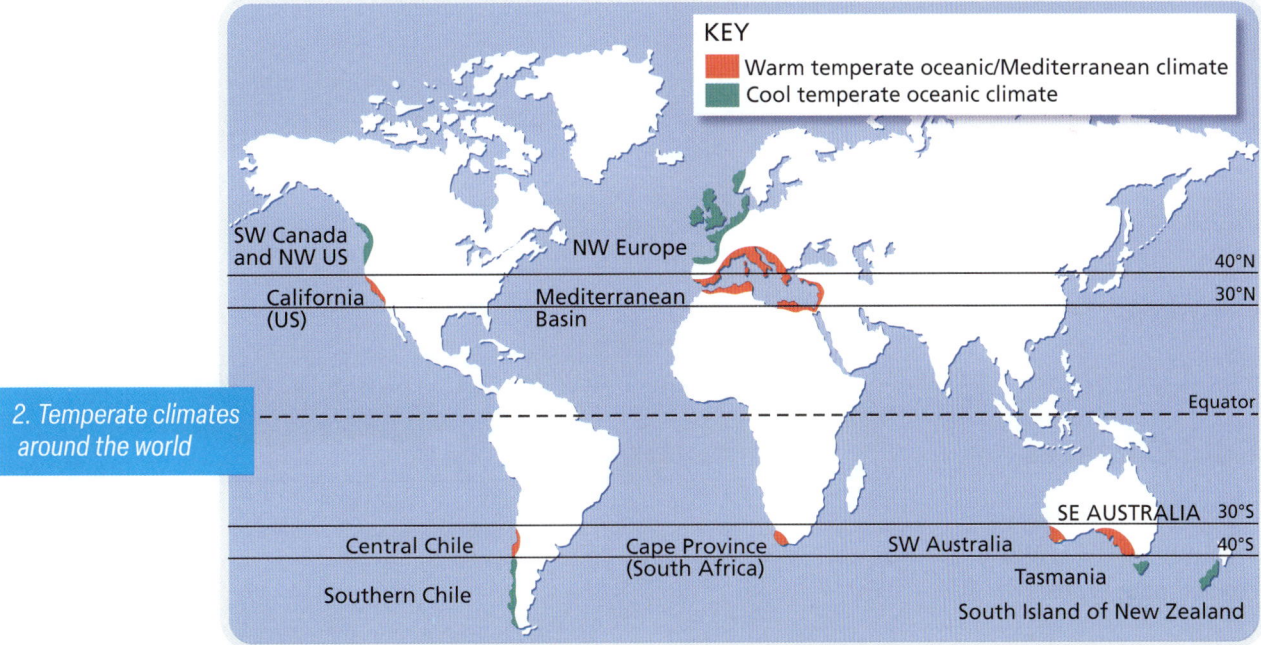

2. Temperate climates around the world

Characteristics	Warm Temperate Oceanic/Mediterranean	Cool Temperate Oceanic
Location	Located between 30–40 degrees north and south of the equator, on the outer edge of landmasses. Majority of this area found around the Mediterranean Sea, hence its name.	Located between 40–60 degrees north and south of the equator. Found in northwest Europe and on the northwest coast of Canada and the US.
Temperature	• **Summers**: Hot and dry with average temperatures reaching 30 °C, with almost no rain in July and August • **Winters**: Mild and moist with temperatures averaging 8 °C	• **Summers**: Mild with temperatures averaging between 6 and 14 °C • **Winters**: Mild with plenty of precipitation. Frost can occur, meaning temperatures can fall below 0 °C
Precipitation	**Type**: Rainfall. Total yearly average of 400 mm	**Type**: Rain, hail, sleet and snow, with average rainfall 800–2,000 mm per year
Vegetation	Evergreen trees, e.g. pine trees	Deciduous trees, e.g. oak

Question Time

1. Outline the characteristics of a temperate climate you have studied under the following headings: Location, Temperature, Precipitation and Vegetation.
2. Name a country with a:
   (a) Warm temperate climate
   (b) Cold temperate climate (other than Ireland).
3. List the key differences between a warm temperate oceanic climate and a cool temperate oceanic climate.

Cold Climates

Figure 3 shows the locations of the world's cold climate types.

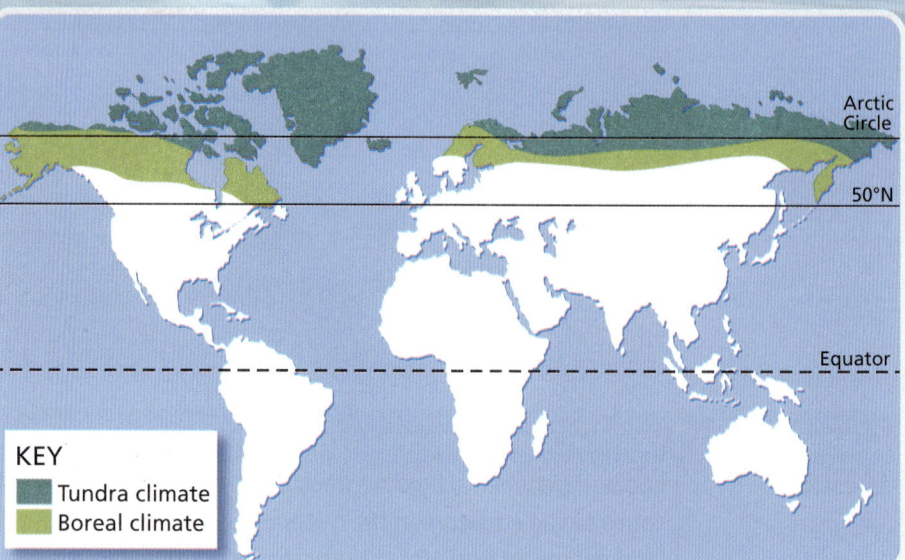

3. Cold climates around the world

Characteristics	Boreal	Tundra
Location	Stretches across the northern hemisphere between 55–66 degrees north of the equator (The word 'boreal' means 'of the north'.)	The North and South Poles, extreme northern parts of Canada and Russia
Temperature	**Summer**: A short season, with average temperatures of 16 °C **Winter**: Average temperatures of −29 °C	**Summer**: 4 °C **Winter**: Average temperatures of −35 °C
Precipitation	**Type**: Rain and snow, with average rainfall less than 400 mm per year	**Type**: Rain and snow, with average rainfall approximately 250 mm per year
Vegetation	Coniferous trees, such as spruce	Mosses

Question Time

1. What type or types of precipitation would you expect in places with a boreal climate? Explain your answer.
2. Explain the characteristics of a cold climate you have studied under the following headings: Location, Temperature, Precipitation and Vegetation.

15.3 Ireland's Climate

Ireland has a cool temperate oceanic climate. Irish winters tend to be cool and windy, while summers are mostly mild and less windy.

Factors that Affect Ireland's Climate

There are a number of factors that influence climate:

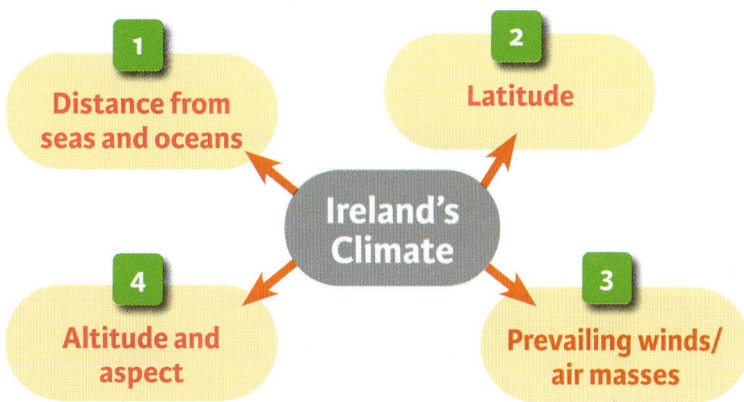

1. Distance from Seas and Oceans

Ireland is an island surrounded by the Atlantic Ocean on our western coast and the Irish Sea to the east. The main influence on Ireland's climate is the Atlantic Ocean and the **North Atlantic Current (NAC)**, also known as the North Atlantic Drift, that flows through it.

The NAC is a warm ocean current – the Gulf Stream – that originates in and flows northeast from the Gulf of Mexico towards northwest Europe. The water at the Gulf of Mexico is heated further by warm water flowing from the equator as it travels towards the northwest of Europe. Without the warming effect of the NAC, Ireland and other places in Europe would be as cold as eastern Canada, which is at the same **latitude** (distance from the equator).

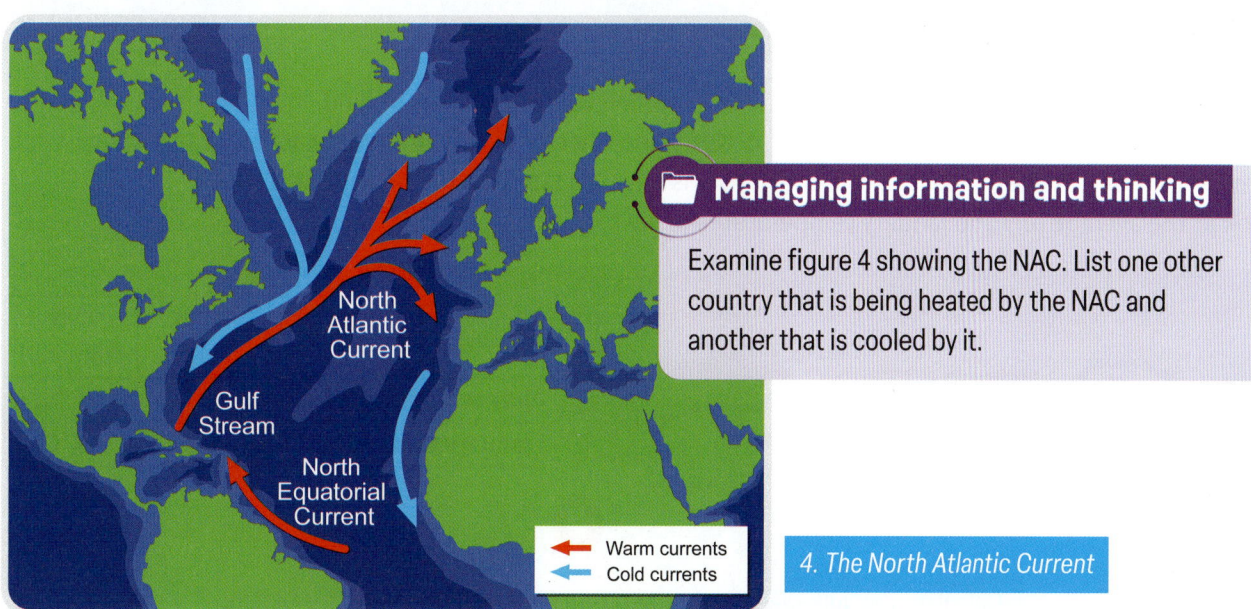

Managing information and thinking

Examine figure 4 showing the NAC. List one other country that is being heated by the NAC and another that is cooled by it.

4. The North Atlantic Current

Our seas heat up slowly in summer, but they also cool down slowly in winter. Because of this, Irish coastal areas with sea winds (e.g. western Mayo) tend not to get very hot in summer or very cold in winter.

In Ireland, land heats up quickly in summer and cools down quickly in winter. This sometimes causes Irish inland areas to have slightly colder winters and slightly warmer summers than Irish coastal areas.

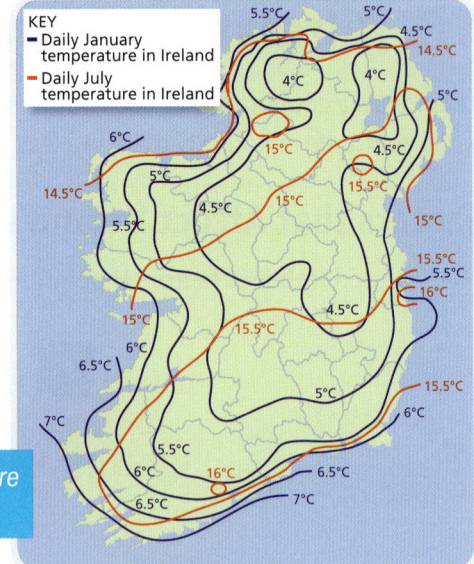

5. The mean average daily temperature across Ireland in January and July

Question Time

1. (a) What is the North Atlantic Current?
   (b) How does it affect temperatures in Ireland's coastal areas?
2. Examine figure 4 showing the North Atlantic Current. Explain how it might affect the climate of a country like Spain.
3. Figure 5 shows the average daily temperature for Ireland's counties in January (blue line) and July (red line).
   Explain how distance from seas and oceans can affect Ireland's climate. Justify your answer.
4. (a) Using figure 5, state the approximate average temperature in your local area in January and in July.
   (b) Calculate the range in average temperature between January and July in your local area (range means difference between the highest and lowest).

Exam hint!

The action verb 'Justify' requires you to give valid reasons or evidence to support an answer or conclusion.

If distance from the sea can affect temperatures in a small country such as Ireland, imagine its effect within a large landmass. Tralee in Co. Kerry and the city of Kyiv in Ukraine are at similar latitudes. While Tralee is on the coast, Kyiv is over 951 km from the nearest sea. Now look at the average July and January temperatures in both cities.

6. Tralee and Kyiv are at similar lattitudes

	Tralee (Kerry)	Kyiv (Ukraine)
July temperatures	15 °C	26 °C
January temperatures	5 °C	−7 °C

Q Calculate the July–January temperature range for Tralee and for Ukraine. Explain the great difference in temperature ranges between the two places.

15. Global Climates

2. Latitude

The latitude of a place is its distance from the equator. The surface of the earth is not evenly heated. The further a place is from the equator, the cooler it is likely to be.

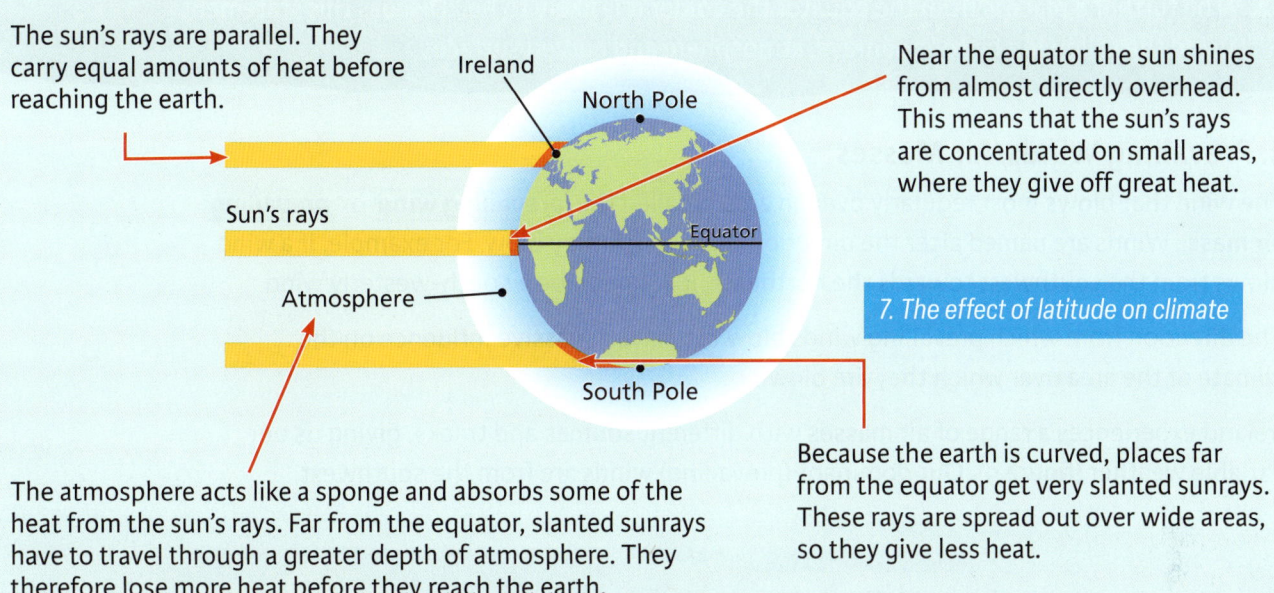

The sun's rays are parallel. They carry equal amounts of heat before reaching the earth.

Near the equator the sun shines from almost directly overhead. This means that the sun's rays are concentrated on small areas, where they give off great heat.

The atmosphere acts like a sponge and absorbs some of the heat from the sun's rays. Far from the equator, slanted sunrays have to travel through a greater depth of atmosphere. They therefore lose more heat before they reach the earth.

Because the earth is curved, places far from the equator get very slanted sunrays. These rays are spread out over wide areas, so they give less heat.

7. The effect of latitude on climate

The sun's rays have a larger surface area to cover when they reach mid-latitude regions, such as Ireland. These regions tend to have moderate temperatures – not too hot and not too cold.

A simple experiment can help us to see how sunlight is more or less concentrated as it reaches different parts of the earth. In a darkened room, shine a torch at different angles on a globe. See how concentrated the light is when it shines directly on area A on earth (on the right below) versus how the light is spread when the torch is held at an angle (area B on earth, on the left below).

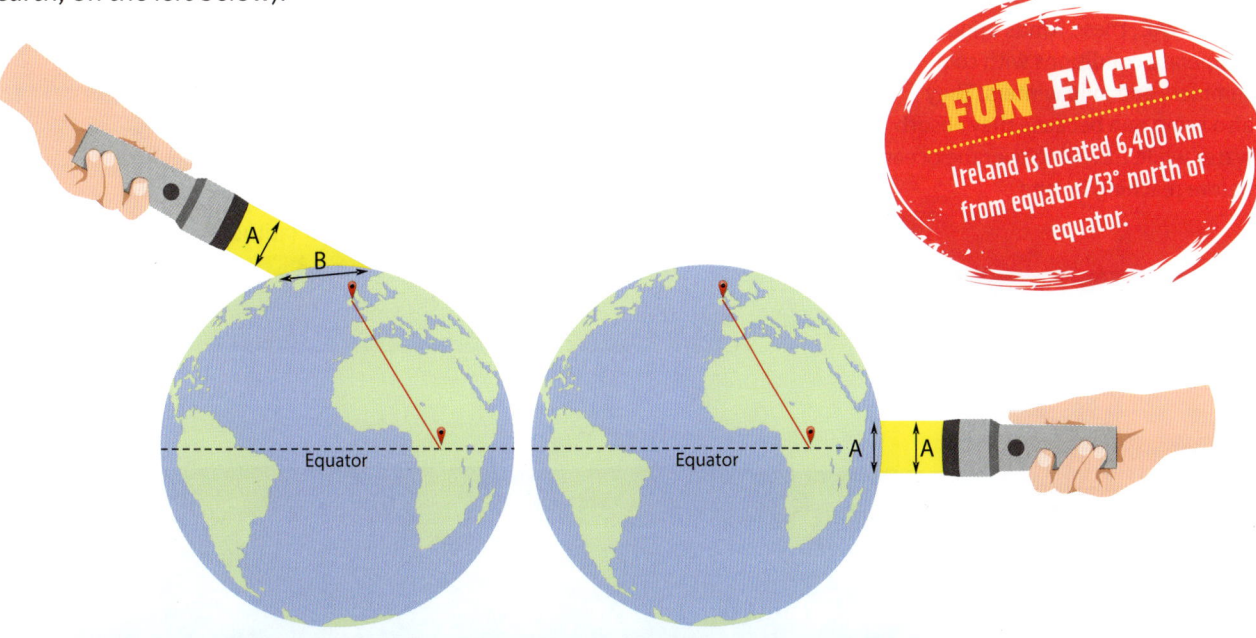

FUN FACT!
Ireland is located 6,400 km from equator/53° north of equator.

8. Experiment to show how the sun's rays are concentrated at different angles

Junior Cycle Geography CYCLONE

Question Time

1. Explain the term 'latitude'.
2. 'The further a place is from the equator the colder it is likely to be.' Discuss this statement using the following factors: sunrays, atmosphere and curved surfaces.

3. Prevailing Winds/Air Masses

The wind that blows most regularly over an area is called the **prevailing wind** or **prevailing air mass**. Winds are named after the direction from which they blow. For example, if a wind blows from the southwest towards the northeast, it is said to be a south-westerly wind.

The direction from which prevailing winds blow can have a massive influence on the climate of the area over which they are blowing.

Ireland experiences a range of air masses with different sources and tracks, giving us our variable weather (figure 9). Our dominant (prevailing) winds are from the southwest.

Northerly winds (polar winds): These winds come from the colder, higher latitude areas (further from the equator). They are dry until they absorb moisture as they move downwards into lower warmer latitudes.

Easterly winds (coming from Eastern Europe): These winds are warm in summer and very cold in winter as they are travelling across large areas of land. They are dry because they absorb little water while travelling over land.

South-westerly winds: Ireland's prevailing wind blows from across the Atlantic Ocean. Heated in winter by the NAC, these winds bring temperate conditions with cool summers and mild winters. They also bring rain because they absorb water while travelling over the ocean before reaching Ireland.

Southerly winds (tropical winds): These are warm winds because they come from warmer, lower latitude areas (closer to the equator). They may bring rain if they give off moisture as they approach cooler, higher latitudes.

9. The air masses that affect Ireland's climate

Question Time

1. What does the term 'prevailing wind' mean?
2. Name the four winds that affect Ireland's climate.
3. Describe Ireland's prevailing wind.

15. Global Climates

4. Altitude and Aspect

Smaller areas may experience special climatic conditions of their own. These **local climates** are influenced by **altitude** and **aspect**.

Altitude and Local Climate

Altitude means height above sea level. The higher a place is, the colder its climate will be. Temperatures decrease by roughly 1 °C for every 150 m climbed. This decrease in temperature is called the **lapse rate**.

Ireland's landscape is a mixture of reasonably flat land in the midlands and mountainous areas on the coastline. Compare the landscape map of Ireland on the left with the map displaying average temperatures across the country on the right. We can see a connection between altitude and temperature.

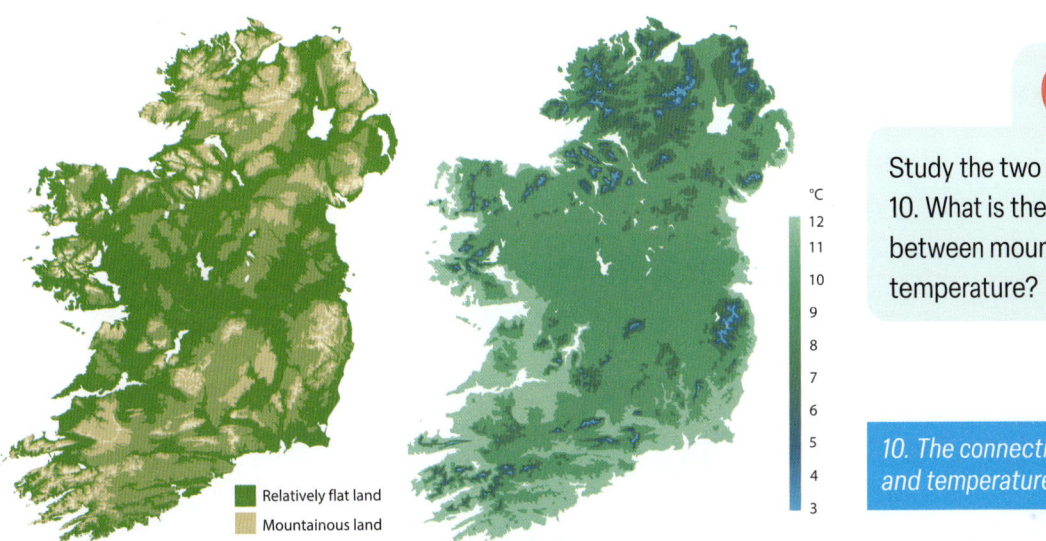

Q Study the two maps in figure 10. What is the connection between mountainous areas and temperature?

10. The connection between altitude and temperature

Increased (higher) altitude also increases chances of rainfall. Annual amounts of precipitation increase by between 100 mm and 200 mm on exposed slopes for every 100 m rise in altitude.

Aspect and Local Climate

Aspect is the direction in which a slope faces.

The sun rises in the east, travels in a southerly direction and sets in the west. For the northern hemisphere, this means that south-facing slopes are warmer than north-facing slopes.

Aspect will also affect precipitation levels. Slopes facing the sea experience sea winds and therefore higher levels of rainfall called **relief rainfall**. The slopes on the **leeward** (sheltered) sides are drier.

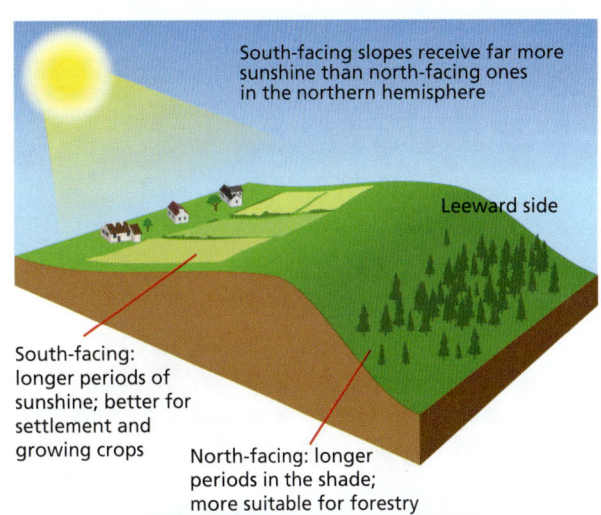

11. The effect of aspect on climate

Mountainous areas along the west coast and particularly their west-facing slopes tend to experience a lot of relief rainfall. This is why parts of counties Kerry, Galway and Donegal are among Ireland's wettest areas.

The southeast of Ireland, e.g. Wexford, is well removed from the effects of the wet westerly winds. It is therefore usually the warmest and driest part of Ireland. On average, it gets more than seven hours of sunshine a day in early summer, whereas the rest of Ireland gets between 5 and 6.5 hours. For this reason, it is known as the 'sunny southeast'.

12. The mountainous areas of Ireland

Question Time

1. Explain the terms 'altitude' and 'aspect'.
2. Carrauntoohil, Co. Kerry, is Ireland's highest point at over 1000 m above sea level. How do you think altitude affects the local climate of any settlement located nearby?
3. Describe how the local climate of a settlement on the leeward side of the MacGillycuddy's Reeks differs from that on the sea-facing slopes?

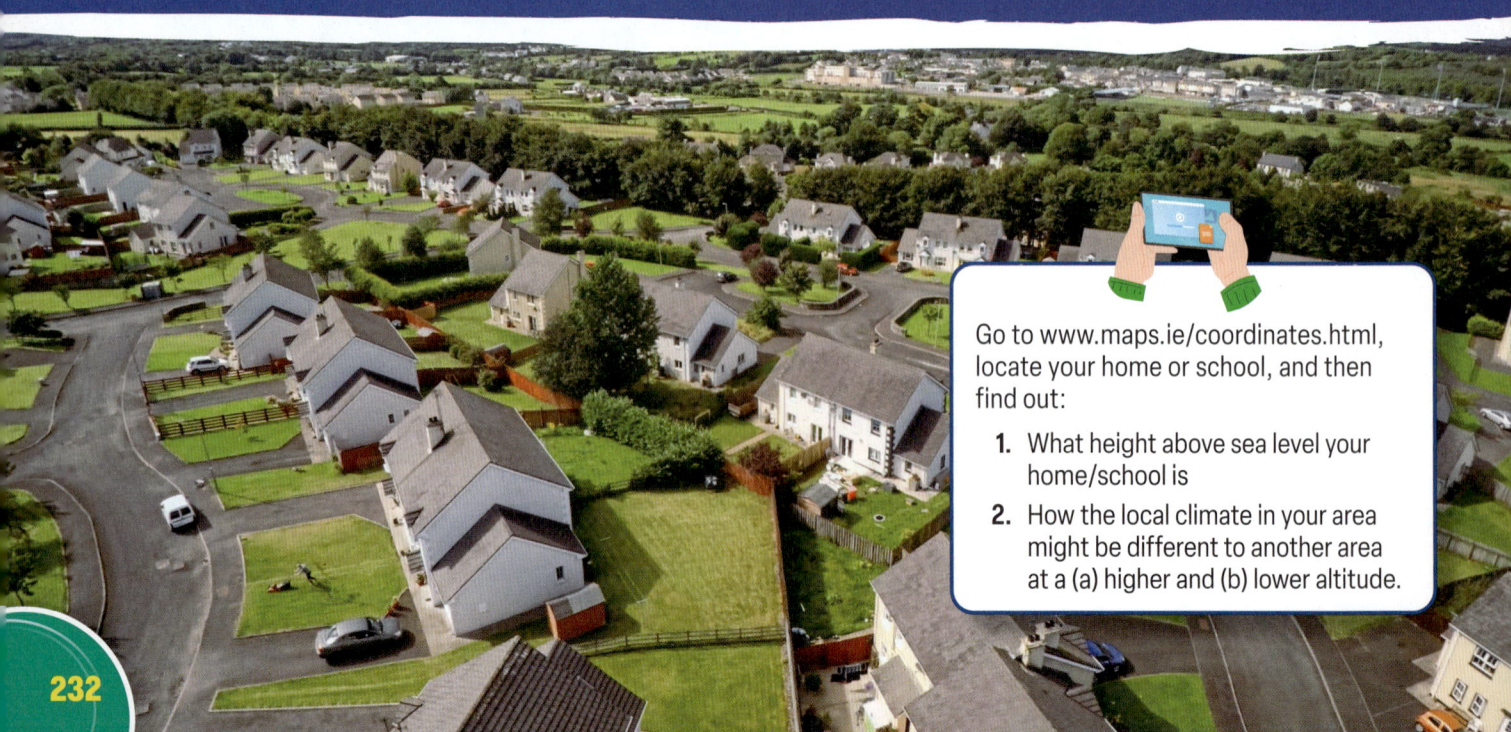

Go to www.maps.ie/coordinates.html, locate your home or school, and then find out:

1. What height above sea level your home/school is
2. How the local climate in your area might be different to another area at a (a) higher and (b) lower altitude.

15. Global Climates

Sample questions

1. Answer the short questions below:

   (i) One factor which influences the temperature of a place on the earth's surface is its distance from the equator. This distance north or south of the equator is called:
   - Altitude • Latitude • Aspect • Relief

   (ii) Which of the following statements below are true?
   - Hot climates are found 50 degrees above and below the equator.
   - A savanna is an example of a hot climate.
   - Hot climates can experience precipitation in the form of rainfall.
   - Deserts have little to no rainfall annually.
   - Equatorial climates have twice the annual rainfall of Ireland.

   Exam hints!

   In the exam, short questions can take many forms. Parts (ii) and (iii) are just two examples. With True/False statements, read each one carefully and focus on the information (facts/figures) presented when deciding.

   (iii) Copy and complete the table below describing cold climates, using the following terms:

   Tundra   Rain (less than 400 mm per year) and snow
   4 °C   −35 °C   66   −29 °C   Temperature

Characteristics	Boreal	
Location	Found between 55 and ____ degrees north of the equator	Northern parts of Canada and Russia and the North Pole
	Summers: average temperature 16 °C	**Summer**: average temperature ____
	Winters: average temperature ____	**Winter**: average temperature ____
Precipitation		Approximately 250 mm per year

2. Examine the map and in your copy label each letter as either a hot, temperate or cold climate.

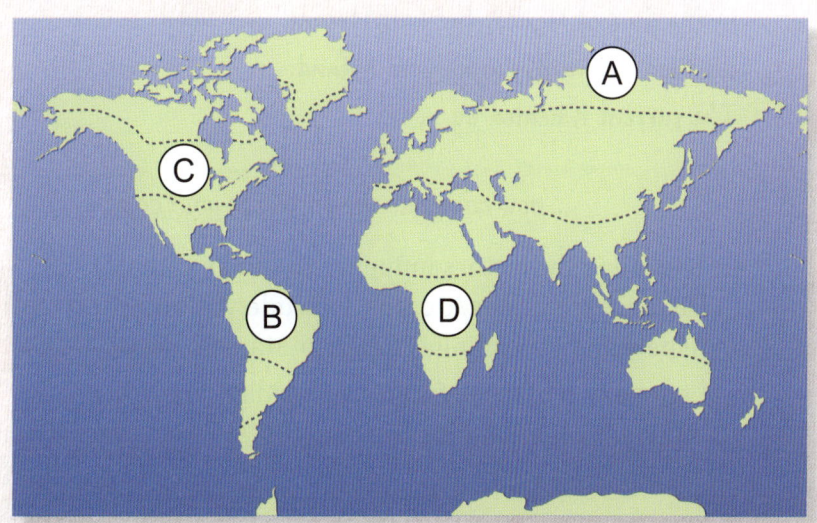

3. (a) Study the image below and answer the questions that follow.

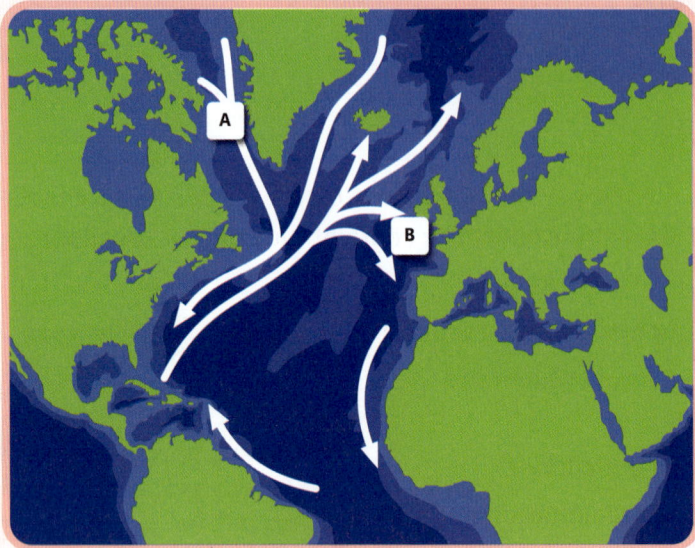

(i) Name the current in this image.
(ii) State whether this current is warm or cold at points A and B.
(iii) Where does this current originate?

(b) Explain how this current influences Ireland's climate.

Exam hints!

For this question, start by describing the origins of the current. Next explain the effect the equator has on this current. Finish your answer by describing the effects this current has on Ireland and other places in Europe.
Complete the sample below that has been started for you.

Let's get started! Sample starter

> Ireland is surrounded by water on all sides. The Atlantic Ocean is the main influence on Ireland's climate. The [insert current mentioned in 3a (i) here] flows through the Atlantic Ocean. This current begins in the Gulf of Mexico, the Gulf Stream, and flows northeast towards Europe and Ireland.
> The current is heated by ...

... now complete the answer in your copy.

4. The following factors influence Ireland's climate: Distance from Seas/Oceans, Latitude, Altitude/Aspect and Prevailing Winds/Air Masses.

Choose **two** of these factors and explain their influence on Ireland's climate.

When answering a question like this, choose the factors you feel most confident about. You do not have to write about each factor you choose in the order asked here. For example, you could start with Altitude/Aspect.

» State your chosen factor clearly in your answer.
» Explain how each of the factors you have chosen influences Ireland's climate.

Learning Outcome: 2.6

16

Climate Change

Learning Intentions

You will be able to:
- Discuss human causes of climate change
- Explain the greenhouse effect
- Describe some of the implications of climate change.

key words

- Climate
- Climate change
- Temperature
- Thermometer
- Ice core
- Global warming
- Greenhouse gases
- Carbon dioxide
- Methane
- Nitrous oxide
- Greenhouse effect
- Desertification
- Heatwave
- Drought
- Flooding
- Erosion

16.1 How Do We Know That Our Climate Is Changing?

As we learned in Chapter 15, **climate** is the average weather across a large region measured over a period of time (30 years). **Climate change** means there has been a large change in the climate records of temperature, rainfall, wind speeds, etc., lasting decades or longer.

The earth's climate has changed many times during its history. However, this current period is different, as there is evidence that human activity is speeding up natural climate change. In this chapter, we will look at the evidence that shows our climate is changing.

Temperature pre-1900

You will remember from Chapter 13 that **temperature** is measured using a **thermometer**. However, humans began to use thermometers to record temperature data only in the last 140 years. This means that there are no reliable weather records before the 1880s detailing the earth's weather patterns, such as its temperature.

In order to piece together weather data from before this time, scientists analyse sediment cores taken from ocean and lake floors and from frozen **ice cores**.

- Ice cores are cylinders of ice drilled out of an ice sheet or glacier.
- As layers of fresh snow fall and sediment becomes buried, they trap and preserve (store) evidence of the global temperature experienced at that time of the burial. Investigating the physical and chemical make-up of the ice core can tell us of past changes in climate.
- Most ice core records come from Antarctica and Greenland. The oldest continuous ice core records are from Greenland and go back 130,000 years.

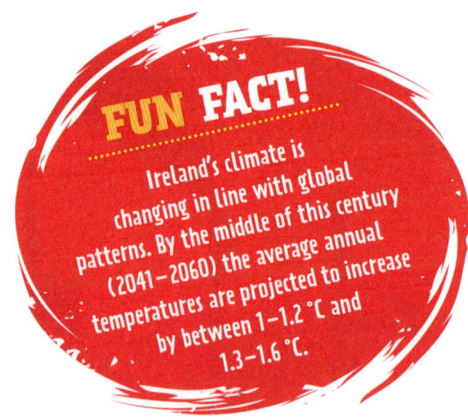

FUN FACT!
Ireland's climate is changing in line with global patterns. By the middle of this century (2041–2060) the average annual temperatures are projected to increase by between 1–1.2 °C and 1.3–1.6 °C.

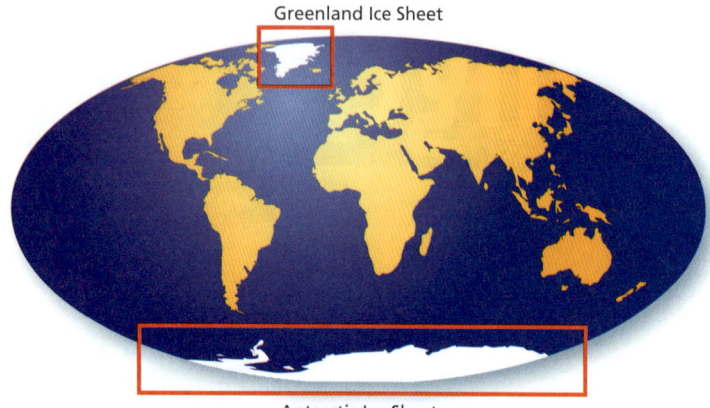

1. The location of ice sheets used for evidence of past changes in climate

Alternating light and dark bands storing evidence of past climates.

Light and dark bands alternate like tree rings in this ice core from Greenland. The longest ice cores can be more than 3 km deep.

Go to YouTube and look up 'National Ice Core Lab Stores Valuable Ancient Ice – Science Nation' (2:32) to see how ice cores are extracted and stored.

Each silver tube on these shelves contains a 1-metre-long section of an ice core. They are stored at a temperature of –36 °C.

16.2 Changes in Global Climate

When compared with the average temperatures of the years 1880–2000, average global temperatures have increased over the past couple of decades. This phenomenon is known as ' global warming '.

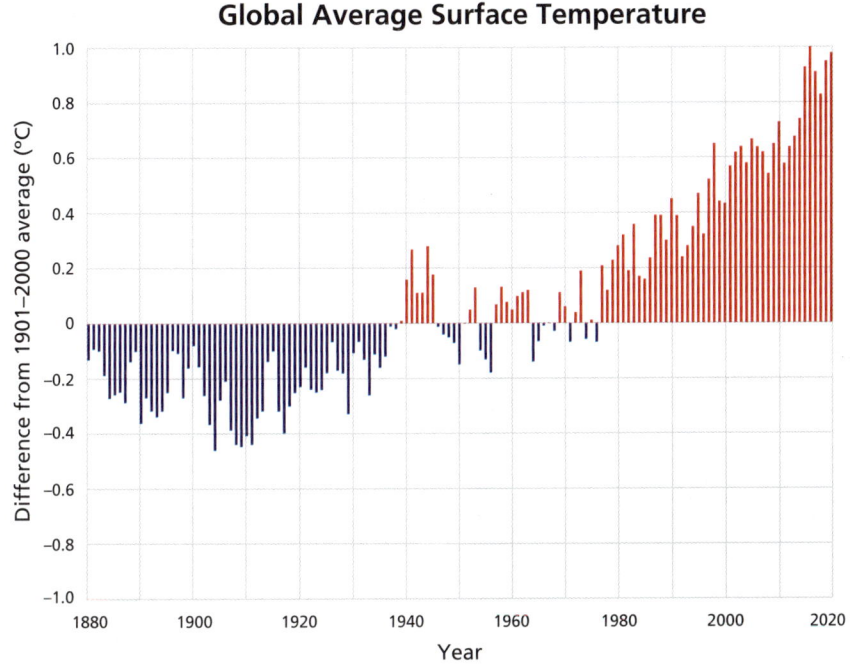

> **⅓² Being numerate**
>
> Looking at the period 1920–2020 in figure 2, how many years have had temperatures above average?

Search online for 'NASA Global temperature anomalies from 1880 to 2021'.

2. Changes in global average surface temperature, 1880–2020

Question Time

1. Describe how scientists discover what temperatures were like prior to the 1900s.
2. What is climate change?
3. What evidence is there to show that global warming is occurring?

16.3 Climate Change and Human Activity

Scientists now believe that human activities have had a significant impact on global warming. Since the start of the Industrial Revolution from around 1760, human activity and industry have released large amounts of greenhouse gases (carbon dioxide , methane , nitrous oxide) into the atmosphere. To understand why this is harmful, we must look at the ' greenhouse effect '.

Industrial Revolution: A period of dramatic change in which manufacturing by hand was increasingly replaced by manufacturing by machines.

The Greenhouse Effect

The greenhouse effect is the warming of the earth's surface and the air above it. It is caused by gases in the air that trap energy from the sun. The heat-trapping gases are known as greenhouse gases.

During the day, the sun's solar energy penetrates the earth's atmosphere. The earth's surface then warms up. At night, the earth's surface cools down. It releases the heat back into the air, but some of the heat gets trapped by greenhouse gases in the atmosphere. That is what keeps our earth at an average temperature of 15 °C.

3. The greenhouse effect

The greenhouse effect is now too strong, and the earth is getting warmer and warmer. There are now too much greenhouse gases, created by human activities, contained in the atmosphere, trapping solar energy.

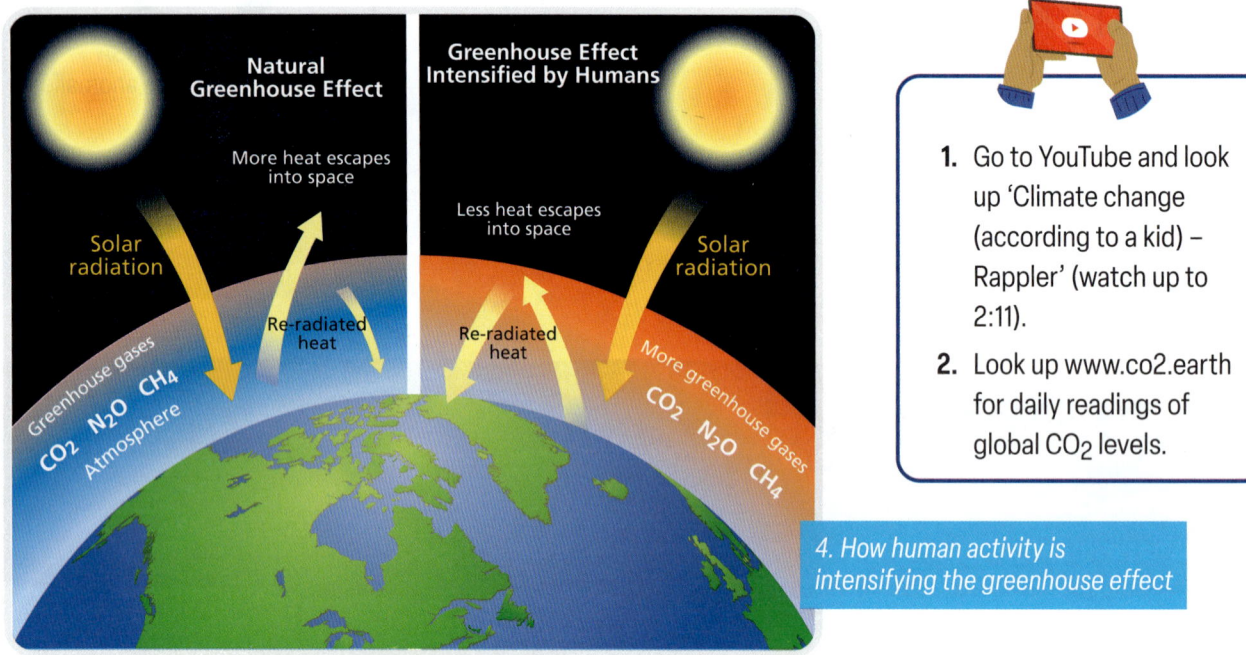

1. Go to YouTube and look up 'Climate change (according to a kid) – Rappler' (watch up to 2:11).
2. Look up www.co2.earth for daily readings of global CO_2 levels.

4. How human activity is intensifying the greenhouse effect

Causes of Climate Change

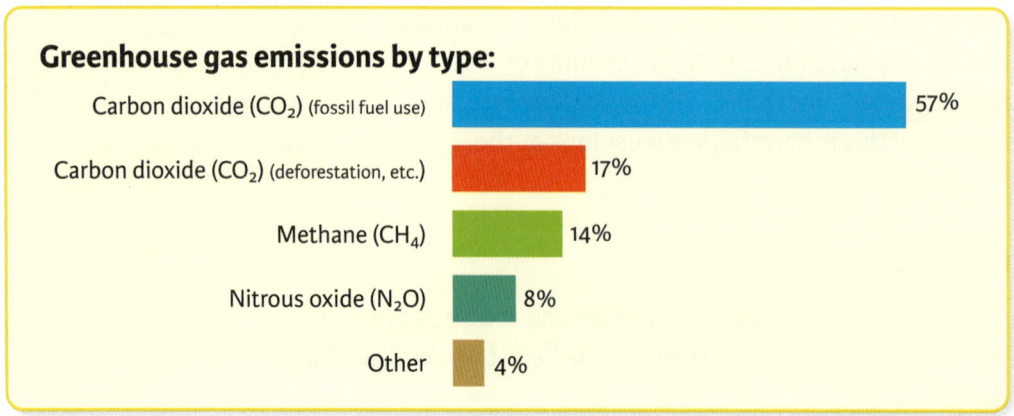

Greenhouse gas emissions by type:

- Carbon dioxide (CO_2) (fossil fuel use) — 57%
- Carbon dioxide (CO_2) (deforestation, etc.) — 17%
- Methane (CH_4) — 14%
- Nitrous oxide (N_2O) — 8%
- Other — 4%

16. Climate Change

1. **Carbon dioxide (CO_2)** makes up around three-quarters of the warming impact of current human greenhouse gas emissions.

- Burning fossil fuels such as coal and oil releases energy, which is turned into heat, electricity or power for transportation, e.g. in power plants, cars and aeroplanes.
- Deforestation is the permanent removal of forests and has a huge impact on greenhouse gas emissions. Forests in many areas have been cleared for timber or burned and cleared to change the land use. Trees pull in carbon dioxide and release oxygen. When forests are cleared, large amounts of stored CO_2 are released into the atmosphere.

2. Methane (CH_4) accounts for around 14% of current human greenhouse gas emissions. Methane is very effective at absorbing heat.

- Methane emissions are produced whenever fossil fuels are extracted from the earth. Methane is the main ingredient of natural gas. Leaks arising from extracting gas release methane straight into the atmosphere.
- Animals such as cows create large amounts of methane during their normal digestion process, releasing it either through belching or flatulence (gas). In countries like Brazil, more greenhouse gas emissions come from cows than from motor vehicles.
- Methane is created by decomposing (rotting) solid waste in landfills. This also happens with animal and human waste.
- Biofuels are fuels made from living things such as crops. Each year, biofuels produce 12 million tonnes of methane, making them a major source of the gas. An example is biodiesel made from vegetable oil.

3. Nitrous oxide (N_2O) accounts for around 8% of current human greenhouse gas emissions. Nitrous oxide is up to 300 times more effective at trapping heat than carbon dioxide.

- Artificial fertilisers are the biggest source of nitrous oxide in the atmosphere. These are chemicals added to the soil to make it more fertile. They increase the release of nitrous oxide into the atmosphere.
- Increasing global energy demands have led to greater emissions from coal-burning energy plants. This generates more nitrous oxide emissions.

We can all play an important role in tackling climate change. Using a carbon calculator can help us to reduce our carbon count number (also known as carbon footprint) and to take responsibility for our greenhouse gas emissions. Search online for 'carbon footprint calculator'. Use the carbon calculator to find out your carbon footprint and discover what changes you can make to improve it.

Question Time

1. What is the greenhouse effect?
2. Using a diagram, illustrate the difference between the natural greenhouse effect and the greenhouse effect intensified by humans.
3. Outline the main sources of each of the following greenhouse gases:
   (a) Carbon dioxide (CO_2)   (b) Methane (CH_4)   (c) Nitrous oxide (N_2O)

16.4 Implications and Effects of Climate Change

Climate change affects all regions around the world.

1. Polar ice sheets are melting, causing the level of the sea to rise.
2. In some regions, extreme weather events and very heavy rainfall are becoming more common.
3. Other regions are experiencing **desertification**, with extreme **heatwaves** and **droughts**.
4. Not only are these changes impacting humans, they are also having a serious effect on wildlife.

All of these impacts are expected to get worse in the years ahead.

1. Melting Ice and Rising Seas

- When water warms up, it expands. At the same time, global warming causes polar ice sheets and glaciers to melt. The combination of these changes is causing sea levels to rise, resulting in **flooding** and **erosion** of coastal areas.
- Many fresh water supplies could become contaminated with sea water, resulting in a lack of clean water in some regions.

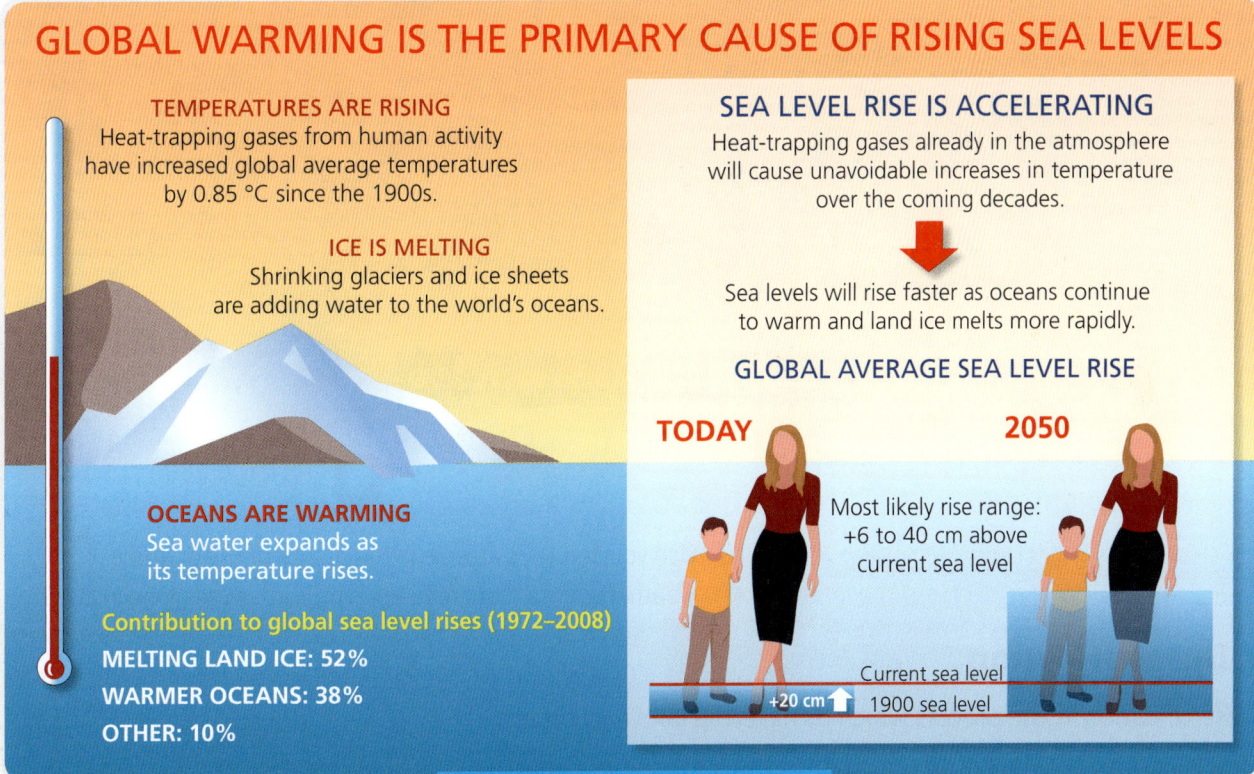

5. Global warming and sea level rise

Being numerate

1. By how many centimetres have sea levels risen from 1900 to today?
2. By how many centimetres are sea levels expected to increase by 2050?

2. Extreme Weather Events

Extreme weather events, such as droughts, floods and heatwaves, are becoming more frequent. These events can have devastating impacts on large areas.

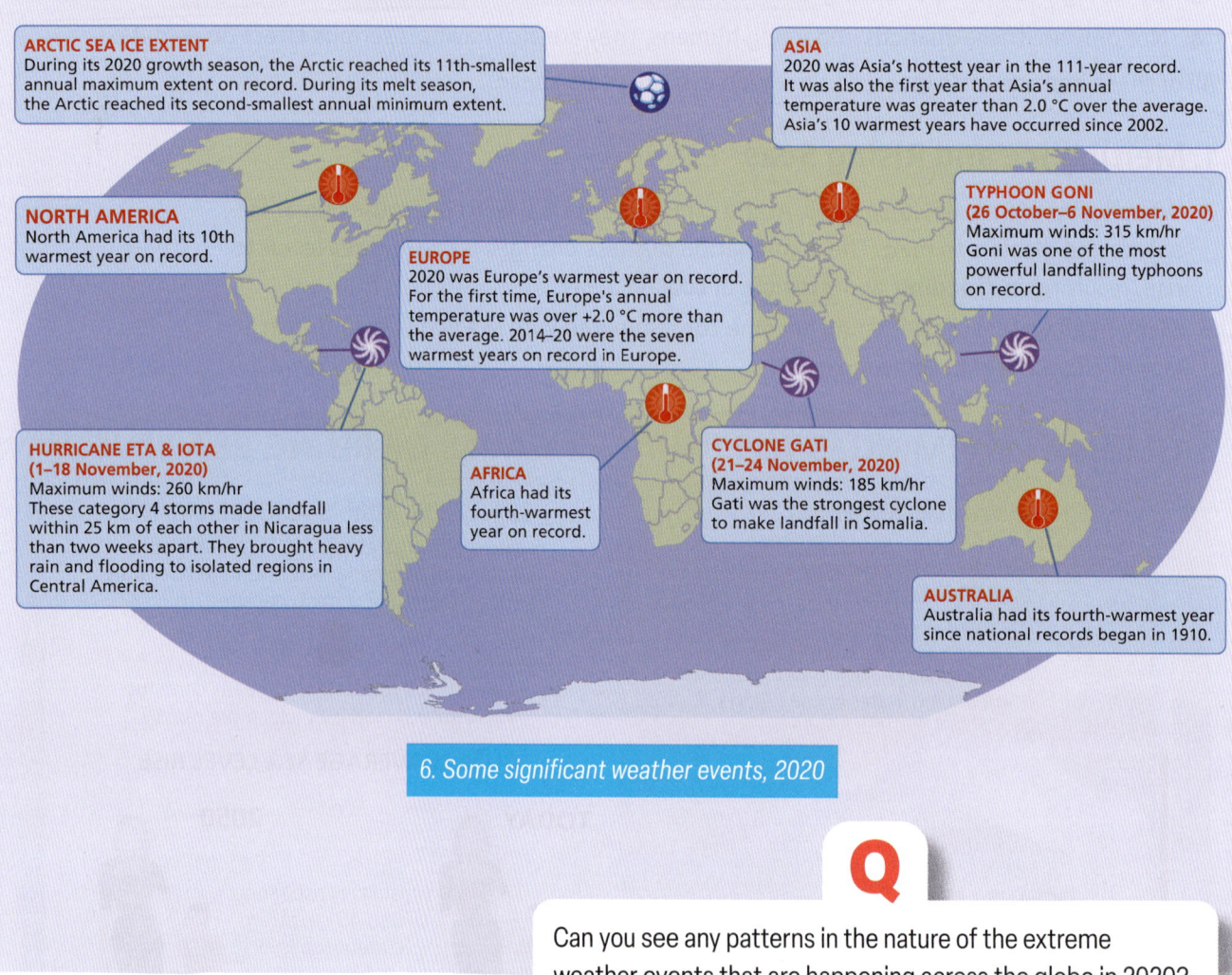

MAJOR WEATHER EVENTS CITED IN THE WORLD METEOROLOGICAL ORGANIZATION STUDY

ARCTIC SEA ICE EXTENT
During its 2020 growth season, the Arctic reached its 11th-smallest annual maximum extent on record. During its melt season, the Arctic reached its second-smallest annual minimum extent.

ASIA
2020 was Asia's hottest year in the 111-year record. It was also the first year that Asia's annual temperature was greater than 2.0 °C over the average. Asia's 10 warmest years have occurred since 2002.

NORTH AMERICA
North America had its 10th warmest year on record.

TYPHOON GONI
(26 October–6 November, 2020)
Maximum winds: 315 km/hr
Goni was one of the most powerful landfalling typhoons on record.

EUROPE
2020 was Europe's warmest year on record. For the first time, Europe's annual temperature was over +2.0 °C more than the average. 2014–20 were the seven warmest years on record in Europe.

HURRICANE ETA & IOTA
(1–18 November, 2020)
Maximum winds: 260 km/hr
These category 4 storms made landfall within 25 km of each other in Nicaragua less than two weeks apart. They brought heavy rain and flooding to isolated regions in Central America.

AFRICA
Africa had its fourth-warmest year on record.

CYCLONE GATI
(21–24 November, 2020)
Maximum winds: 185 km/hr
Gati was the strongest cyclone to make landfall in Somalia.

AUSTRALIA
Australia had its fourth-warmest year since national records began in 1910.

6. Some significant weather events, 2020

Q Can you see any patterns in the nature of the extreme weather events that are happening across the globe in 2020?

3. Desertification

'Desertification' is the process which turns fertile land into desert.

Global warming can cause air temperatures to increase and rainfall to decrease. This results in drought (lack of water) and inhibits the growth of plants. The desert regions then spread, with many areas at risk of being overcome by them.

The loss of fertile land means many agricultural communities struggle to feed themselves, while others may have to migrate (move) to other regions to find work.

16. Climate Change

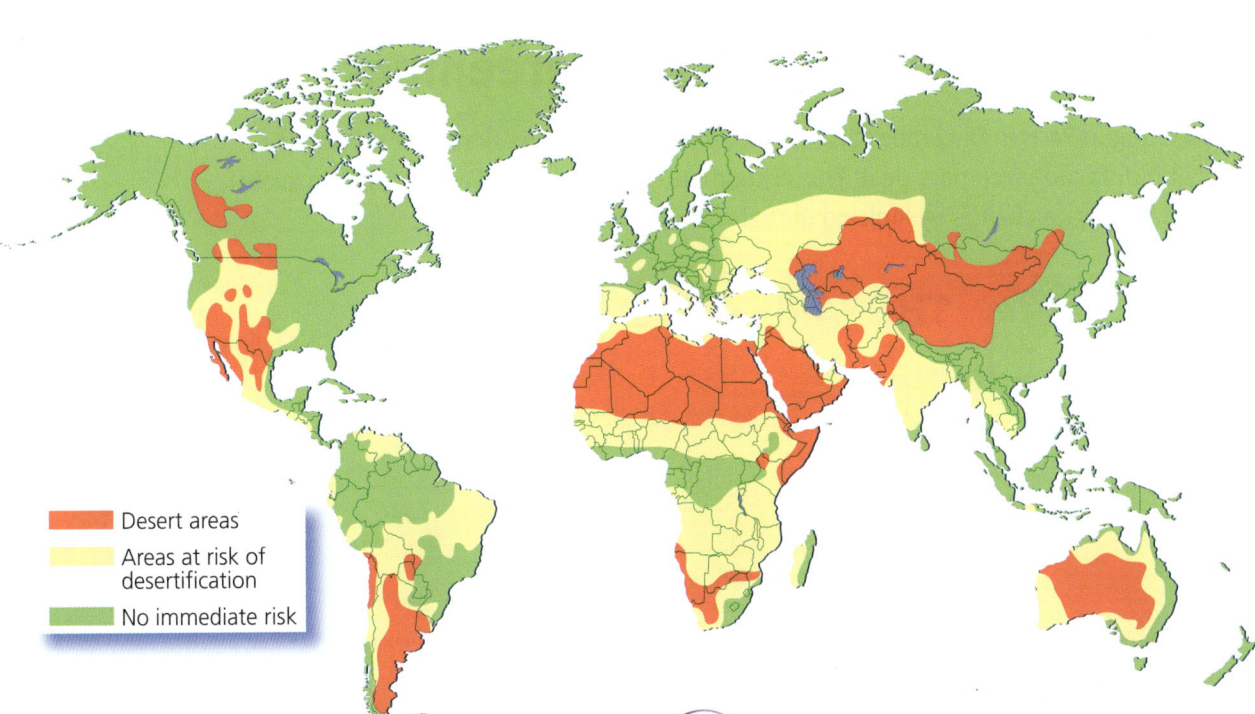

- Desert areas
- Areas at risk of desertification
- No immediate risk

7. Desertification around the globe

Managing information and thinking

Using an atlas/Google Maps and figure 7, list two countries that contain:

1. Desert areas
2. Areas at risk of desertification

4. Threats to Wildlife

Global warming is a real threat to many species of animal. It is estimated that an average rise in global temperature of 1.5 °C could put 30% of species at risk of extinction.

Rising seas and stormy weather will affect sea turtles by eroding or destroying many of the beaches where they lay their eggs.

The Arctic is warming about twice as fast as the global average, causing the ice that polar bears depend on to melt away.

Question Time

Select any three implications of climate change from the list below and describe what is likely to happen in each situation:

1. Melting ice and rising sea levels
2. Extreme weather events
3. Desertification
4. Threats to wildlife

16.5 Solutions: A Climate Agreement

At a meeting in Paris in December 2015, 195 countries signed the first-ever worldwide, legally binding global climate deal. This agreement set out a global action plan to avoid dangerous climate change by limiting global warming to well below 2 °C.

THE PARIS CLIMATE AGREEMENT

TEMPERATURE 2100
- Keep warming 'well below' 2 °C
- Continue all efforts to limit the rise in temperatures to 1.5 °C

EMISSIONS OBJECTIVES 2050
- Aim for greenhouse gas emissions to peak 'as soon as possible'

BURDEN-SHARING
- Wealthier, developed countries to provide financial resources to help the poorer developing countries

REVIEW MECHANISM
- To come together every five years to monitor and set new targets with the help of the latest science available
- Report to each other and the public on how well they are doing to put their targets into action

The UN Climate Change Conference (COP27) was held in Sharm El Sheikh, Egypt, in 2022. At this conference, 120 world leaders came together to agree to continue to work towards the Paris Climate targets. Some important decisions were taken:

- Agreement to provide 'loss and damage' funding for vulnerable countries hit hard by climate disasters
- Reaffirmed commitment to limit global temperature rise to 1.5 °C above pre-industrial levels
- The launch of a new five-year programme to promote climate technology solutions in developing countries
- Young people were given a greater voice at COP27 than had occurred in previous summits. The first ever youth-led climate forum took place to hear the voices, suggestions and solutions on climate issues.
- The announcement of a US$3.1 billion plan to ensure everyone on the planet is protected by early warning systems within the next five years.

Go to the YouTube and search for 'Ever wondered: What is the Paris Agreement, and how does it work?' to watch a video about this agreement.

Being literate

> 'If working apart we are a force powerful enough to destabilise our planet, surely working together, we are powerful enough to save it.'
>
> Sir David Attenborough speaking at COP 26, November 2021

Use the above quote to inspire you to write a speech to make at a school assembly on global warming and the greenhouse effect:

What have we done to cause climate change and what can we do to save our planet?

16. Climate Change

EXAM FOCUS

Sample questions

1. The following table has been sourced from data shared by NASA.

Carbon Dioxide	Global Temperature	Arctic Ice	Ice Sheets	Sea Level
Carbon dioxide levels in the air are at their highest in 650,000 years	Nineteen of the 20 warmest years on record have occurred since 2001	In 2012, Arctic summer ice shrank to the lowest extent on record	Satellite data show that earth's polar ice sheets are losing mass	Global average sea level has risen nearly 178 mm over the past 100 years

(a) **Choose** TWO of the following and **outline** possible causes for the change in record over time.

Carbon Dioxide   Global Temperature
Arctic Ice   Ice Sheets   Sea Level

Exam hints!

This is a two-part question. You must start by naming the two areas that you have selected to answer on. They have been listed for you. Write about each area one at a time and outline possible causes for why there has been a change in that area over time. A sample answer has been started for you below. Complete the answer in your copy.

Let's get started! Sample starter

> I have chosen:
> 1.   Carbon dioxide level   and
> 2.   Carbon dioxide levels: These have risen to record high levels. During the last century, human activities such as the burning of fossil fuels (e.g. coal, oil and gas) have increased the concentration of CO_2 in the atmosphere. In many areas, deforestation to change the land use in certain regions has also had a negative impact. When forests are cleared, large amounts of stored CO_2 greenhouse gases are released, which increases carbon dioxide levels.

... now complete the answer in your copy.

(b) **What** are the consequences of climate change in your **local area**?

(c) In **what** ways could it be possible for people to increase or add to climate change?

(NCCA assessment items, Sample 8 Q4 (a)–(c))

Your local area can refer to your town, county or country. Try to use local examples from the area in which you live if possible, as you may be most familiar with these.

245

2. Read the article below and answer each of the following questions.

Humanitarian Crisis in Somalia Driven by Climate Change

Somalia has a hot desert climate with high temperatures and little rainfall. This makes any rain that falls very important. Climate change is leading to less rainfall in certain parts of the world, including Somalia. Between 2016 and 2018, Somalia experienced below average rainfall. This led to many problems for farmers including crop failure, widespread livestock deaths and loss of assets.

As well as farmers, the wider population of Somalia was affected as there was large scale population displacement, hunger and malnutrition. By 2018 the number of people in need of urgent humanitarian assistance was estimated at 4.2 million.

(i) Between what years did Somalia experience below average rainfall?

(ii) **Name** two effects of this below average rainfall on Somalia.

(iii) Study the two climate graphs below which show the monthly precipitation and average monthly temperature for two different locations. Which climate graph shows the monthly precipitation and average monthly temperature for Somalia?

Exam hints!

For reading comprehension pieces such as this, you would expect to find the answers to the initial questions within the article. Use a highlighter to highlight the key points asked in the questions.

This question is asking you to draw on your knowledge across the topics of climate, weather and climate change. What do you know about precipitation levels in Somalia already? Use information given to you in the article if you are unsure. Then look at the climate graphs carefully to see what is shown on each axis, and use the key provided.

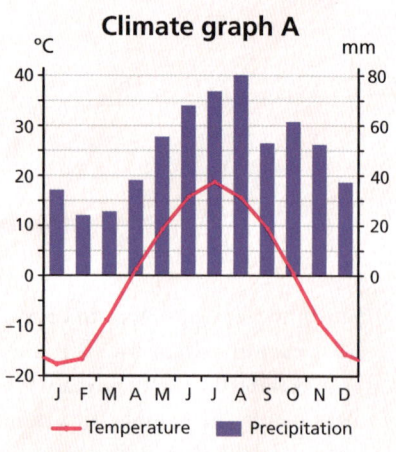

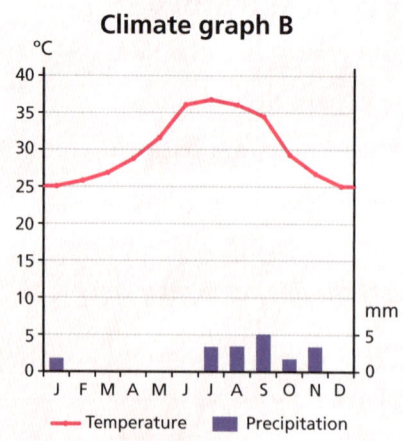

(SEC sample paper, Q5 (c) (i)–(iii))

16. Climate Change

CBA 1 Geography in the News

Devastating Impact of Australian Bush Fires

Nearly 3 billion animals were killed or displaced by Australia's unprecedented 2019–20 wildfires in 'one of the worst wildlife disasters in modern history', according to a newly released report.

The fires ravaged more than 115,000 square kilometres of drought-stricken bushland and forest across Australia in late 2019 and early 2020, killing more than 30 people and destroying thousands of homes.

Scientists say global warming is lengthening Australia's summers and making them increasingly dangerous, with shorter winters making it more difficult to carry out bushfire prevention work.

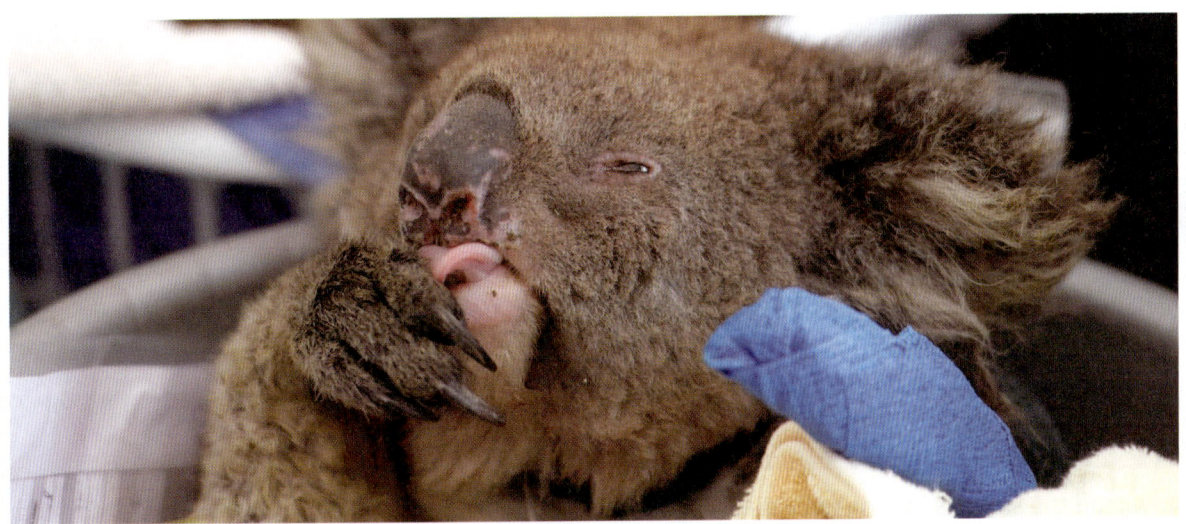

1. **When** did the Australian bushfires mentioned in the article occur?
2. **State** one environmental and one social impact of the bushfires.
3. **How** has global warming had an impact on Australian bushfires, according to the article?
4. You must **write** a report on this significant geographical event. Some suggested articles are provided below, which are reliable sources of further information. You may also conduct your own research. Your report must fulfil the success criteria on the next page.

Go online and search for:
- Almost 3 billion animals affected by Australian bushfires, report shows – Guardian
- The size of Australia's bushfire crisis captured in five big numbers – ABC news
- Climate 'overwhelming' driver of Australian bushfires – study – RTE news

247

Junior Cycle Geography CYCLONE

I Must

- Select to present my report in one of the following ways:
  - Poster
  - PowerPoint/Keynote presentation
  - News video recording
  - Radio/podcast recording
  - Typed/handwritten
- Include information on:
  - Where the bushfires took place
  - When the event occurred
  - What the cause of the event was
  - Who was impacted by the event
  - How people responded to the event
- Include suitable images.

I Should

- outline any sustainability concerns relating to the bushfires.

I Could

- Draw some conclusions on this event and suggest suitable measures to help limit the impact into the future.

Self-Assessment

Review your piece of work and write down two things that you feel you have done well and one thing that you think you can improve on.

Redrafting

Now read the success criteria once again to make sure you have met all requirements. Take into account your self-assessment notes and make any changes you think are needed. When you are happy, prepare to perform/present your piece of work to the class.

Learning Outcome: 1.4

17

Soils

Key words

- Fertile
- Mineral particles
- Water
- Permeability
- Air
- Organic matter
- Micro-organisms
- Plant litter
- Humus
- Soil profile
- Horizon
- Topsoil
- Subsoil
- Bedrock
- Leaching
- Percolates
- Hardpan
- Humification

Learning Intentions

You will be able to:
- Explain what soil is
- Describe what soil is made of and how it is formed
- Explain how leaching and plant litter impact soil
- Discuss four different types of soils in Ireland
- Study soil in your local area in relation to its composition and vegetation.

17.1 What Is Soil?

Soil is the thin, uppermost layer of the earth. Without soil, plants could not grow and there would be no food for land animals. Therefore, soil is an essential natural resource (something created by nature and a useful resource to humans). Soil uses include:

- **Agriculture**: Soil is **fertile** and has the vital nutrients needed to support plant growth for human and animal needs.
- **Building**: Soil can be used to build. A lot of building materials (bricks, etc.) can be made from soil. It is also mixed with cement to offer support since it is compact and can hold material.
- **Storing harmful greenhouse gas emissions**: Plants and soils together absorb about 30% of the CO_2 emitted by human activities each year. This will be explained further in Chapter 24.

Junior Cycle Geography CYCLONE

> **Working with others**
>
> With the person sitting next to you, suggest two other possible uses of soil.

The Composition of Soil

Soil is composed of five main ingredients. These ingredients work together to make the soil fertile.

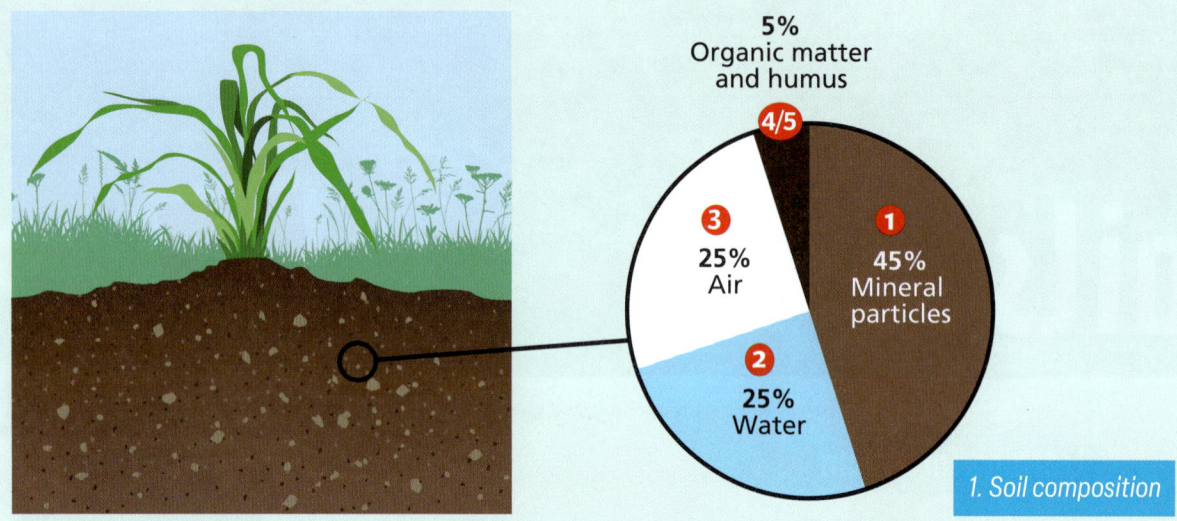

1. Soil composition

① **Mineral particles** of sand, silt and clay make up most of the soil. These particles vary in size, with clay being the smallest and sand the largest. Smaller-sized particles pack more closely together and slow down the flow of **water** through the soil. The composition of a soil can affect the ability of water to flow through the soil. The speed at which water moves through soil or rock is referred to as its **permeability**.

② Water binds (sticks) the soil together. It also dissolves any minerals and nutrients in the soil, so plants can soak them up in liquid form through their roots.

③ **Air** is found in the pores (spaces) between the soil particles. It provides the necessary oxygen and nitrogen for plants, and oxygen for the living organisms found in soil.

④ **Organic matter** is made of any living thing, such as earthworms and beetles, as well as **micro-organisms** (bacteria only visible with a microscope) found in the soil. As worms move through the soil, they also help air and water to circulate through it.

⑤ Plant remains, or **plant litter**, such as leaves and twigs, turn into **humus** when they decay in the soil. Humus is a dark, gel-like material that increases soil fertility (ability of soil to grow plants). Dead animals also form humus. Organisms such as beetles and earthworms also aid in the formation of humus as they eat the plant litter or dead animals and then excrete them, which fertilises the soil. The darker the soil, the more humus it has and the more fertile it is.

FUN FACT! There are more micro-organisms in a handful of soil than there are people on earth.

Being numerate

In your copy, use the information provided in figure 1 to create a bar chart labelled 'What soil is made of'.

Question Time

1. Define soil.
2. Explain two uses of soil.
3. Name and give a short description of each of the ingredients of soil.

17.2 How Soil Is Formed

A number of factors contribute to the formation of soil. Factors such as climate, topography (the physical features of an area of land), parent material, organisms, time and human activity all play a key role in the formation of soil. Slight changes to any of these factors will result in different types of soil forming.

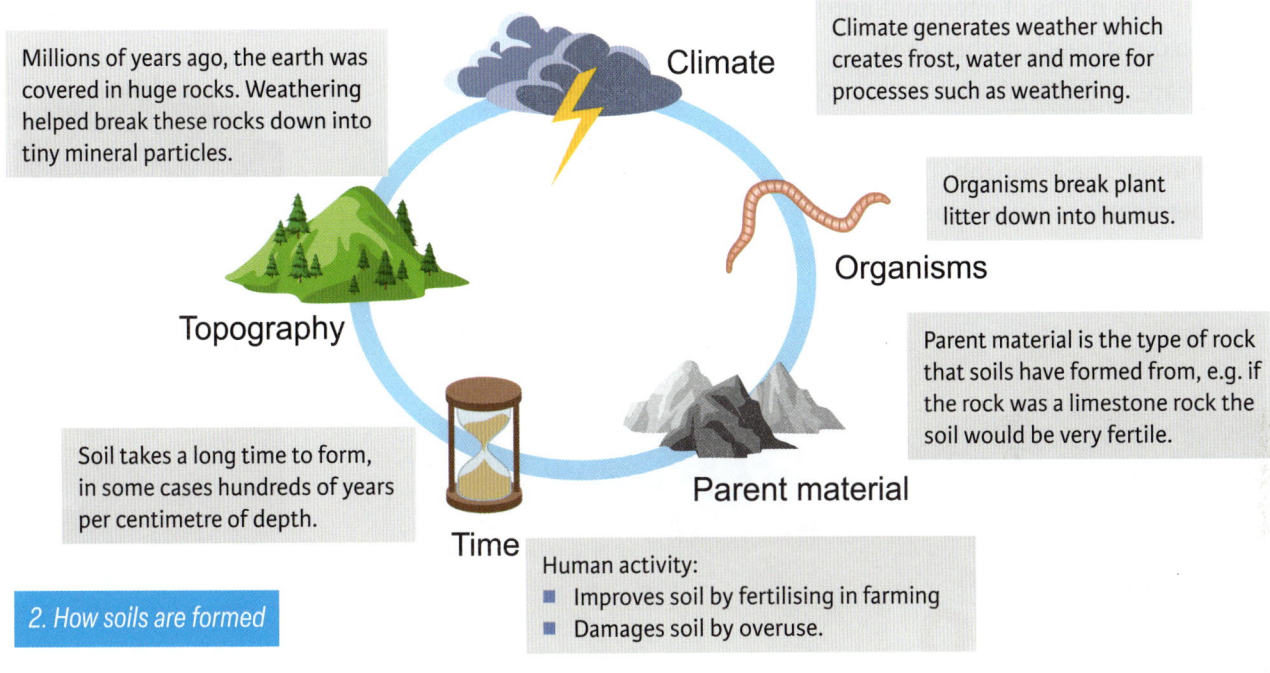

Millions of years ago, the earth was covered in huge rocks. Weathering helped break these rocks down into tiny mineral particles.

Climate generates weather which creates frost, water and more for processes such as weathering.

Organisms break plant litter down into humus.

Parent material is the type of rock that soils have formed from, e.g. if the rock was a limestone rock the soil would be very fertile.

Soil takes a long time to form, in some cases hundreds of years per centimetre of depth.

Human activity:
- Improves soil by fertilising in farming
- Damages soil by overuse.

2. How soils are formed

17.3 Soil Profiles

A **soil profile** is a vertical section of soil from its surface downwards. Soils have layers called **horizons**. Each layer is different from the one above and below it.

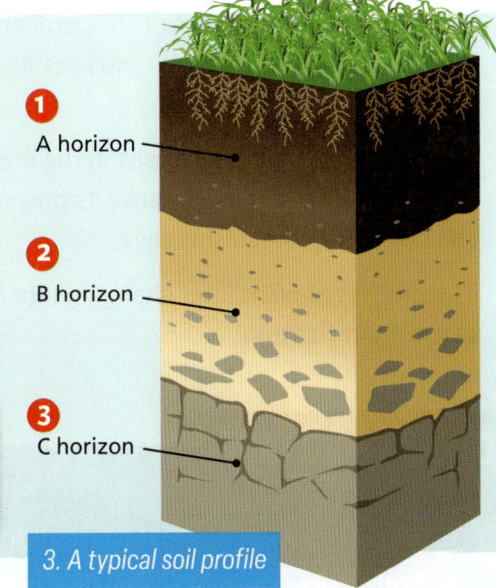

① The A horizon is sometimes called the **topsoil**. It is the closest to the surface and most fertile layer of soil. It is usually dark in colour because it contains the most humus and plant litter.

② The B horizon is called the **subsoil**. It is found just below the topsoil, has less humus and is lighter in colour than the topsoil. It may contain more stones than the topsoil as it is closer to the **bedrock** (C horizon).

③ The C horizon is also called the bedrock. It is a mixture of solid rock at its base and broken rock at its top.

Q Draw your own labelled diagram to describe the layers in a soil profile.

3. A typical soil profile

251

Two Impacts on Soil Formation

1. Leaching

Leaching occurs when there is a high amount of rainfall. The water **percolates** (soaks) down through the soil and washes important minerals and nutrients down into the B horizon.

Severe leaching is very bad for soil.

- It deprives plants of important nutrients and minerals when it washes them out of reach of their roots.
- It creates a **hardpan** when iron oxide minerals are washed from topsoil particles. These minerals then collect at the bottom of the A horizon just above the B horizon. This turns into a thin crusty layer that is impermeable to water (does not allow water to flow through it). This causes the A horizon to become waterlogged and flooded.

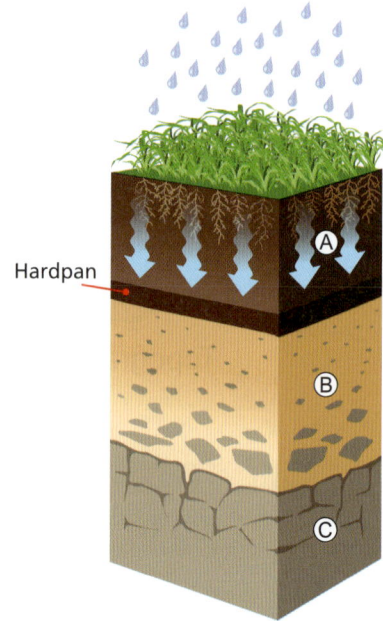

4. The impact of severe leaching on soil

Q Can you identify the A horizon, the hardpan and the B horizon in this photo?

2. The Breakdown of Plant Litter/Humification

Plant litter is the dead leaves and twigs that have fallen on top of the A horizon. It is broken down by insects, such as earthworms and beetles, and micro-organisms into humus.

Oxygen is also needed to assist in the breakdown of plant litter into humus, because its presence in the soil helps keep organisms alive.

The process of plant litter changing into humus is called **humification**.

Question Time

1. What is a hardpan? What causes it to form?
2. Explain how the breakdown of plant litter helps to make soil fertile.

17.4 Ireland's Soil Types

The amount of humus present in soil and the movement of water through soil combine to create many different types of soil.

Examining figure 5, what is the most common soil type in Ireland? Name one county that has each of the soil types indicated (use an atlas or Google Maps if necessary). What is the main soil type(s) in your own county?

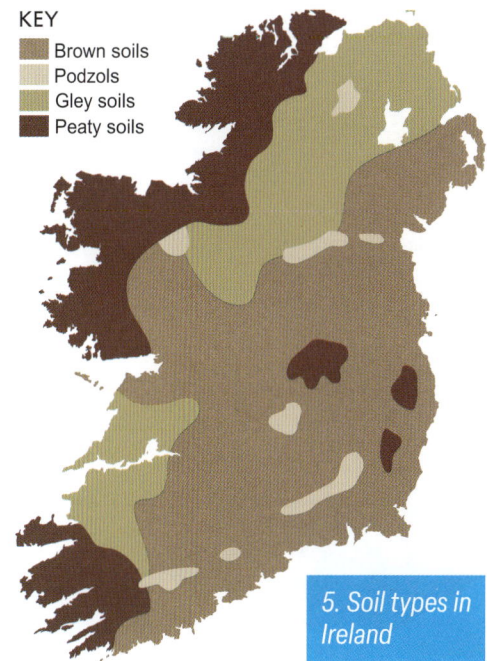

5. Soil types in Ireland

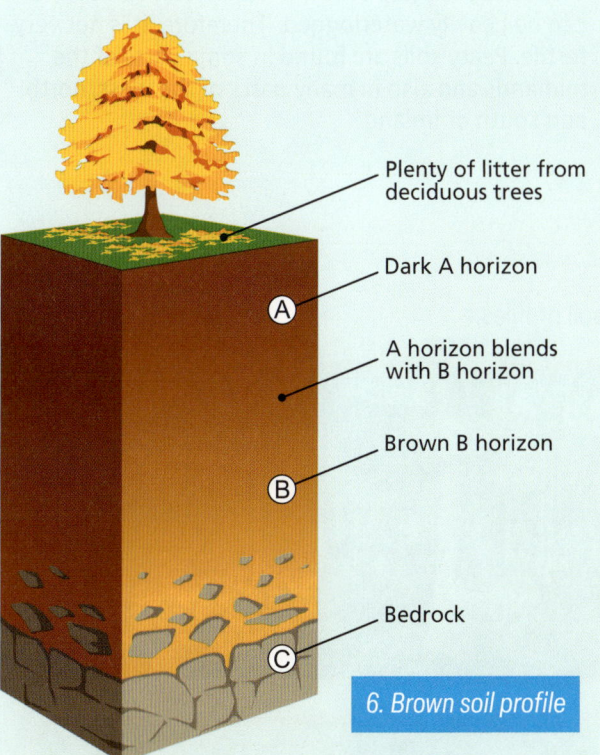

6. Brown soil profile

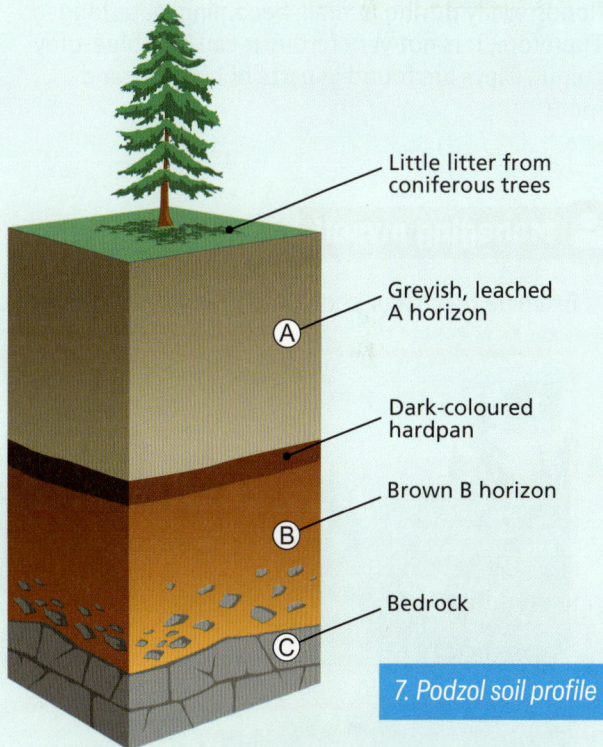

7. Podzol soil profile

Brown soil is Ireland's most common soil type, meaning that most areas are suitable for farming. As the name suggests, this soil is brown in colour. It is found in areas where rainfall is limited. It is also found mainly in areas of deciduous (trees that shed their leaves) forest. This soil is very fertile and excellent for farming. Brown soil is found in the midlands and in the southeast.

Podzol soil is found more in highland areas of Ireland where there is more rainfall. The presence of a hardpan causes the A horizon to turn grey. Podzol soil is found in areas of coniferous (trees that do not shed their leaves) forest. This means there is little plant litter on the forest floor to turn into humus. As a result, the soils are less fertile. They can be found in Clare, Derry and Antrim.

Junior Cycle Geography CYCLONE

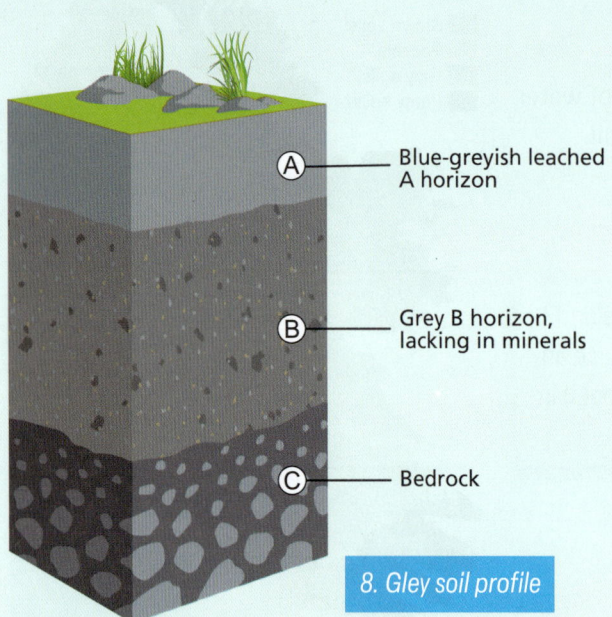

- A — Blue-greyish leached A horizon
- B — Grey B horizon, lacking in minerals
- C — Bedrock

8. Gley soil profile

- A — Dark brown/black flooded A horizon
- B — Light/dark brown saturated B horizon
- C — Saturated C horizon

9. Peaty soil profile

Gley soil forms on areas of rolling lowland or gently sloping hills. It is not deep and because of this floods easily during rainfall, becoming waterlogged. Therefore, it is not very fertile. It can be a blue-grey colour. Gleys are found in parts of Limerick and Clare.

Peaty soil is also called bogland. This soil takes thousands of years to form. It is dark in colour and can be heavily waterlogged. Therefore, it is not very fertile. Peaty soils are found in some parts of the midlands and also in many parts of the west, north and south of Ireland.

Managing myself

Examine these four photographs and match them to their soil profiles.

1

2

3

4

17. Soils

EXAM FOCUS

Sample questions

1. Answer these short questions:

   (i) Use the terms in the box below to complete the definition and description of soil in your copy.

water 25%	natural resource	horizons
A horizon	thin uppermost layer	organic matter 5%

   Soil is the _____ of the earth. Without soil, plants could not grow and there would be no food for land animals. Therefore, soil is an essential _____ _____ (created by nature and a useful resource to humans).

   Soil has four main ingredients, which are mineral particles 45%, _____, air 25% and _____. Soils have layers called _____. The _____ is the most fertile layer.

   (ii) List three uses of soil and briefly explain two of these uses.

2. Copy this incomplete soil profile.

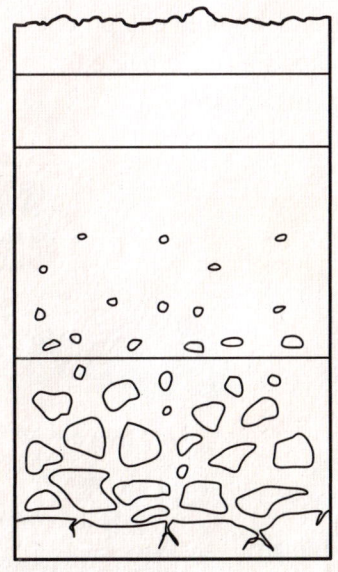

Exam hint!

When sketching, use a ruler for straight lines for neatness. Make your sketch big enough to be clear and easy to label. Label all parts of the sketch needed. Include necessary details only. Use shading or colours if you have time.

   (i) **Label** each horizon.
   (ii) Use this diagram to **explain** how leaching occurs.
   (iii) **Explain** two impacts leaching can have on soil.

3. (i) **Name** one Irish soil type that you have studied.
   (ii) **Give** two examples of where this soil type is most common.
   (iii) **Describe** the structure/composition of the soil you have chosen under the following headings: Colour, Fertility, Leaching (present or not), Uses

Junior Cycle Geography CYCLONE

Let's get started! Sample starter

(i) The soil sample I have chosen is a peaty soil.
(ii) Peaty soils are found in some parts of the midlands and also in many parts of the west, north and south of Ireland.
(iii) ...

... now complete the answer in your copy.

Exam hint!

This sample answer is a guide to answering a question like Q3. Whichever soil you choose when answering part (iii), include any characteristics which make it different from other soils, such as colour, how it is formed, how fertile it is and what vegetation grows on it.
If including a diagram, draw it neatly, label it correctly and, if time allows, shade it using colours.

CBA 2 My Geography Moment

The Soil in My Locality

Identify the soil types in your locality. Instructions and required equipment are given for the five tasks listed. Sample results maps are also provided. Record results in your copies. Present your findings on a poster.

This is a group task.

SUCCESS CRITERIA

We Must

- Extract soil samples from three different sites.
- Carry out Tasks 1 and 2, plus one other task, to determine the soil type of our samples.
  - **Task 1**: Examine the vegetation of our three chosen sites.
  - **Task 2**: Examine the texture of the three soil samples.
  - **Task 3**: Examine the humus content of one soil sample.
  - **Task 4**: Measure the moisture and air content of one soil sample.
  - **Task 5**: Measure the permeability of one soil sample.
- Create a poster describing and naming the type(s) of soil in our area.

We Should

- Carry out one further experiment on each of the three samples to help determine soil type.

We Could

- Taking precautions and under supervision, extract a soil sample from a peatland/bog area and study it using Tasks 3, 4 or 5.

Choosing Suitable Sites and Extracting Samples

First you need to choose three sites from which to extract soil samples.

Choosing your sites

- Study the weather in the days leading up to taking your soil samples. Avoid taking your samples after long periods of rain, as the soil may be flooded, which will give you a false profile. To this end, download a weather app on your phone and monitor the weather for the week ahead.
- Choose sites that have grass or plant cover.
- Identify three contrasting soil profile sites from which to take your samples. Site A could be a garden, Site B could be a forest and Site C could be a coastal/river area.
- Use a measuring tape or metre stick to map out a 2 m square area in each of these three sites from which to take your samples.

To find more information on soil surveys in Ireland, visit gis.teagasc.ie/soils.

You will need:

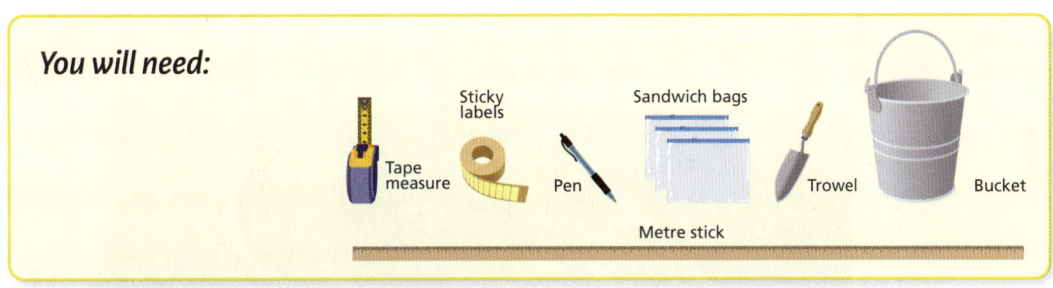

Tape measure, Sticky labels, Pen, Sandwich bags, Trowel, Bucket, Metre stick

Extracting samples

1. Insert trowel to desired depth.

2. Re-insert trowel next to first opening to create a slice.

3. Remove trowel with soil slice intact.

4. Cut a 2.5 cm-wide strip from centre of the soil slice.

5. Remove grass and add the soil to your bucket.

6. Return remaining soil to the opening.

7. Gather five random soil samples in total from the specific site using the same process.

8. Mix the samples and fill the sandwich bag.

- Repeat this process for each of the three sites identified: extract and mix five soil samples from each specific area and place each in a sandwich bag.
- Label the sandwich bags A, B and C.

You need five soil samples to carry out each of the tasks below.

TASK 1

Examine the vegetation of your three chosen sites

You will need:

Quadrat

Pen
Paper

Instructions

1. A quadrat is a square frame which is placed on the ground to look at or count the number of plants or animals within it. Place the quadrat in a number of random areas inside your mapped-out 2 m square.
2. Make a note of the vegetation found in each quadrat, e.g. grass, leaves, plant life.
3. Record your findings.
4. Compile your results to see the main vegetation types for your study sites. (Refer to the results map, p. 261.)
5. Take photographs of the chosen sites.

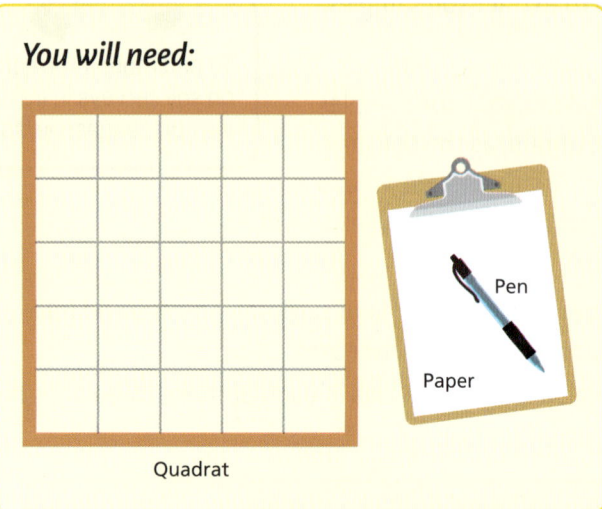

TASK 2

Examine the texture of the three soil samples

Instructions

Take a small piece of each of the samples in turn and carry out the three steps.

You will need:

Small water bottle

1. One by one, wet each of the three samples with a small amount of water and rub the wet sample between your fingers.

2. Record your observations. (Refer to the results map, p. 261.)
3. Take photographs of the wet samples.

TASK 3 — Examine the humus content of one soil sample

You will need:

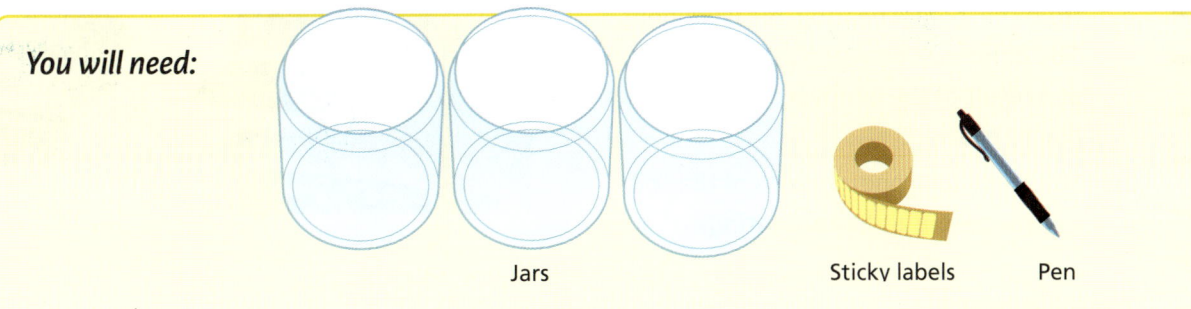

Jars   Sticky labels   Pen

Instructions

1. Choose one soil sample, A, B or C.
2. Place 10 cm of the soil into a clear jar and fill to the top with water. Put a lid on the jar.
3. Shake the jar well and leave to settle for 24 hours.
4. After this period, the soil will have settled so that the heaviest material is at the bottom and lighter material is at the top.
5. The lightest material (humus) will be on top.
6. Measure this material to discover humus content.
7. Record your measurements and draw a diagram/take a photograph of your soil jar. (Refer to the results map, p. 262.)

Shaken and settled
Soil sample
Humus
Clay
Silt
Sand

Cloudy water in the top of the jar indicates plenty of clay

TASK 4 — Measure the moisture and air content of one soil sample

You will need:

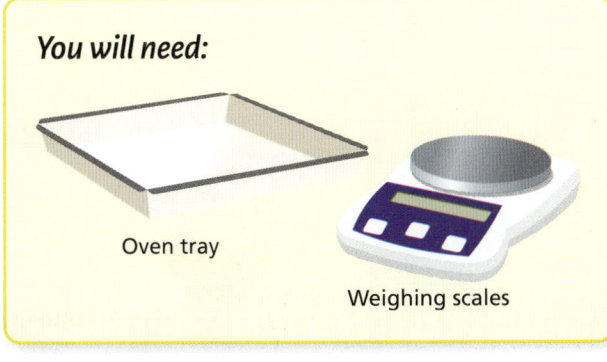

Oven tray   Weighing scales

Instructions

1. Choose one soil sample, A, B or C.
2. Weigh your soil sample and record the result. (Refer to the results map, p. 262.)
3. Place the soil sample on a tray and put it into an oven at a temperature of about 110 °C for 10 minutes. (If you do not have access to an oven, place the sample near a radiator for a few hours to give it time to dry out.)
4. Remove the sample from oven and weigh it again.

5. Record the new weight.
6. Subtract the 'after heating' measurement from the 'before heating' measurement. The difference between the weights is as a result of the removal of moisture/air due to heating.
7. You could take photographs of the soil sample before and after heating.

TASK 5 — Measure the permeability of one soil sample

You will need:

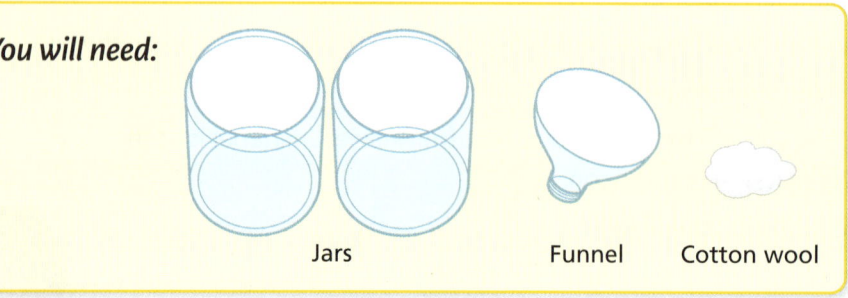

Jars   Funnel   Cotton wool

Instructions

1. Choose one soil sample, A, B or C.
2. Plug the funnel (which can be the top of an empty 2-litre drinks bottle) with cotton wool, place it into an empty jar and place the sample soil in the funnel.
3. Fill another jar with 250 ml of water, and then slowly pour this water onto the soil in the funnel.
4. Using a stopwatch, time how long it takes the water to pass through the soil into the jar.
5. Does all of the water pass through the soil?
6. Record your observations. (Refer to the results map, p. 263.)

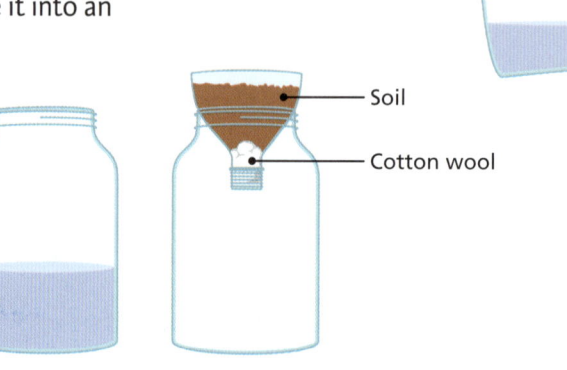

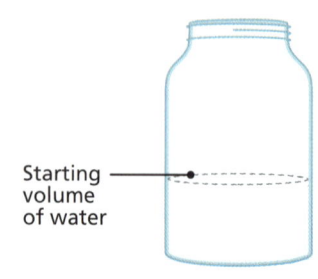
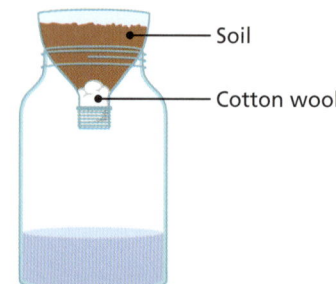

Results Maps

Task 1: Examine the vegetation of your three chosen sites

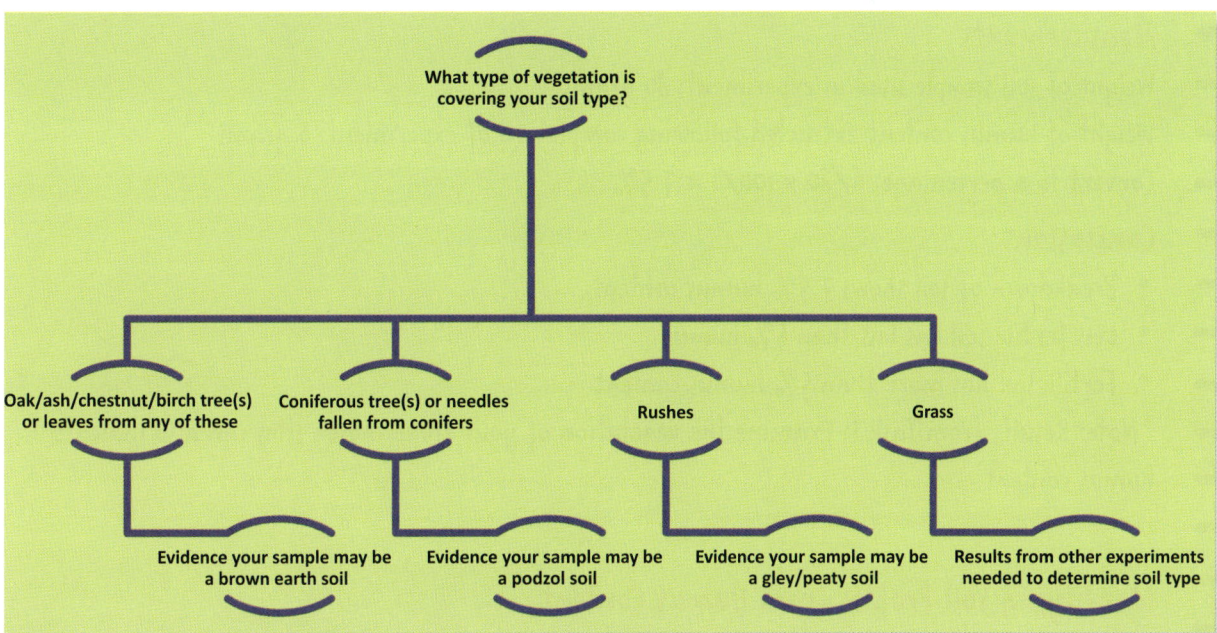

Task 2: Examine the texture of the three soil samples

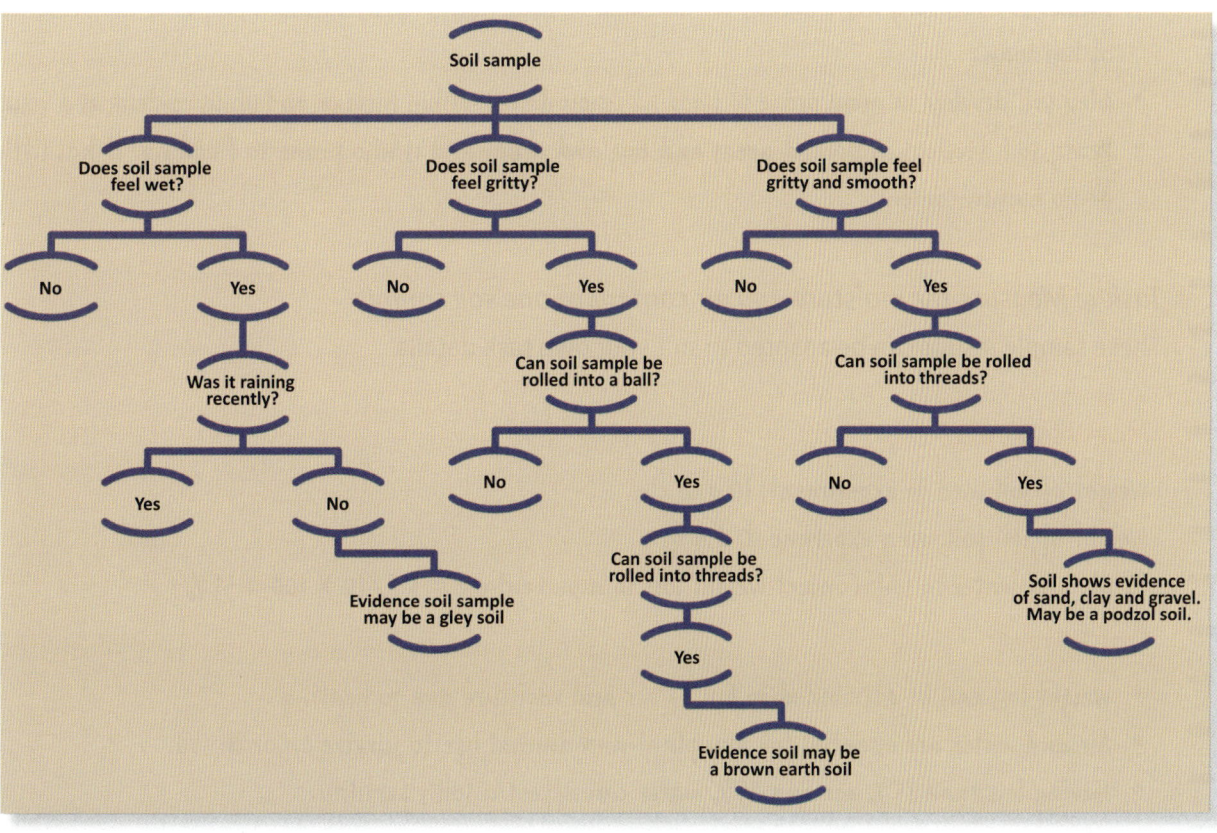

Task 3: Examine the humus content of one soil sample

These sample results can be adapted to suit your own task details.

Measurements
Weight of soil sample used in experiment: 80 grams
Weight of humus content extracted following completion of experiment: 6 grams
Convert to a percentage: $6/80 \times 100/1 = 7.5\%$

Conclusions
- Breakdown of soil shows 7.5% humus content.
- Less fertile soil has less than 5% humus.
- Fertile soil has more than 5% humus content.

*Note: Results from Task 1: Examine the vegetation of your three chosen sites can also indicate humus content.

Identifying Soil Profiles from Humus Content
- **Brown soil** is rich in humus content due to presence of vegetation; humus content gives this soil its brown colour.
- **Podzol soil** has very little humus content due to such characteristics as leaching and the presence of hardpan.
- **Gley soil** develops in areas prone to flooding (mountains). It has little or no humus content as a result.
- **Peaty soil** develops in upland areas and bogland areas and is also prone to flooding. It has little or no humus content.

Task 4: Measure the moisture and air content of one soil sample

These sample results can be adapted to suit your own task details.

Measurements
Weight of soil used in experiment: 10 g
Weight of soil following experiment: 6.5 g
% air and water in soil: Convert weight loss to a percentage = $3.5/10 \times 100 = 35\%$

Conclusions
- Weight loss can be attributed to loss of air and moisture due to heating.
- Air and water are essential for any plant and animal life to survive on or in soil.
- Having less than 25% air and 25% water can affect a soil's fertility.

Identifying Soil Profiles from Moisture and Air Content

- **Brown soil** has high air content and high water permeability. It is relatively fertile.
- **Podzol soil** has low air content (leaching causing hardpan), high water permeability and low fertility.
- **Gley soil** has low air content, low water permeability (water table usually present) and low fertility. The water table is the level beneath which the ground is filled with water.
- **Peaty soil** has low air content, low water permeability and low fertility.

Task 5: Measure the permeability of one soil sample

Amount of water that goes through the soil (ml)	Length of time (seconds/mins)

Identifying Soil Profiles from Soil Permeability

- **Brown soil** generally has good soil structure with adequate space for air and water. Water would flow through quite easily. It has good fertility.
- **Podzol soil:** Water will also flow through this type of soil quite easily, but it has poor fertility due to leaching and presence of hardpan.
- **Gley soil** may have a poor soil structure. These soils may be sticky and hard to farm, as water does not flow through them easily, which can result in flooding.
- **Peaty soil** can often be saturated, because peat is good at retaining water. This can cause peat soils to flood. Permeability and fertility are poor.

Peer Comparison

Compare the soil profiles from the experiments carried out by your group with those of another group.

- What experiments did the other group choose?
- Did the results from their experiments confirm similar soil types to yours?
- Were any results considerably different to your group's results? If so, what factors may have contributed to that (weather, choice of location for samples, etc.)?

18 Population

Learning Outcomes: 3.1, 3.3, 3.9

key words

- Demography
- Birth rate
- Death rate
- Natural increase
- Natural decrease
- Population cycle
- Demographic transition model
- Globalisation
- Interconnected
- Population pyramid
- Age–sex structure
- Expansive
- Constrictive
- Stationary

Learning Intentions

You will be able to:
- Describe how a country's population changes as it moves through each stage in the demographic transition model
- Describe the characteristics displayed by countries at each stage of the demographic transition model
- Describe how the population has changed in Ireland and in a developing country over time.

18.1 World Population

The world's population has increased over time. Population growth has occurred at uneven and fluctuating (up and down) rates. There have been times when the population has increased and times when it has decreased. In 2022, the world's population reached 8 billion, and is growing steadily.

1. The projected world population until 2100

World population
Projected world population until 2100

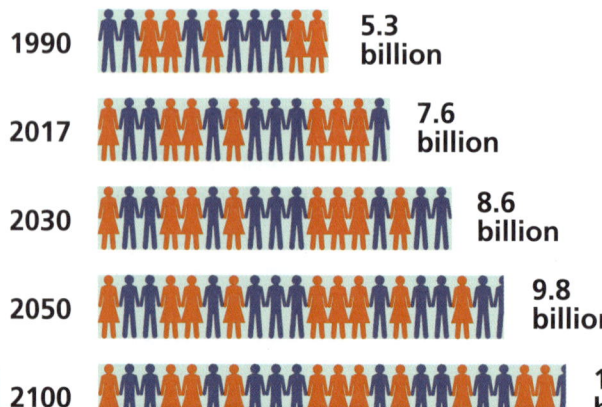

1990	5.3 billion
2017	7.6 billion
2030	8.6 billion
2050	9.8 billion
2100	11.2 billion

18. Population

 Managing information and thinking

Look at the projected world population in 2030 and compare it to 2100. What impact do you think this population change will have on our planet?

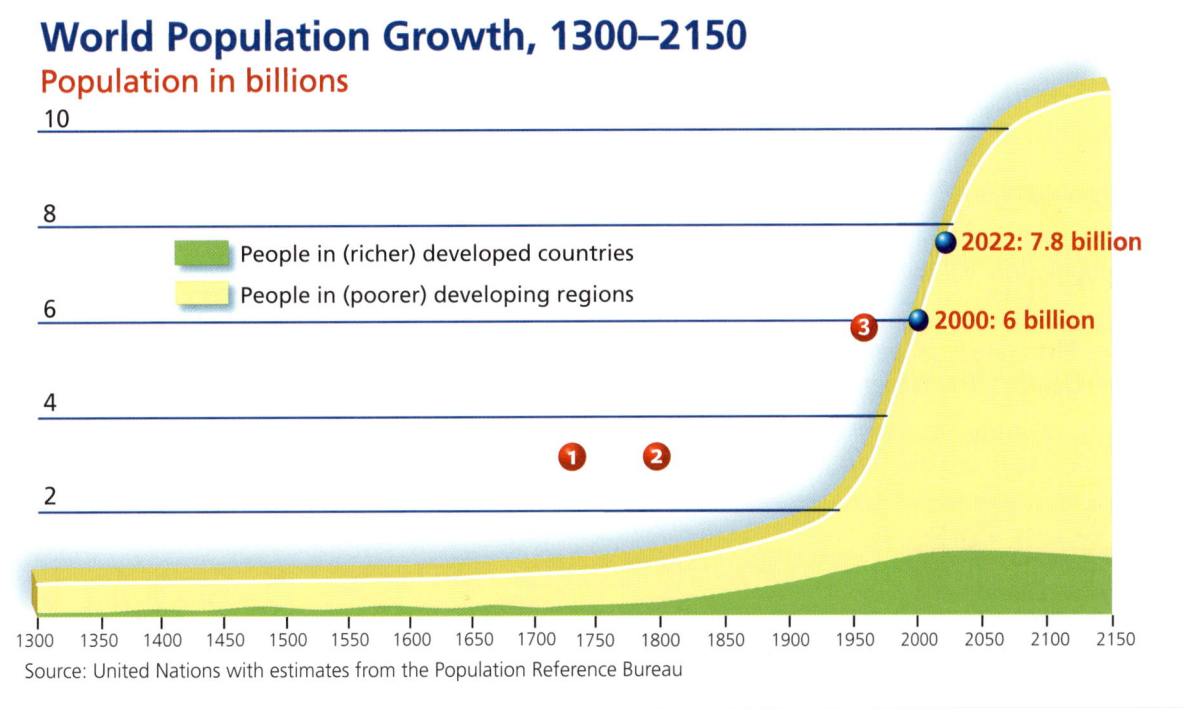

2. World population growth, 1300–2150

❶ Before 1750, the world's population was very slow to grow. This was a time of fluctuation, as population went up and down due to wars, famines and plagues, such as the Black Death.

❷ From 1750 onwards, improved medical care and a better supply of food allowed the population to grow rapidly in Europe.

❸ Over the last 100 years, population has grown rapidly. This is especially the case in developing countries. The population explosion is the result of better healthcare and food supplies generally throughout the world.

Being numerate

1. Based on current estimates, roughly what will the population be by 2050?
2. Calculate the change in population from 2000 to 2022.
3. True or false?
   (i) Our planet's population fluctuated in the period between 1300 and 1750.
   (ii) The populations of rich countries are growing much faster than those of poor countries.
   (iii) The world's population grew more in the last 100 years than in any period before that.

FUN FACT! In 1970, there was roughly half the number of people in the world as there are now.

265

18.2 Population Change and Density

The study of population is known as **demography**. We calculate world population change by measuring the **birth rate** (the number of live births per thousand of the population in one year) against the **death rate** (the number of deaths per thousand of the population in one year).

When the birth rate is higher than the death rate, the population will have a **natural increase**.

Example:
Birth rate is 25 (per thousand)
Death rate is 13 (per thousand)

Natural increase = 25 – 13
= **12 per thousand or 1.2%** (12/1,000 × 100/1 = 1.2)

When the death rate is greater than the birth rate, a **natural decrease** will occur.

Example:
Birth rate is 14 (per thousand)
Death rate is 19 (per thousand)

Natural decrease = 19 – 14
= **5 per thousand or 0.5%** (5/1000 × 100/1 = 0.5)

Being numerate

Calculate the natural increase or natural decrease for the following countries. The first one has been done for you.

Country	Birth Rate	Death Rate	Natural Increase	Natural Decrease	% Change Per Year
US	12.0	9.0	3	–	0.3%
Ireland	11.2	6.4			
Germany	9.3	11.9			
Israel	19.4	5.3			
Italy	6.8	12.6			

Population Density

Population density is the average number of people per square kilometre. It changes across different countries, regions and cities. To calculate population density, divide the total population by the total area. As of April 2022, Ireland's population was 5,123,536 and the area of Ireland is 70,273 km². The population density is calculated below:

$$\text{Population density} = \frac{\text{Total population}}{\text{Total area}} = \frac{5,123,536}{70,273} = 73 \text{ people per km}^2 \text{ (rounded up)}$$

Go to the website https://www.worldometers.info/ to investigate the global population. Make a note of:
1. The world population today
2. The number of births today and so far this year
3. The number of deaths today and so far this year
4. Population growth today and so far this year

18. Population

Question Time

1. Why did the world's population fluctuate before the 1700s?
2. What caused the rapid growth in world population from 1900 onward?
3. Define each of the following terms:
   (a) Birth rate (b) Death rate (c) Natural increase (d) Natural decrease (e) Population density

18.3 The Population Cycle

As the economy of a country develops, the population of that country grows very slowly, then very rapidly and finally very slowly or not at all. This is known as the **population cycle** or the **demographic transition model**.

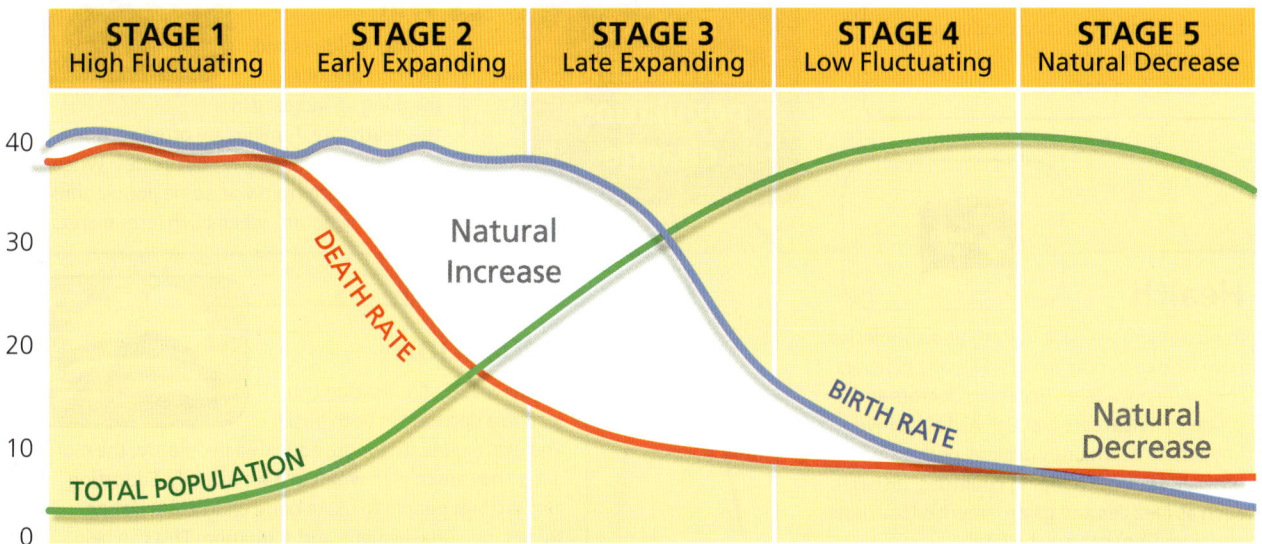

STAGE 1 High Fluctuating	STAGE 2 Early Expanding	STAGE 3 Late Expanding	STAGE 4 Low Fluctuating	STAGE 5 Natural Decrease
The birth and death rates are both very high. A country is not economically developed. Countries may be experiencing war, famine and diseases. Population can fluctuate or grow very slowly.	Birth rates remain high, but the death rates now begin to fall because of better food, medical care and supply of clean water. The economy is starting to grow. Population now grows very quickly. There is a natural increase.	People begin to plan the size of their families, so, the birth rate now begins to fall. The gap between birth and death rates gets smaller. The economy is continuing to grow. The rate of population growth slows down.	Most people are living long lives, but they are having fewer children. Both the birth and death rates are now low. The economy is now considered developed. Population now fluctuates and only increases very slowly.	Women are having very few children, and most people are living into old age. The population is now in decline. This is a natural decrease. The economy is very well developed.
Example: Currently no country is at this stage.	**Example:** Nigeria	**Example:** India	**Example:** Ireland	**Example:** Germany

3. The demographic transition model

Q

Using the demographic transition model in figure 3, answer the following questions:

1. Why do you think Stage 1 is known as the 'High Fluctuating' stage?
2. In what stage does the death rate decline most rapidly? Why is this the case?
3. Compare and contrast Stages 2 and 4. Name two differences between them.
4. What stage do you think Ireland will enter in the future? Give one reason for your answer.

18.4 Factors Influencing Population Change

There are several factors that influence population change.

1 Food

The agricultural revolution in the eighteenth and nineteenth centuries had a direct impact on population growth. Machinery and new farming methods, such as crop rotation and selective breeding, meant farmers could produce more food. The use of fertilisers improved crop production, leading to an increase in food supply and fewer famines.

2 Globalisation

Globalisation is the process by which the world is becoming more **interconnected** (linked) as a result of increased trade and movement of people. People in the poorest parts of the world are becoming more aware of the wealth and lifestyle of people in richer countries and want to move there. In developing countries, many people are being forced to move because of natural disasters and war. This can cause population decreases in some regions and increases in others.

3 War

In wars, soldiers and civilians are killed, meaning an increase in the death rate. Families are often separated, which can result in lower birth rates. When soldiers return from war, some populations see a 'baby boom', when birth rates increase.

4 Education

Higher levels of education can lead to a decrease in both birth and death rates. The more educated people are, the more likely they are to plan when to have a family. Education also enables people to make good choices in relation to diet, personal hygiene and sanitation. This can help children to grow up healthy and live for longer.

5 Place of women in society

When there is equality between men and women, both sexes have equal power in making decisions about their lives. In developed countries, many women will have careers outside of the home. In developing countries, women often marry at a young age and have larger families. As these countries become more developed, girls will receive better education and make choices about family planning and contraception. In turn, birth rates may decline.

6 Technology

Advances in medical technology can help save lives. New medicines can assist people in living longer. Advances in farm machinery have helped with increasing food supplies. Water treatment equipment has helped to provide clean and safe drinking water.

7 Health

Improvements in vaccinations and antibiotics have helped people to survive many illnesses that were once fatal. Access to doctors and proper medical care has also had a positive impact. As death rates decline, populations grow. In some developing countries, diseases such as measles and gastroenteritis (diarrhoea and vomiting) can still be fatal.

Question Time

Select any five of the seven factors listed and describe in your own words how each factor influences population change.

18.5 Population Structure

Population Pyramids

The population structure of a country can be displayed using a **population pyramid**. Population pyramids show the differences between the numbers of males and females and the number or percentage of people in each age category (known as the **age–sex structure**).

Population pyramids show:

- A series of bars stacked on top of each other. The length of the bar shows the percentage of the population in a certain age group (sometimes it can show the number of people).
- The scale bar at the bottom allows you to measure these percentages.
- The youngest age group, 0–4, is found at the base of the pyramid, with the oldest group, 100+, at the top.
- Figures for males are shown on one side of the population pyramid, with figures for females on the other.

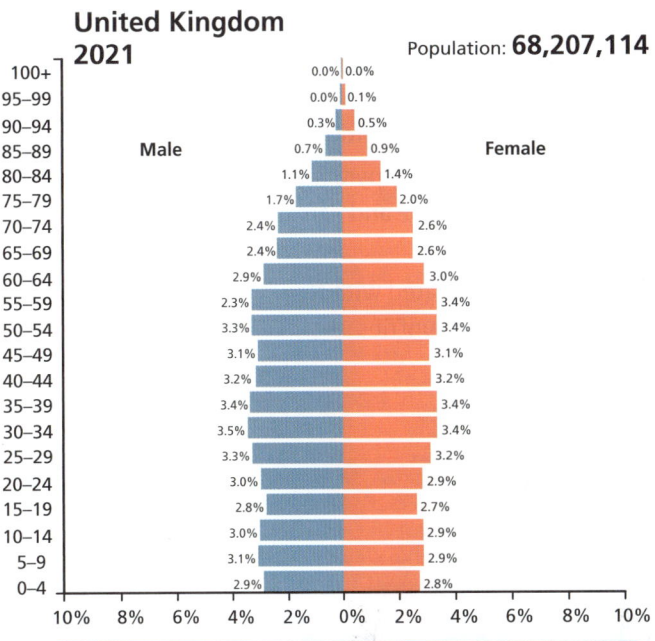

4. A population pyramid for the United Kingdom, 2021

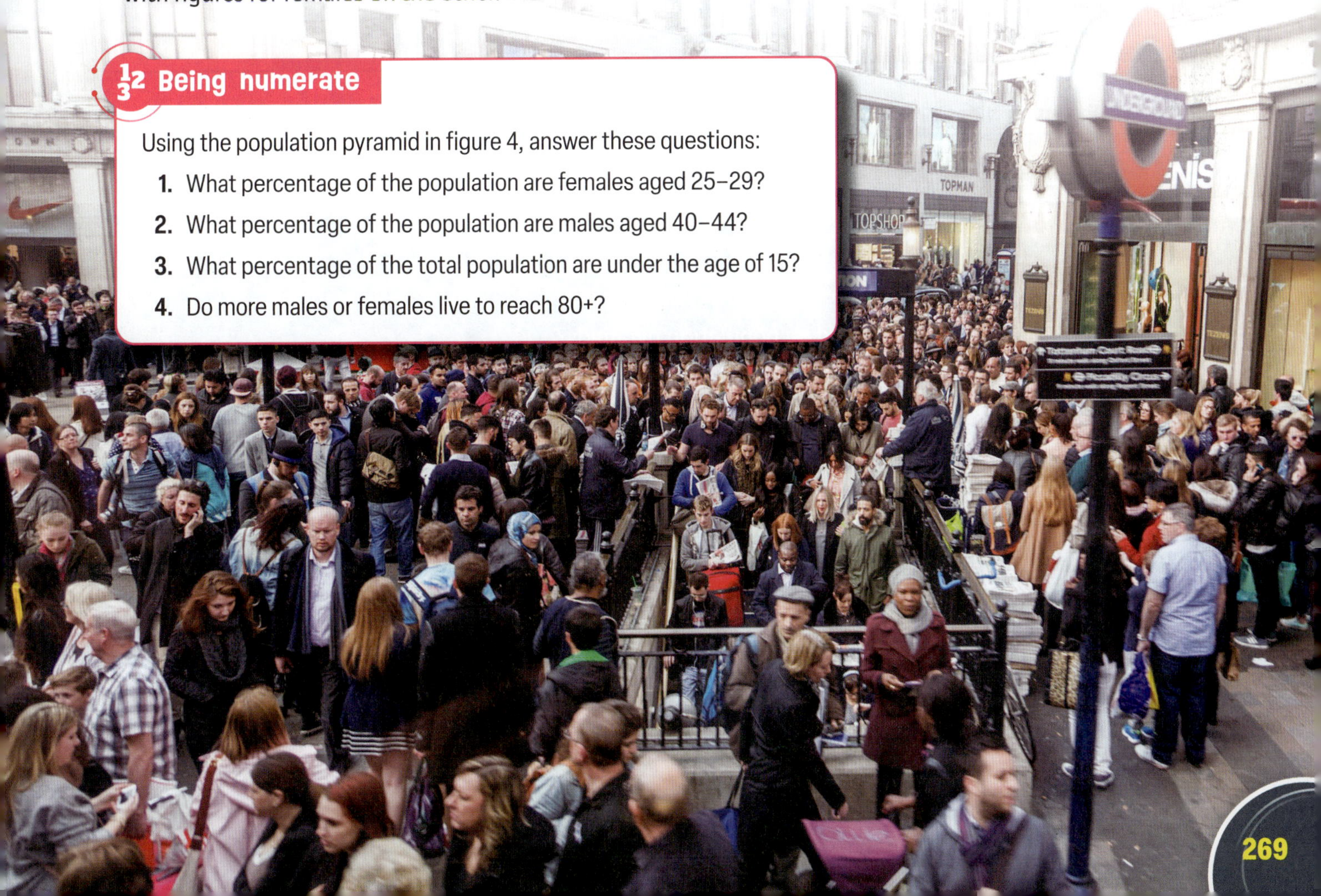

Being numerate

Using the population pyramid in figure 4, answer these questions:

1. What percentage of the population are females aged 25–29?
2. What percentage of the population are males aged 40–44?
3. What percentage of the total population are under the age of 15?
4. Do more males or females live to reach 80+?

Types of Population Pyramid

There are three types of population pyramid: expansive, stationary and constrictive.

Expansive	Stationary	Constrictive
• Used to describe populations that are young and growing • Look like a typical 'pyramid', with a broad base and narrow top • Usually associated with slowly developing countries	• Used to describe populations that are not growing • High life expectancy and low birth rate • Narrow towards the top • Usually associated with developed countries	• Used to describe populations that are becoming old and shrinking • Low birth rate • Can look like beehives • Usually associated with developed countries
Example: Nigeria	**Example:** Ireland	**Example:** Germany

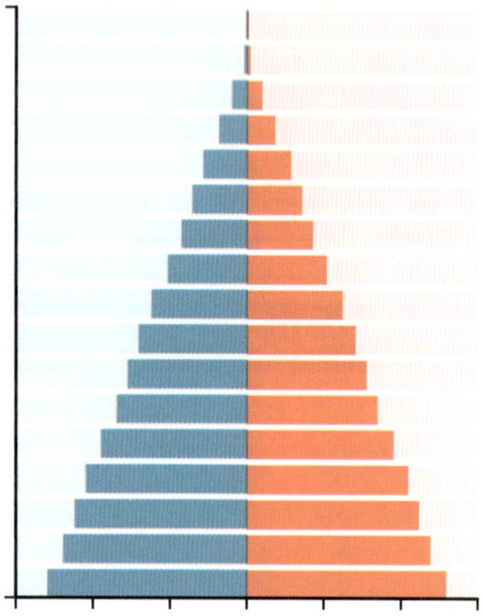

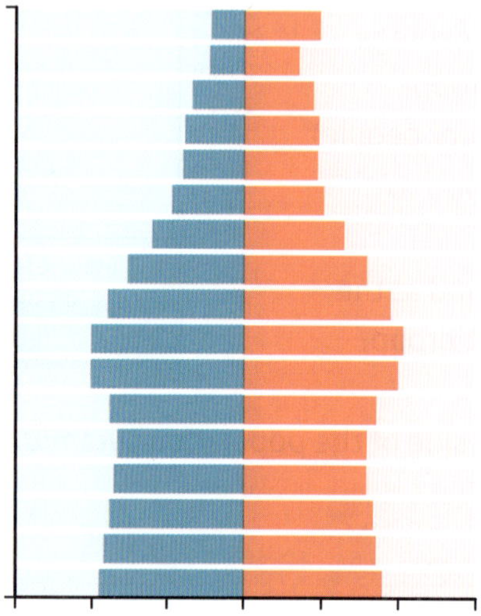

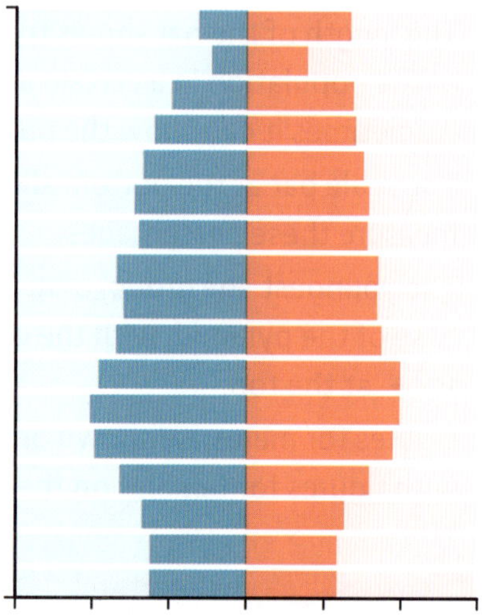

Question Time

1. What do population pyramids tell us about a population?
2. How is the information displayed on population pyramids?
3. Draw a rough diagram of the following types of population pyramids, giving a brief description of what type of population structure each is displaying:
   (a) Expansive
   (b) Stationary
   (c) Constrictive

18. Population

Case Study: Population Change in a Developed Country – Ireland

History: The Great Famine

The Great Famine had a huge impact on Ireland's population.

- In 1841, the population of the Republic of Ireland was 6.5 million. However, in the years 1845–52, Ireland experienced the Great Famine. The potato crop failed because of blight.
- The food supply was cut off. As a result, more than 1 million people died, and a further 1 million people emigrated to escape hunger, disease and poverty.

After the Famine

- Many people left Ireland for the United States, Britain, Canada and Australia.
- Those who left were mainly young, and more females than males left. This had an impact on marriage and birth rates.

Ireland's Population in the Modern Era

In the 1960s, the Republic of Ireland's population began to grow for the first time in 120 years. The government worked hard to encourage large multinational companies to set up in Ireland. This created thousands of jobs and brought people into the country. The population continued to grow after Ireland joined the European Economic Community (now the European Union) in 1973.

Between 1995 and 2008, Ireland experienced a period of great economic growth known as the 'Celtic Tiger'. By 2008, the population had grown to almost 4.4 million. However, in 2008, Ireland went through a period of economic decline. Large numbers of young people left Ireland to seek work abroad, but population growth remained steady due to natural increase.

In the 2022 census, Ireland's population had reached 5.1 million – the first time since the Great Famine that the population was over 5 million.

A Famine memorial, Co. Sligo. The family are comforting one another, and the young child points to the New Land to which they will sail.

Irish Population, CSO 1996–2022		
Census year	**Population**	**Change**
1996	3,626,087	100,368
2002	3,917,203	291,116
2006	4,239,848	322,645
2011	4,588,252	348,404
2016	4,761,865	173,613
2022	5,123,536	

Being numerate

Looking at the table:

1. What was the population of Ireland according to the 2016 census?
2. By how much did the population increase between 1996 and 2022?
3. Calculate the change in population from 2016 to 2022.

Current Population

Ireland is a developed country. The population is growing slowly. The population pyramid has a stationary shape.

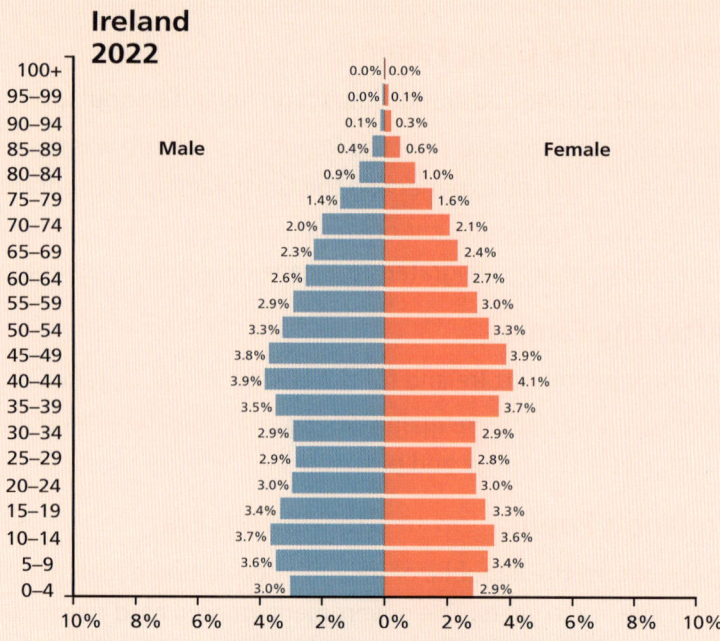

5. Population pyramid for the Republic of Ireland, 2022

> **$\frac{1}{3}\frac{2}{}$ Being numerate**
>
> Looking at figure 5:
> 1. What percentage of the population is aged 0–4?
> 2. Which age group contains the largest percentage of Ireland's population?
> 3. What percentage of the population is 80+?

Birth Rate

In 2021, Ireland had one of the highest birth rates in the European Union, with 12 births per 1,000 people. Ireland has experienced what is described as a 'baby boom' in recent years. There is access to good healthcare and low infant mortality (the number of deaths of infants under one-year-old per thousand live births).

Death Rate

In 2021, Ireland had one of the lowest death rates in the European Union, with 6.3 deaths per 1,000 people. People are living longer due to good healthcare, high standards of living and better care for the elderly. Life expectancy (how long a person is expected to live) in Ireland is 81.

The Future for Ireland's Population

Ireland's population is projected to rise to 5.2 million by 2030. The largest increase is predicted to occur in Dublin and the mid-east of the country. This important information helps the government to plan for the growing need for housing, for employment and for essential facilities such as schools and hospitals.

1. What was Ireland's population before the Great Famine?
2. Describe the impact the Great Famine had on the Irish population.
3. Describe Ireland's current birth and death rates.
4. Ireland's population is projected to increase in the future. Explain two ways that this will impact on services.

18. Population

Case Study: Population Change in a Developing Country – Nigeria

History: Colonialism and the Slave Trade

- Nigeria is a country located in West Africa. It was a colony of the British Empire until 1960, when it gained its independence. (A colony is a country that is taken over by another country, which occupies it with settlers and takes advantage of it economically.)
- Up to the late 1700s, there was an active slave trade operating out of Nigeria. European and Nigerian rulers traded enslaved Nigerians. In 1807, the British stopped their slave trade. As well as the horrible abuse of humans, the slave trade had a negative impact on population in Nigeria.
- Writing in 1849, British naval officer Lieutenant Patrick Forbes estimated that, during a period of 26 years, ships carrying 1.8 million enslaved people from West Africa managed to slip past naval barriers. These people were taken to the United States and forced to work on plantations there.

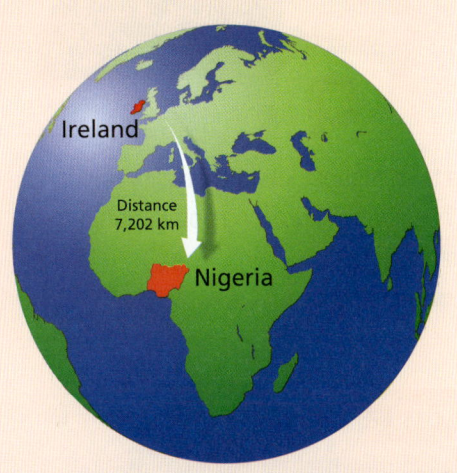

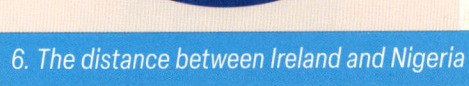

6. *The distance between Ireland and Nigeria*

Q Describe what you can see in this image. What impact do you think the slave trade had on the population of Nigeria?

Nigeria's Population in the Modern Era

- In 1950, the population of Nigeria was less than 40 million.
- There has been a rapid and large population increase over the last 70 years.

Nigeria Population 1990–2022		
Census Year	**Population**	**Change**
1990	95,617,345	
1995	108,424,822	12,807,477
2000	122,876,723	14,451,901
2005	139,611,303	16,734,580
2010	159,424,724	
2015	182,201,962	
2022*	216,746,933	

*Estimated due to Covid-19

Being numerate

Examine and copy the table in your copy. Calculate the change in Nigeria's population from one census to the next and fill in the blank spaces.

Factors Influencing Population Change

- **Religion**: Islam is the main religion of Nigeria. It encourages large families and early marriage. Polygamy, where a man takes more than one wife, is common.
- **Male child preference**: In many cultures in Nigeria, male children are more highly valued than females. Males can carry on the family name and are considered to have greater strength for physical work. This results in women having many babies to try to have male children.
- **Security in old age**: In many cultures in Nigeria, children are the only form of support for the older generations. Children are expected to take care of parents in their old age.
- **High infant mortality**: Nigeria has one of the highest infant mortality rates in the world. Parents feel they need to have many babies in the hope that some will survive to support them in old age and provide an income for the family.

Current Population

Nigeria is a slowly developing country. The population pyramid is expansive, like a typical pyramid. The population is growing rapidly.

Q Despite having the third highest infant mortality rate in the world, Nigeria has a rapidly growing population. Why do you think this might be?

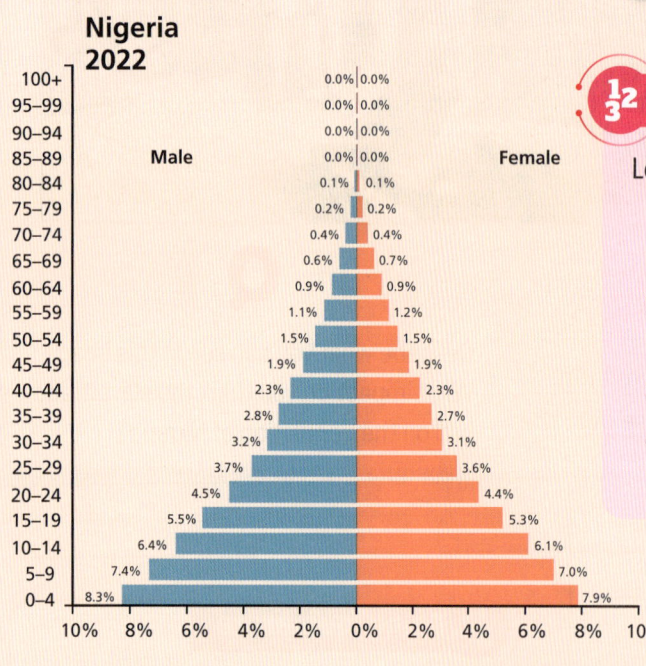

Being numerate

Looking at figure 7:

1. What percentage of the population of Nigeria is aged 0–4?
2. Which age group contains the largest percentage of Nigeria's population?
3. What percentage of the population is 80+?
4. How do these figures compare to Ireland's? Give two comparisons.

7. The population pyramid for Nigeria, 2022

Birth Rate

In 2021, the birth rate in Nigeria was 36.8 births per 1,000 people. Nigeria still has a very high infant mortality rate, with many children dying before reaching their first birthday.

Death Rate

In 2021, Nigeria had a death rate of 11.3 deaths per 1,000 people. The main causes of death in Nigeria are malaria, gastroenteritis and HIV/Aids. Life expectancy is just 54 in Nigeria.

18. Population

The Future for Nigeria's Population

It is estimated that by 2030, the population of Nigeria will reach 262 million, making it one of the fastest-growing countries in the world.

Question Time

1. Nigeria was once a British colony. What is a colony?
2. Describe Nigeria's population change from 1950 to 2022.
3. Explain three factors that have influenced Nigeria's population growth.
4. Compare Nigeria's birth, death and life expectancy rates to Ireland's.
5. (a) What is expected to happen to Nigeria's population by 2030?
   (b) How might this impact on services and the economy in Nigeria?

Sample questions

1. (a) **Examine** the demographic transition model below and decide whether each of the following questions is true or false.

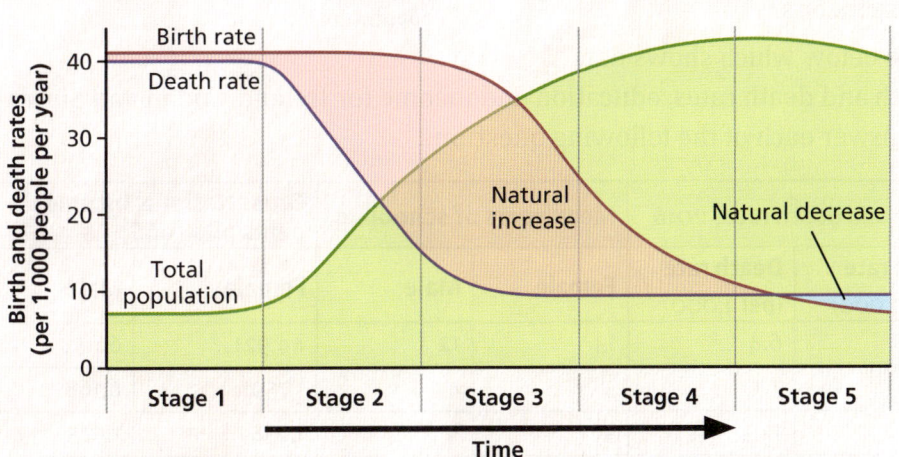

Demographic Transition Model

> **Exam hint!**
>
> Read each statement below and look at the stage that it refers to on the diagram. Next, look carefully at all labels and colour coding on the graph for that section.

(i) The birth rate is increasing during stage 3.

(ii) The death rate is highest in stage 4.

(iii) Total population growth in stage 4 is slower than in stage 2.

(iv) The death rate being lower than the birth rate is the only reason a population increases.

(2022 SEC exam paper, Q4 (a) (i)–(iv))

(b) **Outline** how the birth rate changes between stage 3 and stage 4 of the demographic transition model.
(NCCA assessment items, Sample 1 Q2)

(c) According to the demographic transition model, birth rates and death rates are decreasing in stage 3. **Outline** three reasons for this possible decrease.
NCCA assessment items, Sample 1 Q3

Exam hints!

To answer part (c), you should start by stating your three reasons. Next, take each reason one at a time and outline in detail why it causes the decrease birth and death rates. The answer has been started below for you. Complete it in your copy.

Let's get started! Sample starter

> Three reasons for the possible decrease in birth and death rates in stage 3 are:
> 1. Family planning   2. Improved medical care   3.
> Birth rates drop as people begin to plan family size more.
> Birth rates are decreasing in stage 3 because the economy of the country is doing well and women's role in society may be impacted by this. Women are well educated in family planning and have greater access to contraception.
> The death rate drops because of the impact of improved medical supports ...

... now complete the answer in your copy.

(d) **Explain** two ways in which a government may use the information in the demographic transition model to support its citizens.

What might governments need to plan for if there is a large number of young people? What if there is a large older population?

2. Study the table below which shows data relating to birth and death rates, education and income for Ireland, Sudan and Sierra Leone. Then answer each of the following questions.

	Birth and death rates 2016		Mean years of schooling		Gross National Income per capita (US$)	
	Birth rate (per 1,000)	Death rate (per 1,000)	Female	Male	Female	Male
Ireland	13.4	6.4	13	12	44,921	66,583
Sudan	32.9	7.3	3	4	1,759	6,168
Sierra Leone	34.5	12.3	3	4	1,238	1,525

(i) **Which** country from the table has the highest mean years of schooling for males?
    Ireland   Sudan   Sierra Leone

(ii) What was the Gross National Income per capita (in US$) for females in Sierra Leone?

(iii) **Calculate** the natural increase in population (per 1,000) in Sudan for 2016.

Look back on p. 266 to see an example of how this is calculated. Remember to always show how you have worked out the answer. You should show all rough work.

(2022 SEC exam paper, Q4 (b) (i)–(iii))

19 Migration

Learning Outcomes: 3.2, 3.9

key words

- Internal migration
- International migration
- Migrant
- Immigrant
- Emigrant
- Refugee
- Asylum seeker
- Internally displaced
- Forced migration
- Individual migration
- Organised migration
- Push factor
- Pull factor
- Globalisation
- Network factors
- Multi-ethnic

Learning Intentions

You will be able to:
- Explain the key terms associated with migration
- Describe the push and pull factors that can influence migration
- Describe a range of causes of individual, group or organised migration
- Explain a range of consequences of migration
- Discuss the impact of globalisation on migration.

19.1 People on the Move

Throughout history, people have moved or migrated from one place to live in another. This migration has taken place on different scales or levels. Some people move from one place to another within a country, e.g. a young person leaving Dublin to go to college in Galway. This is known as **internal migration**. Others leave one country to go live in another. This is known as **international migration**, e.g. a person who leaves Ireland to work in Australia.

Key Definitions

- A **migrant** is a person who moves from one place to another to find work or better living conditions.
- An **immigrant** is a person who comes to live in a foreign country. The country to which an immigrant has come is called the host country.
- An **emigrant** is a person who leaves their own country to live in another. The country from which the emigrant has left is called the donor country.
- A **refugee** is an immigrant given special permission to live in another country. They have fled their own country because of war, natural disasters or persecution (bad treatment).
- An **asylum seeker** is an immigrant looking to be accepted as a refugee in another country.
- An **internally displaced** person is a person forced to flee their home but who remains within their own country's borders.
- **Forced migration** occurs when someone is forced to move due to war, famine or persecution.
- **Individual migration** is when one person moves from one country or district to another.
- **Organised migration** involves planned migrations of people carried out by governments or other agencies.

Ukrainians fleeing war were welcomed to Ireland in 2022

Melissa Carthy, an Irish woman living in Australia

Managing information and thinking

Photograph A shows support for Ukrainian refugees arriving in Ireland. They are fleeing war in their home country and are seeking permission to stay in Ireland. Photograph B shows Irish woman Melissa Carthy, who moved to Australia in search of a better life.

1. Which of the nine terms given on the right best describe the people shown?
2. Has anyone in your family emigrated or immigrated? If they did, where did they go and why?

19.2 Why People Migrate: Push and Pull Factors

People migrate for a variety of reasons. These reasons are divided into push factors and pull factors.

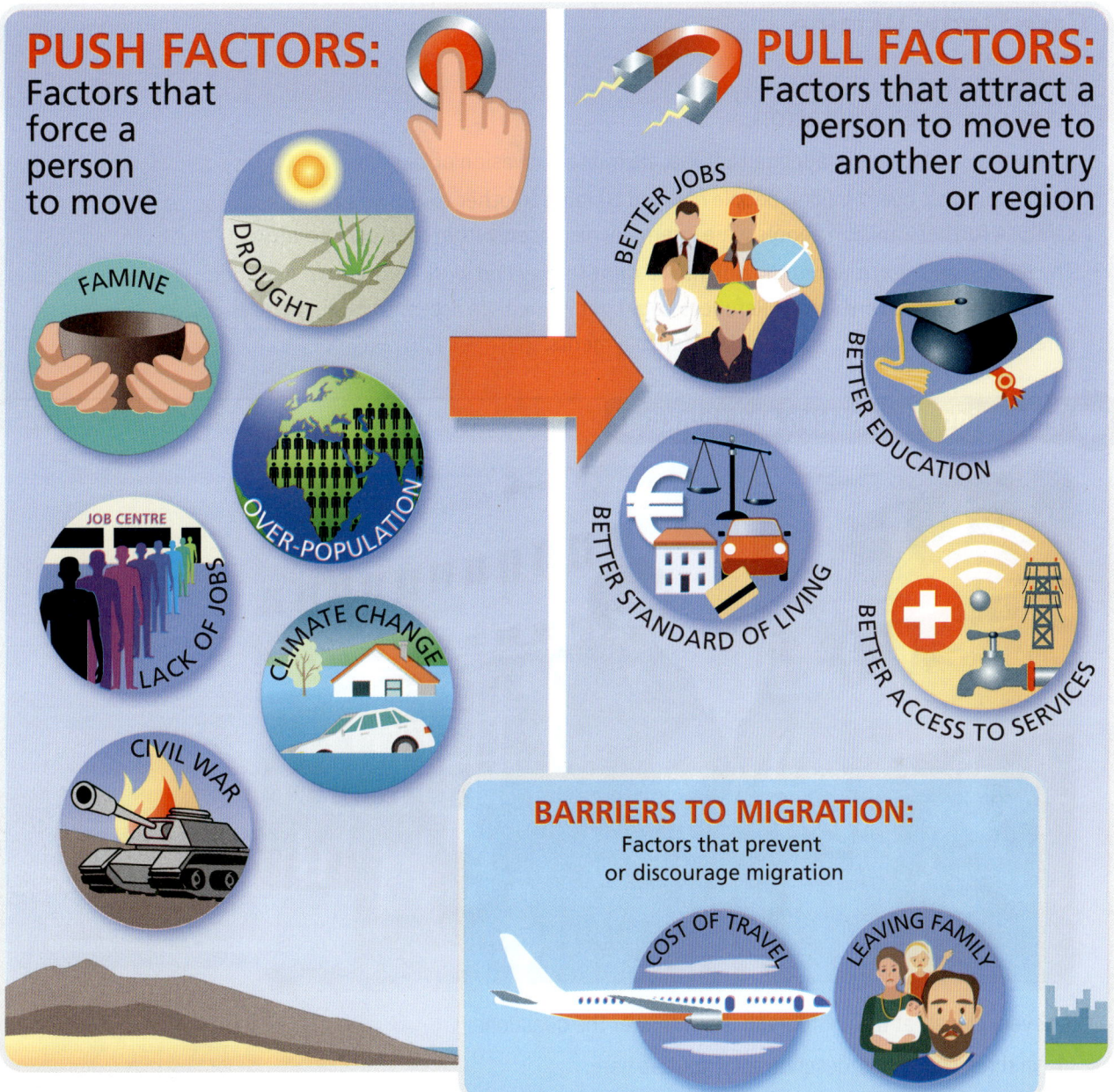

Question Time

1. What is a migrant?
2. Explain each of the following terms: (a) immigrant, (b) emigrant, (c) refugee, (d) asylum seeker.
3. 'Migration can be caused by push and pull factors.'
   (a) Describe what is meant by push and pull factors and explain how they influence migration.
   (b) Give two examples of push factors and two examples of pull factors.

19.3 Forced Migration

Case Study: Ukraine

Causes of Forced Migration

- In February 2022, Russia launched a full-scale military invasion of Ukraine.
- Within months, over 7 million people had fled Ukraine to other countries in search of safety. A further 8 million people were internally displaced within their own country.
- The European Union granted Ukrainians the right to stay and work throughout member countries for up to three years. Refugees could get immediate access to food, medical care, education, housing and social welfare payments.

Managing information and thinking

Ukrainian refugees arriving into a processing centre at Dublin Airport

Look very closely at the photograph. Then answer the questions:

1. In three sentences, describe what you see here.
2. Write a description of what you think life might be like for someone your age who has arrived into a new country seeking asylum from war.

Consequences of the Forced Migration

- Millions of people have had to leave their homes and families behind to seek safety in new and often unfamiliar places. Some people have lost their lives in an attempt to get to safety.
- Many homes, schools, hospitals and care facilities have been completely destroyed by the war.
- The majority of internally displaced people are women and children. Many of them are lacking the basic items such as medication and food that they need to look after themselves.

19. Migration

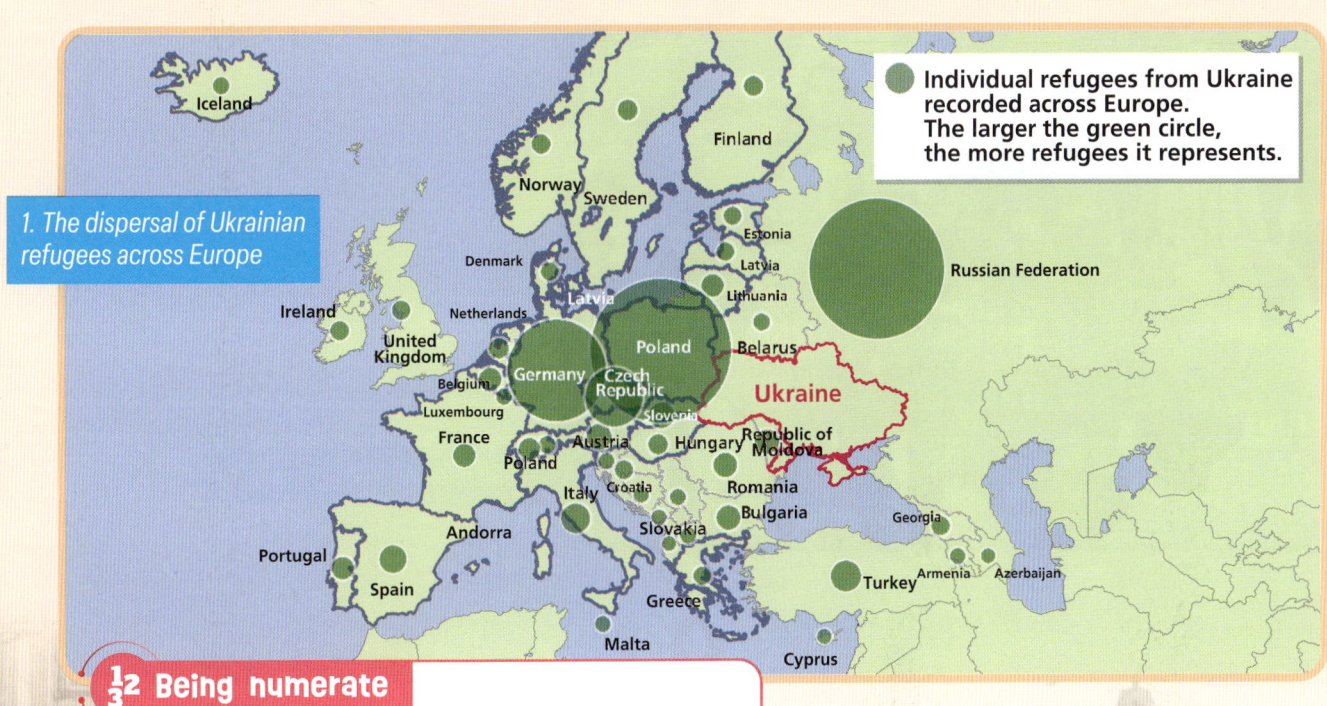

1. The dispersal of Ukrainian refugees across Europe

Individual refugees from Ukraine recorded across Europe. The larger the green circle, the more refugees it represents.

⅓² Being numerate

Examine figure 1. Which neighbouring country has the largest number of individual Ukrainian refugees?

Push Factors
- War
- Fear for personal safety
- Destruction of schools, hospitals, water supplies, sanitation
- Economy destroyed, lack of employment

Pull Factors
- Hope of a better life, safety
- Employment
- Education for young people
- Safe housing
- Access to food and clean drinking water

Barriers to Migration
- Lack of finance to fund the move
- Leaving family behind
- Great danger involved in travelling to safety

Destruction of buildings and vehicles in a Ukrainian city

281

Junior Cycle Geography CYCLONE

A Ukrainian family fleeing to safety

1. What is the cause of the large-scale migration of people from Ukraine?
2. Describe two consequences of Ukrainians fleeing and seeking asylum in other countries.
3. Briefly describe two push factors, two pull factors and two barriers to migration for Ukrainian refugees.

19.4 Individual Migration

Case Study: Edelle's Story

Edelle McCrudden is an ambitious and energetic woman. Born and raised in Co. Monaghan, she was a school prefect in her post-primary school, took a leading role in her local youth group and achieved over 500 points in her Leaving Certificate examination.

Edelle migrated from Monaghan to Dublin to study Science Education. She found shared accommodation in Dublin.

'I was so nervous and excited at the same time. Dublin offered many fantastic universities, along with social opportunities and a diverse culture that was more than I had access to in my quiet, rural village. Although I would miss my family and friends in Monaghan, my educational goals would only be achieved by going to university. I knew that financially it would be a struggle, so I would have to balance study and social events with a part-time job to afford my four-year degree.'

Having graduated with both Bachelor's and Master's degrees, Edelle began her employment as a teacher in a post-primary school in Dublin. She moved further away from the city to be able to afford rent.

19. Migration

From Monaghan ...

To Dublin City ...

To Sydney

After eight years of teaching, Edelle's fiancé was offered a job in Sydney, Australia. The lifestyle and salary levels were a strong draw.

'Although it is as expensive to live in Sydney as it is in Dublin, the salaries we both earn here in Australia are much higher. It was not an easy decision to move. We miss our family and friends. But we are lucky that we can return home at any stage. The money we're earning here will allow us to save for a deposit for a house in Ireland – something that we're unsure we could have done had we stayed at home.'

1. In pairs, discuss each of the areas shown in the photographs and list one positive and one negative for migrating to live in each of the two areas she moved to.
2. Edelle migrated twice. Where did Edelle move from and to each time? Which type of migration did each move involve?
3. Draw this table into your copy and explain two push factors, two pull factors and any possible barriers Edelle faced for both migrations.

Migration 1	Migration 2
Push Factors	**Push Factors**
1.	1.
2.	2.
Pull Factors	**Pull Factors**
1.	1.
2.	2.
Possible Barriers to Migration	
Dublin:	
Sydney:	

19.5 Organised Migration

Case Study: The Plantation of Ulster

In 1609, King James I of England started sending English and Scottish Protestants to confiscate and settle on land taken from the native Irish. Almost 4 million acres of land in the province of Ulster were divided up into estates and rented cheaply to the English and Scottish settlers.

2. The land taken over by settlers in the plantation of Ulster

Q Use a map of Ireland to identify the counties in which each of the groups was settled.

Causes of the Organised Migration

- King James I wanted to rule over the entire island of Ireland and to put a stop to rebellions. Ulster had been the strongest of the four provinces in the struggle against British rule.
- King James I wanted to 'civilise' Ulster and to introduce the Protestant religion and English customs to the region.
- The English and Scottish settlers were promised good farmland and a better life in Ulster.

Consequences of the Organised Migration

- The plantation introduced Protestantism and Presbyterianism to Ulster and they became the majority religions. This contributed to unrest and a divide between Catholics and Protestants in Northern Ireland that still exists today.
- Before the plantation, Irish was spoken in Ulster. During the plantation, the English language was introduced.
- English farming methods were introduced, with more focus on crop-growing and less on cattle farming than before.
- English and Scottish customs and cultures were introduced to Ulster during the plantation.

19. Migration

Managing information and thinking

What do these two photographs tell you about life in Ulster long after the initial plantation?

Push Factors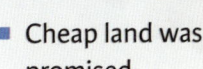
- At the start of the 1600s, unemployment was rising in Scotland. Tenant farmers were desperate enough to leave their homes to travel to Ulster in the hope of a better life.
- England and Scotland were becoming overpopulated.

Pull Factors
- Cheap land was promised.
- Some planters wanted to introduce their religious beliefs and customs to help 'civilise' the native Irish.

Barriers to Migration
- There was fear of attacks by the native Irish.
- New planters were expected to build large castles to protect themselves, which was costly and took time.

1. Explain two causes of the Ulster plantation.
2. Briefly explain one push factor, one pull factor and one barrier to the plantation of Ulster.
3. Describe two long-term consequences of the Ulster plantation for Northern Ireland.

19.6 Globalisation and Migration

> *Today, the number of people living outside their country of birth is larger than at any other time in history. International migrants would now make up the world's fifth most populated country if they all lived in the same place.*
>
> United Nations Population Fund

Globalisation is the process whereby the world has become more interconnected. It is the increasing interaction of peoples, states or countries through the growth of the international flow of money, ideas and culture.

Globalisation has introduced factors known as '**network factors**' that have influenced migration. Network factors include:

- The free flow of information
- Improved global communications
- The faster and lower cost of moving from one place to another.

The Impacts of Migration for Migrants

The impacts of migration can be positive and negative for the migrant.

Positives

- Immigration can provide a supply of low-cost workers for some countries. This is positive provided these migrant workers' rights are protected and that they are paid a fair wage.
- Migrants also bring qualifications and skills that can fill a gap in the jobs market, like engineers, nurses, doctors and teachers.

- Upon return, migrants bring new skills to their country of origin (home country) such as the ability to speak foreign languages. These new skills can help to improve the economy.
- Workers may send money home to families in their home nations, which can be an important source of income for these countries.
- The creation of a **multi-ethnic** (culturally mixed) society increases understanding and tolerance of other cultures.

Negatives

- Immigrants can face discrimination, or be accused of lowering wages or of being associated with crime.
- When people emigrate from their home country, it can lead to a loss of well-educated young people (known as a 'brain drain'). This can have negative impacts on population growth and development in the country of origin.
- The migration of young families can cause a loss of cultural leadership and traditions in the country of origin.

 Go to YouTube and search for 'Pitching Up – The Guardian' (15:28) to watch an interesting short documentary on what is called 'Ireland's most ethnically diverse town'. Before you watch it, guess which county this town is in.

Question Time

1. Name two network factors that have had an impact on migration.
2. Explain two positive and two negative impacts of migration for the migrant.

Junior Cycle Geography CYCLONE

Sample questions

1. Natural disasters can force people to move away from their homes.

    (i) A natural disaster is one example of a push factor. The table below lists other factors that affect migration. **Choose** the three push factors listed in the table.

    | Job opportunities | Shops and services are nearby | Persecution |
    | Unemployment | Famine | Good schools |

    (ii) A donor region is the place that a person moves away from when they migrate. **Discuss** one impact of outward migration on donor regions.

    (2022 SEC exam paper, Q9 (c) (i) & (ii))

2. Read the article below and answer the questions that follow.

This Syrian refugee family fled their home in Aleppo in 2012. When bombing struck their town, 'within 24 hours the city was destroyed,' Ahmad recalls. They fled to Lebanon, where they shared a small flat with Ahmad's three siblings and their children. Ahmad was able to find some work from time to time, but knew his future was bleak, especially as his 6-year-old son Abdullah had developed hearing problems. The family was eventually accepted to resettle in Germany under the government's Humanitarian Assistance programme. They travelled to Germany in September 2013 and were among the first group of Syrians to arrive in the country under this programme.

19. Migration

(i) **Define** the term 'refugee'.

(ii) **What** is the impact of families like the one above relocating to host countries such as Germany?

(iii) **List** two positive impacts of families like the family above locating to your local area.

(iv) The data presented below represents child migrants hosted in a country. Use the data in the table to **draw** a bar graph. (In this data set 'child migrants' are persons aged 19 and under.)

Year	1990	1995	2000	2005	2010	2015
Percentage of migrants	21%	20%	19%	15%	14%	14%

(v) **Outline** one pattern that the table demonstrates.

(NCCA assessment items, Sample 7 Q2, Q4(a), Q4(b), Q5 & Q6)

Exam hints!

To answer part (ii), you should consider what each country would need to do to prepare and to look after refugees arriving (e.g. housing, food, medical care, education). Also consider what the family can contribute to the community they join (e.g. culture, language, skills and employment).

For part (iv), carefully draw the x-axis and y-axis in your copy. Make sure to label each axis to state what it is showing. Always use a ruler and pencil to draw a bar graph.

For part (v), consider whether the percentages go up, down or stay constant. In your answer, outline the pattern that you see.

20

Geographical Skills: Aerial Photographs, Charts, Graphs and Infographics

Element: Geographical Skills

You will:
- Develop your graphicacy through reading and interpreting visual stimuli and data sets
- Investigate geographical information, by suitably interpreting, analysing and presenting data.

Learning Intentions

You will be able to:
- Use and interpret information on aerial photographs
- Draw sketch maps of aerial photographs
- Compare OS maps with aerial photographs
- Understand why charts, graphs and infographics are used and how to create them.

FUN FACT!
Every two minutes, we snap more photographs than the whole of humanity did in the 1800s.

20. Geographical Skills: Aerial Photographs, Charts, Graphs and Infographics

Part 1: Aerial Photographs

key words
- Vertical
- Oblique
- Foreground
- Background
- Urban planning

20.1 Aerial Photographs

Types of Aerial Photograph

Aerial photographs display the surface of the land. They are taken from the air and can be divided into two types: **vertical** and **oblique**.

Vertical photographs
Taken with the camera pointing straight down on the area being photographed

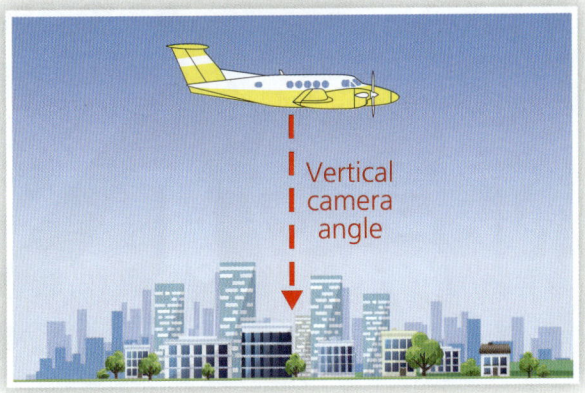

↓

Oblique photographs
Taken with the camera looking down at an angle on the area being photographed

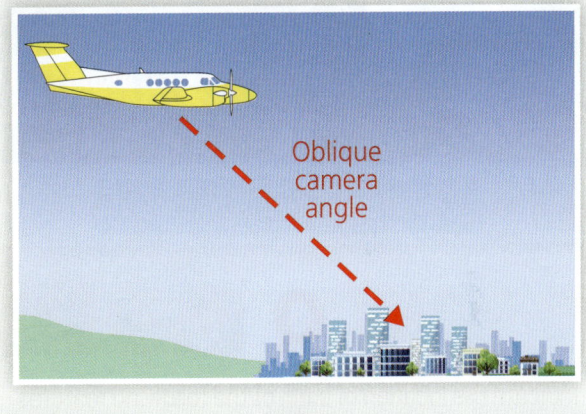

↓

 Using the camera on your phone, practise taking oblique and vertical photographs.

Locating Places on Photographs

Vertical photographs

To locate features on a vertical photograph, divide it into nine equal squares (figure 1). Give the location of the feature using the name of the square in which the feature is found.

1. Dividing up a vertical photograph

Oblique photographs

To locate features on an oblique photograph, you must first divide it into nine equal squares (figure 2). Give the location of the feature using the name of the square in which the feature is found.

- The areas nearest the camera are called the **foreground**. They cover smaller areas but will appear larger in the image.
- The areas furthest from the camera are called the **background**. They cover very large areas of land but appear smaller.

2. Dividing up an oblique photograph

 Q Indicate the location of each of the following features in figure 2: the bridge, a row of houses and a factory.

20. Geographical Skills: Aerial Photographs, Charts, Graphs and Infographics

Determining Direction on Aerial Photographs

Some vertical photographs include a north arrow (figure 3). In such cases, use compass directions when asked to identify the location of features.

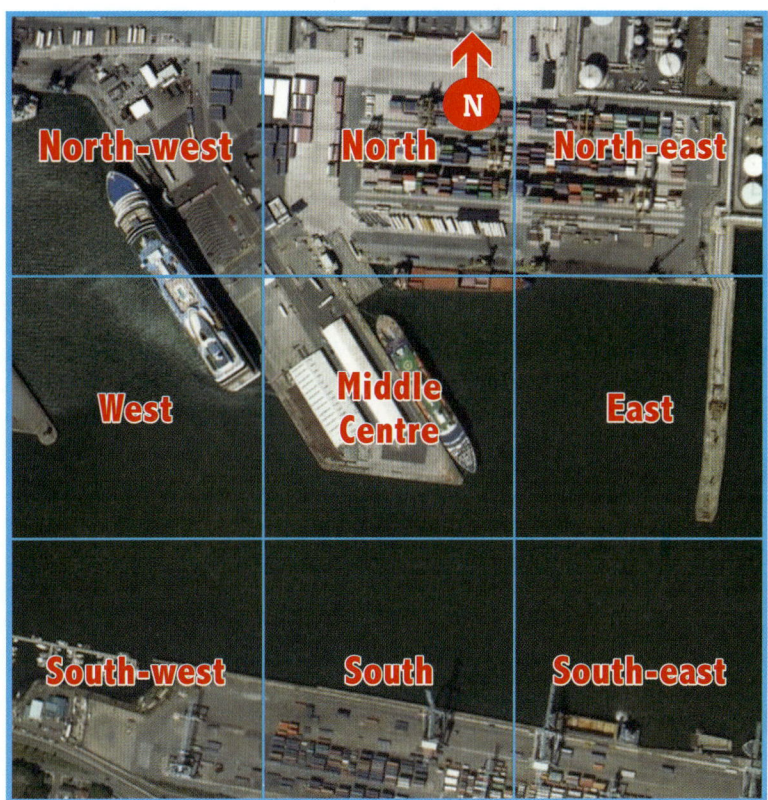

3. A vertical photograph showing direction

If there is no north arrow on the aerial photograph, use the following steps to determine direction.

1. Correctly position the Ordnance Survey (OS) map so the north arrow symbol can be read correctly like in the image above.
2. Choose a dominant feature in the foreground of the aerial photograph, such as a church, bridge or beach.
3. Pick another dominant feature in the background of the photograph and use a pencil to join the two with a straight line.
4. Now locate both features you chose on the OS map and join them with a straight line.
5. Use the north arrow on the OS map to determine the direction.

This method can be used on vertical and oblique aerial photographs.

Junior Cycle Geography CYCLONE

How to Sketch an Aerial Photograph

Use the following steps to correctly sketch an aerial photograph:

① Draw a four-sided border or frame for your sketch. This must be the same shape as the aerial photograph you are sketching. Unless instructed otherwise, make the frame half the scale of the original (half the length and half the width).

② Divide the photograph you are sketching into sections. Do likewise with your frame (figure 4). These sections will help you to better locate and sketch features of the aerial photograph.

③ Always draw important features such as the coastline, skyline and horizon (the line in the background where the sky touches the land or water) on your sketch map. This will also help you to sketch features in their correct location on your map.

④ Title your sketch map, e.g. 'Sketch of [*Name of location*] Aerial Photograph'.

⑤ Label all features you are asked to sketch in the question.

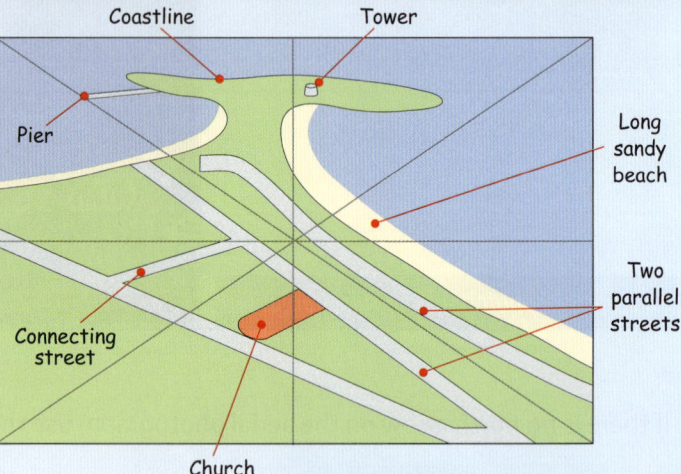

4. Segmenting a photograph and drawing a sketch map

Tips!

» Always sketch with a pencil.
» Insert all features asked for.
» If asked to draw houses, buildings or industrial estates, draw the outline shape of them.
» Allow plenty of room to properly label your features.
» Draw double lines for roads and rivers.
» Use colours sparingly.
» Never attempt to trace a photograph you are asked to sketch.

294

20. Geographical Skills: Aerial Photographs, Charts, Graphs and Infographics

Sligo Town

1. Examine this aerial photograph of Sligo Town. In your copy, match each of the features labelled A–H with the descriptions below.

Bridge	Meander	
Recreational area	Shopping centre	
Factory	School	
Church	Residential	

2. Draw a sketch map of the aerial photograph of Sligo Town. On your map, sketch and label the following:
   - The river
   - The bridge
   - The graveyard
   - The church
   - Three streets converging on a roundabout
   - A factory
   - A row of houses.

295

20.2 Examining Aerial Photographs: Colour and Shape

Much like on an OS map, the different colours on an aerial photograph can help us to identify the possible uses of land and other features.

- **Trees** appear dark or light green during spring and summer.
- **Rivers, lakes and seas** can be different shades of blue or possibly silver depending on the light they reflect.
- **Roads** will be light grey.

Identify the location of the following features on this photograph of Dungarvan, Co. Waterford:

(a) trees, (b) a body of water, (c) roads

- **Cereals** (crops such as barley or wheat) appears yellow or golden in colour during the autumn or spring.
- **Pastureland** (grassland left for animals to graze) appears green or dark green during summer months.

Identify the location of the following features in this photo:

(a) An area of land used for crop growth

(b) An area of land used for grazing.

Identifying Time of Year on Aerial Photographs

There are clues to look out for that may help us to identify the approximate time of the year an aerial photograph was taken.

Summertime	Wintertime
• **Trees:** In summertime, deciduous trees will be covered in leaves (May to October). • **Farm animals**: You might see farm animals grazing in the fields. • **Cereal growth**: You may see fields of ripe cereals (July to September).	• **Trees**: In wintertime, deciduous trees will be bare (November to April). • **Farm animals**: Absent from the fields, as they are housed in sheds during cold winter months. • Freshly ploughed fields (November to April).

20. Geographical Skills: Aerial Photographs, Charts, Graphs and Infographics

Rural Settlement and Land Use on Aerial Photographs

Aerial photographs offer evidence of different types of settlement. We can also see how land is used.

Historical Settlement

Defensive sites:
Look for ringforts, castles and round towers as evidence of defensive settlement throughout the years.

Ringfort

Burial sites:
Graveyards, standing stones and dolmens are indirect evidence that settlement existed in an area.

Present-Day Settlement

Houses and villages are evidence of present-day settlement. They are easily identified on an aerial photograph.

Land Use

Many kinds of rural land use can be identified on photographs, such as the following:

Pastoral farming	Look for animals grazing and green fields.
Arable farming	Look for brown fields that show ploughing has occurred. Light brown fields might indicate crop growth or newly cut crop.
Woodland	Look for large areas of deciduous and coniferous trees.
Buildings	Look for individual houses, housing estates, farm sheds, factories and industrial estates.
Transport	Look for roads and railway lines.
Recreational	Look for sports grounds, playgrounds and golf courses.

297

Junior Cycle Geography CYCLONE

20.3 Urban Settlement: Towns and Cities on Aerial Photographs

The Locations of Towns

When looking at OS maps in Chapter 7, we learned that the location of towns has been influenced by:

1. The presence of flat or gently sloping land
2. The meeting of transport routes
3. Bridging points on rivers
4. The coast.

We can also look out for these factors on aerial photographs.

Q Examine this aerial photograph of Ennis, Co. Clare. Remembering the four factors listed above, offer three explanations as to why the town of Ennis developed where it has. Refer to evidence from the photo to support your answer. At what time of year do you think this photo was taken? Use evidence from the photo to support your answer.

20. Geographical Skills: Aerial Photographs, Charts, Graphs and Infographics

Town Functions and Services

The functions of an urban area and the services it provides help it to develop. Most towns have multifunctional (more than one) purposes and offer several services to the people who live there.

Functions
- **Residential**: houses, apartments
- **Industrial**: factories, industrial estates

Functions that can also be services
- **Retail**: shops, shopping centres
- **Commercial**: office blocks
- **Religious**: churches, abbeys and cathedrals
- **Transport**: roads, railway lines, ports, canals, airports
- **Recreational**: stadiums, playing fields, golf courses, parks, marinas
- **Tourism**: beaches, hotels, caravan parks
- **Medical**: hospitals
- **Educational**: schools, colleges

Past/Historical functions
- **Defensive**: castles, towers, town walls
- **Commercial**: old warehouses, canals or docks, town squares used for cattle fairs

 Working with others

With the person sitting next to you, identify as many functions, functions/services and historical functions your town or nearest town may have.

Economic Activities on Aerial Photographs

Most of the functions and services of an urban area are directly related to economic activities. It is from these activities that people make a living. Economic activities can be divided into three categories.

	Explanation	What to Look for on an Aerial Photograph
Primary Economic Activities	The extraction and use of natural resources from the land or sea	Look for fields, farm buildings, fishing boats, piers and quarries (a large, deep pit, from which stone or other materials have been taken).
Secondary Economic Activities	Include factories or businesses that make things	Look for large buildings and industrial estates.
Tertiary Economic Activities	Provide useful services, e.g. transport, shopping, educational, religious, healthcare, recreational and tourism	Look for railway stations, shopping centres, office blocks, schools, colleges, hospitals, sports grounds, car parks, marinas.

Junior Cycle Geography CYCLONE

Q

1. Examine this aerial photograph of Wicklow town. List evidence of any primary, secondary and tertiary economic activities you can see.
2. Identify and name a method of coastal protection you can see in the photograph.
3. Suggest one reason why risk of flooding during storms and high rainfall might not be a concern for the people of Wicklow town.

20. Geographical Skills: Aerial Photographs, Charts, Graphs and Infographics

Identifying Traffic Problems and Traffic Solutions on an Aerial Photograph

The management of traffic as it flows through Irish towns and cities is a key concern for people who live in them.

Traffic Trouble Spots
- **Central Business District (CBD)**: the part of a town or city made up of the main shopping streets
- Places where streets become narrow
- Places where streets converge (meet), e.g. near a bridge

Traffic Solutions
- **Traffic lanes**: using arrows or other markings
- **Traffic lights**: placed at busy junctions
- **Yellow boxes**: used to regulate the flow of traffic at junctions
- **Roundabouts**: used to reduce potential delays where busy roads meet
- **Parking restrictions**: single and double yellow lines along sides of streets to prevent or limit parking
- **Off-road car parks**: multistorey car parks placed off street to prevent parked cars from cluttering a town
- **Pedestrianised streets**: for people and not vehicles
- **Bypasses/Ring roads**: used to take traffic and heavy goods vehicles around towns and city centres rather than through them

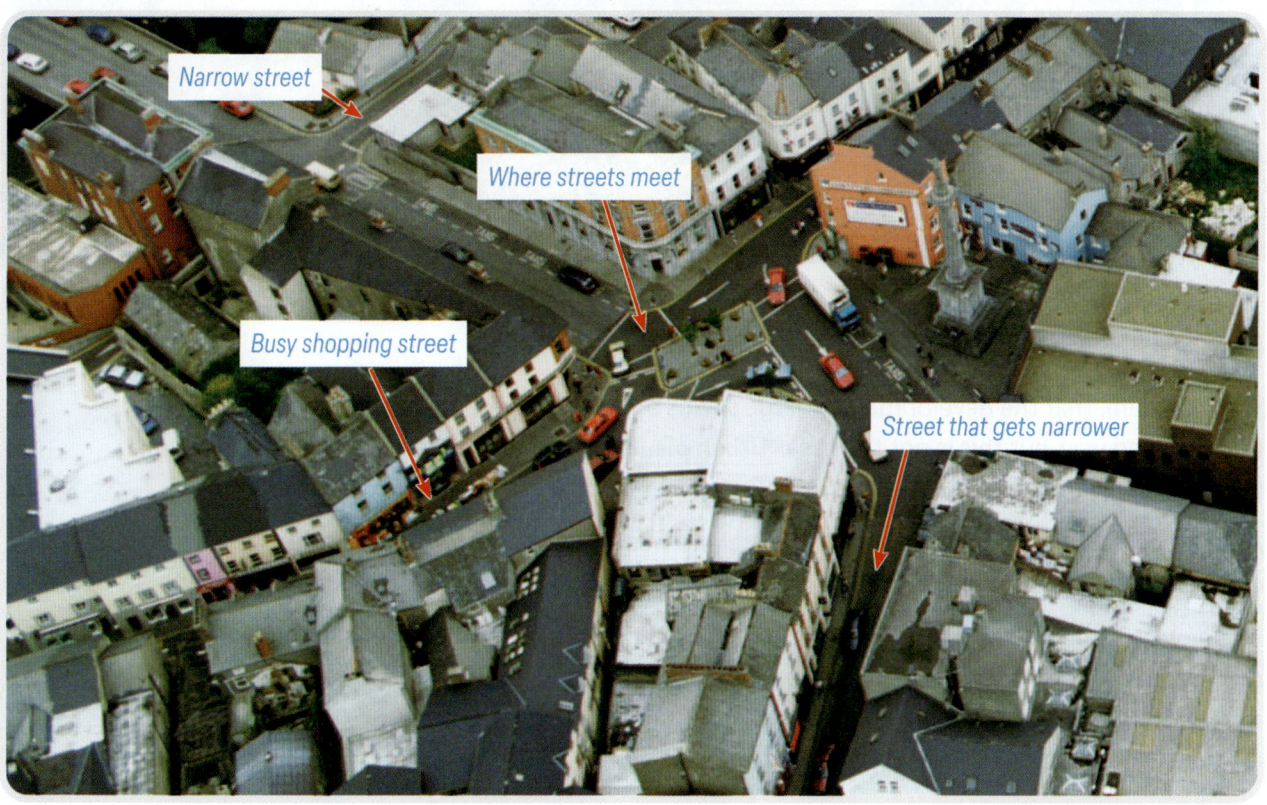

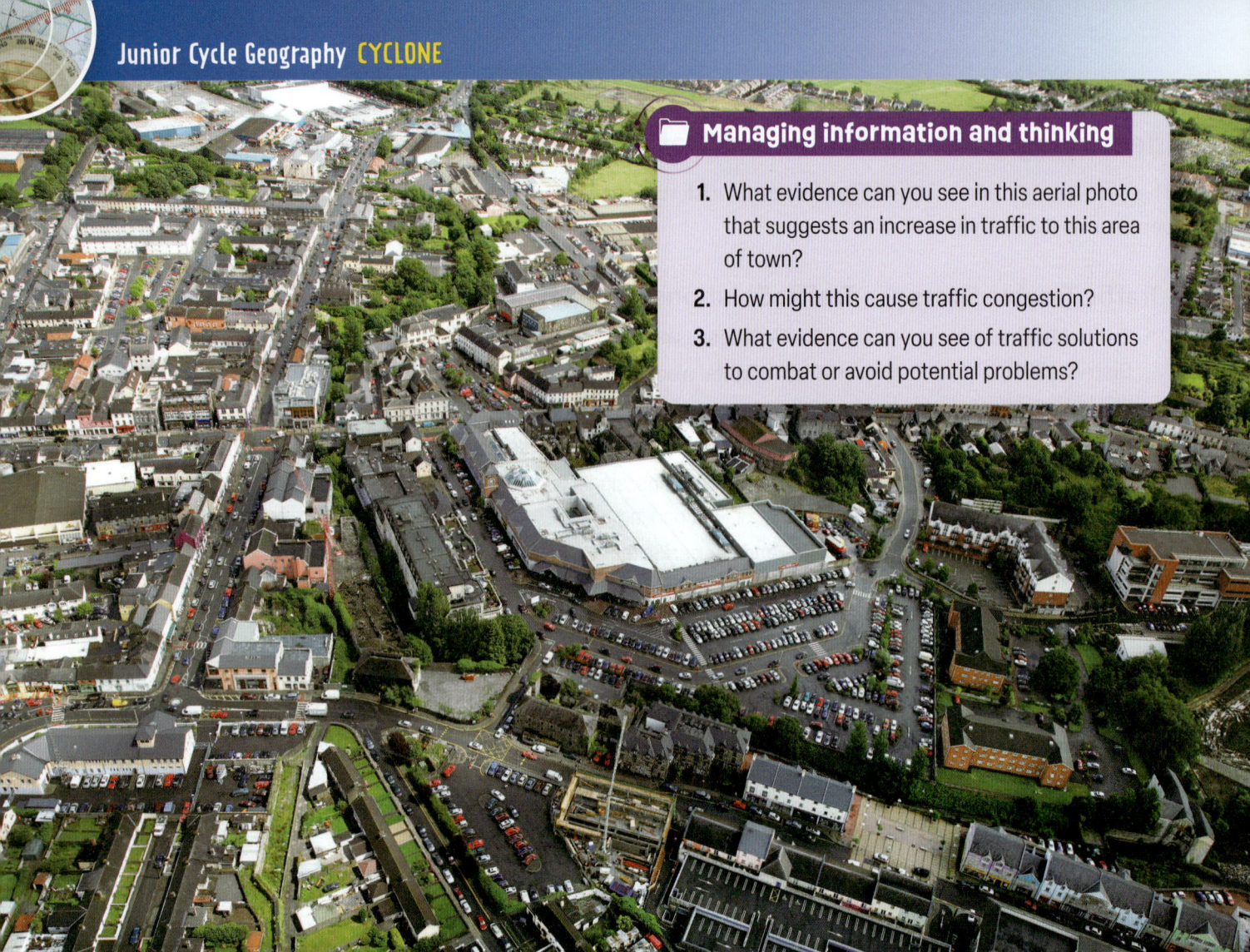

Managing information and thinking

1. What evidence can you see in this aerial photo that suggests an increase in traffic to this area of town?
2. How might this cause traffic congestion?
3. What evidence can you see of traffic solutions to combat or avoid potential problems?

Urban Planning

You may think that the development of large towns or cities happens by chance and in an uncontrolled way, but this is not the case. **Urban planning** is the process of designing and shaping urban areas, such as towns and cities.

A very recent example of urban planning in Ireland was the choice of location for the new children's hospital in Dublin. Photo A on the next page shows the site before building began.

The new children's hospital is being built on the existing St James's site. It is expected to be completed by November 2023. Photo B on the next page shows what the new development is expected to look like.

You can identify urban planning when you see evidence of:

- Traffic management
- Special zones for housing, business and shopping
- Streets meeting at right angles; they can form an orderly grid, or square-like patterns
- Specially designated green areas, parks and sports fields
- Central areas such as town squares
- Trees, planted to offer greenery and shade.

20. Geographical Skills: Aerial Photographs, Charts, Graphs and Infographics

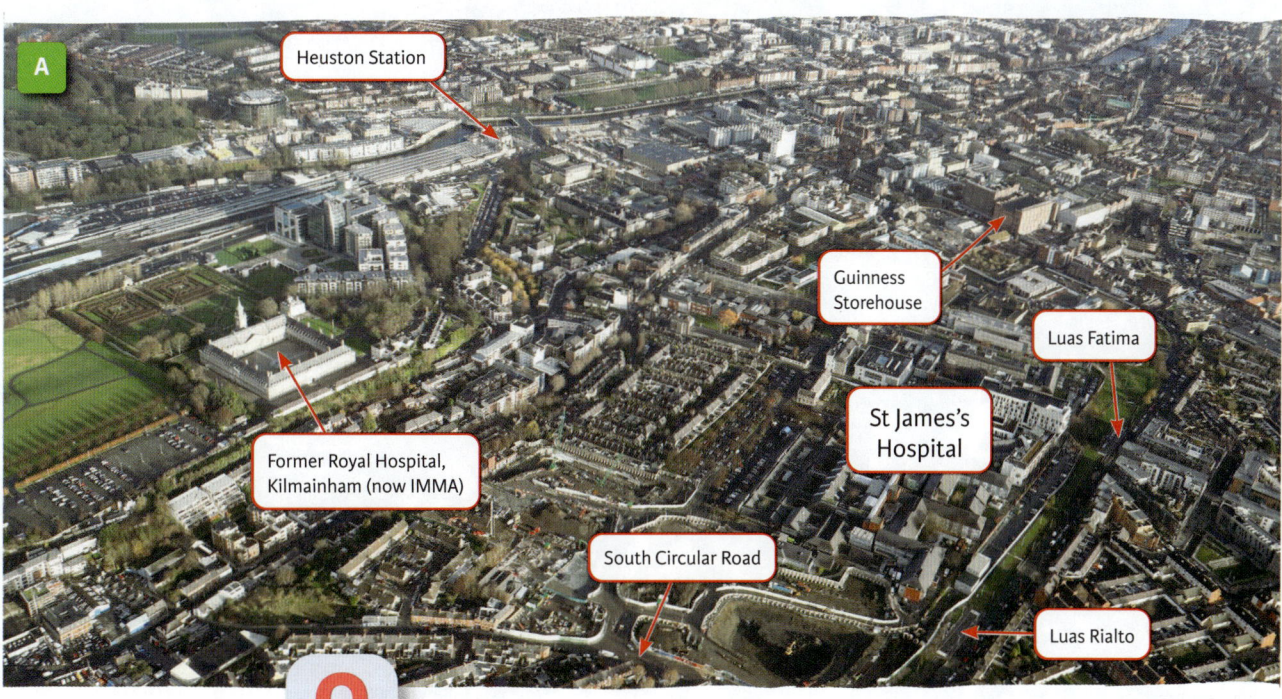

A

Labels: Heuston Station; Guinness Storehouse; Luas Fatima; St James's Hospital; Former Royal Hospital, Kilmainham (now IMMA); South Circular Road; Luas Rialto

Q What services can you see in this image of the existing St James's Hospital site that show evidence of urban planning?

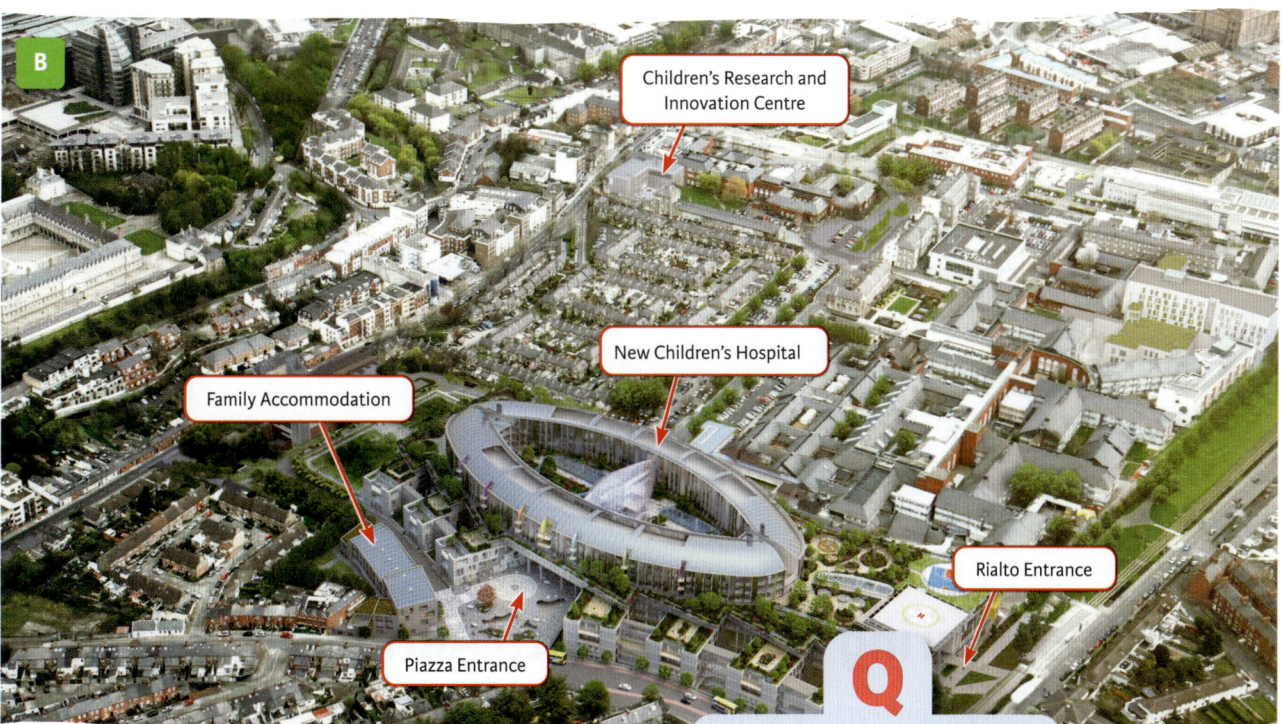

B

Labels: Children's Research and Innovation Centre; New Children's Hospital; Family Accommodation; Piazza Entrance; Rialto Entrance

Q What future services planned for the new children's hospital can you identify in this image?

303

Junior Cycle Geography CYCLONE

Working with others

Examining this aerial photo of Galway, identify as many examples as you can of evidence of urban planning. Compare your list with that of the person sitting next to you.

Recreation and Tourism on Aerial Photographs

You can identify recreation and tourism areas on aerial photographs when you see evidence of:

- Scenic areas such as oceans, rivers, lakes, sandy beaches, cliffs and other features that would attract people to a specific area
- Services such as hotels, campsites and car parks
- Shops and shopping centres
- Green areas such as parks and playing fields
- Historical attractions such as castles.

Being creative

Make a list of the tourist attractions your local area provides. Then draw a sketch map of your local area and include the location of the recreational and tourist areas your local area provides. Use this website for access to aerial photographs: maps.scoilnet.ie

20.4 Comparing OS Maps with Aerial Photographs

OS maps and aerial photographs each have unique ways to display the physical landscape. The aerial photograph and the OS map shown in figure 5 are of the same area. As we see, both have different advantages when it comes to examining an area.

All OS maps are drawn to a single scale. This means we can calculate distance on them. Vertical photographs are to scale but oblique photographs are not.

20. Geographical Skills: Aerial Photographs, Charts, Graphs and Infographics

Q Can you think of any advantages or disadvantages of OS maps and aerial photographs?

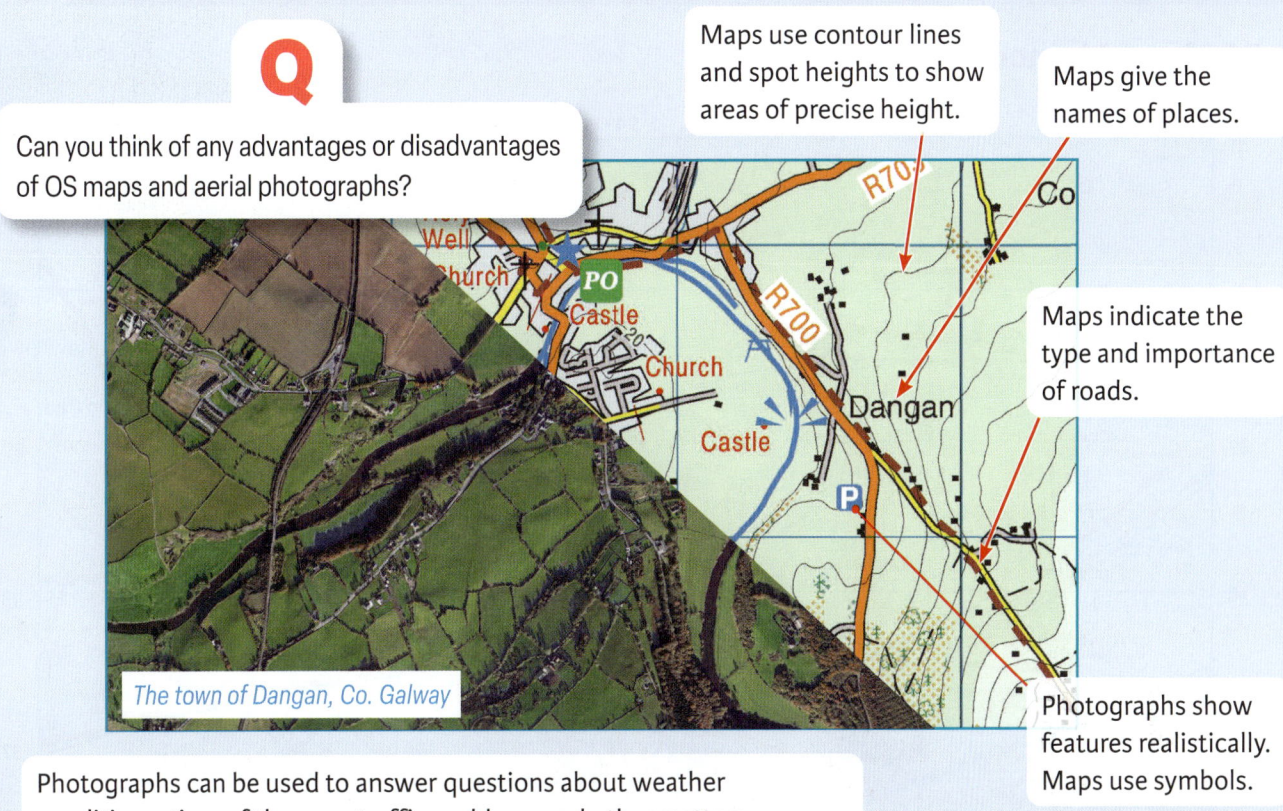

Maps use contour lines and spot heights to show areas of precise height.

Maps give the names of places.

Maps indicate the type and importance of roads.

Photographs show features realistically. Maps use symbols.

Photographs can be used to answer questions about weather conditions, time of the year, traffic problems and other matters.

The town of Dangan, Co. Galway

EXAM FOCUS

Sample questions

1. Examine the aerial photograph and map extract below and answer each of the following questions:

 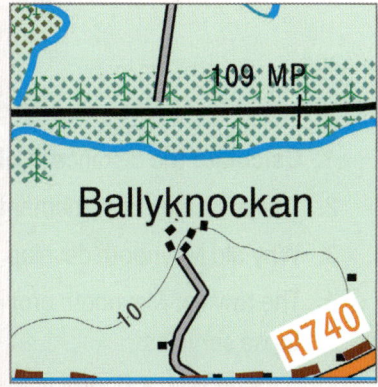

   (i) What type of forest is the most common in the OS map extract shown above? (Hint: use the legend on p. 85 to help you.)

   (ii) Name two possible primary economic activities that are visible on the photograph and the OS map.

305

OS Map 1: Maynooth, Co. Kildare

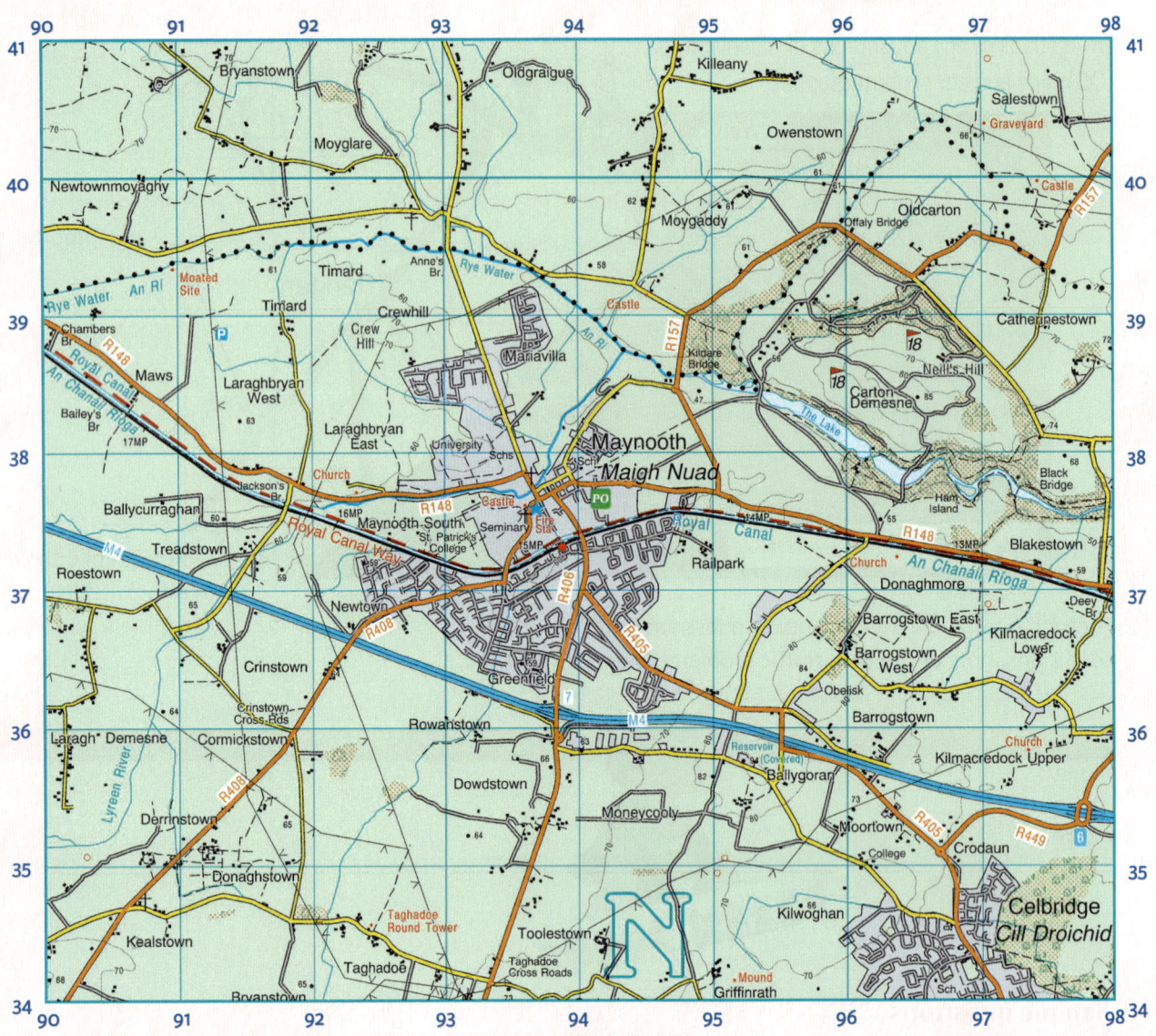

Q

1. **Calculate** the distance in kilometres of the section of the R148 road shown on the map.
2. There is a lack of settlement at N 95 37. **Give** one possible reason for this.
3. **Why** did Maynooth develop where it has? Give three reasons explaining its location.
4. The town of Maynooth provides a number of services. With reference to the OS map, **describe** three of these services.
5. With reference to the OS map, **list** the six-figure grid references for three pieces of evidence of historical settlement.

20. Geographical Skills: Aerial Photographs, Charts, Graphs and Infographics

Photograph 1: Maynooth, Co. Kildare

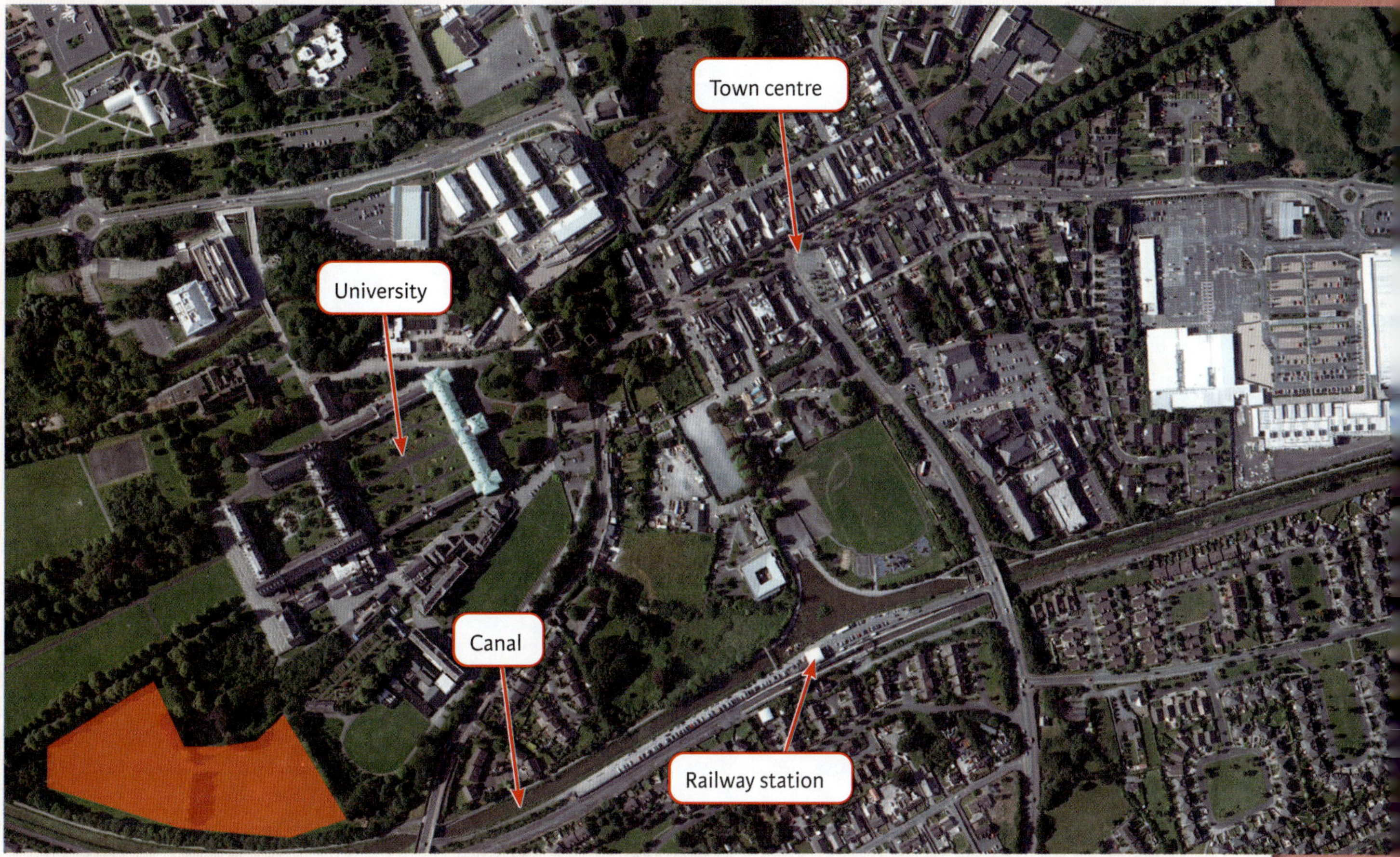

1. Is this photograph vertical or oblique?
2. **Explain** two ways in which the use of colour helps you to understand land use on this photograph.
3. **Draw** a sketch map of this aerial photograph. **Show** and **name** the following on your map:
   - Dual carriageway road
   - The canal
   - A hotel
   - A main road
   - A greenfield area
   - The university
4. Using the correct terms (those you saw in figure 1 on p. 292), locate the following on this photo:
   - The university
   - A housing estate
   - The train station
   - Tennis courts
5. Using the OS map and aerial photograph of Maynooth, determine in which direction you would be travelling if you went from the university to the railway station.
6. Using the correct terms, locate one place on the photo where traffic congestion may be an issue for the town. **Name** two steps that you think have been taken to manage traffic at this location.
7. The area highlighted in red on this photo is located at N 924 367 on the corresponding OS map of Maynooth. It is the proposed site for a new factory. Using evidence from both the OS map and the aerial photograph, **suggest** four reasons why this is a suitable location.

OS Map 2: Limerick City

Q

1. **What** natural river feature is found at R 55 55?
2. **Use** six-figure grid referencing to locate an antiquity on this map.
3. **Identify** the features at each of the following grid reference locations:
   - R 589 548
   - R 576 575
4. **Calculate** the area of this OS map.

20. Geographical Skills: Aerial Photographs, Charts, Graphs and Infographics

Photograph 2: Limerick City

Q

1. Is this photograph vertical or oblique?
2. **Describe** three possible ways in which the River Shannon may have influenced the development of Limerick City.
3. **Draw** a sketch map of the area shown in the photograph. **Show** and **name** the following on your sketch map:
   - The river
   - Two bridges
   - An area for leisure activities
   - Trees
   - A factory
   - A block of apartment buildings.
4. **What** evidence from this photograph shows that flooding is not a major concern for the city?
5. Do you think the factories located in the left background of this photograph are suitably located? Referring to the photograph, **give** three reasons to support your answer.
6. This photograph was taken over the point R 57 57 on the OS map of Limerick City on the previous page. In what direction is the camera pointing?

Photograph 3: Rosslare, Co. Wexford

Q

1. Is this photograph vertical or oblique?
2. (a) A multinational company has decided to build a new factory in the location marked X. In which of the following sections of the photograph are industrial buildings visible?
   (i) North-west, (ii) South-east or (iii) South-west
   (b) **Explain** two reasons why the area marked X would be a good location for a factory. Use evidence from the photograph to support each reason.

Photograph 4: Aviva Stadium, Dublin

Q

1. Is this photograph vertical or oblique?
2. Using the correct term (refer back to figure 3 if you need to), **give** the location of the Aviva Stadium on this map.
3. **What** evidence can you identify that supports the argument that this is a good location for the Aviva Stadium?
4. Using evidence from the photograph, **what** potential objections may have been made against locating the stadium in this location?
5. **What** evidence of urban planning can you see in the photograph?
6. **Give** one piece of evidence of flood control measures visible in this photograph.

Part 2: Charts, Graphs and Infographics

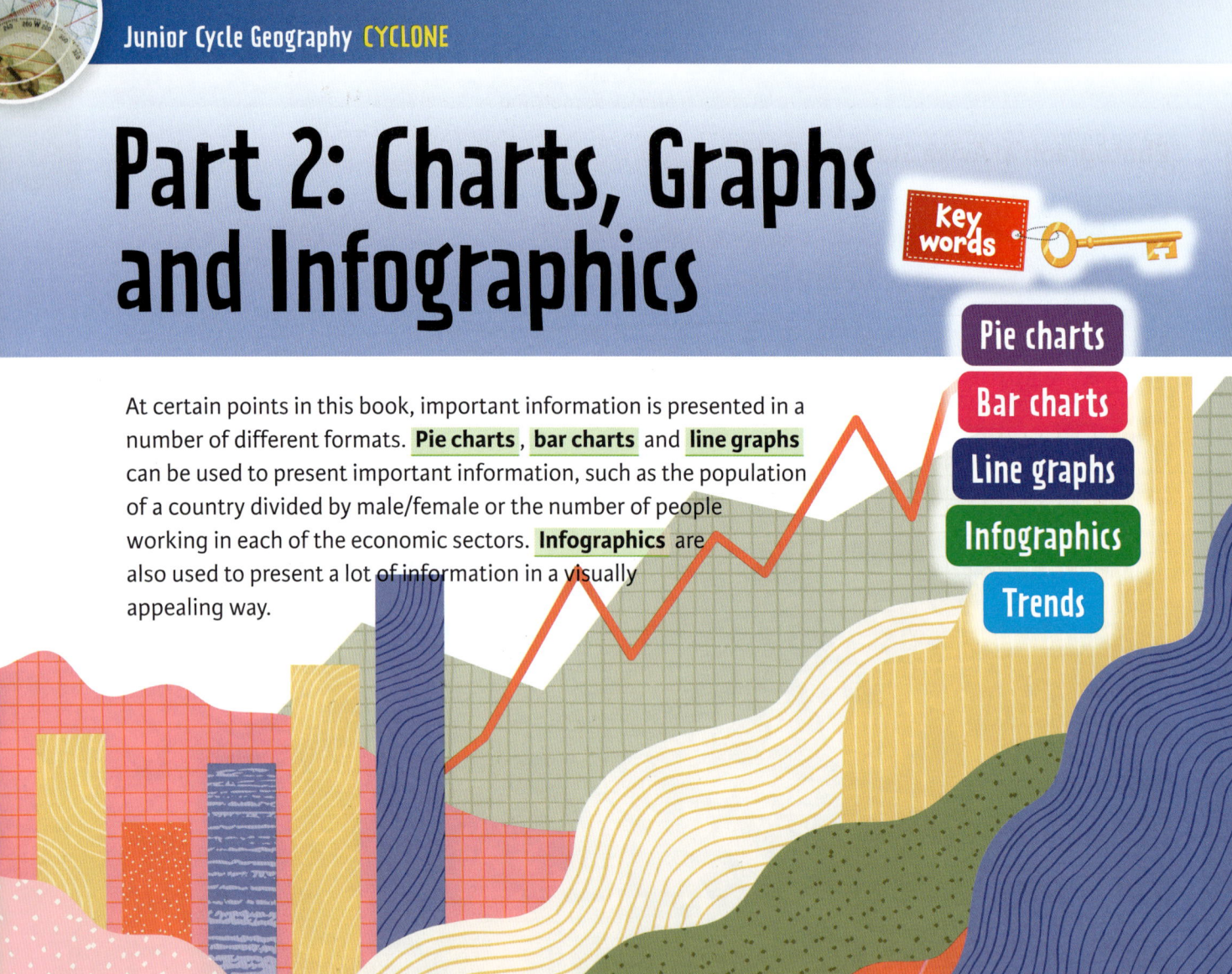

key words
- Pie charts
- Bar charts
- Line graphs
- Infographics
- Trends

At certain points in this book, important information is presented in a number of different formats. **Pie charts**, **bar charts** and **line graphs** can be used to present important information, such as the population of a country divided by male/female or the number of people working in each of the economic sectors. **Infographics** are also used to present a lot of information in a visually appealing way.

20.5 Line Graphs

Line graphs are often used to display information in a visually appealing way. Unlike pie charts or bar charts, line graphs can track changes to information over periods of time, e.g. the number of students choosing a particular subject in 5th Year over a five-year period.

The following table displays the information accurately:

Subject	2012	2013	2014	2015	2016
Geography	34	30	24	35	38
Chemistry	18	20	19	20	18
Engineering	16	20	20	22	24
French	14	16	12	10	10

However, it is not very visually appealing. **Trends** (the way something is developing) can also be difficult to see in a table. A line graph displays the same information in a far more appealing way. Trends can be seen clearly, and information is easier to understand.

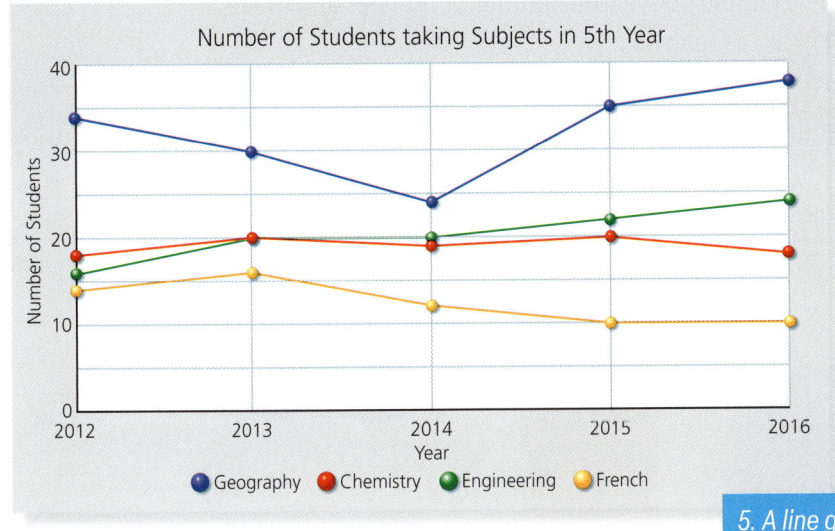

5. A line graph

Q Examine the line graph in figure 6 and answer these questions:
1. According to the line graph, which subject is the most popular?
2. What trends can you identify for each of the subjects over this five-year period?

20.6 Pie Charts

A pie chart displays information using a circle divided into segments and is useful for showing how a total number is divided up. Each segment displays a piece of information as a percentage of a total.

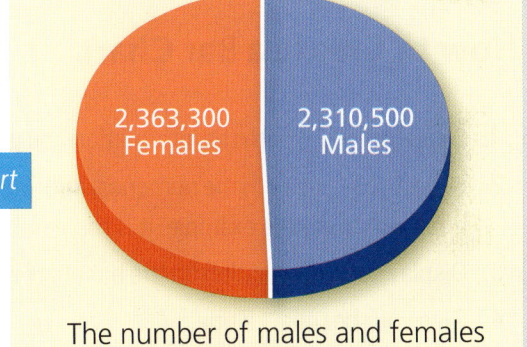

6. A pie chart

The number of males and females in Ireland in 2016

How to Create a Pie Chart

The information displayed in pie charts is numerical. To accurately display numerical information on a pie chart, you must convert the numbers into degrees (°).

Converting numbers into degrees

Let's say your class has 24 students: 15 girls and 9 boys. To convert that information into data that can be displayed on a pie chart, do the following:

1. Put the numbers of females and males over the total number in the class: $\frac{15}{24}$ and $\frac{15}{24}$
2. Multiply each fraction by 360 (the number of degrees in a circle) to change the information into degrees:

    $\frac{15}{24} \times \frac{360}{1} = 225°$

    $\frac{9}{24} \times \frac{360}{1} = 135°$

3. Now use a compass to draw the circle and a protractor to accurately measure the degrees you just worked out.
4. Give your pie chart a title/label.

20.7 Bar Charts

Bar charts display information using bars of different heights and are suitable for making comparisons. The information from the example above (the number of females and males in a class) would be displayed on a bar chart as follows:

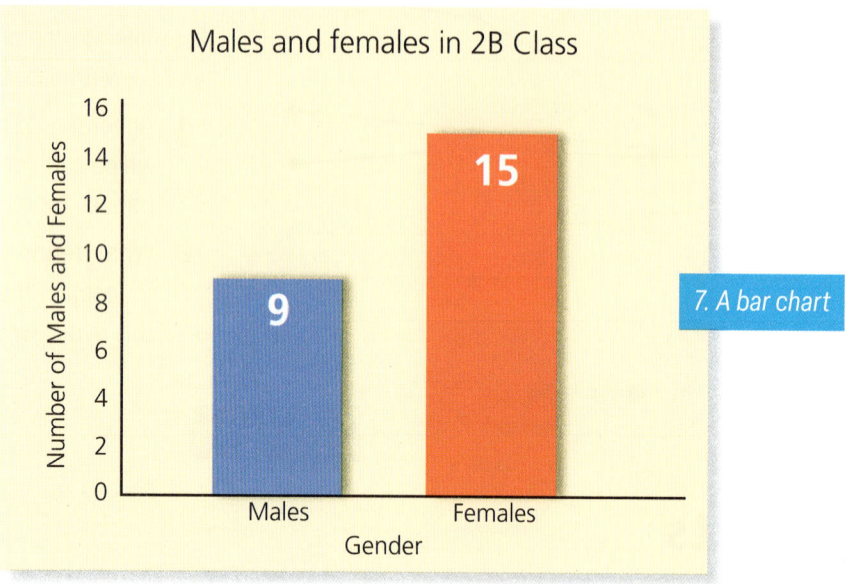

7. A bar chart

How to Create a Bar Chart

1. Label the x-axis with the information you wish to display – in this case, gender.
2. Label the y-axis with the numerical information you wish to display and add numbers in multiples as appropriate. In this example, it is the number of males and females.
3. Display your information on the bar chart using different coloured bars.
4. Give your bar chart a title.

20.8 Infographics

Infographics display a lot of information in a visually appealing and easy-to-understand way.

How to Create an Infographic

- Keep it simple.
- Keep it focused.
- Show things visually.
- Make it a manageable size.
- Check your facts and figures.

20. Geographical Skills: Aerial Photographs, Charts, Graphs and Infographics

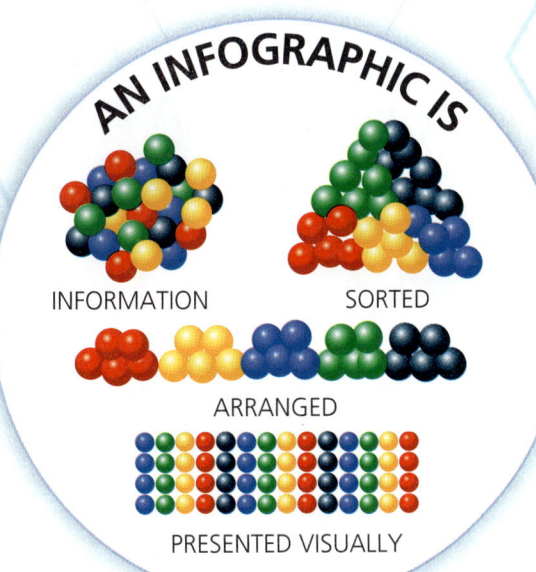

AN INFOGRAPHIC IS
INFORMATION
SORTED
ARRANGED
PRESENTED VISUALLY

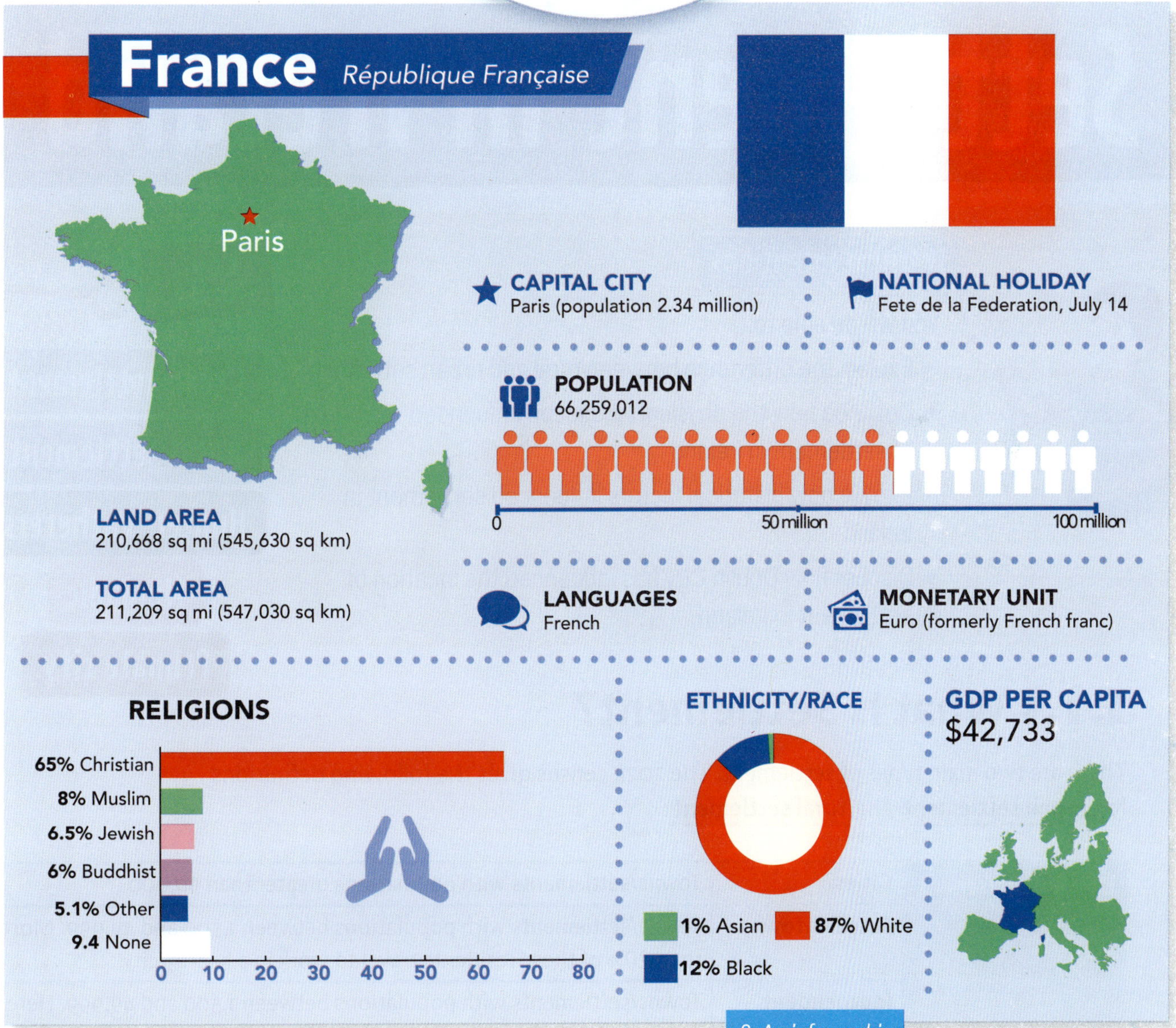

France *République Française*

CAPITAL CITY
Paris (population 2.34 million)

NATIONAL HOLIDAY
Fete de la Federation, July 14

POPULATION
66,259,012

0 — 50 million — 100 million

LAND AREA
210,668 sq mi (545,630 sq km)

TOTAL AREA
211,209 sq mi (547,030 sq km)

LANGUAGES
French

MONETARY UNIT
Euro (formerly French franc)

RELIGIONS
- 65% Christian
- 8% Muslim
- 6.5% Jewish
- 6% Buddhist
- 5.1% Other
- 9.4 None

0 10 20 30 40 50 60 70 80

ETHNICITY/RACE
- 1% Asian
- 87% White
- 12% Black

GDP PER CAPITA
$42,733

8. An infographic

315

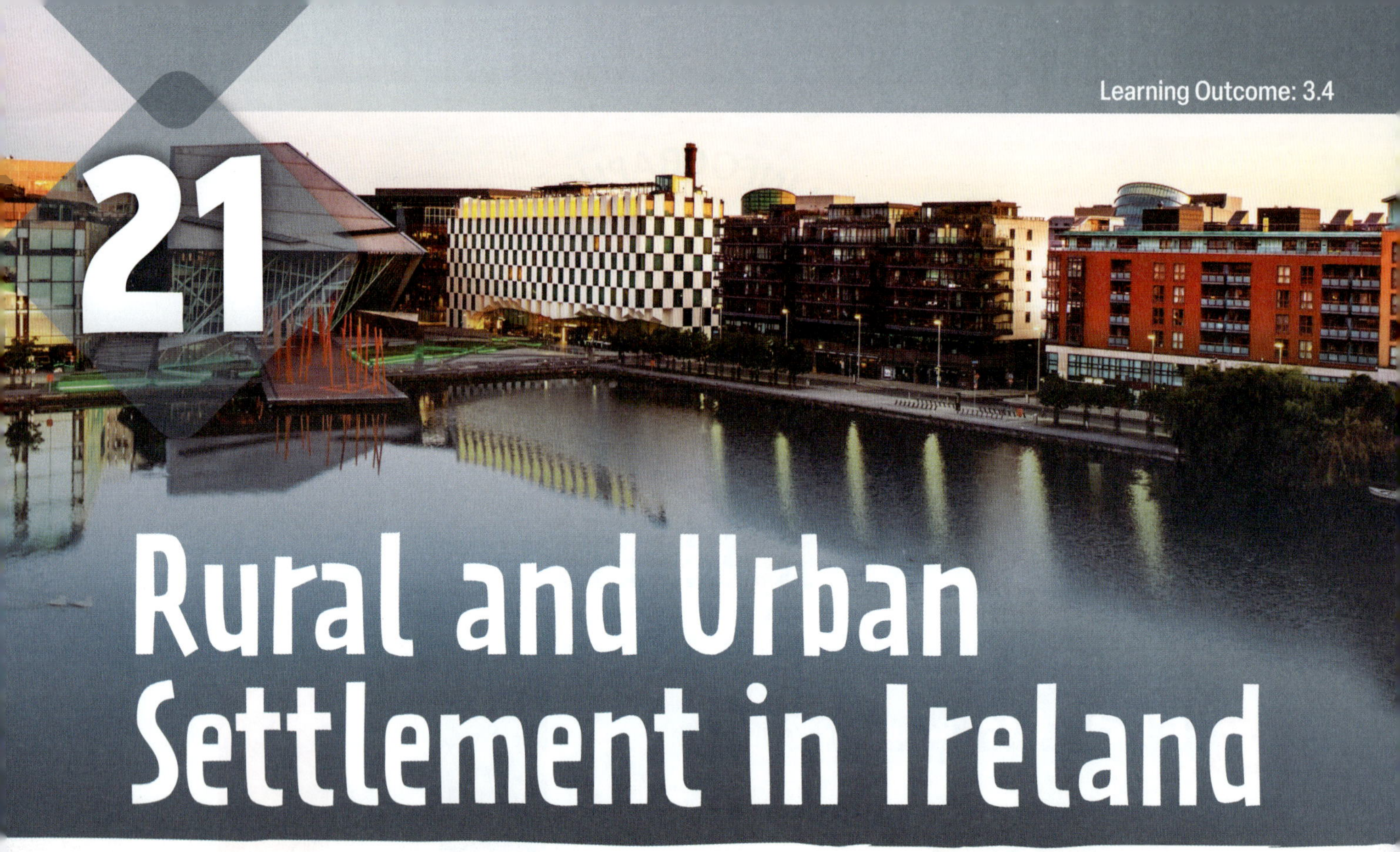

Learning Outcome: 3.4

21 Rural and Urban Settlement in Ireland

Learning Intentions

You will be able to:
- Explain the difference between rural and urban settlement
- Describe how the physical landscape influenced the origin of settlement in Ireland
- Explain how historical factors influenced settlement in Ireland
- Discuss how Dublin City has influenced the location of settlement in Ireland.

Key words
- Urban settlement
- Rural settlement
- Hunter-gatherers
- Nomadic
- Primate city

21.1 What Is Settlement?

There are two main types of settlement. The 2022 census gives the following definitions for **urban settlement** and **rural settlement**:

Urban Settlement	Cities	Towns/settlements with populations greater than 50,000
	Satellite towns	Towns/settlements with populations between 1,500 and 49,999. More than 20% of these residents travel to work in a nearby city.
	Independent urban towns	Towns/settlements with populations between 1,500 and 49,999. Here less than 20% of these residents travel to work in a nearby city.
Rural Settlement		Areas with a population of less than 1,500 people

21. Rural and Urban Settlement in Ireland

Working with others

With the person sitting next to you, debate the pros and cons of living in towns vs living in the countryside. Use the images here to help.

FUN FACT! Data from 2020 shows that almost 64% of the population of Ireland now live in urban areas.

Question Time

1. (a) Define the three types of urban settlement in Ireland.
   (b) Describe the key differences between each type of urban settlement in Ireland.
2. Classify the location of your home, school and next nearest settlement according to the urban and rural definitions.

The following factors have influenced settlement in Ireland:

Physical landscape

Historical factors

The importance of Dublin City

21.2 The Physical Landscape

The first people came to Ireland to live approximately 10,000 years ago during a period of the Stone Age known as the Mesolithic. The people did not live in large settled communities. They were **hunter-gatherers** who lived a **nomadic** lifestyle. They lived in small groups, and moved from place to place, sometimes along river valleys, to find food.

Ireland's first farmers, who raised animals and grew crops, came later. They settled in places and started to build more permanent dwellings. This period is known as the Neolithic.

What Influenced Where They Chose to Build Settlements?

Altitude and relief

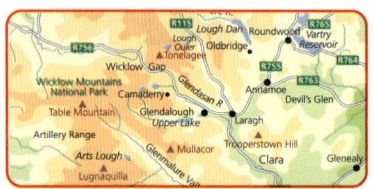

Altitude means a location's height above sea level. Relief is the difference between its highest and lowest points and what the land is like in between (flat or mountainous). Flat or gently sloping lands are:

- Easier to build on and to farm. They are also better for building infrastructure, such as road networks, on.
- Usually warmer and less windy than upland areas.

Drainage

Most Irish towns and villages are found in areas that are well drained by rivers. Areas of land that are marshy are avoided because they are likely to flood.

Settlement

Rivers

- For Ireland's first settlers, rivers provided routes through the country and back to the coastline (see Chapter 10).
- Rivers provided water for drinking, cooking and other domestic purposes.
- Settlements were built on rivers at dry point sites. These were places where the river did not flood during times of heavy rainfall.
- Towns where bridges were built became bridging points. Roads usually converged on these points.
- Port activities, such as trading, have led to the growth of urban settlements at the mouths of rivers.

Fertile soil

Fertile land, found in flat lowland river valleys, is better for growing crops.

21. Rural and Urban Settlement in Ireland

Athlone Town and the River Shannon

Q

1. Looking at the aerial photograph of Athlone, describe three ways in which the River Shannon may have contributed to the growth of the town.

2. Draw a sketch map of the aerial photograph. On your sketch, show and name the following:
   - The river
   - A bridge
   - A car park
   - A housing estate
   - An area of potential flooding
   - A place facilitating boating activities

3. Examine the photograph of Athlone for one minute. Then turn to the person next to you and discuss why settlement has developed in the foreground and centre of this photograph. Try to come up with two reasons.

Managing information

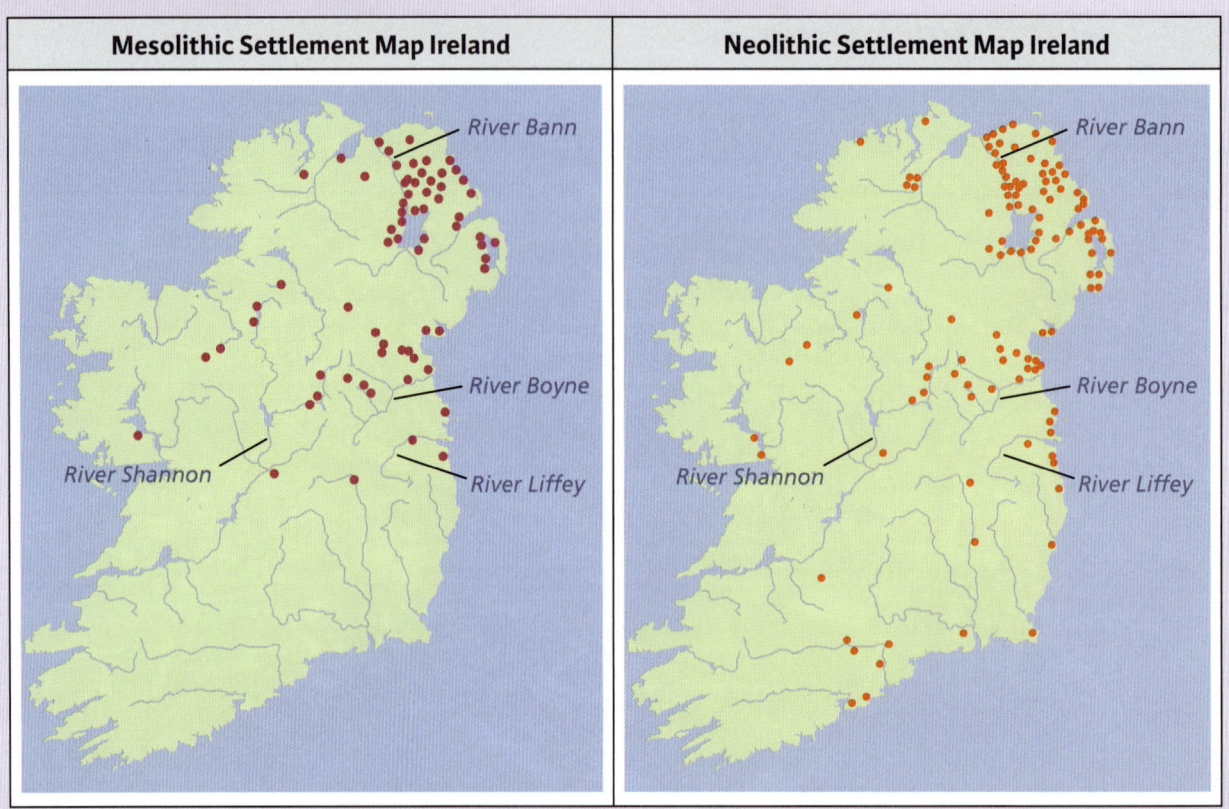

Study both maps above. Write a paragraph explaining:
- The difference in the amount of settlement between each map
- Three possible reasons for this difference.

Managing information and thinking

Examine the OS map of Killybegs. Give three pieces of evidence that show how the physical landscape influenced the location of Killybegs. Provide grid references to support your answers.

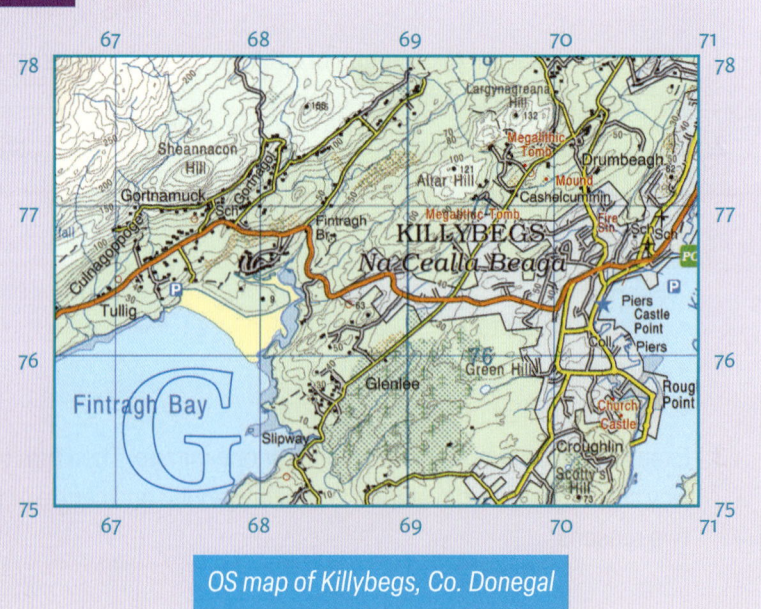

OS map of Killybegs, Co. Donegal

21. Rural and Urban Settlement in Ireland

21.3 Historical Factors

Throughout history, groups of people have come to Ireland and settled in different parts of the country.

Viking Settlement

- Viking settlement in Ireland took place between 795– 1169 CE.
- Irish cities such as Dublin, Cork and Limerick can trace their origins back to Viking settlement.
- The Vikings settled near the coast so they could trade and return home (to Norway, Sweden and Denmark), if necessary.
- Buildings consisted mainly of wooden forts, built on top of hills.

Norman Settlement

- The history of the Normans in Ireland stretches from 1169 to 1500 CE.
- The Normans further developed already existing Viking settlements.
- Towns such as Kilkenny can trace their origins to Norman settlement.
- Normans began their settlement in the south-east of the country. They used rivers such as the Slaney to move to further inland.
- Norman settlements often consisted of stone castles.

Present-Day Settlement

- This map shows the location of urban centres with populations over 15,000 (2022 figures).
- As we will see in Chapter 22, more and more people now live and work in urban areas.
- The result is a huge demand for housing in cities and urban areas, which has led to an increase in house prices.
- This has created urban sprawl, forcing many people to live in areas surrounding cities, especially Dublin. These people often have to commute (travel to and from) long distances to work.

Managing information and thinking

Examine the map showing present-day settlement of populations over 15,000 in Ireland. Out of the places pinpointed, how many are in the county of Dublin? After Dublin, in which county are the most centres of population over 15,000? What do you notice about the location of populations over 15,000 in Ireland?

Question Time

1. Rivers are one physical factor which influenced the location of Viking settlement. Explain one other physical factor which influenced the location of their settlements.
2. Explain why the Normans began their settlement in the south-east of the country.
3. Present-day settlement is influenced by other factors. Explain two of these factors in detail.

Historical Settlement on OS Maps and Aerial Photographs

The Neolithic structures built by Ireland's first settlers to bury their dead provide evidence of where they settled.

Study the OS map and answer the questions.

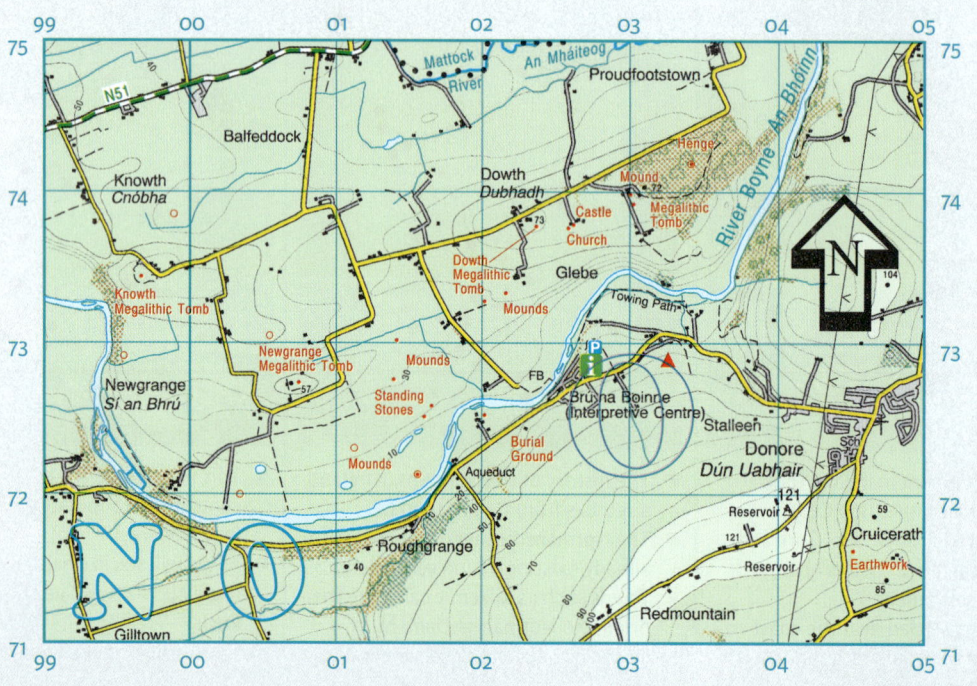

1. What structures built by Neolithic people to bury their dead can be seen on this map? (Hint: a famous attraction in Meath.)
2. Give a six-figure grid reference for two of these structures.

21. Rural and Urban Settlement in Ireland

Managing information and thinking

Examine the OS map of Kilkenny and answer the following questions:

1. With reference to the map and photograph, give three reasons for the development over time of Kilkenny City.
2. What evidence can you see in the photo that suggests Kilkenny was once a Norman town?
3. Locate the castle shown in the photo on the OS map and give its six-figure grid reference.

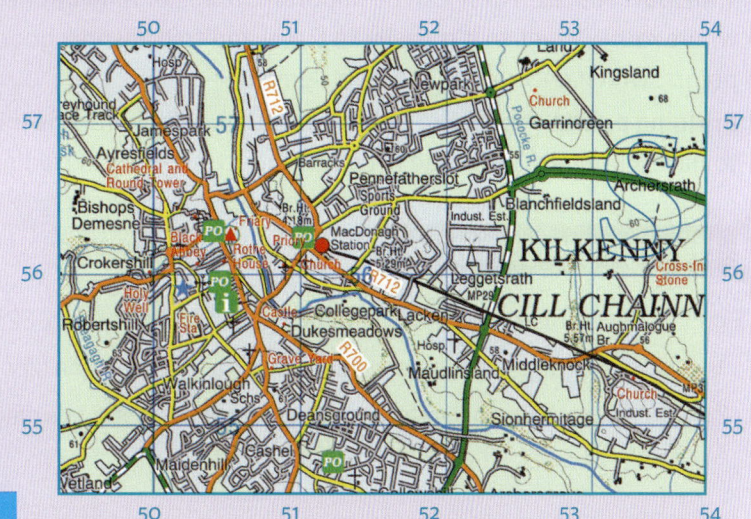

OS map (above) and aerial photograph (below) of Kilkenny City

See pp. 98–101 in Chapter 7 for information on settlement patterns and OS maps.

Managing information and thinking

Revise the functions of settlement on p. 102 of Chapter 7. List the functions provided by the settlement you live in or near.

Attempt to chart the increase/decrease in the functions provided by this settlement over time.

The functions of a settlement can depend on a number of factors. The settlement's location is important, e.g. those on a coastline may have a port function. Norman towns would have had a defensive function with a castle or town walls to protect against attack.

In Chapter 22, we will examine how the change in settlements over time can lead to the development of more functions and to increased population.

Factors Which Influenced the Development and Location of Towns

The location of urban settlements such as towns can also be influenced by the following factors:

- **Transport routes**: These can be important roads or railway lines. People like to live close to where routes meet so it is easier to travel and to transport goods. Trade can develop at these points, which encourages more settlement.
- **Bridging points**: A bridging point is a settlement where a river is narrow or shallow enough to have a bridge. The bridge becomes a route centre and trading centre.

Managing information

Examine the OS map and aerial photograph of Kilkenny City on p. 323.
1. Give the six-figure grid reference for a bridging point in the city.
2. Give the six-figure grid reference where two different roads meet in or near the city.

21.4 The Importance of Dublin City

Dublin is Ireland's **primate city**. A primate city is a city with a population of more than twice the size of the next largest city in the same country (or containing over one-third of a nation's population).

Urban settlement has developed in clusters around Dublin City, as can be seen in figure 1. Urban settlement has developed outside of Dublin because of its close proximity to the city.

- **Satellite towns**: Swords is an example of a satellite town. It is located just over 17 km away from Dublin city centre. In 2022, it had a population of approximately 39,000 people. Swords is located in Fingal north Co. Dublin. According to census data, Swords is the largest feeder town to Dublin City. Nearly 8,000, or 40% of the 20,000 workers in Swords, commuted to work there.
- **Independent urban towns:** Kildare town in Co. Kildare is an example of an independent urban town. It is located 60 km from Dublin city centre. In 2022, the population of Kildare town was approximately 8,600. Of the town's 3,520 workers, 646 or 18% commute to Dublin for work.

1. Urban settlement around Dublin City and county

Using a map of Ireland or Google Maps, identify the counties in which each of the towns labelled in figure 1 is found. Give two reasons why settlement has clustered around Dublin.

Being Ireland's primate city means that Dublin:
- Provides employment opportunities across all economic sectors. For example, multinational corporations (MNCs) such as Google have their headquarters in Dublin. It is also where the Irish government, Dáil Éireann, sits.
- Is the setting for many sports and cultural events, e.g. the All-Ireland hurling and football finals in Croke Park, major art exhibitions in the National Gallery of Ireland.
- Has transport infrastructure – Dublin Bus, the Luas and the DART – linking the city with surrounding urban towns.

Factors Affecting the Location of Dublin City

Drainage

At one point, Dublin had over 60 rivers flowing entirely above ground. Over time, many of these rivers disappeared underground as Dublin grew and there was more construction. These rivers controlled flooding in the Dublin area. Today the Rivers Tolka, Liffey and Dodder keep large parts of Dublin City from flooding.

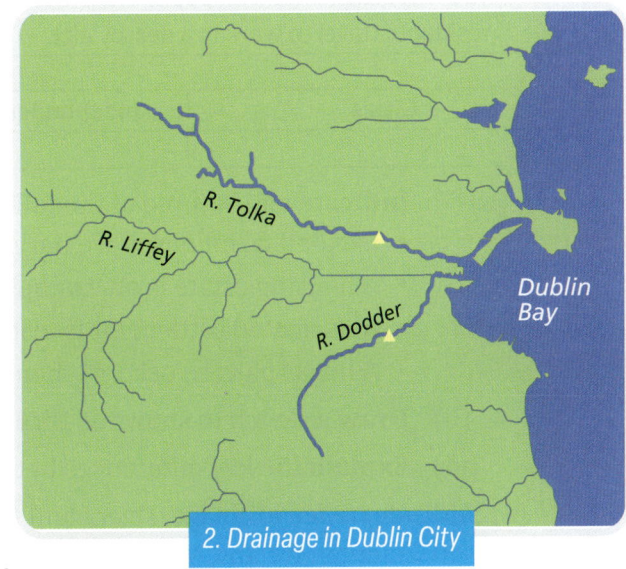
2. Drainage in Dublin City

Altitude and Relief

As you can see from the map below, the altitude of Dublin city centre (light blue to blue) ranges from 3 m to 9 m above sea level in the inner-city areas. The surrounding suburbs (light green to green) have a higher altitude of between 30.5 m to 60+ m above sea level. Towns and infrastructure built on land above 200 m experience more wind, rain and colder temperatures. It is easier to build infrastructure on lowland areas.

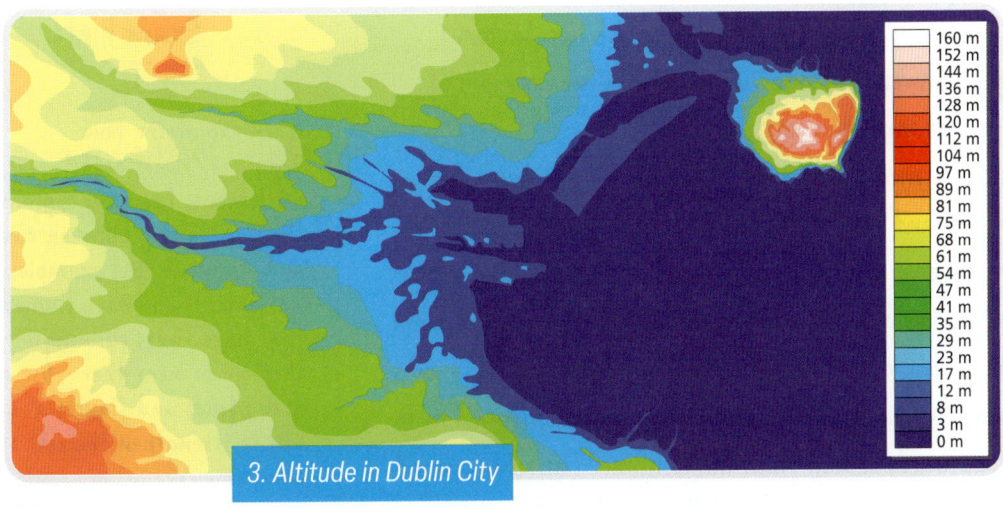
3. Altitude in Dublin City

Managing information and thinking

Using Google Maps or an atlas, list a settlement found on or close to each of the three main rivers of Dublin. Investigate what type of settlement this is (satellite town or independent urban town).

Sample questions

1. (a) In your copy, **match** the type of urban settlement with the correct definition:

A	Towns/settlements with populations between 1,500 and 49,999. Here less than 20% of these residents travel to work in a nearby city.
B	Towns/settlements with populations greater than 50,000
C	Towns/settlements with populations between 1,500 and 49,999. More than 20% of these residents travel to work in a nearby city.

Cities	Independent Urban Towns	Satellite Towns

(b) (i) **Indicate** which three of the following are true regarding settlement patterns:
- Isolated farmhouses are part of a dispersed settlement pattern.
- A housing estate is an example of a linear pattern of settlement.
- A nucleated settlement is also called a clustered settlement.
- A linear pattern occurs when houses form a line along a road or river.

(ii) **Draw** a sketch to show the three types of settlement patterns.

(iii) **Explain** the key differences between the three types of settlement pattern.

2. (a) Altitude/relief, drainage/rivers and _____ are contributing factors to settlement in Ireland. What is the missing factor?

(b) **Choose** any two of the factors above and **describe** their impact on a settlement in Ireland.

Let's get started! Sample starter

> 2. (b) Altitude/relief and drainage/rivers are the two factors I have chosen. In my answer I will use Dublin City as an example of how both these factors impacted settlement here.
> Altitude/relief: Altitude means the height above sea level. Relief is the shape and slope of the land between places. The more flat or gentle sloping land is, the easier it is for settlement to start there. Viking settlers looked for land with a low altitude and gentle relief. This protected them from harsh weather conditions. The gentle relief of the land would have made farming easier. Dublin City's altitude moves gradually from 3 m to 23 m as you move outwards or west from the coastline. The relief around Dublin City is gently sloping land. The low altitude and gentle relief allowed the settlement of Dublin City to reach what it is today. …

Exam hint!

When explaining how drainage/rivers affect settlement, begin by explaining the relationship between rivers and drainage. Use examples of settlements you have studied that have rivers running through or near them. Use as many facts and figures as you can to strengthen the rest of your answer.

… now complete the answer in your copy.

Learning Outcome: 3.5

Urban Change in Dublin

22

Learning Intentions

You will be able to:
- Outline what urban change is
- Examine the causes of urban change in an Irish city
- Describe the effects of urban change on an Irish city.

key words

- Urbanisation
- Urban change
- In-migration
- Immigration
- Urban sprawl
- Urban decay
- Migration
- Recession
- Urban redevelopment
- Urban renewal
- New towns

22.1 Urban Change

The growth of both the size and number of urban areas (towns and cities) is sometimes referred to as **urbanisation**. Urban areas around the world are constantly undergoing **urban change**. This is the process whereby cities, towns and other built-up areas develop, usually to meet the changing needs of their populations. This process can result in changes to how the urban land is used, to the types of buildings constructed and to the expansion of urban areas.

A

B

Q These two photographs show Shenzhen, one of the fastest-growing cities in China. This was a small fishing village as recently as 40 years ago (image A). Today it has a population of over 12 million. What do you think caused so many people to move to this area?

327

Cities such as Dublin undergo constant and various changes. The greatest of these changes is their growth in terms of area and population.

The table below outlines some causes and effects of the growth of Dublin.

Causes of Population Change
- Relatively high natural increase in population
- In-migration
- Land use within a city

Urban Change/ Urbanisation →

Effects
- Urban sprawl
- Urban decay
- Urban redevelopment and urban renewal
- New towns

22.2 Population Change in Dublin

Dublin county is divided into four council regions. These regions are called South Dublin, Fingal, Dún Laoghaire-Rathdown and Dublin City. Together, these areas make up the Dublin metropolitan area. A metropolitan area is a city and its surrounding areas.

Relatively High Natural Increase

A relatively high natural increase in population occurs when the birth rate is higher than the death rate (see Chapter 18: Population). In 2022, the Dublin metropolitan area had a total population of 1,256,000, which was a 1.13% increase from 2021. The population is expected to continue to grow.

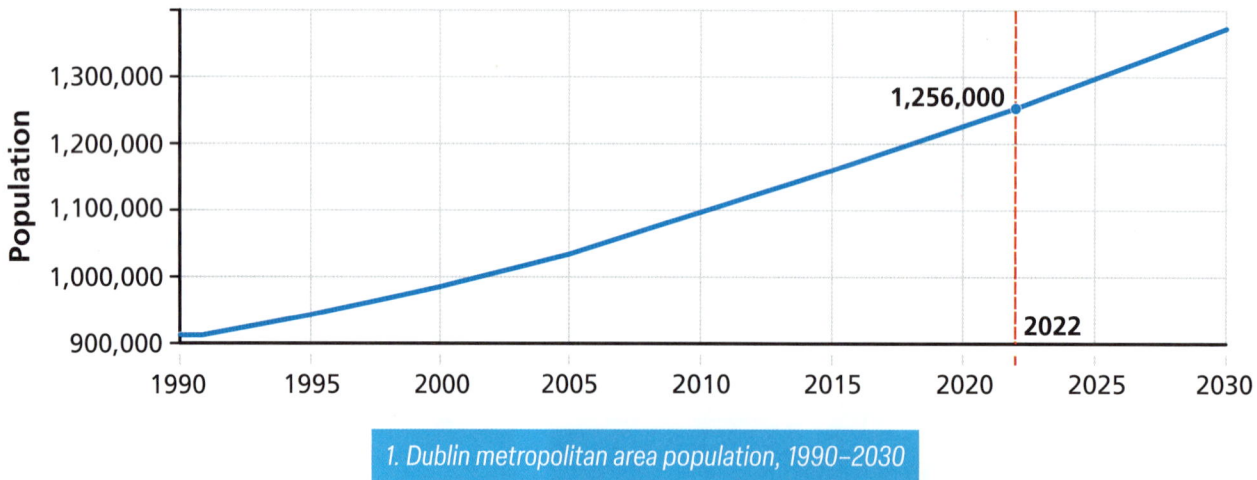

1. Dublin metropolitan area population, 1990–2030

A natural population increase tends to occur in areas with a young adult population. Fingal in north Dublin has the lowest average age in the country at 34.

Being numerate

Examine the graph in figure 1 and answer the following questions.
1. What is the projected population for the Dublin metropolitan area in the years 2025 and 2030?
2. Calculate the difference in population for each year as a percentage.

22. Urban Change in Dublin

Immigration

Immigration occurs when people choose to move from their country to a new country for economic reasons (see Chapter 19: Migration). Immigrants often choose to settle in urban areas where there are good job prospects, strong transport links and an existing community or network of people from their homeland.

Land Use Within a City

In Chapter 20, we learned about the functions or services of towns. The more functions a settlement has, the more attractive a place it is for people to live. The functions of an urban area change to meet the needs of its increasing population.

One in six of Fingal residents is a non-Irish national. Fingal also has the most diverse population nationally with 24.5% of its 330,000 population from an ethnic group other than white Irish.

The population of Dublin City has steadily increased during the last 25 years. In 2021, the city's population was 554,554. In 2016, non-Irish nationals made up 17% of the population.

In 2021, the population of South Dublin was 278,767. Several different nationalities live in the region, including people from Poland, Lithuania, other EU countries and the UK. These nationalities make up 30,000 of the region's population.

In 2016, non-Irish nationals made up 10.5% of the 218,000 population of Dún Laoghaire-Rathdown. The nationalities include people from the UK, Poland, Lithuania, Nigeria and Brazil.

2. Dublin metropolitan area's councils

1. List three nationalities other than Irish found in any of Dublin's four councils.
2. Which council has the highest percentage of non-Irish nationals in its population?
3. Give one reason why you think Dublin county is divided into four separate councils.

Case Study: Dublin's Functions Over Time

Dublin City provides a wide variety of functions. Over time, these functions have evolved. Some functions have remained part of Dublin for centuries. Others have developed more recently.

Defensive

Norman walls

3. Viking and Norman settlement shown on a modern-day map of Dublin

Defence was one of the earliest functions of Dublin City. Viking settlers arrived in 841 CE. They built a castle alongside the River Poddle, which is a tributary of the Liffey. The castle protected settlers from attacks.

The Normans walled the city over 300 years later. More people wanted to live in or near cities with defences as they offered protection.

Port

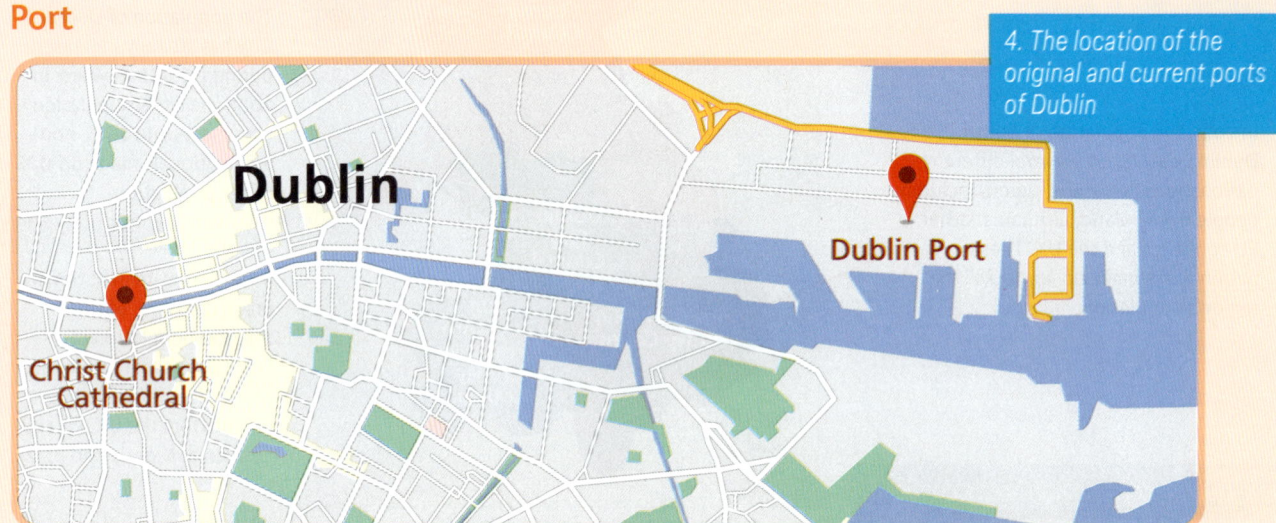

4. The location of the original and current ports of Dublin

The original Port of Dublin was situated further inland along the Liffey, close to Christchurch Cathedral.

Through Viking and Norman times, Dublin's trade with England and Europe increased. In the early part of the nineteenth century, the city's docklands became an important source of work. Workers lived in the city to shorten the commute to work. The development of Dublin Port is linked to the development of Dublin City.

Transport

Ireland's first railway line opened in 1834. It went from Dublin (Westland Row) to Kingstown (Dún Laoghaire) and had nine stations. The daily number of passengers using the train ran into the thousands. This made it the first ever commuter railway. The railway line allowed the city to expand southwards, creating new suburbs and increasing the city's population.

22. Urban Change in Dublin

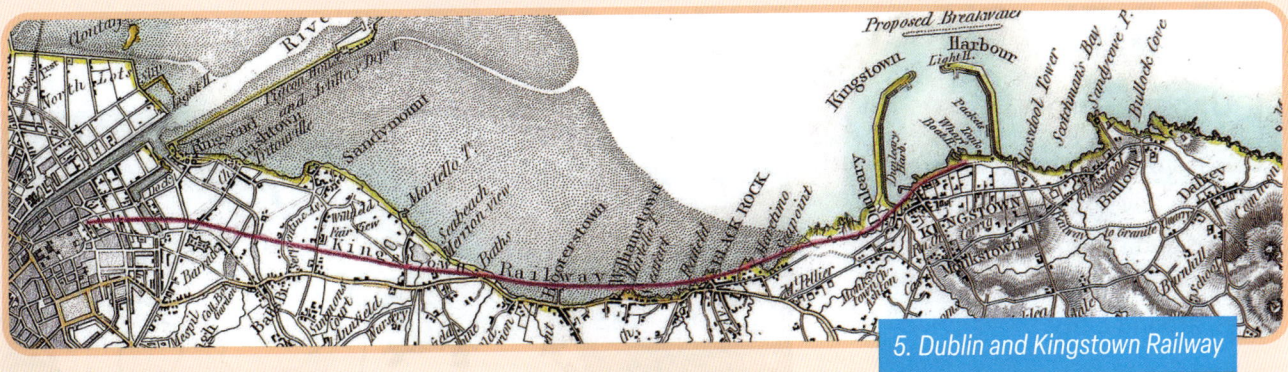

5. Dublin and Kingstown Railway

Present-Day Dublin

- Dublin's more recent functions are located in the city centre, which is dominated by large office buildings particularly in the International Financial Services Centre (IFSC).
- Industrial estates located around the city centre provide space for a variety of economic activities.
- Businesses such as retail, digital media and headquarters of MNCs are located in Dublin City. This pushed settlement westwards, with the creation of suburbs such as Tallaght and Swords.

📁 Managing information and thinking

Look back to Chapter 20 and list examples of the following from the map:

- A residential area
- A type of economic cluster/industrial estate
- A primary economic activity
- A secondary economic activity

Key
- Metropolitan area
- City
- City core
- Metropolitan centre
- Economic clusters
- Other clusters
  A – Docklands and Dublin Port
  B – Grangegorman
  C – Heuston
  D – Digital Hub and St James's Campus

6. Functions located in and around Dublin City

Q

1. (a) List one historical and one present-day function of Dublin City.
   (b) Briefly explain how each function helped to change Dublin.
2. Examine figure 6. Compare and contrast one function of Dublin's core, city and metropolitan area.

22.3 The Effects of Urban Change

1. Urban Sprawl

Urban sprawl is the rapid spread of urban developments (such as houses and shopping centres) onto undeveloped rural land (greenfield sites) around a city.

Dublin experienced rapid urban sprawl during the 'Celtic Tiger' years 1995–2008. This was a time of great economic growth in Ireland. During this time, many people from rural areas and from abroad moved to Dublin in search of work and higher living standards. The resulting population growth led to greater demands for suburban housing and services. This in turn led to the outward spread of the city in the form of urban sprawl.

Urban sprawl has affected both Dublin's inner city and its outer areas.

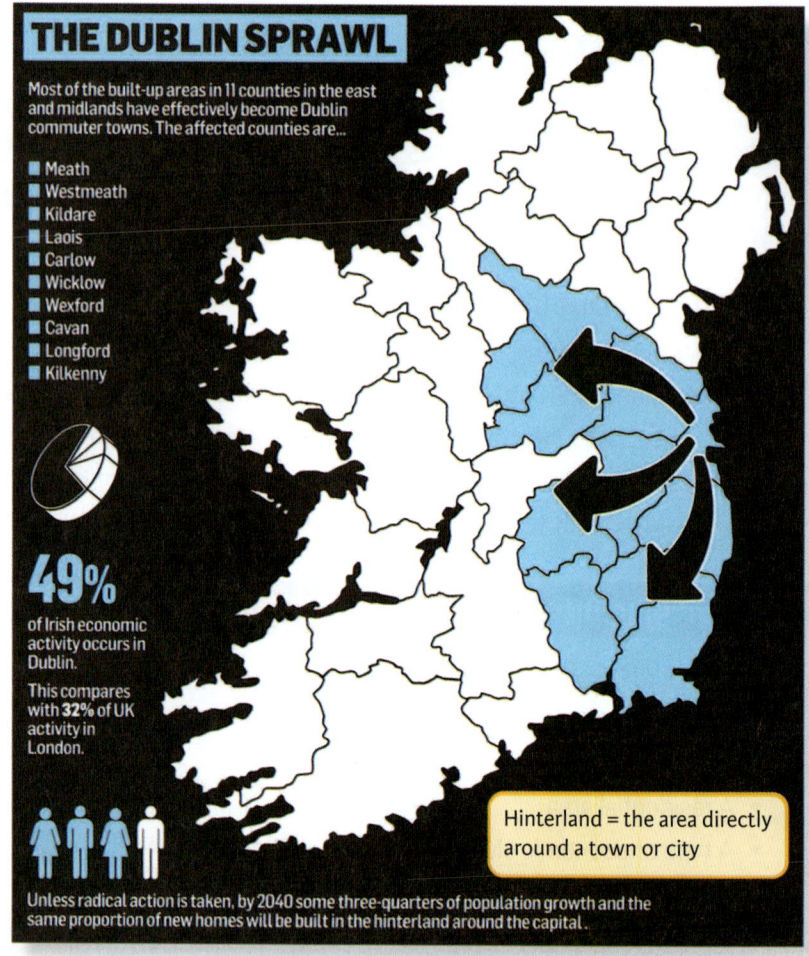

7. Expected urban sprawl in Dublin City and surrounds by 2040

Managing information and thinking

Examine figure 7. It shows the expected level of urban sprawl in Dublin by 2040.

1. How many counties have commuter towns for Dublin?
2. Name a large town in any three counties of your choice which you think might be a commuter town to Dublin.

Effects of Urban Sprawl

- Valuable farmland, such as fertile land in North County Dublin, has been used up for building purposes.
- Old villages such as Dundrum were swallowed up by urban growth. These villages lost their unique identities and became part of a large, impersonal city.
- Some large suburbs are considered boring, 'soulless' places with relatively few recreational or social amenities. They are almost deserted by day, when many of their inhabitants are either commuting or at work.

22. Urban Change in Dublin

Case Study: The Cost of Urban Sprawl – House Prices

Urbanisation has affected house prices in Dublin and surrounding urban areas. If we compare 2021 and 2022 prices in Dublin, there is a clear yearly increase (given as a percentage).

A lack of affordable accommodation in Dublin City and county has contributed to an increase in housing costs in urban areas serving Dublin, e.g. Maynooth, Co. Kildare, and Ashbourne, Co. Meath.

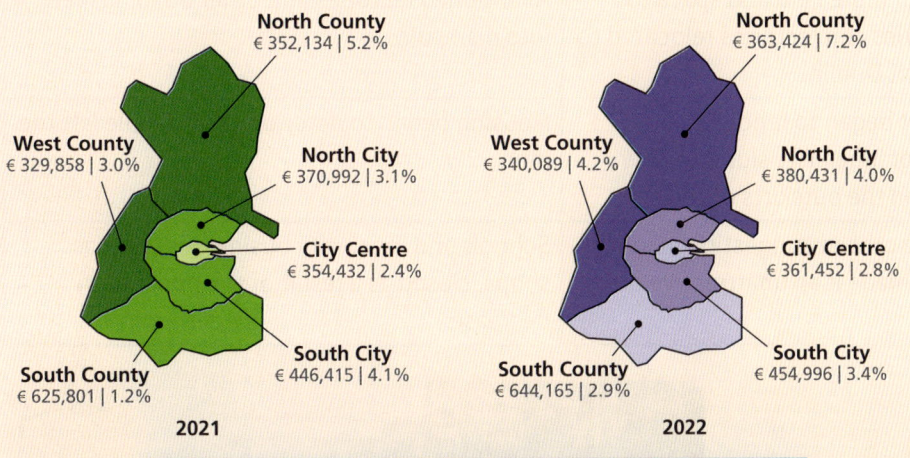

2021
- North County: €352,134 | 5.2%
- West County: €329,858 | 3.0%
- North City: €370,992 | 3.1%
- City Centre: €354,432 | 2.4%
- South City: €446,415 | 4.1%
- South County: €625,801 | 1.2%

2022
- North County: €363,424 | 7.2%
- West County: €340,089 | 4.2%
- North City: €380,431 | 4.0%
- City Centre: €361,452 | 2.8%
- South City: €454,996 | 3.4%
- South County: €644,165 | 2.9%

½² Being numerate

Create a bar chart comparing 2021 house prices in Dublin to those in 2022 (figure 8). Label the x-axis 'Dublin areas' and the y-axis 'House prices'.

8. Average Dublin City and county house prices, 2021 vs 2022

Question Time

1. Define 'urban sprawl'.
2. List two effects of urban sprawl and explain one of them.
3. Think of one possible way in which urban sprawl might have affected the counties surrounding Dublin and briefly explain it.

2. Urban Decay

Inner-city **urban decay** is the deterioration of an urban area due to neglect or age. It can result in buildings being left empty and becoming run down.

Urban decay is a significant problem in cities all over the world, including Dublin. It can be a result of **migration** from an area, which results in a change in the number and age of people living in that area.

Communicating

Talk to the person beside you. Can you think of specific examples in your town or a town near you where there is urban decay?

Case Study: Dublin Inner-City Urban Decay 1960–1990

Between 1961 and 1981, the population of Dublin's inner city grew rapidly. There was a need for urban development and to reduce overcrowding in the inner city. Urban decay took hold of the inner city. The causes and effects are explained in the table below.

	Causes	Effects
Industrial development	Due to a need for more land and space to develop, industries and factories relocated to the outer areas of Dublin.	Abandoned factories and other buildings fell into disrepair.
Rehousing projects	The government began to rehouse people from poor inner-city slums in new developments on the outskirts of the city.	Housing began to deteriorate as people left the inner city.
Traffic congestion	Increased traffic caused congestion in inner-city Dublin and made it an unattractive place to live.	Schools and inner-city businesses, such as shops, closed as a result of the population decrease.

Q Look at the photo of Ballymun in the 1960s, which was developed to rehome inner-city residents. Can you think of two advantages and two disadvantages of this scheme?

Dublin Inner-City Urban Decay Post-Celtic Tiger

During Ireland's Celtic Tiger period, inner-city Dublin was redeveloped. Some of the buildings in the inner-city area were improved.

In 2008, the Celtic Tiger boom ended suddenly. This had several effects:

- Ireland went into **recession** (its economy declined for a sustained length of time).
- This had an effect on inner-city Dublin. People had less money to spend, which decreased profits for shops in the inner city.
- Shop owners could no longer afford to rent buildings in the expensive inner-city area. These buildings were left vacant for long periods of time. Many shops and businesses also closed for good after the Covid-19 lockdowns in 2020 and 2021.

Dublin's famous department store, Clerys, closed down in 2015

22. Urban Change in Dublin

Working with others

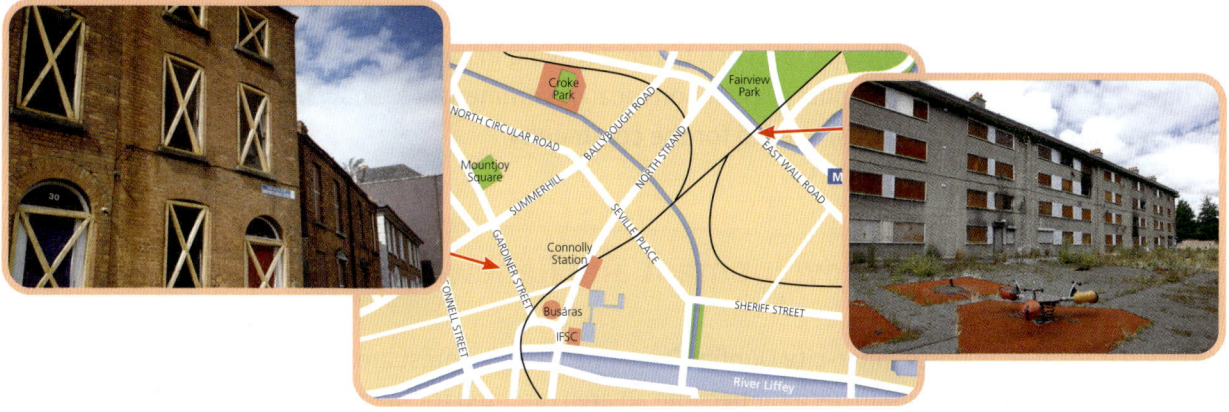

Take one minute to examine these photos of inner-city Dublin. Then, with the person next to you, list the evidence you can see of urban decay.

Question Time

1. Explain what is meant by urban decay.
2. Explain why local authorities moved people out of Dublin's inner city in the 1960s. Do you think this was a good idea? Why?
3. How do you think inner-city areas suffer when their population decreases?

3. Urban Redevelopment and Urban Renewal

The solutions to urban decay are urban renewal and redevelopment.

- **Urban redevelopment** is when old, run-down houses and buildings are demolished. The original tenants are moved to new suburbs, and new shops and businesses are built on the valuable inner-city land instead. The redevelopment of the land near Connolly Station to create the IFSC is an example of this.
- **Urban renewal** is when old buildings are restored or replaced by new buildings. Facilities, such as community centres and Astro Parks, are built to encourage existing residents to remain in the area, e.g. the renewal of Dublin Docklands in the 1990s and 2000s. Urban renewal is sometimes called urban regeneration.

Q The original Ballymun flats – an example of urban redevelopment in the 1960s – were demolished because they had become run down (photo A). Since then, there has been urban renewal in the area (photo B). Why do you think it was decided to renew this area rather than move the residents to a new area?

Urban redevelopment and urban renewal are linked to population change. When people leave inner-city areas and move to suburbs, old buildings can be redeveloped completely, changing their purpose from residential to commercial.

Urban renewal is stimulated when people wish to remain living in inner-city areas that may have become run down. This puts pressure on local governments to renew buildings and to develop amenities (facilities) needed for the community.

Case Study: Urban Redevelopment and Renewal in Smithfield

Smithfield is located in north inner-city Dublin. Once one of the poorest areas in Dublin, experiencing high levels of urban decay, it became an area targeted for massive urban renewal and redevelopment. The area underwent significant commercial and residential development. This led to it becoming newly fashionable in the first decade of the twenty-first century.

Managing information and thinking

Go to the RTÉ Archives website, www.rte.ie/archives, click on 'Search' and look up 'Urban Renewal in Smithfield Square'. Watch the video and then answer the following questions regarding the renewal plan for Smithfield:

1. What was Smithfield Market known for?
2. What four areas of Smithfield did the plan focus on?
3. What concerns did the existing residents have for any planned redevelopment of the area?

Question Time

1. Explain the key differences between urban redevelopment and urban renewal.
2. Using one of the urban areas discussed in this chapter, explain how urban redevelopment/renewal has benefitted the area and the people living there.

22. Urban Change in Dublin

4. New Towns

The creation of **new towns** is a solution to urban sprawl. New towns are built outside of but close to the city. These large, planned towns provide homes, services and employment for many thousands of people.

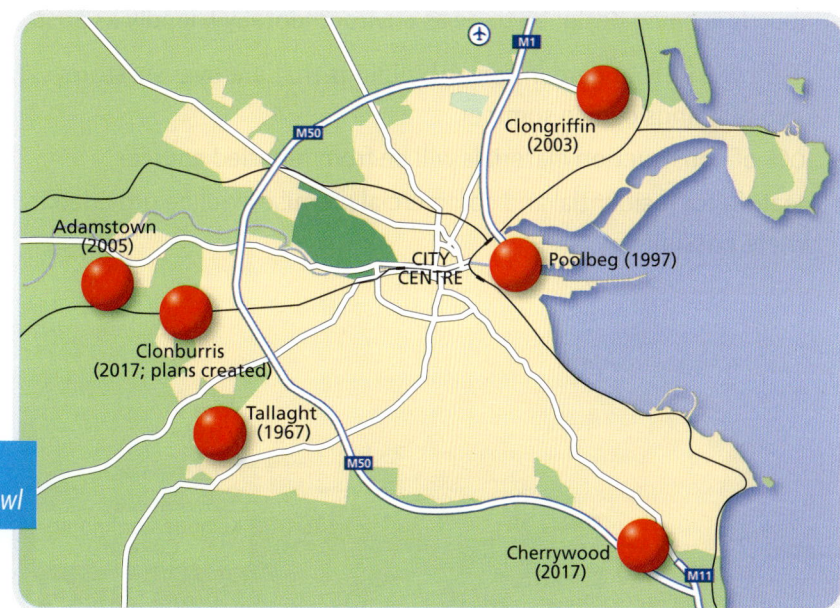

9. Some examples of new Dublin towns, built to reduce the spread of urban sprawl

Case Study: The New Town of Clonburris

In 2019, planning was approved for a new town submitted by South Dublin County. Clonburris is a new urban neighbourhood which, when complete, will deliver more than 8,000 homes for 23,000 people. It will also include retail and commercial hubs supporting up to 10,000 jobs.

The Development of Clonburris

Clonburris is located 12 km from the city centre and 21 km from Dublin Airport. It is also 23 km from the M50, which goes around Dublin City making it easier to access the city and suburbs. The new town will also benefit from being located on the Kildare–Dublin railway line.

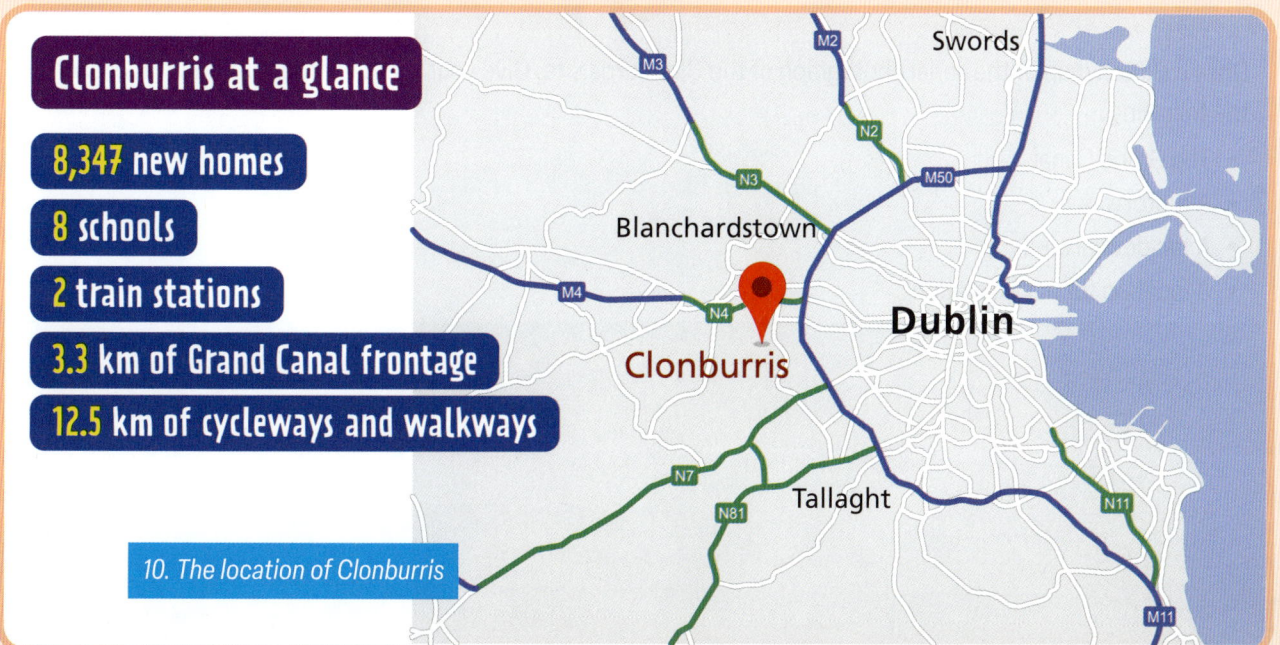

Clonburris at a glance
- 8,347 new homes
- 8 schools
- 2 train stations
- 3.3 km of Grand Canal frontage
- 12.5 km of cycleways and walkways

10. The location of Clonburris

Clonburris is being designed to be as environmentally friendly as possible:

- There will be roof gardens on all buildings that are over six storeys. This will help to keep buildings cool in summer.
- 10% of construction materials will be from recycled sources.
- Heating for residential and commercial premises will be provided by solar power where possible.

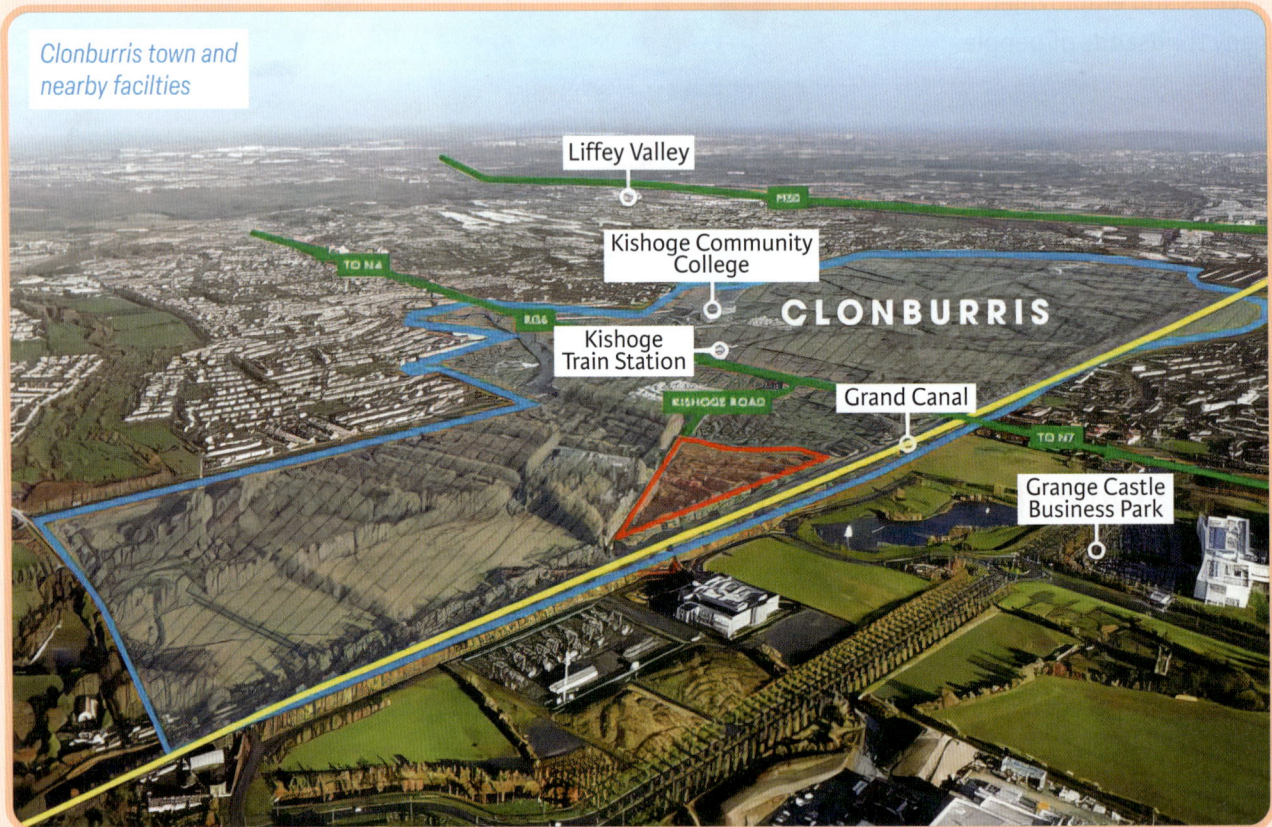

Clonburris town and nearby facilties

Draw a sketch map of the aerial photograph of the Clonburris site. Give your sketch a title. Draw and label each of the following:

- The Grand Canal
- A motorway
- A residential area
- A location for a train station

22. Urban Change in Dublin

Sample questions

1. **Examine** the infographic, which shows the percentage of the population of each continent living in urban areas in 1950 and in 2020.

   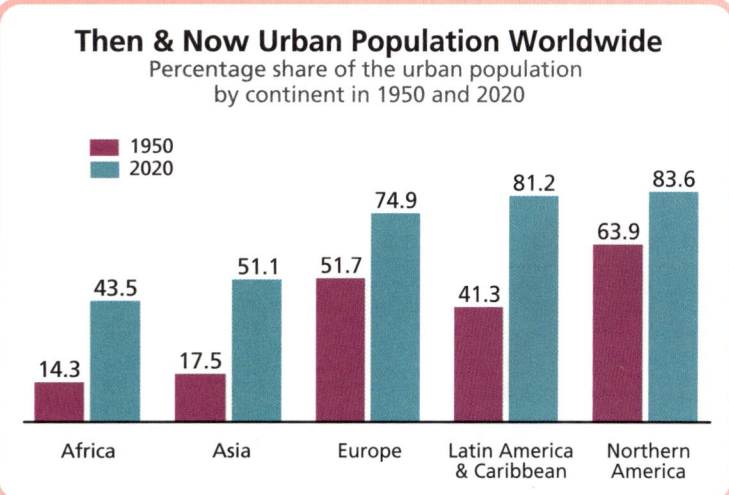

   (i) **Which** continent has shown the greatest change in urban population between 1950 and 2020?

   (ii) **Which** continent has shown the least change in urban population between 1950 and 2020?

   (iii) **Describe** the trend indicated by the bar chart.

2. 'Population change is a driving force behind urban change.'

   Explain two causes of population change in an area/region you have studied.

3. Cities can be great places to live, but they are not without their problems.

   Study the photographs and answer the questions below.

   (i) **Explain** two problems associated with urban change in Irish cities.

   (ii) **List** a solution for each of the problems and explain one.

Exam hint!

For Q3(i), clearly state the two problems you have chosen. Explain each problem individually and use named examples of urban areas you have studied. For part (ii) list two solutions — one for each problem mentioned in part (i). You can use solutions mentioned in your case studies to help.

Junior Cycle Geography **CYCLONE**

CBA 2 My Geography Moment

Examining Urban Change

Identify whether the city or town you live in or the nearest city or town to you has recently undergone urban change. Examine evidence of population change and the causes and effects of this urban change. You can display your findings on a poster or as a presentation.

This task can be done individually or in groups.

I/We Must

- Name the town or city I have chosen that has experienced urban change
- Find out the current population of the urban area and how it has changed over time (whether it has increased or decreased in population between censuses)
- Explain two causes of urban change for the chosen town or city (push-pull factors)
- Explain two effects of urban change on the chosen town or city
- Include photos to showcase the urban area and illustrate the points made.

I/We Should

- Use Scoilnet maps to compare old and new maps of the chosen urban area
- Explain three ways in which the urban area has changed over time, using evidence from the maps.

I/We Could

- Identify an area of urban decline/decay in the chosen urban area
- Put together a plan for regenerating this area and include sketches for what the regenerated area would look like
- Explain how the regenerated area would serve the needs of the local community.

Peer Assessment

Swap your report with a classmate's/another group's report. Write down two things you feel they did well and one thing they could improve on.

Redrafting

Review the success criteria again to make sure you have met all the requirements. Take into account the peer assessment notes and make any changes you think are needed. When you are happy, you can prepare to show your work to the class.

Learning Outcome: 2.3

23

Primary Economic Activities

Learning Intentions

You will be able to:
- Explain the term 'physical landscape'
- Explain the factors that influence the physical landscape
- Describe the three types of economic activity
- Outline what primary economic activities are
- Realise the effects the physical landscape has on primary economic activities (using case studies).

key words

- Physical landscape
- Natural features
- Climate
- Soils
- Relief
- Human activity
- Primary, secondary and tertiary economic activities
- Natural resource
- Farming
- Pastoral
- Arable
- Mixed
- Forestry
- Cool temperate oceanic climate

23.1 The Physical Landscape

The **physical landscape** is the visible **natural features** such as bodies of water (seas, rivers and lakes), mountains, flat lands called plains, deserts and soils.

 Communicating

In pairs, discuss the following:
1. Describe the physical area in which you live, e.g. is the sea nearby? Are there mountains or rivers?
2. What work do the members of your family undertake?

FUN FACT!
Approximately 70% of the earth's surface is covered in water.

Landscapes Found Around the World

Field landscape

Mountain landscape

Bodies of water

Question Time

1. Define the term 'physical landscape'.
2. Name three landscapes found around the world.

23. Primary Economic Activities

What Factors Influence the Physical Landscape?

Climate is the average weather across a large area over a period of 30 years.

Soils are the thin, uppermost layer of the earth. The type of climate and relief in an area will determine the type of soil it has.

Relief/height/slope is how high or flat the land is.

The shape of the coastline is greatly influenced by the sea.

Human activity is anything that humans do that changes the physical landscape or the earth's surface, e.g. establishing cities.

23.2 Introducing Economic Activities

People work in all kinds of different jobs in order to earn money. Jobs and industries are also known as economic activities. They can be broken down into three groups: **primary**, **secondary** **and tertiary economic activities**.

1. Primary economic activities are those in which people get **natural resources** from the land or sea, e.g. **farming**.

2. Secondary economic activities are those in which people process or manufacture products, e.g. car manufacturing.

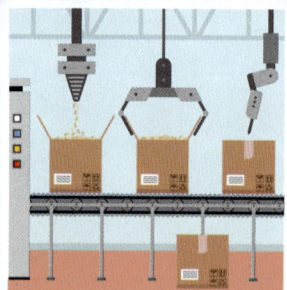   

3. Tertiary economic activities are those that provide useful services for people, e.g. driving taxis.

Working with others

In pairs, write down:
1. Four primary economic activities
2. Four secondary economic activities
3. Four tertiary economic activities.

23.3 What Are Primary Economic Activities?

Ask yourself the following question: How do we extract (take) the natural resources from the earth?

The answer is: through primary economic activities.

Primary economic activities involve the taking and using of the earth's natural resources. Examples of these activities include mining, farming, fishing and forestry.

Working with others

A

1

B

2

C

3

Working with your classmate, match each of the natural resources (images 1, 2 and 3) with the primary activity that extracted them (images A, B and C).

23.4 The Physical Landscape: Farming and Forestry

Farming is the raising of livestock and the growing of crops. There are many types of farming.

Pastoral farming	The farming of animals, e.g. dairy farming, raising beef cattle and raising sheep
Arable farming	Growing crops in fields, which have usually been ploughed before planting
Mixed farming	Elements of both pastoral and arable farming
Forestry	The planting, management and maintenance of trees. Some greenfield areas in Ireland are used for forestry, while others are used for farming.

The physical landscape has a huge influence on what our land is used for. In Chapter 17, we studied local soil and looked at what vegetation grows on it and why.

Some of the factors that determine if land is used for farming or forestry include how fertile the soil is, what the relief is like and what the climate is like.

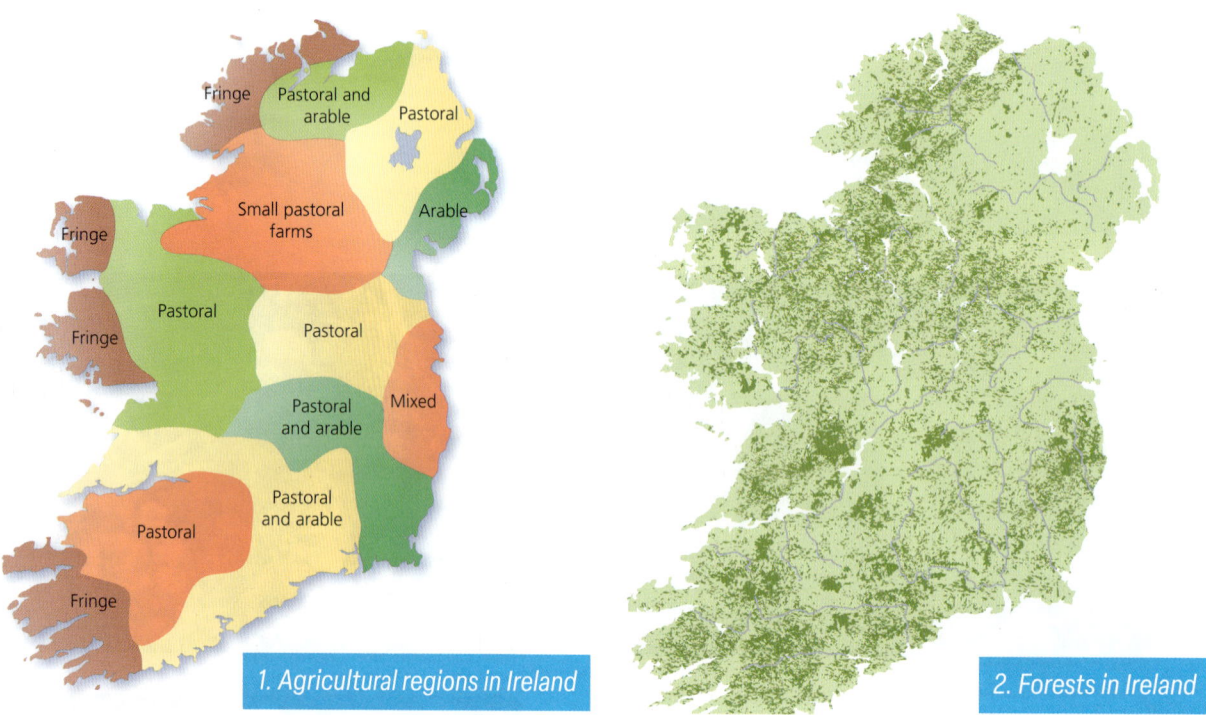

1. Agricultural regions in Ireland
2. Forests in Ireland

Question Time

1. Looking at figure 1, identify the main type of agriculture in Connacht. Why do you think this is the case?
2. Looking at figure 2, name three counties that have a heavy covering of forest. Give a reason why forests might be grown in these counties more than in others.
3. Describe how the land in your local area is used.
4. Has the physical landscape had an influence on the types of jobs people have in your area?

23. Primary Economic Activities

In the next section, we will identify how the physical landscape influences primary economic activities in two very different areas. We will look at mixed farming in Laois and forestry in the West of Ireland (Galway, Mayo and Roscommon).

Case Study: Mixed Farm in Co. Laois

Andrew Mulhare runs his own farm outside the village of Ballybrittas in Co. Laois. It is a mixed farm. It is also a very good example of how a farm operates as a system, with inputs, processes and outputs. We will look at systems in more detail in Chapter 25.

Inputs	Processes

• **Labour**: This is the people who work on the farm. On this mixed farm the majority of the work is done by Andrew himself. He hires seasonal workers at busy times of the year.
• **Capital**: This is money used to run the farm. Selling cattle and crops is the source of capital for Andrew's farm.
• **Machinery**: Seed drills and harvesters are used to sow and harvest the barley grown on the farm.
• **Farm buildings**: A large shed houses cattle during the winter. A meal bin stores animal feed for cattle and a grain store stores the harvested barley.
• **Stock**: Andrew has 38 cows, which usually produce 38 calves each year.
	• Summer and winter **barley** are grown on the farm.
	  – Winter barley is sown towards the end of September and harvested in April or May.
	  – Summer barley is sown in mid-March and harvested in July.
	• **Grass** is grown year-round and harvested in summer months.
	• **Cattle**: Calves are born in early spring. They are reared on the farm and sold to markets 16 months later.

	Outputs
	Examine figure 3 below and identify the outputs of Andrew's farm.

Average farm size in the Midlands is 39 hectares (ha). Andrew's farm is bigger, which allows him to engage in mixed farming.

1 ha = 0.01 of a km^2

Suckler animals suckle/feed their young. Andrew's cattle are reared from calves in the autumn and sold 16 months later.

3. Andrew's farm

The soil on the farm is of good quality and rich in nutrients. The soil helps to produce high crop yields. Yields mean the amount produced.

The main crops grown are spring and winter barley. The soil is fertile so yields are high in both winter and summer. Summer barley is used for malting (used for alcoholic drinks).

Andrew also grows turnips, which are sold at local markets, and silage. The silage is used to feed his cattle.

Sustainability on Andrew's Farm

Andrew adopts a number of practices to reduce the impact of his farming on the environment. He raises cattle from calves to beef in under 16 months. This reduces the amount of methane produced by his cattle, which is harmful to the ozone layer.

Andrew uses the manure produced by his cattle stock to fertilise the land for his crops. He also uses a particular type of grass which has lots of clover in it. More clover can increase grass growth. Both of these actions reduce the need for artificial fertiliser.

Andrew prevents soil poaching (soil breaking down due to heavy weight on top of it) by moving feeding troughs around the fields.

How Does the Physical Landscape Influence the Farm?

Managing information and thinking

Examine the photo of Andrew's farm. Identify fields you think may be used for each of the following:

- Growing barley
- Growing grass for silage
- Cattle grazing.

- **Soil**: The rock underlying the soil in most of Andrew's farm is limestone. Limestone contains sediments deposited by glacial deposition, which help to keep the soil fertile. This supports good quality mixed farming.
- **Drainage**: The River Barrow is a large river found in the north and east of Co. Laois. It flows eastwards through the county. The river helps to prevent flooding on the farm, which would destroy crops and prevent cattle from grazing the land.
- **Relief**: The land on the farm is generally flat. This makes it easier to use farm machinery for planting and harvesting crops and for cutting grass for silage.

1. Briefly explain how a farm operates as an example of a system.
2. How is Andrew's farm an example of a mixed farm?
3. Explain how the sustainable practices used on this farm may impact the local environment.
4. Explain two ways that the physical landscape influences Andrew's farm.

23. Primary Economic Activities

Physical Landscape and Forestry in the West of Ireland (Galway, Mayo, Roscommon)

In the past, farms in the West of Ireland were small and were not very productive because of the physical landscape. Some farmers have changed to forestry as the landscape is more suited to growing certain types of tree.

FUN FACT! Approximately 11% of Ireland is covered in forest (the lowest in Europe). Around 64% of our land is used for farming.

- **Climate**: The West of Ireland has a **cool temperate oceanic climate**. Usually described as dull, wet and windy, this part of the country gets over 250 days of rainfall per year. Average temperatures are similar to those in Dublin (5–14 °C).
- **Soil**: As the West of Ireland is mountainous, soils are generally of poor quality, e.g. peat and gley soils. The regular rainfall can cause flooding due to poor drainage. This greatly affects the quality of the soil.
- **Relief**: As you can see in figure 4, the West of Ireland has lots of mountains, e.g. Nephin Beg in Co. Mayo. It also has lowlands and lakes such as Lough Corrib. Major rivers include the Shannon, the Corrib and the Moy. Much of the land is bog, which is not suited for farming crops because it is does not drain.

Forestry in the West of Ireland

- As of 2020, there were 190,000 hectares of forestry in the West of Ireland.
- Forestry is a type of agriculture that is suitable in areas of poor land where successful arable or pastoral farming is not possible. The Irish government and the EU have given farmers in the West grants and subsidies to encourage them to move into forestry.

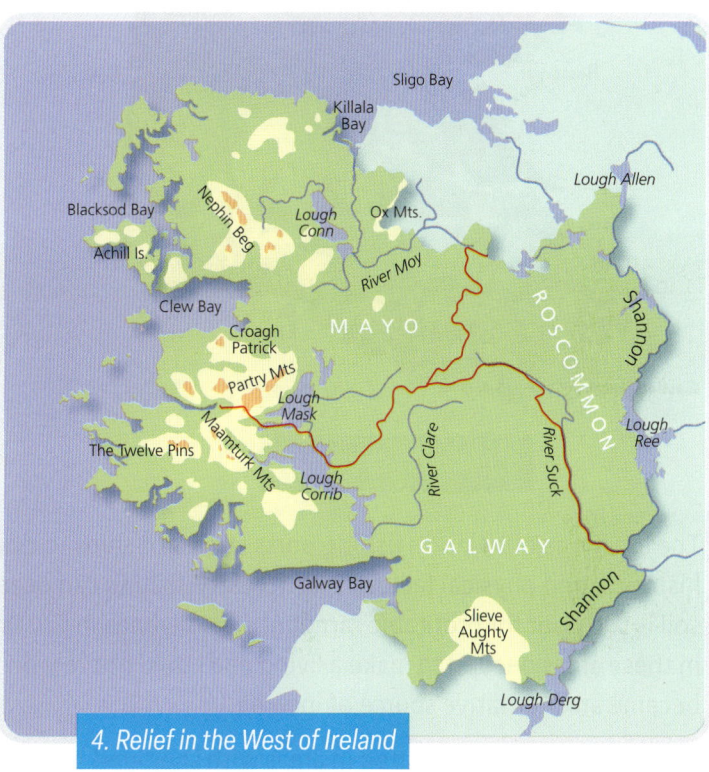

4. Relief in the West of Ireland

What Is Forestry Worth to the Economy?

- Forestry is a vital resource for the Irish economy.
- The forestry sector contributes €2.3 billion to the Irish economy annually and employs 16,000 people.

Question Time

1. List an economic benefit of forestry to Ireland.
2. Explain how soil, climate and relief contribute to the growth of forests in the West of Ireland.

23.5 The Physical Landscape: Fishing

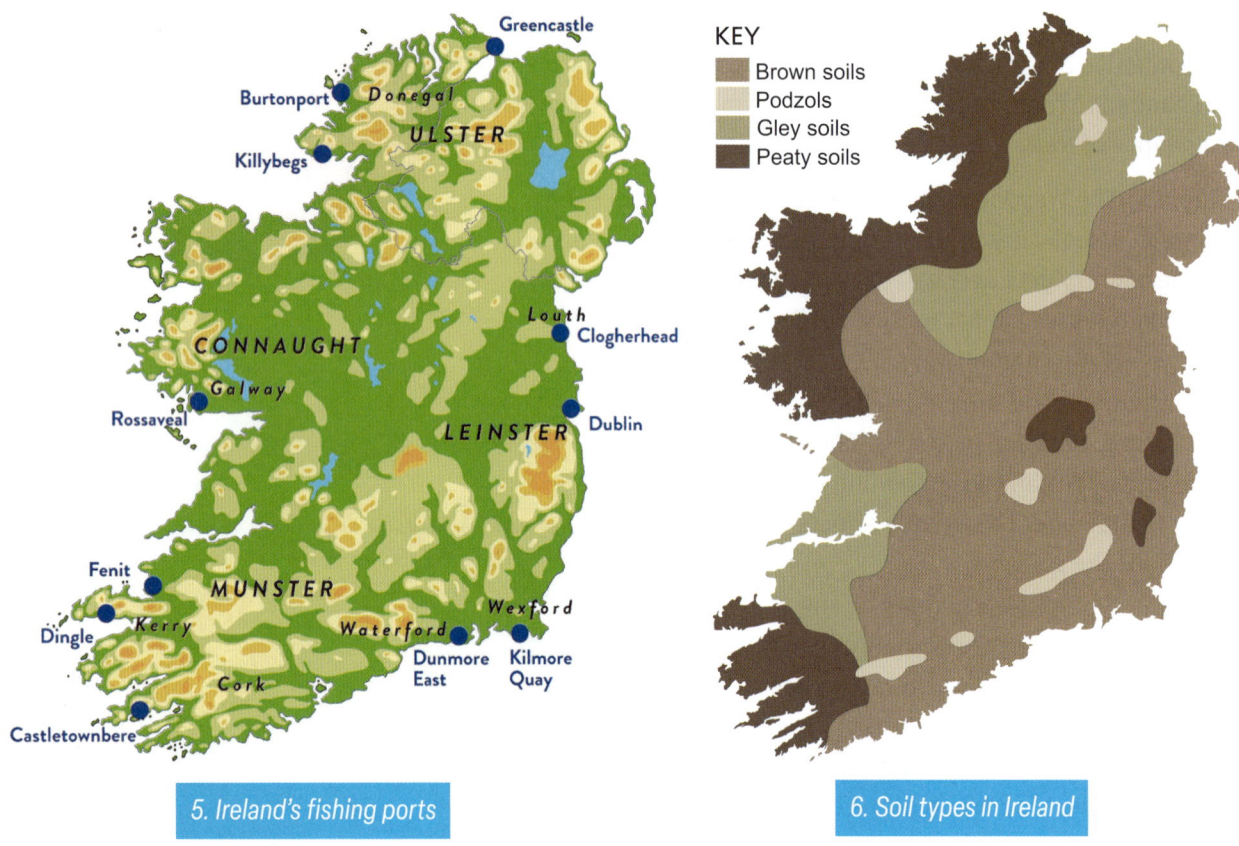

5. Ireland's fishing ports

6. Soil types in Ireland

The locations of Ireland's fishing ports (areas for ships to dock in) are influenced by historical and physical factors. Like forestry, fishing can occur in counties where the soil is too poor for profitable farming, e.g. Killybegs in Co. Donegal. Historically, people in these areas needed to make a living and could not rely on farming to do so. Fishing became an alternative source of income.

Fishing ports are also located in areas with sheltered bays, e.g. Dunmore East in Waterford. Bays offer protection from storms and waves that can damage ships. Towns and cities located on the coastline have better access to the sea.

Communicating

Work with the person sitting next to you.
1. On figure 5, locate two ports that are found in areas with poor soil (figure 6). Name the ports, identify the types of soil there and say why fishing ports may have been set up there.
2. On the same map, locate two ports found in areas with good soil. Name the ports, identify the soil types and explain why fishing ports may have been set up there.

23. Primary Economic Activities

Case Study: Killybegs, Co. Donegal

7. The location of Killybegs

- **Soil**: The land in and around Killybegs has a mixture of gley and peaty soils. These soils are not fertile enough for productive farming.
- **Relief**: The Crownarad Mountain to the west of Killybegs is over 490 m high. To the north and east of the town the land is less mountainous but still has hills between 100 m and 135 m high. It is difficult to use farm machinery on hilly land. Landslides can also occur during bad weather, removing soil needed for farming.
- **Location**: Killybegs Harbour is located in a deep inlet, making it one of the safest, most sheltered, deep-water harbours on the Irish coast.

FUN FACT! Ireland's territorial waters for fishing measure 41,000 km².

Managing information and thinking

Examine this map, which shows Ireland's territorial fishing waters. What advantages does extending our waters into the Atlantic Ocean have for the Irish fishing industry?

8. Ireland's territorial fishing waters

What Is Fishing Worth to the Irish Economy?

- According to Bord Iascaigh Mhara, the Irish seafood industry was worth €1.26 billion in 2021.
- Some 9,200 people are employed by the fishing industry, mainly in coastal counties.

EXAM FOCUS

Sample questions

1. (a) **Examine** the list of workers in the box below and categorise each into primary, secondary or tertiary economic activities.

   | Brewer | Hairdresser | Teacher | Farmer |
   | Forestry worker | Baker | Dentist | Miner | Mechanic |

   (b) **Define** primary, secondary and tertiary economic activities.

2. (a) (i) Examine the photographs. Label each one as either pastoral or arable.

   (ii) **Define** arable farming and pastoral farming.

   (b) **Examine** the infographic below showing the structure of Irish farming.

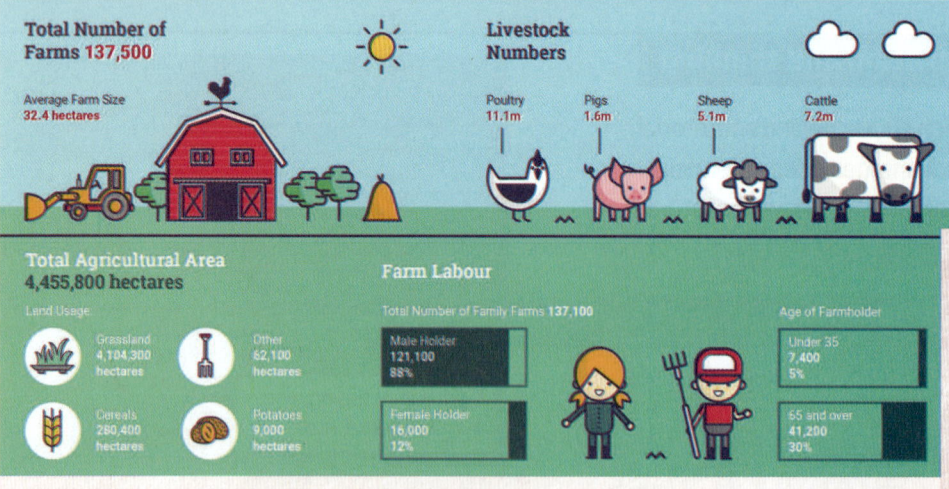

   (i) **Which** type of pastoral farming has the largest number of animals in Ireland?

   (ii) **Which** type of arable farming uses the greatest number of hectares in Ireland?

   (iii) **List** an economic benefit of farming from this infographic.

Exam hint!

When answering questions on an infographic, take a few minutes to examine it carefully. Read the heading first (if there is one), then scan through the main content. Underline anything that may help you to answer the questions.

23. Primary Economic Activities

3. **Examine** the section of OS map below and answer the question.

Location A on the map is used for controlled forest farming. Location B is used for mixed farming. Location C has a fishing port. Using evidence from the map, explain how the physical landscape has influenced the location of each type of economic activity. You may include factors not visible on the map also.

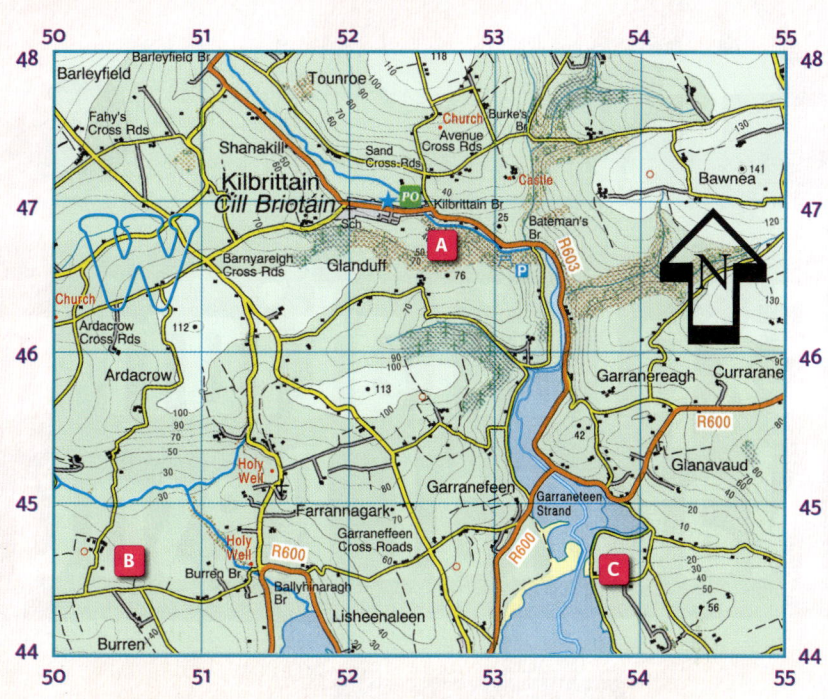

Let's get started! Sample starter

The physical landscape can influence the location of economic activities in Ireland.

Location A: Controlled forest

Drainage: The forest area located at A is alongside a river. There is very little settlement located on that side of the river (W 527 467). While providing drainage to the area, the river may also flood in times of heavy rain. This would destroy most crops grown on a farm if placed here....

Relief: [Mention here the spot height nearby and the slope the trees are growing on. How would this make things difficult for farm machinery or farm animals?]

Soil type: [To answer this part of the question, you will need to revise the information in Chapter 17: Soils. What soils encourage good growth of vegetation? Do trees need very fertile soil to grow?]

Location B: Mixed farming

River drainage: [Settlement nearby, what does this imply for drainage in the immediate area?]

Relief: [How flat is the land in the area?]

Location C:

[Is the fishing port located in a bay? What might this provide?]

Relief: [Is the land flat or gently sloping?]

> **Exam hint!**
>
> Use four- and six-figure grid references to strengthen each part of your answer. Locate spot heights and describe the contour lines.

... now complete the answer in your copy.

Learning Outcome: 2.4

24 Exploitation of Natural Resources

Learning Intentions

You will be able to:
- Describe what natural resources are
- Explain the term 'exploitation'
- Describe the impacts of using natural resources in a sustainable way.

Key words

- Natural resources
- Renewable
- Non-renewable
- Exploitation
- Overexploitation
- Sustainable exploitation
- Fresh water
- Drought
- Irrigation
- Overfishing
- Super-trawlers
- Fishing quotas
- Aquaculture
- Fish farming
- Felled
- Seedlings
- Timber-based products
- Non-timber-based products
- Deforestation
- Agribusiness
- Coillte
- Saplings
- Desertification
- Soil sealing
- Brownfield sites
- Overgrazing

24. Exploitation of Natural Resources

24.1 Natural Resources

Natural resources are materials from the earth that are used to support life and meet people's needs. Any natural substance that humans use can be considered a natural resource. Oil, coal, natural gas, metals, stone and sand are natural resources. Other natural resources are air, sunlight, trees, fish and water. These resources can be divided into two types: **renewable** and **non-renewable**.

> **Managing information and thinking**
>
> How many natural resources can you see in this picture?
>
>

24.2 What Is Sustainable Exploitation?

Exploitation of resources is making use of and benefitting from them (e.g. water, fish, forestry and soil).

The exploitation of resources is necessary for people's survival. The **overexploitation** of resources has serious consequences. Overexploitation happens when too much of a resource is used. This may result in resources becoming depleted (seriously reduced in quantity) or exhausted (wiped out completely).

The world's resources must also benefit future generations of people. To slow down overexploitation, the world's natural resources must be exploited in a fair and equal way, so that all can benefit now and into the future.

Sustainable exploitation is the use of resources in such a way that they will be available and usable in the future.

How to Think Sustainably

> **Q** The Christmas trees on the left are real, while the one on the right is artificial. Does your family put up a real or an artificial tree at Christmas? Which do you think has more of an impact on natural resources?

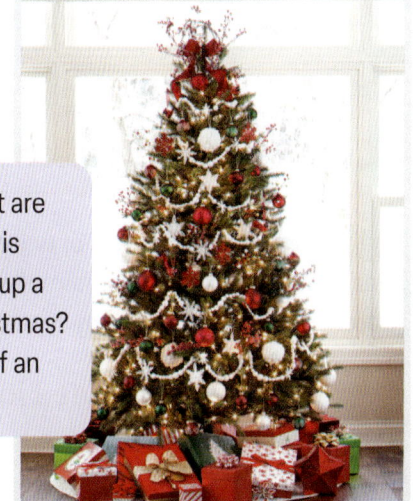

355

24.3 Sustainable Exploitation of Water: Egypt and Ireland

Even though water is a natural renewable resource, there are some parts of the world that suffer from a shortage of **fresh water**. Fresh water is not salty, so does not include sea water.

There are a number of factors which explain why people cannot get access to fresh water. They can be both physical and human.

NOT SO FUN FACT! Only 0.003% of water on earth is fresh water that is available for human consumption.

Physical Factors

- **Rivers**: River systems move huge quantities of water from where it falls as rain towards seas and lakes. River systems can dry up in areas with little or no rainfall, making them unreliable as sources of water.
- **Climate**: The changing climate affects the amount of rainfall and snowfall and rates of evaporation. Climate can vary over time, with wetter and drier periods and hotter and colder periods. This can affect water availability. Areas with low rainfall have access to less water. An example is Egypt, where desert conditions mean that there is little accessible fresh water. To cope with this, water is taken from the ocean and the salt is removed. This is a very expensive process.

Human Factors

- **Pollution of water supplies**: Fresh water sources around the world are threatened by water pollution. Some examples of how human activity pollutes water resources are:
  - Human littering in rivers, oceans, lakes and other bodies of water
  - Agricultural run-off (rainwater that flows over farmland and can pick up chemicals used in fertilisers) that makes its way into our rivers and streams and groundwater sources.
- **Overexploitation of reservoirs**: Taking more water from reservoirs than is falling as rain can lead to **drought**. In hot summer months, there can be less rainfall so people need to be careful about their water use. Councils can impose hosepipe bans to stop people watering their gardens.
- **Politics**: Countries that share water supplies need to cooperate to make sure water is shared fairly and sustainably. The River Nile, for example, is the primary water source for both Egypt and Sudan. The Nile is classed as an international river and flows through nine countries before reaching the Mediterranean Sea.

Case Study: Irrigation in Egypt

Irrigation is the transport of water from one area that has a plentiful supply to another area that has low levels of rainfall or cannot access water easily to help crops grow. This water can be transported by pipes and canals.

Why Is Irrigation Needed in Egypt?

The need for irrigation in Egypt is driven by physical and human factors.

24. Exploitation of Natural Resources

Physical Factors

- Egypt is surrounded by desert to the east, west and south, and bordered by the Mediterranean in the north. It receives little annual precipitation. The Nile therefore is the primary source of fresh water and the only source of water for agriculture.

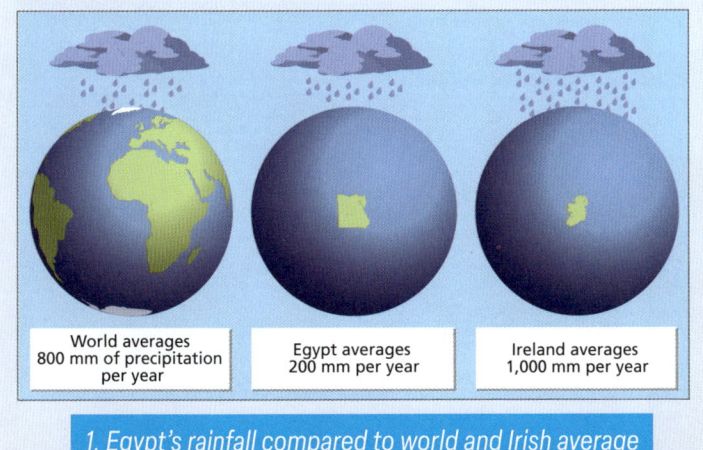

1. Egypt's rainfall compared to world and Irish average

Human Factors

Egypt's population is generally rising at an extremely fast rate, which causes the government concern. In 2022, the population was approximately 106 million. It is predicted to reach 160 million by 2050. Some 90% of the country's population lives close to the River Nile. The Nile is under extreme pressure to provide water for agricultural, household and business purposes.

The Farm-level Irrigation Modernization Project (FIMP)

In 2010, the Egyptian government, with funding from the World Bank, set up Farm-level Irrigation Modernization Project (FIMP) to modernise the irrigation system throughout the Nile region in Egypt.

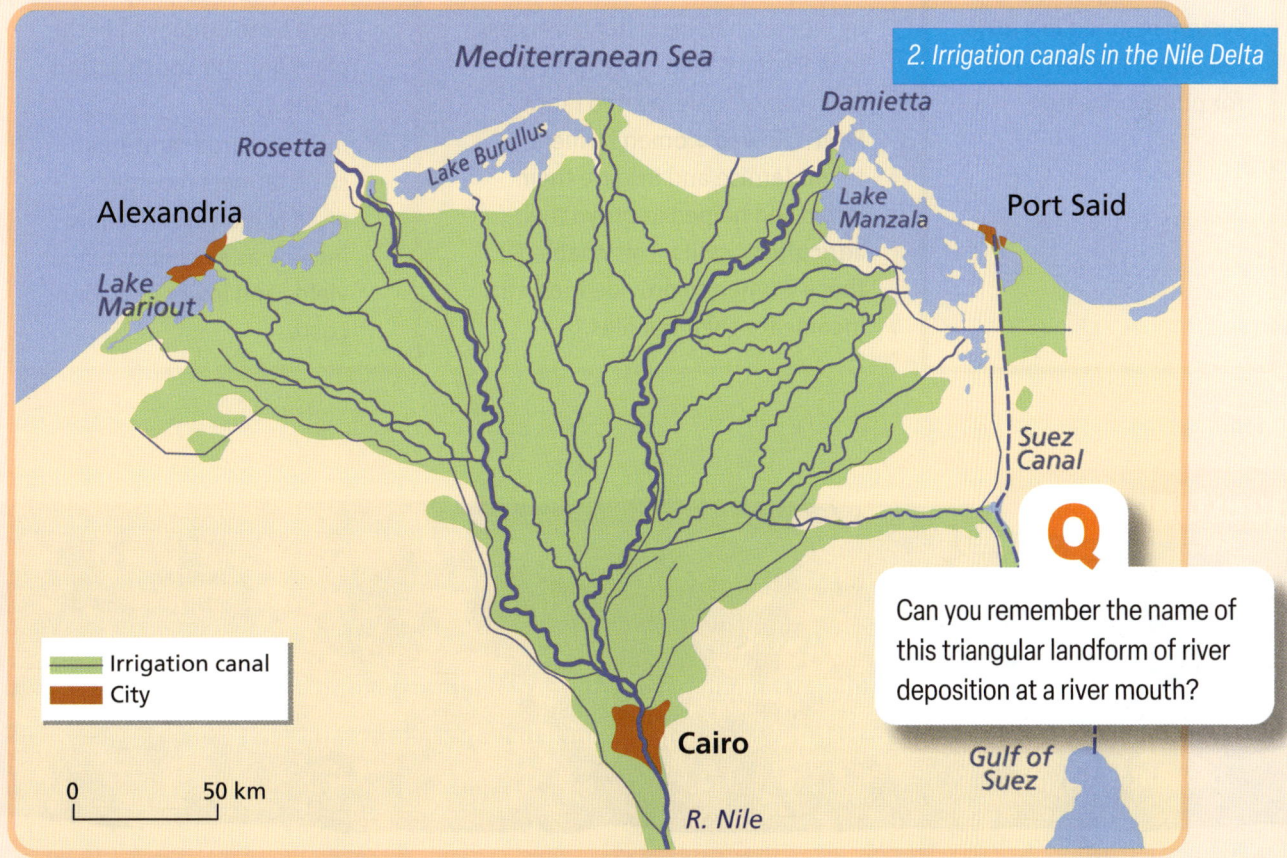

2. Irrigation canals in the Nile Delta

Q Can you remember the name of this triangular landform of river deposition at a river mouth?

357

How the Scheme Worked?

The scheme modernised the irrigation system from overground channels into buried canals with brick lining and pipes. This meant that less water was lost to evaporation. The project will also save 5 billion m³ of water per year. At pumping stations, diesel pumps were replaced with new electric pumps.

Q What advantages do you think modern irrigation canals have over open canals in a country like Egypt? (Think about the climate.)

Environmental Impact

The new irrigation system resulted in a 46% reduction in energy costs for pumping water.

Social Impact

In the past, farms owned and run by women had been denied access to water provided by irrigation in favour of male-owned farms. FIMP allowed women farmers to modernise and irrigate their farms and helped improve their agricultural skills. This enabled these women to provide food and income for their families.

Economic Impact

- To date 10,000 people have been employed in the planning and construction of the system.
- In 2020, large- and small-scale farmers recorded an 18% increase in family income as a result of higher yields and lower irrigation costs.

Question Time

1. What is overexploitation?
2. Explain what irrigation is and how it works.
3. List and explain one physical and one human reason for the need for irrigation in Egypt.
4. Describe one economic, one social and one environmental benefit that irrigation has brought to Egypt.

24. Exploitation of Natural Resources

Sustainable Exploitation of Water in Ireland

Irish Water is a company that is part owned by the Irish government. It is responsible for delivering water and gas infrastructure and services for Ireland. It also operates the water treatment plants across the country. These plants make waste water safe for use in our homes.

Every day in Ireland, we lose about 38% of our treated water through leaks. This is an unsustainable waste of water. It is estimated that between 2022 and 2030, an average of €250 million will be needed each year to fix leaks in the system.

How Is Irish Water Tackling the Issue?

Irish Water aims to reduce the number of leaks in the water system. This Leak Reduction Programme has a number of benefits.

A more reliable water supply	Improved water quality	Reducing high levels of leakage	Individual water connections
Fixing and replacing old, damaged pipes prone to bursts, reducing disruption to supply	Reducing health risks and contamination by replacing aging, damaged cast iron and lead pipes	Reducing the amount of clean drinking water lost into the ground by replacing old, damaged pipes	Reducing disruptions by replacing shared and backyard services with new, individual supply connections

Case Study: Reducing Water Wastage in Wicklow

Since 2018, Irish Water has delivered over 17 km of new water mains (pipes bringing water to businesses and homes in Wicklow). It has done this in partnership with Wicklow County Council as part of the Leakage Reduction Programme.

How Has This Benefitted Co. Wicklow?

- Improved water supply for over 70,000 people
- 225 leaks repaired at no cost to homeowners
- Reduction in supply issues in the county

Question Time

1. How has reducing water wastage improved the sustainability of water for Co. Wicklow?
2. Explain one possible economic benefit from Irish Water repairing leaks around the country.
3. Suggest one step Irish Water could take to ensure it will not have to spend €250 million a year after 2030 on repairing leaks.

24.4 Sustainable Exploitation of Fish: Ireland

Even though fish are a renewable resource, they are being rapidly overfished in many parts of the world. **Overfishing** occurs when fish are taken from the sea faster than they can reproduce. When this happens, fish stocks decrease quickly and become depleted (used up).

Why Is Overfishing Happening?

There are a number of causes of overfishing:

- **Efficient ships**: **Super-trawlers** have increased the amount of fish caught each year. These ships can catch fish and then store their catch in huge refrigerators to preserve them. Smaller ships take the catch back to land, while trawlers remain at sea. The super-trawler in figure 3 below is called the *Margiris*. It is 142 m long. In 2015, it was investigated for illegal fishing in Irish waters.
- **Improved technology**: Super-trawlers have improved capability to locate fish. They use echo sounders and sonar equipment to locate shoals (large groups of fish). Powerful motorised winch cranes can haul in huge nets full of fish.
- **Netting**: Monofilament nets are hard for fish to see. They can stretch to many different lengths (depending on the laws of the country in control of the fishing waters) and hang like massive curtains in the water. Trawl nets scoop fish from the bottom of the sea floor.
- **Different types of commercial fishing**: Developments in modern commercial fishing have led to increased catches. Countries are allowed to catch more fish to meet the needs of their growing populations.

NOT SO FUN FACT! Overfishing means that three-quarters of the world's fish stocks are being harvested faster than they can reproduce. If current trends continue, the world's food fisheries could be completely eliminated by 2050.

Sydney Opera House: 185 m

Airbus A380: 72.75 m

Queen Mary II: 311 m

Margiris: 142 m

Titanic: 269 m

Super-trawler *Margiris*'s net is 600 m long and can open to 200 m wide. This is large enough to hold eight jumbo jets.

3. The size and fishing range of super-trawler Margiris

Purse Seining	Bottom Trawling	Traps and Pots
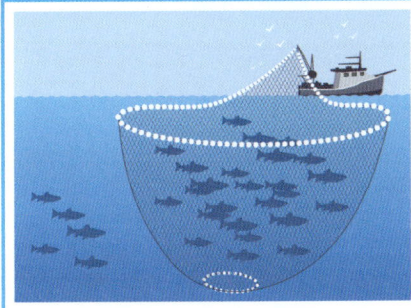		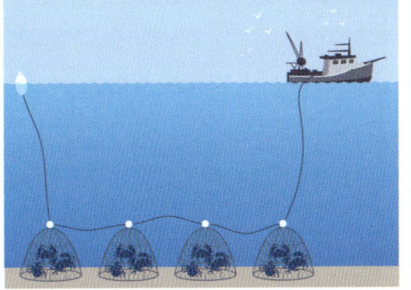
This involves catching fish which inhabit the mid-water regions or upper layers of water (pelagic fish), e.g. tuna.	Bottom trawling catches fish that inhabit the bottom layers of water near the ocean floor (demersal fish), e.g. flounder.	Traps and pots are mainly used to catch invertebrates like crab and lobster.

4. Modern fishing techniques

Sustainable Exploitation of Fish Stocks

There are a number of ways to prevent overfishing, so that fish stocks can recover or at least stay the same while still being fished. These include:

- Ensuring fishing trawlers fish only within their own waters. All countries with a coast have designated fishing zones off their coastlines.
- Reducing the amount of fish (**fishing quotas**) that are allowed to be caught. In Ireland, quotas are set by the government and the European Union and are calculated based on the EU's total allowable catches (TACs). TACs are limits on the amount of fish a country is allowed to catch over one year.
- Surveying fish stocks to record if they fall too low. The catching of certain fish can be stopped completely if stocks are too low.
- Reducing fishing fleets (number of super-trawlers) in our seas.
- Restricting fishing of certain types of fish at specific times of the year so as not to interrupt breeding seasons.

FUN FACT! According to the Food and Agriculture Organization (FAO), there are 38 million fishermen or fish farmers in the world today. The infographic on the next page gives a number of reasons why fishing must be done in a sustainable way.

> **Communicating**

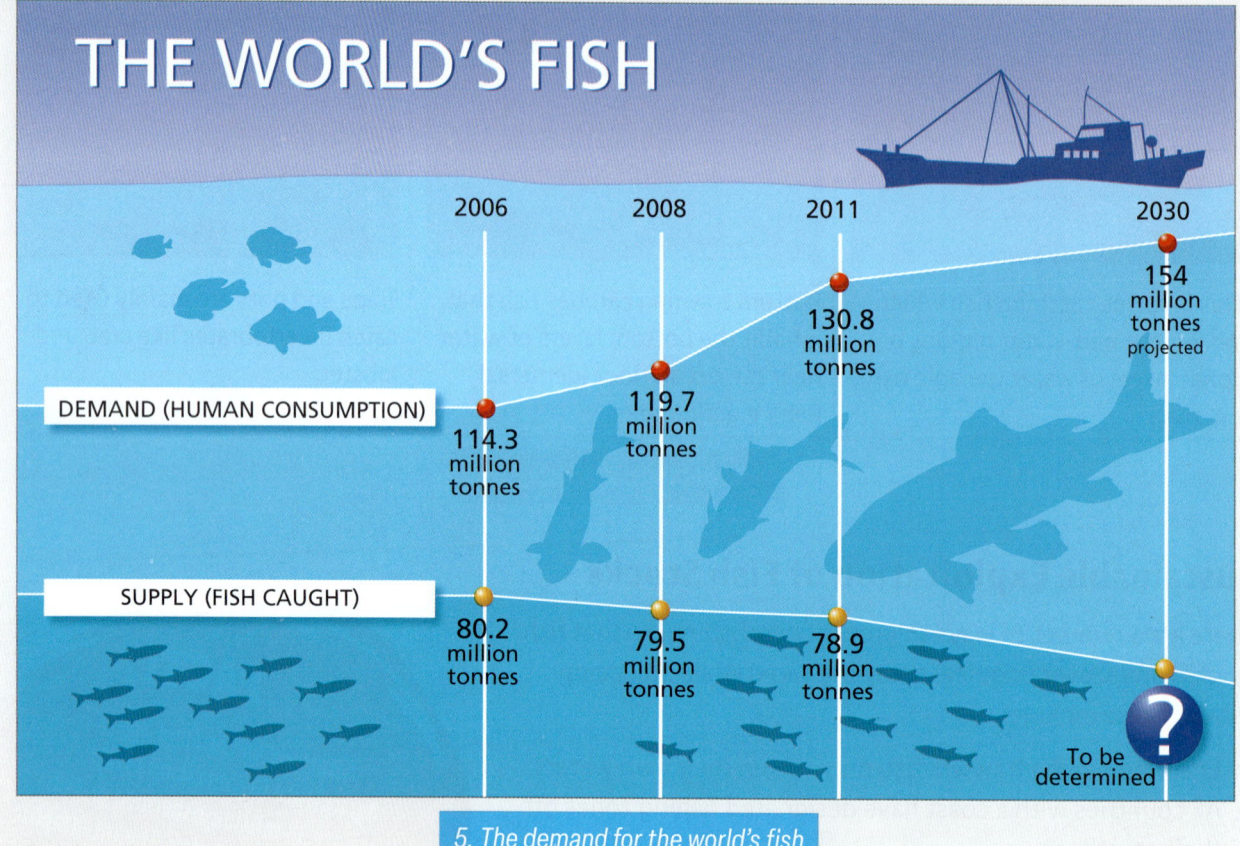

5. The demand for the world's fish

In pairs, examine the infographic. It compares the amount of fish being caught each year with the demand for fish for human consumption (for eating). What is the main point this image is trying to get across? (Hint: Does supply meet demand?)

Sustainable Fishing in Ireland

A practice called **aquaculture** helps to provide sustainable fish supplies in Ireland. Aquaculture is the breeding, rearing and harvesting of animals and plants in water. **Fish farming** is Ireland's most common type of aquaculture.

Fish farming and aquaculture raise fish and shellfish in tanks or enclosures. This ensures their numbers can be maintained while not being threatened by overfishing or pollution.

Fish farming and aquaculture in Ireland can have positive and negative impacts.

A fish farm

24. Exploitation of Natural Resources

Environmental Impact

- Fish farming and aquaculture have helped fish stocks (numbers) to recover and even increase for some types of fish, e.g. salmon.
- However, there are concerns that the environment can suffer as a result of these practices. For example, releasing farmed salmon into the wild could potentially spread disease to wild salmon.
- There is also a concern that wild fish caught and processed into food for farmed fish will reduce wild fish stocks.

Social Impact

- Fishing is worth hundreds of millions to the Irish economy each year. In 2020, the total value of Ireland's seafood economy was just under €1.1 billion.
- Starting up a fish farm can be expensive.

Economic Impact

- Figure 6 shows the numbers of people employed in sea fishing and aquaculture. Both provide an income for thousands of people.
- Fish farming and aquaculture jobs are often low paid. There can also be a lack of trained staff to fill all jobs needed on a fish farm.

Aquaculture

1,841 Total employed
995 Total FTE*

Sea fishing

3,217 Total employed
2,426 Total FTE

FTE* Full time employed

6. *The numbers of people employed in aquaculture and sea fishing in Ireland*

Question Time

1. List four causes of overfishing.
2. Explain how technology has contributed to overfishing.
3. Describe two fishing methods and the type of fish they catch.
4. What is fish farming/aquaculture?
5. Explain one environmental impact, one economic impact and one social impact of aquaculture in Ireland. The impacts can be positive or negative.

24.5 Sustainable Exploitation of Forestry: Ireland and Malaysia

A sustainable forest is one that is carefully managed. Trees that are **felled** (cut down) are replaced with **seedlings** (young trees) that eventually grow into mature trees. This is a carefully and skilfully managed process.

Forests provide a number of essential products. These can be split into two different types: **timber-based products** and **non-timber-based products**. Timber-based products are all wood-based, e.g. planks and hurleys. Non-timber-based products are made up of anything else a forest can provide, e.g. medicines and maple syrup.

Threats to Forestry: Illegal Deforestation

There are a number of significant threats to the world's forests. Illegal **deforestation**, or logging, is one of the main risks.

Illegal deforestation occurs when trees are cut down without the permission from those who own or manage the forest, e.g. private owners, companies or government agencies.

- Illegal logging reduces the number of trees, which contributes to global warming.
- It can drive down the price of timber around the world, as illegal timber is often sold more cheaply.
- It also damages the economies of local communities and companies, which rely on legal forestry as a source of income.

One of the main reasons for cutting down trees is to clear the land for **agribusiness**. The table below outlines the differences between agribusiness and traditional farming.

Agribusiness	Traditional farming
A group of businesses purchase land for farming, in large amounts and usually with no interest in sustainability.	A farmer owns or rents their own land.
Businesses hire a farmer to work the land. This farmer is under contract.	The farmer is their own boss.
Profits are returned to the businesses.	Profits are kept by the farmer.

24. Exploitation of Natural Resources

Increased demand for food around the world has increased the rate at which our forests are being cut down and replaced with farms. This is being done both legally and illegally.

This photo shows the results of deforestation in Malaysia. The trees have been cleared to make way for a palm oil plantation. This plantation is an example of an agribusiness.

Case Study: Deforestation in Malaysia

Malaysia is located in Southeast Asia. It is made up of two separate areas of land known as Peninsular Malaysia and East Malaysia. Over two-thirds of land in Malaysia was covered in rainforest. Over the last 30 years, Malaysia has had one of the highest rates of deforestation seen in the world. Now, just over 50% of the land is rainforest.

7. The location of Malaysia

Causes of Deforestation

- **Agribusiness**: Malaysia is one of the largest exporters of palm oil in the world. Plantation owners receive tax incentives, which has led to large amounts of land being converted to palm oil plantations.
- **Population pressure**: The government encouraged the poor to move to the countryside to ease overcrowding in the cities. Many of these settlers then set up plantations, removing forested areas.
- **Subsistence farming**: This type of farming involves producing just enough food to support yourself and your family. Tribal people are usually small-scale, sustainable farmers. They use the 'slash and burn' method, which involves setting fires to clear the land for crops. The fires can get out of control, destroying large areas of land.

Government Interventions

The Malaysian government introduced policies to reduce the levels of deforestation.

- The government launched an education campaign to increase public awareness of the value of tropical rainforests.
- Permanent Forest Estates were created where no change of land use is allowed.
- Local communities were included and involved in forest conservation projects.
- In 2020, Malaysia was awarded the Gold Medal for the world's biggest increase in forested area. So far, over 950,000 hectares have been grown in 10 managed forests spread across the islands.

Environmental Impact	 Social Impact	 Economic Impact
Malaysia hopes to reduce the greenhouse gas emissions across the economy by 45%. Keeping at least 50% of the country covered in forest will help them to achieve this.	Approximately 350,000 people work either directly or indirectly in forestry.	Malaysia exported US$4.4 billion of timber and timber products from January to October 2020.

Sustainable Exploitation of Forestry in Ireland

Coillte (the Irish word for 'forest' or 'woods') is an Irish organisation that owns and manages 7% of the land in Ireland. That is roughly 44,000 hectares, or 440 km² of land. It was set up in 1989 and has helped to increase the amount of forests covering Ireland from 1.5% to 11%.

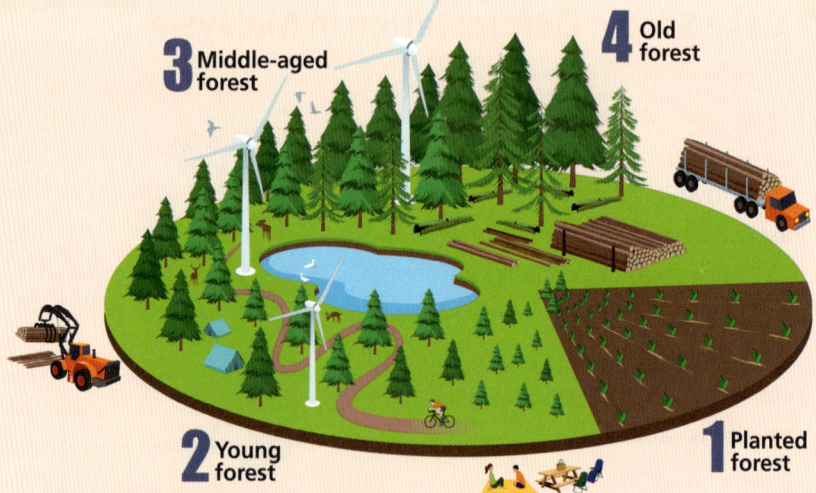

8. Sustainable forestry in Ireland

Coillte makes sure that forestry in Ireland is done in a sustainable way. Figure 8 shows how this is done in stages.

1. **Saplings** (young trees) are planted.
2. When forests reach a young age (0–15 years), they are used for human activities, such as camping and cycling.
3. Middle-aged forests (15–55 years) are used for cutting down trees.
4. Old-age forests (55+ years) are cut down completely.

During stages 3 and 4, young saplings are planted each time a tree is cut down.

The Benefits of Forestry in Ireland

Environmental Impact

Trees convert harmful carbon dioxide into oxygen. In one year, one acre of forest can absorb six tonnes of carbon dioxide and release four tonnes of oxygen. That is the amount of oxygen needed for 18 people in one year.

Social Impact

There are 12,000 people currently employed by Coillte in Ireland.

Economic Impact

Coillte contributes €4.2 billion to the Irish economy each year.

Question Time

1. Explain the difference between timber-based and non-timber-based products.
2. List three timber-based products and three non-timber products that forests provide.
3. How has the increasing need for farmland impacted forests in Malaysia?

24.6 Sustainable Exploitation of Soil: Urbanisation and Desertification

As we learned in Chapter 17, soil is an essential natural resource. Soil enables plants to grow, provides homes for most of the organisms on earth, holds and cleans water and is used for constructing buildings and roads. Soil is a non-renewable resource.

In this section, we will examine how urbanisation and **desertification** can affect soil sustainability.

Changing Land Use Patterns: Urbanisation in Ireland

Urbanisation is the transformation of land, such as agricultural land and forests, into artificial surfaces. These artificial surfaces include areas for buildings, industrial facilities and infrastructure.

Urbanisation causes **soil sealing**. Soil sealing is where the ground is covered with an artificial impermeable layer such as concrete. In Ireland, artificial areas with sealed soil surfaces have increased by 65% since 1990.

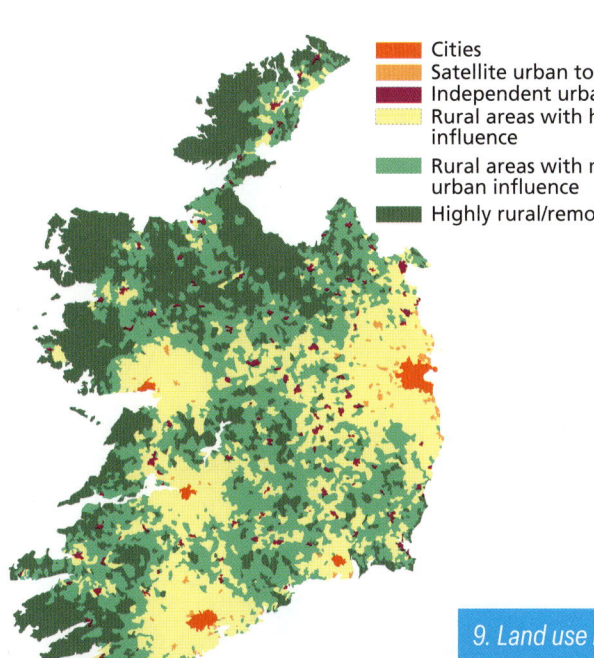

9. Land use in Ireland

Urbanisation can help the economic development of a country as more housing and jobs become available for a growing population. It can also be a serious risk for soils in Ireland. For example, every 20,000 homes built each year in Ireland potentially causes a loss of around 1150 hectares of greenfield agricultural land. Greenfield lands around Ireland's major cities and towns contain high-quality, highly productive soils that, once sealed under new developments, are lost forever.

Project Ireland 2040 is a national strategy outlining how Ireland will plan for its continued population growth. Part of the plan involves renewing and developing existing **brownfield sites** in towns and cities rather than continued urban expansion into the countryside, at the expense of rural Ireland and its soils. A brownfield site is land that was previously used for industrial purposes. Examples of brownfield sites include abandoned factories.

Changing Land Use Patterns: Desertification

As we learned in Chapter 16, desertification is the process which turns fertile land into desert as the quality of the soil declines over time. It occurs mainly in dry areas as a result of changes in climate and human activities, such as deforestation.

Soil degradation is the decline in soil condition. It is caused by its improper use or poor management, usually for agricultural, industrial or urban purposes. Over 75% of the earth's land area is already degraded, and over 90% could become degraded by 2050. Soil degradation is estimated to cost the EU economy tens of billions of euros each year.

Globally, a total area half of the size of the European Union (4.18 million km²) is lost to desertification each year. Africa and Asia are the worst affected. Desertification and climate change are expected to cause a 10% decrease in global crop growth by 2050. Most of this will occur in India, China and sub-Saharan Africa, where crop production could fall by 50%.

Desertification Case Study: China

China has a land area of over 9.5 million km². More than a quarter of China is now turning to desert.

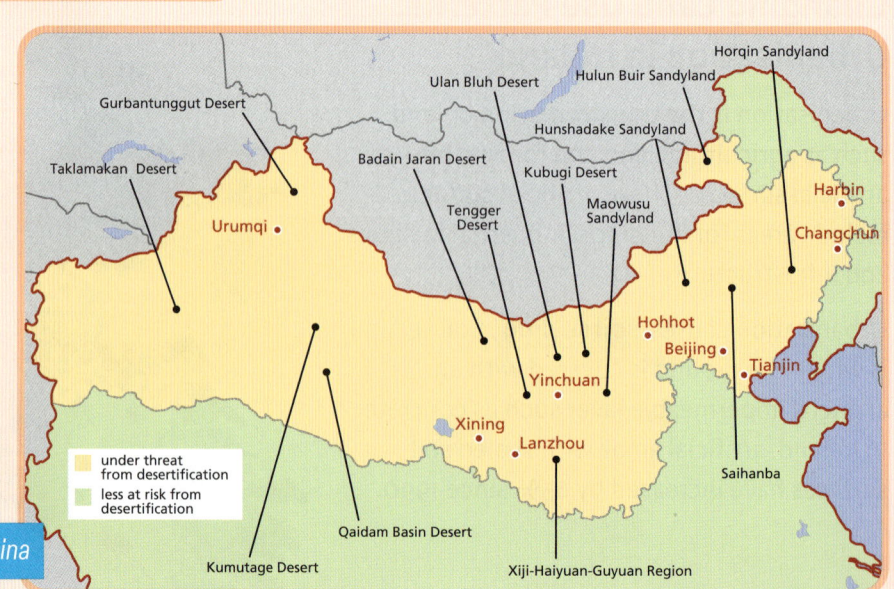

10. The deserts of China

Causes of Desertification in China

Desertification in China is caused by **overgrazing** of livestock, population growth and changes in climate. Each year, the Gobi Desert spreads into 3,600 km² of grassland.

- **Overgrazing**: There are more than 32 million grazing animals in the northern half of China. Sheep make up 70% of the total, followed by goats, cattle, donkeys and horses. Overgrazing has resulted in desertification of almost 80% of the of pastoral land in the northern half of China.
- **Population growth**: There are nearly 1.4 billion people in China. This leads to increased pressure on agriculture to produce enough food to feed the population. Land is farmed intensely to produce rice, maize and wheat. It is difficult for soil to recover and so it erodes faster.
- **Climate change**: China became the world's largest emitter of carbon dioxide in 2006. In 2020, China released 10.67 billion metric tons of CO_2. In comparison, India, a country with almost as large a population, emitted 2.4 billion metric tons. Climate change has led to higher temperatures, which cause drought and evaporation. This hardens the soil. The increase in rainfall washes away the hardened soil.

Solutions to Desertification in China

Overgrazing was a primary cause of land degradation and desertification in China. Sheep are now kept in enclosures, which allows grass to regrow and recover, and keeps soil in place. Farmers received subsidies for sheep sheds.

11. The Great Green Wall

In an effort to combat the loss of its grassland to the Gobi Desert, the Chinese government started the Three-North Shelterbelt Project, also known as the Great Green Wall in 1978.

The plan involved planting a 4,500 km² area along the northern part of China's deserts with 100 billion trees.

So far, over 66 billion trees have been planted.

Question Time

1. What is the name of Ireland's planning strategy to tackle soil loss to urbanisation?
2. What percentage of soil in Ireland has been lost to artificial areas since 1990?
3. What countries/regions are most affected by desertification?
4. List a social impact and an environmental impact as a direct result of increased desertification.

Junior Cycle Geography CYCLONE

EXAM FOCUS

Sample questions

Sustainable Exploitation of Water: Egypt and Ireland

1. Using examples, **explain** the term 'natural resources'.

2. **Study** the poster below which shows the United Nations 2030 Sustainable Development Goals.

SUSTAINABLE DEVELOPMENT GOALS

 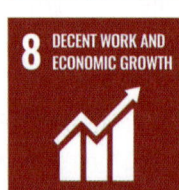  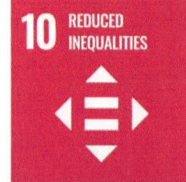  

(i) **Name** three goals from the 17 which you think can directly affect the sustainable use of water around the world.

(ii) **Explain** why two of the goals you chose in part (i) might directly link to sustainable use of water.

Exam hint!

When deciding on which three goals to use to answer parts (i) and (ii), think of how water availability and sustainable use are affected by physical and human factors. When explaining your goals, you can use facts from the regions you examined in the case studies.

24. Exploitation of Natural Resources

3. Study the infographic below on the relationship between society and water.

   (i) **Explain** a physical factor which may explain the scarcity of water for 1/5 of the world's population.

   (ii) **Choose** a fact from the infographic that may explain how developed countries also have issues with sustainable water use.

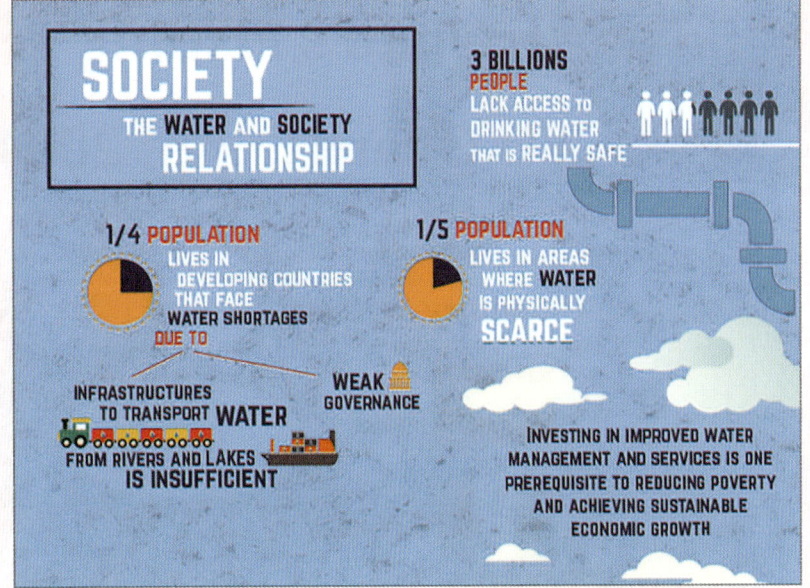

4. (i) **Name** a region that you have studied that exploits water sustainably.

   (ii) **Explain**:
   - Why this region needs to use water sustainably
   - How it was achieved.

   (iii) **State** one economic, one social and one environmental benefit.

Exam hint!

To answer Q4, state your region clearly. Write three clear paragraphs, one for each part of (ii). Explain why your chosen region needs to use water sustainably. Include any unique things about the region's sustainable water usage that you can recall.

1. **Examine** the infographic below showing the value of fishing to the Irish economy in 2021.

   (i) **Which** type of fishing is more valuable to the Irish economy, aquaculture or seafood?

   (ii) **List** one economic and one social benefit of fishing shown in this infographic.

   (iii) **Name** Ireland's main export market for fish.

   (iv) **Describe** the similarities in the location of Ireland's two main fishing ports.

371

2. Examine the infographic and answer the question below.

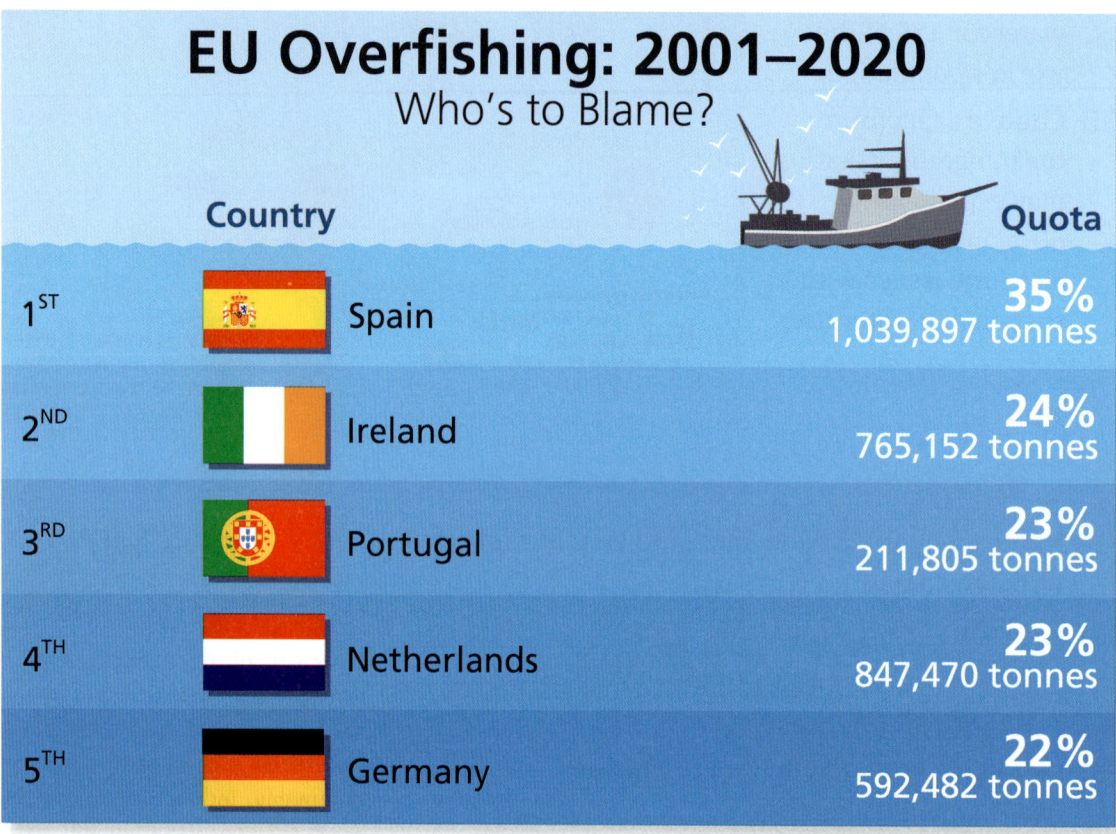

A study of fishing trends between 2001 and 2020 places Ireland second behind Spain in the list of countries that have fished over their quota. In this 19-year period, Ireland caught almost 770,000 tones (24%) more than was allowed.

Write a letter to Bord Iascaigh Mhara, the Irish fish trade and marketing body, in which you explain how sustainable fishing might be one possible solution to this issue. Explain what fish farming/aquaculture are and provide one social, one economic and one environmental benefit of sustainable fishing.

24. Exploitation of Natural Resources

Sustainable Exploitation of Forestry: Ireland and Malaysia

1. (i) **Explain** how a forest can be managed sustainably.
   (ii) **What** is illegal deforestation?
   (iii) **Explain** two differences between traditional farming and agribusiness.

2. **Compare** and **contrast** how forests are managed sustainably in two different areas you have studied.

Sustainable Exploitation of Soil: Urbanisation and Desertification

1. (i) Circle the correct answer.
   Soil is a renewable / non-renewable resource
   (ii) **Describe** what a brownfield site is.
   (iii) **Illustrate** the link between urbanisation and soil degradation in a region you have studied.

2. (i) **Name** one human activity, apart from urbanisation, which can cause desertification.
   (ii) As human population grows, the pressure on soil to meet the demands of this grows also. Using an area you have studied:
   - **Name** three ways soil has been impacted by population growth
   - **Explain** one way to tackle the impact of population growth on soil.

Exam hints!

When asked to 'compare and contrast', you must describe what is similar and what is different about two or more things. For Q2, highlight the reason(s) why your chosen areas need to manage their forests (look back at pp. 365–366). Mention any similarities or differences in how their forests are managed.

The action verb 'Illustrate' requires you to include examples in your answer. Name a region you have studied where urbanisation has affected soil. Explain the terms 'urbanisation' and 'soil degradation' to show how both are linked (see pp. 367–368). Explain how soil usage is affected once infrastructure like buildings and roads are built on top of it.

To answer Q2(ii), you must name three clear and separate ways soil has been impacted by population growth (see pp. 368 and 369). Then explain one way to tackle the impact of population growth on soil. Use facts and figures to support your answer.

Learning Outcome: 2.5

25 Secondary Economic Activities

Learning Intentions

You will be able to:
- Explain what a secondary economic activity is
- Explain the factors that influence the location of industry
- Explain the different types of industry
- Explain why industries might change location over time.

key words

- Raw materials
- Inputs
- Process
- Outputs
- Finished products
- Semi-finished products
- Transport
- Market
- Services
- Labour
- Linkages
- Capital
- Government and EU policy
- Corporation tax
- Personal preference
- Common Market
- Industrial estate
- Heavy industry
- Light industry
- Multinational corporation
- Footloose industry

25.1 What Are Secondary Economic Activities?

Secondary economic activities involve the processing of **raw materials**. These raw materials can sometimes be sourced from primary economic activities. Secondary economic activities function as a system. The raw or resource materials are the **inputs**, or ingredients, needed to make the products. They are **processed** (changed) to become new products. These new products are called **outputs**. They can be either **finished products** or **semi-finished products**.

Finished products can be heavy or bulky, e.g. concrete structures, or light, e.g. medicines. Semi-finished products have had some processing, but will undergo more changes before they become ready for use. Examples include steel, computer chips and sugar.

Think back to Chapter 24 where we learned about the sustainable exploitation of forests. Managed forests can be used as part of a system.

Inputs	Processes	Outputs
 Inputs/Raw materials are trees. Trees are cut down.	 The cut-down trees are moved to a saw mill, where they are processed. The branches and bark are removed. Then the wood is cut into boards. They are stored and allowed to fully dry out.	 The finished boards or planks are an example of a semi-finished product. The boards can be processed further to make doors or furniture.

Systems and the Environment

During each stage of a system, waste or waste products can be created. Companies must ensure that any waste materials or waste products are managed correctly to protect the environment.

Examples of waste products from the system used to make planks of wood are sawdust and wood chippings.

Sawdust

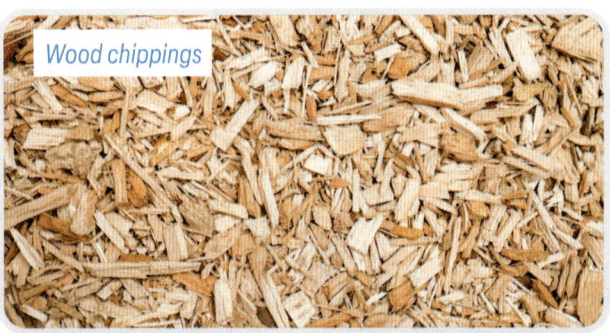
Wood chippings

Junior Cycle Geography CYCLONE

Waste products from manufactured boards can be used:
- As wood chips for bedding for farm animals in sheds
- As sawdust to soak up spillages
- To make compost.

Functions of Secondary Economic Activities

Secondary economic activities:

1. Bring together the raw materials and resources needed to make a new product
2. Process the raw materials and resources. Sometimes this is done through a number of different activities.
3. Create a product that is usually sold for profit and that is useful for consumers.

Question Time

1. Explain what secondary economic activities are.
2. How do secondary economic activities function as a system?
3. (a) Explain how producing timber planks works as a system with inputs, processes and outputs.
   (b) What can be done with the waste material from producing timber planks?

25.2 Location of Industry

There are a number of factors that must be considered when a company is choosing a location for its factory/industry. Depending on the product a manufacturer makes or the service they provide, some factors are more important than others.

Influences on the Location of Industry

Raw Material		The type of raw (resource) materials used by the factory will help decide on its location. If the raw materials are bulky (heavy), the factory will need to locate close to them to cut down on transport costs.
Transport		A factory needs to be able to get its raw materials in from its suppliers and its finished products out to its market. To do this, it may need to be close to transport links, e.g. air, sea, road and rail.
Services		Factories need services such as water, waste disposal, electricity, Wi-Fi and broadband in order to do business. Factories will locate where these services are easily accessible.
Labour		Factories need workers. For that reason, factories will locate near towns and cities, which provide a larger workforce. Depending on the type of work involved, the workers will need to be skilled (with third-level qualifications) or semi-skilled (trained on the job).

25. Secondary Economic Activities

Market		The market is anywhere the finished product is sold. This can include online sales. Depending on the product, the factory may or may not be located near its target market (the particular group of customers for whom the product is made).
Linkages		When a factory uses and relies on products made by another company, it makes sense to locate the factories near to each other – to link up. This will cut down on transport time and costs, e.g. setting up a bakery close to a flour mill.
Capital		A company needs money (capital) to cover the costs of building the factory, buying the raw material, paying the workers, covering transport costs and much more. This money can come from banks, investors or the government. In the past, factories often located near cities so that they could get access to banks. This made it easier to borrow money to help finance the running of the factory.
Government and EU Policy		The Irish government has a number of agencies that help companies setting up in Ireland, or Irish companies setting up abroad. They do this by giving loans or grants (money that does not have to be paid back) and advice to companies. Since 2023, Ireland has **corporation tax** rates (a tax on companies' profits) of just 12.5% and 15%. The rate paid depends on the size of a company's profits. This attracts companies to set up in Ireland. The EU provides funding to the agencies which help set up Irish companies at home and abroad. It also provides funding for infrastructure such as roads, which benefits industry.
Personal Preference		A person setting up a new business or building a new factory may wish to do so where they live or in a location they prefer. They do this for a number of reasons, e.g. to bring employment to their local area.
Common Market		The United Kingdom left the EU at the end of January 2020, meaning it left the EU Common Market. Countries within the EU Common Market can trade freely with each other, which means that there are no tariffs (taxes paid on imports or exports) on good and services traded between them. Many companies who wanted to remain in the Common Market left the UK. Ireland was the most popular place for companies to relocate in 2021.
Industrial estates		An industrial estate is an area of land developed as a site for a number of factories and businesses. It may include existing buildings ready for industries to move into and use. Industrial estates are usually located on the edges of towns or cities, close to transport links. They provide good access to services (e.g. electricity, waste disposal, broadband) needed by the various companies that locate in them. An industrial estate can also be an ideal location for linkage industries that rely on each other.
Greenfield or Brownfield Sites		Greenfield sites are underdeveloped areas within or outside a city, typically on agricultural land. These areas have not previously been used for manufacturing industry. Brownfield sites are land that was previously used for industrial purposes.

Being creative

Make up a mnemonic to remember the factors that influence the location of an industry.

A mnemonic uses the first letter in each word to help you remember something. For example, we can use '**R**ichard **O**f **Y**ork **G**ave **B**attle **I**n **V**ain' to remember the colours of the rainbow.

Question Time

1. (a) Identify the nearest industrial estate to where you live.
   (b) List the types of industries found there.
2. (a) Locate a potential greenfield site in or near your local area.
   (b) List three reasons why this greenfield site might be suitable for the location of industry.
3. Explain the link between raw materials and transport when choosing location of industry.
4. Give two reasons why it is an advantage to locate an industry close to a town.
5. How does money in the form of capital and corporation tax influence the location of industry?

Government Supports

There are two Irish government organisations responsible for supporting, promoting and developing Irish industry: the Industrial Development Agency (IDA) and Enterprise Ireland.

IDA Ireland

IDA Ireland partners with overseas companies to secure new investment to set up in Ireland. This investment is known as foreign direct investment. IDA does this by:

- Collaborating with foreign investors in Ireland to help them expand and develop their operations here
- Offering funding and grants to those considering setting up in Ireland.

Enterprise Ireland

Enterprise Ireland is the government organisation responsible for the development and growth of Irish businesses in world markets. It does this by:

- Funding supports – a range of supports for start-ups and business expansion plans
- Providing information about local markets and introductions for Irish companies to international customers
- Helping companies to become more competitive in international markets
- Providing incentives to carry out research and development in new products, services and processes.

FUN FACT!
In 2020, there were 269,208 people employed in foreign-owned companies in Ireland and 216,695 in Irish-owned companies that received assistance from the IDA and Enterprise Ireland.

25.3 Types of Industry

Heavy industry makes heavy/bulky finished products from heavy/bulky raw materials, e.g. ship-building.

Light industry makes small, lightweight finished products from small raw materials, e.g. the pharmaceutical industry.

Multinational corporations (MNCs) may have their headquarters in one country but be present in many countries, e.g. the US company Google has its European headquarters in Dublin and offices around the world.

Footloose industry is not tied to one location. Footloose industries can set up successfully in a wide variety of places. They are usually light industries, as their resource materials are light and easy to transport. Most modern industries are footloose because:

- Improved roads allow for better movement of industrial products over long distances
- Most people have cars, meaning most workers can commute to and from work.

Advantages of footloose industries include:

- They can set up in more than one location, meaning they can have access to a larger market
- Land on the edge of cities is cheaper than in the inner city, which reduces the costs of setting up.

Pharmaceutical companies that make light products such as medical equipment and medicines are examples of footloose industries.

Question Time

1. (a) Name the organisation which helps overseas companies to set up in Ireland.
   (b) Explain one way in which they do this.
2. How does funding help Irish industry to set up and attract overseas industry to come to Ireland?
3. What type of country has its headquarters in one country and locations in many more?
4. What could companies who receive funding use this money for?
5. What is a footloose industry?

Case Study: Heavy Industry – Banagher Precast Concrete

Managing information and thinking

Examine this photo of Banagher Precast Concrete. What evidence can you see that this is a heavy industry?

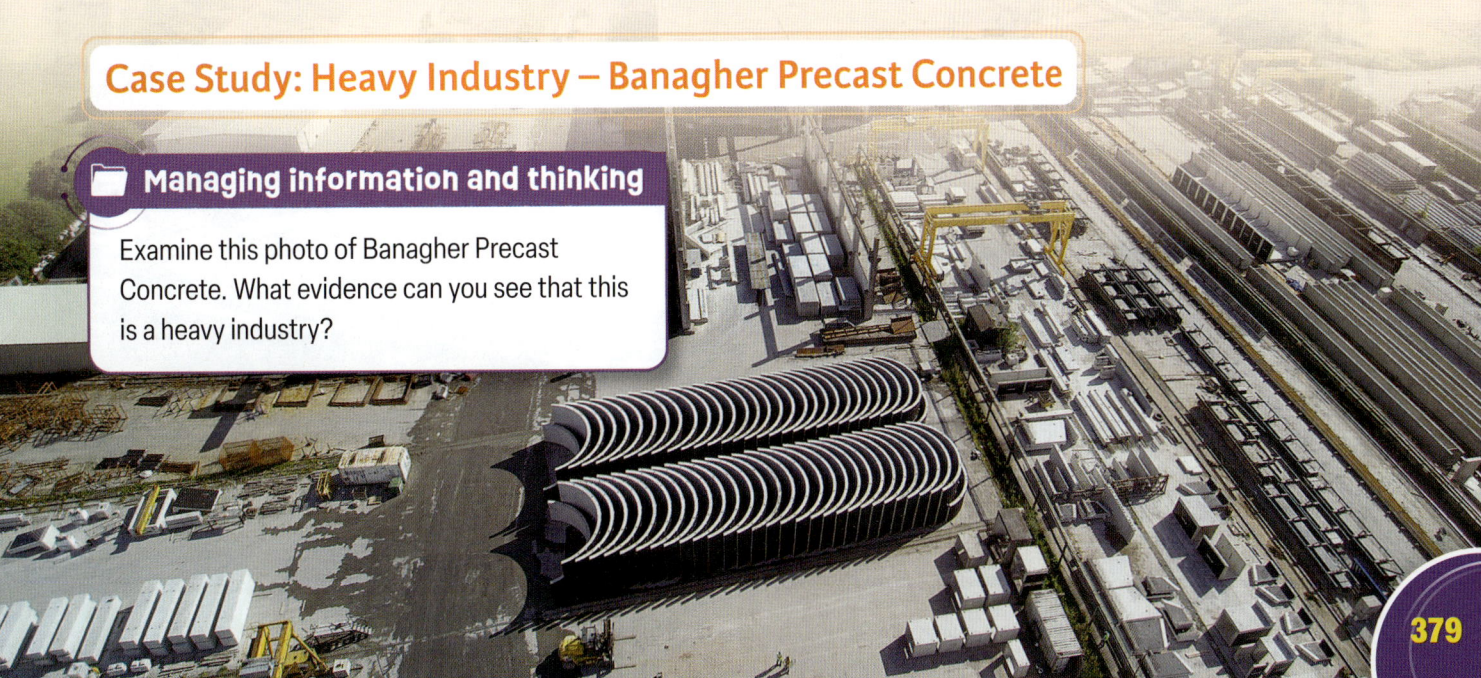

History: The company was originally called Bantile and was first set up in 1949. It became Banagher Precast Concrete in 1976 and is located at Queen Street, Banagher, Co. Offaly.

Finished product: Banagher Precast Concrete's main product is precast concrete. It can shape concrete into many different forms to meet the needs of its market. The concrete is used in the production of bridge beams, cattle slats, car parks, tunnel segments (e.g. Dublin Port Tunnel) and stadiums (e.g. the Aviva, Páirc Uí Chaoimh and the Curragh Racecourse). Banagher Precast Concrete also supplies hurling training walls.

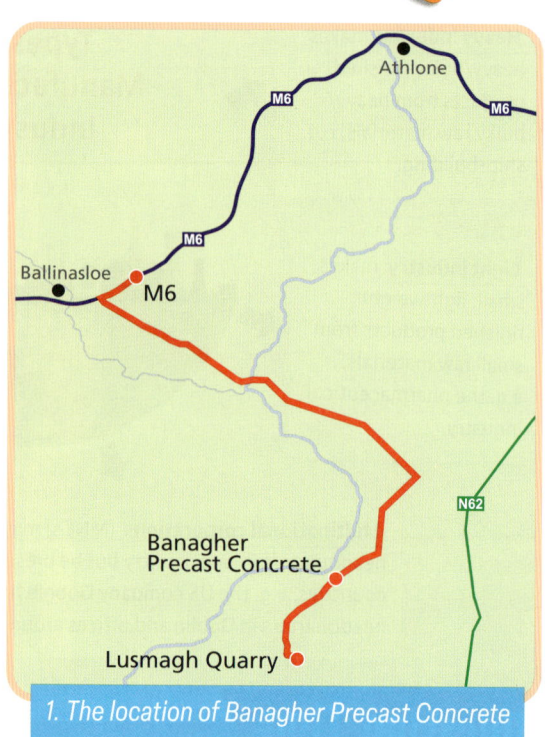

1. The location of Banagher Precast Concrete

Dublin Port Tunnel

Hurling training wall

Raw/resource materials: Banagher Precast Concrete is located 2.4 km away from Lusmagh Quarry. This limestone quarry provides 80% of Banagher Precast Concrete's resource materials, such as sand and gravel. Silica is another important resource material. This mineral is found in the area's soil.

The Aviva

25. Secondary Economic Activities

COUNCIL TO SPEND €200,000 ON LUSMAGH-BANAGHER ROAD FOLLOWING FLOODS

Q Why do you think the local council spent €200,000 on fixing the road between Banagher and Lusmagh?

Labour: Banagher Precast Concrete currently employs approximately 250 people. Much of the workforce lives in the town of Banagher, which has a population of 1,700 people. The local secondary school provides a semi-skilled (not fully trained) workforce that is further trained on site. Other employees live in the nearby towns of Birr, Kilcormac and Tullamore. The good road network between the towns makes it easy for workers to commute to and from work.

Transport facilities: To transport the large bridge beams produced by the factory a bypass (road that goes around the town) was built. This connects to the main road leading out of the town.

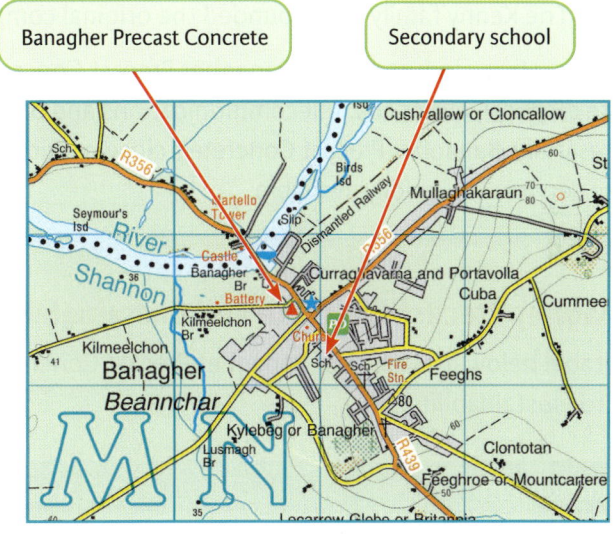

The factory is located 30 minutes from the M6 motorway, which connects Dublin and Galway. This enables Banagher Precast Concrete to transport its products to Dublin Port for exportation within two hours.

📁 Managing information and thinking

Looking at the two photos, A and B, can you identify why a bypass was needed?

A Main Street, Banagher

B Banagher Precast Concrete bridge beam

Markets: Due to the wide variety of finished products made by Banagher Precast Concrete, it has markets throughout the world. For example, Banagher Precast Concrete has manufactured precast breakwaters destined for Mexico and the Maldives, precast civil solutions for London Gateway, HS2, Silvertown Tunnel and prestressed bridge beams for roads across the UK.

Government and EU policies: Banagher Precast Concrete received grants from Enterprise Ireland. Ireland's low corporation tax rate of 12.5% on profits has also helped in the company's success. EU structural funds given to Ireland have contributed towards road-building across the country, which aids Banagher Precast Concrete in distribution.

Personal preference: Banagher Precast Concrete has gone through many changes from its initial set up. The Kenny family, who founded the original company in 1949, had close ties to the town.

Services: Heavy industries like Banagher Precast Concrete need on-site services – such as waste disposal, electricity, telecommunications and broadband – to run their businesses successfully. Banagher Precast Concrete's close proximity to the town of Banagher means that these services are readily available.

Banagher Precast Concrete and Sustainability

The table below lists some of Banagher Precast Concrete's policies for the environment and for sustainability.

Environment	Sustainability
The company promotes environmental and energy awareness to its employees through appropriate training and procedures.	The quarry where materials are sourced is just 5 km from the manufacturing site, which reduces transport and carbon emissions.
The company operates a LEAN organisation (using fewer resources) and aims to reduce, reuse and recycle.	In 2021, the company installed 1,000 PV solar panels above its manufacturing halls to generate green electricity.
It has planted acres of oak woodland locally and created a wildlife habitat, with a fresh water lake stocked with indigenous fish to counterbalance outputs and to leave a positive lasting legacy.	All suppliers to Banagher Precast Concrete are chosen with sustainability in mind.

Question Time

1. Banagher town does not have a large enough population to supply the workforce for Banagher Precast Concrete. How has the company overcome this problem?
2. What natural disaster was caused difficulty for Banagher Precast Concrete in accessing its raw materials? How was this problem overcome?
3. Transporting their finished products from their factory also caused difficulties. How was this overcome?

25.4 Change Over Time in the Location of Industry

Industries can change location over time. Factors that may cause an industry to change its location include:

- A change in demand for the product
- A change in the size of the industry.

Why Do Industries Change Location?

When a product becomes popular and new markets for it are created, there is an increase in demand for the product. The first factory is often unable to meet this increase in demand. New factories in new locations are set up to meet the growing demand.

Case Study: Kerry Group

Kerry Group began as a small dairy company in the south-west of Ireland in 1972 and has grown into a global multinational. It now has locations across Europe, Asia and North America. Kerry Group's headquarters are still located in Tralee, Co. Kerry.

The company started out producing dairy products but has expanded its range to include dairy alternatives, alcoholic drinks and ingredients for food, beverage and pharmaceutical industries.

Kerry Group opened a new facility in South Africa in May of 2022.

Many foods can be made from Kerry Group's range of products

What Factors Caused This Expansion?

Demand

The new facility in South Africa will produce dairy alternatives. Oat milk is already in high demand across the continent of Africa and Kerry Group is the leading supplier there.

Sustainability

As the population in Africa continues to grow, so too does the demand for food. Parts of South Africa are unable to meet the demand for natural milk. Kerry Group works with companies using oat milk to improve their products as an alternative to dairy. The oats used are produced sustainably to have as little impact on the environment as possible.

Q Explain two factors which can cause an industry to locate in other regions.

Junior Cycle Geography CYCLONE

Sample questions

1. (a) Photos A–D show four different products. **Create** a system for each product using the headings Inputs, Processing and Output. One product has been completed for you:

	Inputs	Processing	Output
D	Trees	Cutting, sawing by carpenter	Timber

(b) **Compare** and **contrast** how two different government organisations help industries to set up in Ireland.

(c) (i) **Explain** the term 'corporation tax rate'.
 (ii) How can corporation tax rate affect the choice of location for an industry?

(d) Image A shows a ship being manufactured in South Korea. Image B shows Intel's manufacturing plant in Co. Kildare, which produces microprocessors for computers.

Exam hint!

When asked to compare and contrast, you should explain things that are the same and things that are different. For this question, think of funding and if the industry is Irish or from overseas (see p. 376).

(i) **Which** image shows a heavy industry and which shows a light industry?

(ii) **Explain** the key differences between a heavy and a light industry.

> **Exam hint!**
> To answer part (ii), look at examples of heavy and light industries you are aware of. Think of how big/bulky or lightweight their raw materials and finished products might be.

2. (a) **Examine** the OS map segment below and answer the question.

(i) **Give** the six-figure grid reference for the industrial estate located in the south-west of the town of Mullingar.

(ii) **Explain** how transport, services, labour and market might have influenced the location of this industrial estate.

(b) **Name** an industry you have studied.
Explain the following:
- Whether it is heavy or light
- Five factors that influenced its location
- The steps your chosen company takes to be sustainable.

> **Exam hint!**
> To answer part (b), you can use the case study on Banagher Precast Concrete on pp. 379–382.

385

CBA 2 My Geography Moment

Create a report on a secondary economic activity from your area. Your area can include your nearest town, around your county or another place in Ireland. This task can be done individually or in groups. Begin by stating the secondary economic activity that you will complete the report on.

I/We Must

- Give a brief background to the secondary economic activity, including its location and when it was set up
- Explain what the factory produces. Are the products finished or semi-finished? Are they heavy or light?
- Explain the factors that influenced the location of the secondary economic activity chosen
- Explain how the secondary economic activity works as a system
- Discuss at least two impacts of the secondary economic activity on the area in which it is located, e.g. employment
- Explain how the secondary economic activity tries to reduce its impact on the environment
- Describe how the secondary economic activity acts sustainably.

I/We Should

Locate my chosen secondary economic activity using Scoilnet maps and see how the physical landscape influenced its location.

I/We Could

- Draw a sketch map of the area where the chosen secondary economic activity is located and provide a key for all items shown
- Visit or make contact with the secondary economic activity chosen to gather further information.

Self-Assessment

Reread what you have written and then write down two things you think you did well and one thing you think you could improve on.

Redrafting

Review the success criteria again to make sure you have met all the requirements. Take into account your own self-assessment notes and make any changes you think are needed. When you are happy, you can prepare to show your work to the class.

Learning Outcome: 2.9

26 Tertiary Economic Activities

Learning Intentions

You will be able to:
- Explain the term 'tertiary economic activities'
- Explain what tourism is and understand what is meant by sustainable tourism
- Discuss the connections between tourism and the physical world (landscape), and tourism and transport in Ireland and abroad.

Key words

- Services
- Service sector
- Tourism
- Tourists
- Physical landscape
- Climate
- Transport
- Road network
- International visitors
- Domestic tourism
- Market

26.1 What Are Tertiary Economic Activities?

Tertiary economic activities provide **services** that are useful to communities and people. These activities are also known as the **service sector**. Examples are people working as nurses, taxi drivers or musicians.

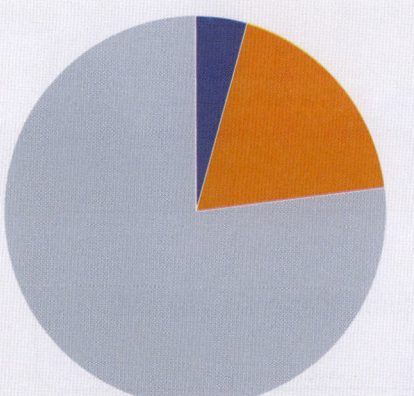

Examine figure 1. Name one example of a job found in each sector. Try to think of an example for tertiary not mentioned above.

1. Employment by economic sector in Ireland, 2020

387

26.2 Tourism, the Physical Landscape and Transport

Tourism is a tertiary economic activity. Tourism is travel for pleasure. It is also the business of of attracting, accommodating and entertaining tourists. **Tourists** are people who travel for leisure outside their local area (town, city or country).

The **physical landscape** stimulates tourism. We will examine how the landscape (notably mountains and beaches) and **climate** influence tourism. We will also look at how tourism relies on and affects **transport**.

Western Europe is one of the world's most developed tourist areas. Let's examine two popular tourist destinations, the Alps and the Mediterranean.

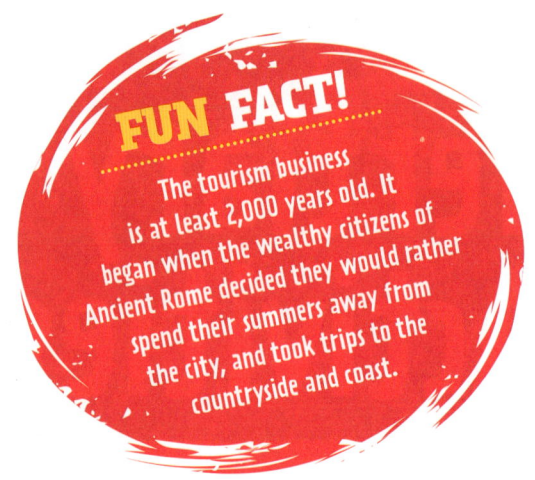

FUN FACT! The tourism business is at least 2,000 years old. It began when the wealthy citizens of Ancient Rome decided they would rather spend their summers away from the city, and took trips to the countryside and coast.

Case Study: The Alps

As we learned in Chapter 4, the Alps are a mountain system located in south-central Europe. They extend for almost 1,200 km, across seven European countries.

2. The Alpine region

Managing information and thinking

Looking at figure 2, identify and name the Alpine countries that are landlocked (surrounded entirely by land). Which of the seven countries has the largest Alpine region?

26. Tertiary Economic Activities

Normally, about 60–80 million tourists travel to the Alps each year. There are over 600 ski resorts in the Alps. France, Switzerland, Austria and Italy provide over 85% of Europe's skiing landscape. The Covid-19 pandemic greatly affected tourism in the region as travel restrictions were in place. However, many local tourism offices said that they had their best summer season in 2020. Governments encouraged the local people from France, Austria, Italy and Switzerland to holiday in the region instead of going abroad.

Tourists want to visit the Alps for a number of reasons.

- The Alps are a 'snow-sure destination'. This means they are likely to have enough snow for skiing during the peak season, which is December right through to May. The peak times are when large numbers of tourists wish to travel to the same places because of ideal weather conditions, school and work holidays or travel deals.
- The Alps provide a wide variety of slopes. This means that experienced skiers can use the steep slopes, while beginners and children can use the gentler slopes.

The Alps and Transport

The physical landscape of the Alps means it is more difficult to travel to and from the region. The cold winter temperatures and heavy snowfalls can make travel conditions dangerous. Sometimes, it can be impossible to drive through parts of the Alps. However, tourists still travel in their millions to the region.

Travelling to the Alps by Air

Many tourists travel to the Alps by air. Airports have been built as close as possible to many of the Alpine resorts. These airports offer flights to and from destinations all over the world, allowing tourism to thrive in the region.

3. Airports servicing the Alps region

Travelling to the Alps by Road and Rail

There are a number of roads and railway lines that travel directly through the Alps. These come from the countries that border the Alps. At over 11 km, the Mont Blanc Tunnel is the longest road tunnel running through the Alps. It links France to Italy and passes near numerous resorts.

The longest railway tunnel running through the Alps is also the longest tunnel in the world. It is called the Gotthard Tunnel and is over 59 km long.

The Mont Blanc Tunnel

The Gotthard Tunnel

How Alpine Tourism Affects the Physical Landscape

With the popularity of tourism in the region increasing year on year, the landscape of the Alps has suffered.

- **Road networks** are being constructed and improved in the region to allow for better access. However, the natural slopes of the mountains are being destroyed to allow for this.
- Modern downhill skiing or snowboarding needs well-prepared and large ski runs to remain attractive to tourists. Creating these ski slopes has also increased the rate of erosion in the Alps.
- Air pollution is impacting the physical world of Alpine regions. Fossil fuels used in transport release emissions. Carbon dioxide and nitrous oxide can sometimes become trapped in the region when weather is calm and winds are weaker.

Search online for the article 'Amazing view and dirty air in the French Alps – BBC News' (7 March 2017). Read the article and then answer these questions.

1. Give two reasons as to why air pollution is so bad in this region.
2. What is the main cause of air pollution, according to the article?
3. List three measures being taken by the local government to decrease levels of air pollution in the region.

26. Tertiary Economic Activities

Q

1. Name the mountain range that provides a source of tourism for the people of north Italy, Switzerland, Austria and the south-east of France?
2. Explain the term 'snow-sure destination'.
3. (a) Describe a difficulty the physical landscape presents for tourism in this region.
   (b) Briefly describe one way the difficulties are overcome.
4. Describe one negative effect tourism has had on this region.

Case Study: The Mediterranean

The Mediterranean region has long been one of the top tourism destinations in the world. The Covid-19 pandemic greatly affected the numbers visiting the region. However, the Mediterranean region recorded its highest figure in 2021 of around 139 million tourists.

The Mediterranean Climate and Tourism

The Mediterranean climate – warm to hot in summer and mild to cool in winter – provides perfect conditions for tourism almost all year round.

- In the summer months, the Mediterranean has an average temperature of 22 °C and between 7 and 11 hours of sunshine per day. This makes the region ideal for beach holidays.
- In winter, the region gets on average 6–10 hours of daily sunshine, and temperatures are usually between 2 and 10 °C.

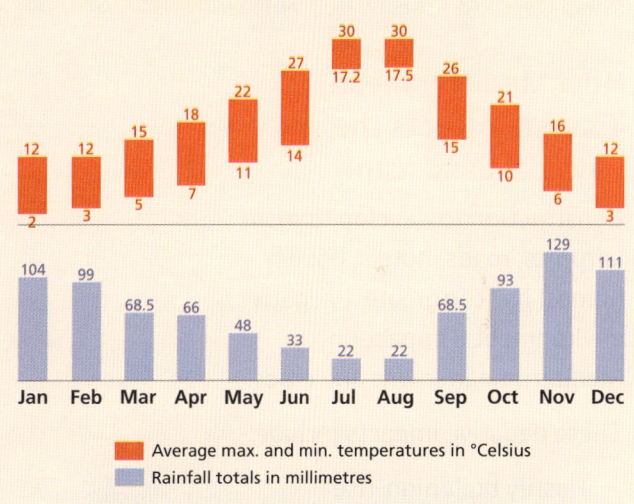

4. Average temperature and rainfall in the Med

The Mediterranean and Transport

The Mediterranean tourism year can be divided into three seasons:

1. Peak season (mid-June through August)
2. Mid-season (April through mid-June and September through October)
3. Off-season (November through March).

The Mediterranean is well serviced by air, sea and road routes. However, high demand at peak times creates an increased need for transport routes.

Being numerate

Figure 4 shows the average maximum and minimum temperatures and average rainfall for each month in the Mediterranean.

1. How many months experience temperatures in excess of 15 °C?
2. What is the total amount of rainfall experienced in the winter months (November–January)?

Increased demand for air travel can raise the cost of flights by hundreds of euro. Airports become crowded, with long queues for check-in, security clearance and passport control.

5. Airports servicing the Mediterranean

Q

Looking at figure 5:
1. Why do you think airports have developed in these locations?
2. Have you ever flown to one of these airports? At what time of year did you fly? Was the airport busy?

How Mediterranean Tourism Affects the Physical Landscape

Construction to develop tourism (airports, roads, hotels, holiday homes) has the greatest negative environmental impact on Mediterranean towns and cities.

These negative impacts include:

- Hastily built high-rise developments, which take away from the natural beauty of the area
- Roads and hotels which take up beach space, leaving less for local people and tourists to enjoy.

High-rise developments in the Mediterranean

Q

1. Describe the Mediterranean's climate that makes it a popular tourist destination.
2. Tourism can have negative effects on the physical landscape of the Mediterranean. Describe two of these.
3. Compare and contrast how the Alps and the Mediterranean rely on different physical features to attract tourists.

26.3 Tourism in Ireland

The number of overseas visitors to Ireland reached record levels in 2016, when 8 million **international visitors** came to the country. **Domestic tourism** (people travelling within their own country for holidays) also reached a record high, with over 2 million Irish people holidaying in Ireland in 2016.

The Covid-19 pandemic had a negative impact on tourism in Ireland as overseas visitors were not able to travel here. To help this sector, the Irish government promoted domestic tourism. People living in Ireland made approximately 5.7 million journeys around Ireland in 2021.

IRELAND'S TOP 10 TOURIST ATTRACTIONS – AS CHOSEN BY YOU!

Slieve League, Co. Donegal

What do you think are our favourite Irish tourist destinations? An online poll has found that the following attractions have been voted as Ireland's top 10 tourist destinations for 2021:

1. The Cliffs of Moher, Co. Clare
2. The Guinness Storehouse, Dublin
3. The Ring of Kerry
4. The Giant's Causeway, Co. Antrim
5. The Wild Atlantic Way
6. Aillwee Cave, Co. Clare
7. Glendalough, Co. Wicklow
8. Hook Head, Co. Wexford
9. Slieve League, Co. Donegal
10. Kilmainham Gaol, Dublin

How many of Ireland's top attractions do you think were influenced by the physical landscape?

Case Study: Tourism and Ireland's Physical Landscape

Ireland relies on its unspoilt physical landscape and stunning natural features rather than its climate to attract tourists.

Some major movie and television production companies have been attracted to Ireland because of its unique landscape. The inclusion of Ireland in international films and television programmes has helped to **market** (advertise) Ireland to the world and attract more tourists to these shores.

Skellig Michael

Skellig Michael (also known as Great Skellig) is a towering sea crag/island rising from the Atlantic Ocean, nearly 12 km west off the Kerry coast.

Star Wars: The Force Awakens (2015) and *Star Wars: The Last Jedi* (2017) featured scenes filmed on Skellig Michael. Boat trips to and from the island have become extremely popular as tourists flock to see where the movies were filmed and to admire the region's natural beauty.

6. The Ring of Kerry and Skellig Ring

A scene from *Star Wars,* filmed on Skellig Michael

Go to YouTube and search for 'Star Wars: The Force Awakens – Behind the scenes in Ireland' (1:08). Write down two reasons why Skellig Micheal was chosen to film scenes for the *Star Wars* movies.

The Cliffs of Moher

In 2009, the Cliffs of Moher were included in scenes from *Harry Potter and the Half-Blood Prince*. The cliffs themselves have long been one of Ireland's most popular tourist destinations. They stretch for 80 km and are among the highest in the country, ranging from 120–214 m in height.

Thinking back to what you learned in Chapter 11: The Sea, what process of erosion helped shape the Cliffs of Moher? What feature of coastal erosion can you see in this photo?

The Cliffs of Moher

26. Tertiary Economic Activities

Settlement and Tourism along Ireland's Coast

Settlement
People have settled along the Irish coast because of its location right next to the ocean. It also provides opportunities for sea transport, tourism and recreation.

Some 1.9 million people (or 40% of the population) live within 5 km of the coast. Many of Ireland's main cities are located on the coast, for example: Belfast, Dublin, Limerick and Galway.

Tourism and Ireland's Coasts
Coastal areas have a variety of scenic landscapes, with historic towns and harbours and lots of options for recreational activities.

The population of some coastal areas can change throughout the year due to tourism. In some resorts, the population can increase threefold in the summer months.

According to Fáilte Ireland, tourism in the coastal counties that stretch from Malin Head to Mizen Head and around to Carlingford is estimated to be worth €2 billion to the economy. This excludes Dublin, which is classified under city tourism.

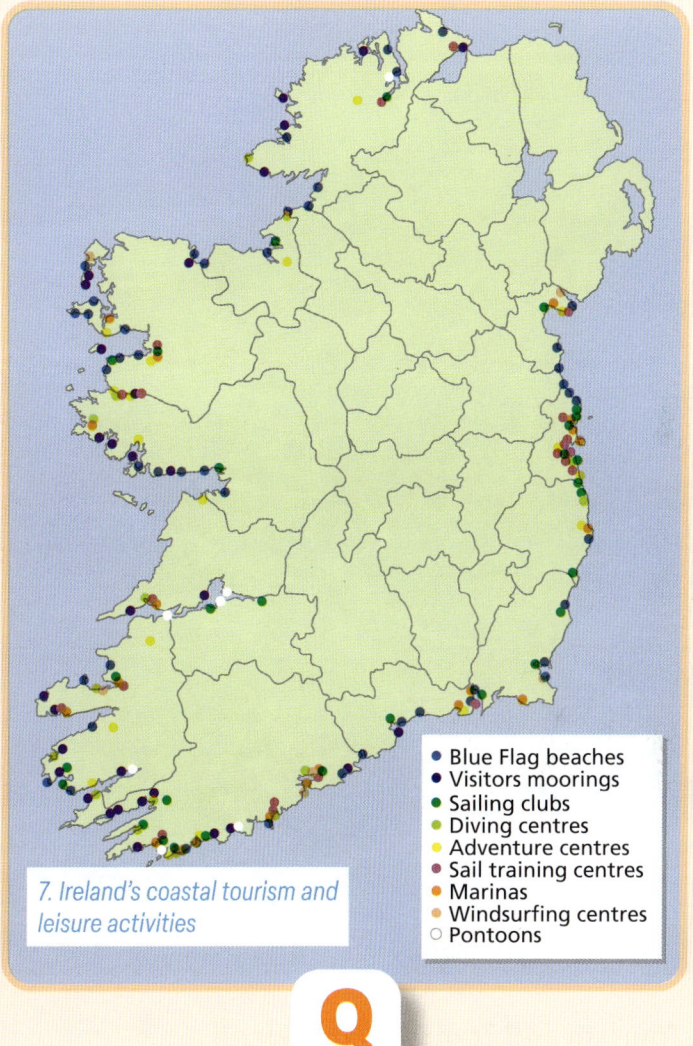

7. Ireland's coastal tourism and leisure activities

- Blue Flag beaches
- Visitors moorings
- Sailing clubs
- Diving centres
- Adventure centres
- Sail training centres
- Marinas
- Windsurfing centres
- Pontoons

1. Using Google Maps or Scoilnet, work out the distance from your home or school to the nearest point on the coast. Record your measurement in kilometres.
2. Find out two tourist activities available in this location. Give the name of the activity, state how much it costs (if anything) and say when it is open. Then select one of these tourist activities and create a poster displaying all of the information you have gathered on it. Present this poster in class.

Q Using the map above, locate and list five coastal tourist attractions between Malin Head and Mizen Head.

Question Time

1. Explain the term 'domestic tourism'.
2. List two physical features that attract tourists to Ireland.
3. Briefly describe the link between tourism and settlement found along Ireland's coastline.

Exam hint!
What happens to the population of coastal towns during summer months?

Ireland's unique landscape and physical beauty are the main attractions for tourists coming to Ireland. However, the increased numbers of tourists visiting Ireland have the potential to damage this natural beauty in the following ways:

- Rising greenhouse gas emissions from increased traffic
- More litter in our rural/coastal areas and unique landscapes
- Activities such as hiking destroying delicate habitats.

Tourism in Glendalough

Q What physical features can you see in this photo that might attract tourists to Glendalough?

Tourism and Transport in Ireland

Whether international or domestic, tourists need a choice of transport routes to reach Ireland and/or to travel around it comfortably.

Ireland's Airports and Ports

International tourists arrive in Ireland by air or sea. A variety of flight routes makes it quick and easy for tourists to reach Ireland from anywhere in the world. Some people might like to travel by ferry, so that they can bring their car, campervan or caravan.

  Working with others

With the person sitting next to you:
- Think about the last holiday each of you went on
- Write down the different types of transport you used to reach your destination
- Write down any types of transport you used while on holiday.

26. Tertiary Economic Activities

Ireland's Airports

Ireland has five international airports:

1. Dublin
2. Cork
3. Shannon
4. Knock
5. Belfast

There are also other airports around the country that fly to specific European destinations or to other parts of Ireland.

Ireland's Ports

Tourists can reach Ireland by sea through the following ports:

1. Larne, Co. Antrim
2. Belfast, Co. Antrim
3. Dublin Port
4. Dún Laoghaire
5. Rosslare, Co. Wexford
6. Cork Port in Ringaskiddy

How Irish Tourism Affects Transport

Airports will provide more flights during their busy 'peak' periods. However, airlines are always concerned about the costs of running flights during off-seasons.

8. Ireland's airports and ports

For example, if fewer than 50 people wished to travel from Spain to Ireland on a particular day, it would be very difficult for an airline to cover its costs or to make a profit from the flight. The airline might decide to fly that route only a couple of times per week.

Contrast this with a busy summer day, when thousands of people might wish to travel from Spain to Ireland. This number might include Irish people returning home and Spanish people visiting Ireland. This high demand would mean that the airlines could offer a choice of flights (different times and different origins/destinations).

Managing information and thinking

The bar chart shows the rounded total number of passengers who arrived into and departed from the Republic of Ireland's main airports (Cork/Dublin/Kerry/Knock/Shannon) in the months of April, May and June between 2019 and 2022.

1. What incident occurred to cause the significant drop in number for 2020 and 2021?
2. Identify the peak month for tourists arriving in 2019 and 2022.

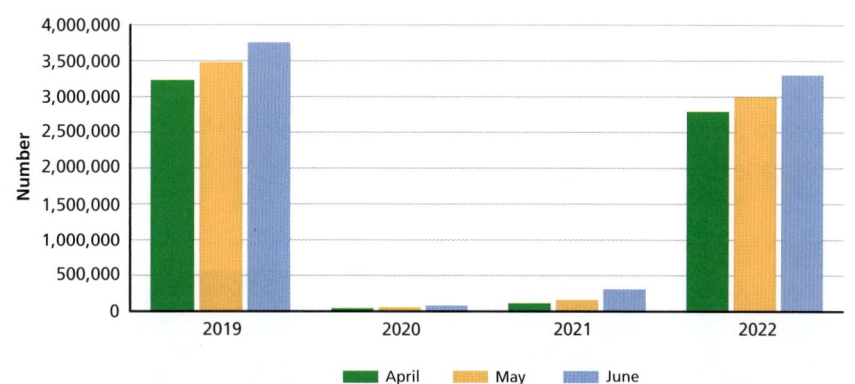

9. Total number of passengers served by the main Irish airports for months of April, May and June from 2019 to 2022

International Trade and the Physical Landscape

Truck drivers travelling between Ireland and Europe go by ferry but drive across the UK. When the United Kingdom left the EU in January 2020, it created difficulties as goods had to be physically checked before entering and leaving the UK. This caused huge delays and backlogs at ports.

To overcome this, ferry companies offered direct services from Rosslare to Dunkirk in France so as to avoid delays in the UK.

Stena Line's tweet about its direct service between Ireland and France

Ireland's Road Network

Once tourists arrive in Ireland, they can travel around the country on a widespread road network.

Roads are really important for tourism. As we can see from the map (figure 10), Ireland has a variety of roads from motorways and primary roads to secondary and regional roads. Each type is influenced by the physical landscape of the area. Less important roads can sometimes be harder to access due to the physical landscape.

Our motorways and primary roads link our cities and large towns. They provide easy access to Dublin City, one of Ireland's main tourist attractions. Secondary and regional roads take longer to travel on, e.g. the N81 from Dublin to Tullow. This is because they do not bypass some towns. Tourists travelling on these roads see more of Ireland's small towns and villages.

FUN FACT! There is over 5,300 km of national road network in Ireland. The US has almost 7 million km of roads!

10. Ireland's national road network

Q Use a ruler to connect Dublin to another major city shown on figure 10. The line of your ruler will show the shortest distance between Dublin and that city. Now, think about why the road might not follow the most direct route ('as the crow flies'). How might the physical landscape have influenced the layout of the road?

Question Time

1. Explain the tourism terms: (a) peak season, (b) mid-season and (c) off season.
2. How did Irish ferry companies adapt to overcome the challenge faced by the UK leaving the EU?
3. Briefly explain the differences between Ireland's motorways, national primary/secondary and regional roads.
4. Describe how each of these roads help develop tourism in Ireland.

26.4 Sustainable Tourism

The quotation on the right is the slogan for sustainable tourism. Sustainable tourism means that tourism can be used to provide both employment and a source of income, but it must do so without damaging the natural environment, so that future generations can benefit from it.

Let's look at how the different regions we have studied in this chapter are trying to achieve this:

Take only photographs and leave only footprints.

The Alps	The Mediterranean	Ireland
• Developing eco-friendlier travel packages transporting guests by rail or bus, thereby reducing carbon footprint • Reducing the number of new ski slopes created each year	• Measuring and monitoring the environmental impacts of tourism activities on the landscape • Identifying, protecting and restoring vulnerable and damaged coastal landscapes • Reducing the number of high-rise hotels built directly on coastlines	• Promoting the Tidy Towns competition, encouraging locals to maintain towns and villages throughout the country • Encouraging travel by rail, bus and bike, and opening up walking routes around Ireland, in an effort to reduce carbon footprint • Encouraging 'staycations' (where people holiday in their own country) to reduce the number of flights • Greenways are trails built to be used exclusively by cyclists, pedestrians and other non-motorised transport

Question Time

1. What is sustainable tourism?
2. List two steps each country/region discussed in this chapter have taken to ensure tourism is more sustainable for the future.
3. Which country/region in your opinion has the best approach towards achieving sustainable tourism? Give a reason for your answer.

Junior Cycle Geography CYCLONE

EXAM FOCUS

Sample questions

1. (a) Read the article below and answer each of the following questions:

Donegal Airport has been voted the most scenic airport in the world for the third year in a row in 2020

Donegal Airport in the northwest of Ireland has been voted as the most scenic airport in the world for the third time in a row. Located at Carrickfinn, Donegal Airport (Aerfort Dhún na nGall) has a runway length of 1500 m and is suitable for small props, small jets, and regional airliners.

(i) **List** the physical features you can see in this image that make Donegal Airport scenic.

(ii) Shannon Airport has a runway length of 3200 m, Dublin Airport's runway is 2600 m, Donegal Airport's runway length is 1500 m. **How** is the physical landscape stopping a longer runway being built at Donegal Airport?

(b) **Describe** how the physical features of a region you have studied have influenced tourism in that region.

Exam hint!

Part (b) is asking you to show your understanding of how the physical features of a region can affect tourism. Choose a region that has a clear link between the physical landscape or its climate and tourism (the Alps, the Mediterranean or Ireland). Describe the feature(s) as best you can. Then use any facts and figures you can remember that show the value to tourism the features bring.

26. Tertiary Economic Activities

2. (a) (i) 'Tourism relies on transport to be successful.'
    Name two ways tourists can arrive into Ireland from overseas.

   (ii) **Explain** why airlines would offer more flights into Ireland from June to August then they might from December to February.

   (b) (i) **Suggest** an idea that could help to make tourism more sustainable for Ireland.

   (ii) 'The physical features of a landscape can make transport to and from a region difficult.'

    Explain how one region has overcome this difficulty.

Let's get started! Sample starter

> The region I have studied is the Alps. They are a mountain range stretching 1200 km across Italy, France, Switzerland and more.
> Driving to and around the Alps can be next to impossible as cold winters and heavy snowfalls make travel very dangerous.
> Travelling by air …
> Travelling by rail …

… now complete the answer in your copy.

3. **Examine** the 1:50 000 Ordnance Survey map extract below of Slea Head in Co. Kerry and the accompanying legend which is available on p. 85. The road shown on the extract is a popular tourist route.

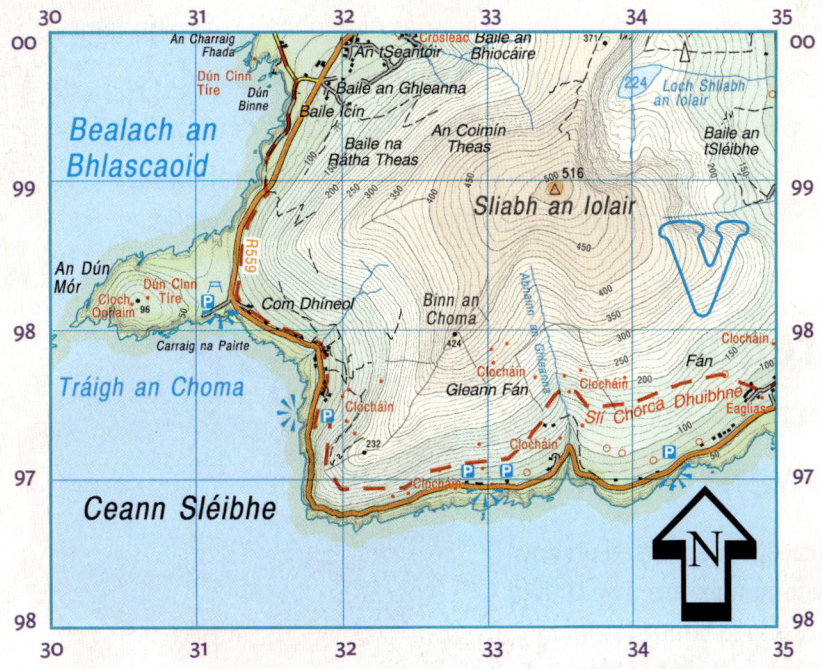

Name two different pieces of evidence from the map extract above that show how this area has developed to support tourism and **state** how each can be used by tourists.

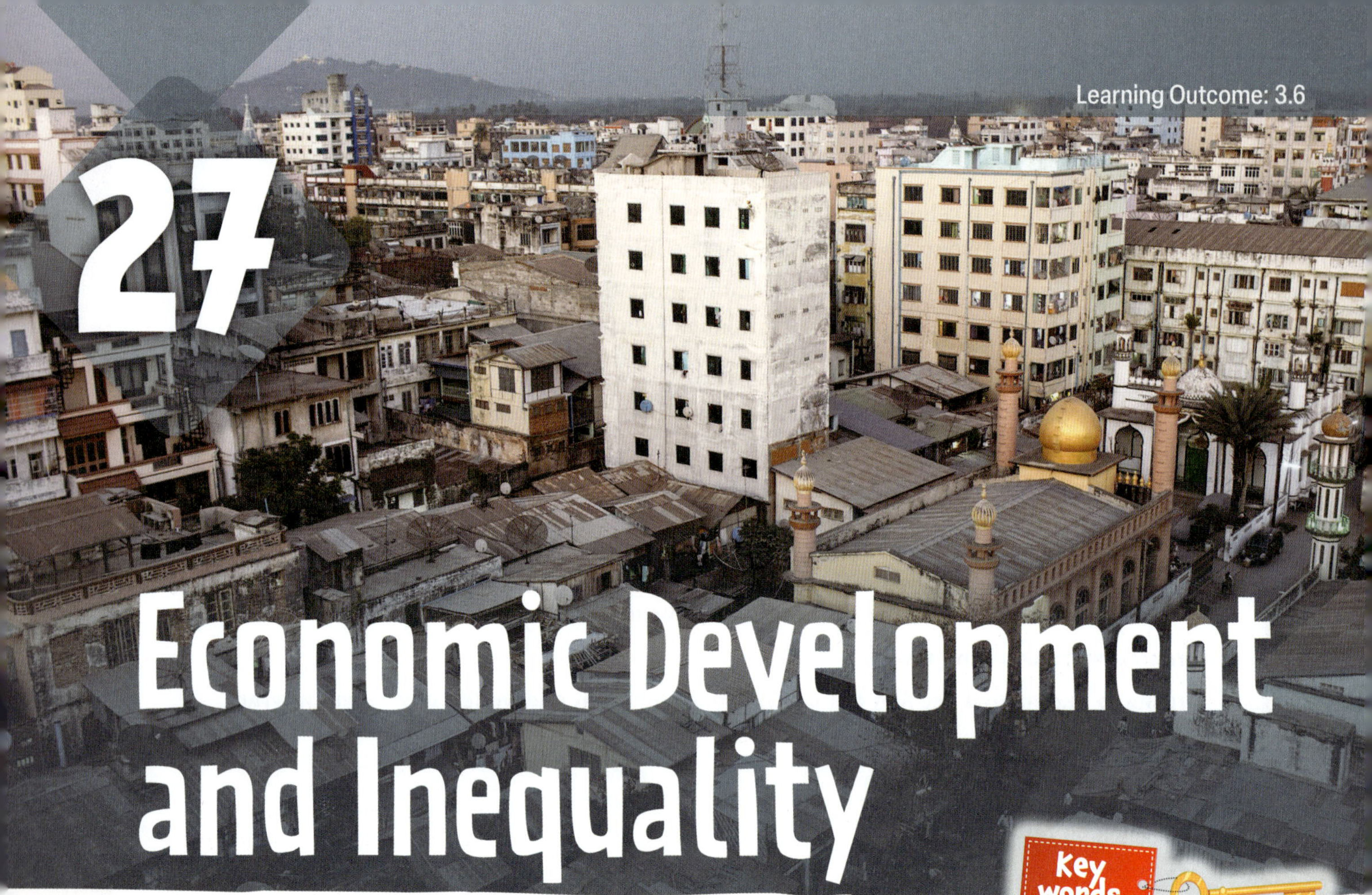

Learning Outcome: 3.6

27 Economic Development and Inequality

Learning Intentions

You will be able to:
* Examine the different categories of economic development
* Describe how economic development is measured
* Examine some causes of unequal economic development
* Examine how economies can grow, as well as rates and patterns of economic development around the world.

key words

- Economy
- Least developed countries
- Developing countries
- Economies in transition
- Developed countries
- Gross national income (GNI)
- Imperialism
- Unfair trade
- Globalisation
- Imports
- Exports
- Multinational companies (MNCs)

27.1 Categories of Economic Development

Countries can be categorised based on how developed their **economies** are. The United Nations (UN) has grouped countries into four different categories: **least developed countries**, **developing countries**, **economies in transition** and **developed countries**.

NOT SO FUN FACT!
The world's 2,153 billionaires have more wealth than the 4.6 billion people who make up 60% of the planet's population.

402

27. Economic Development and Inequality

Countries Grouped by United Nations

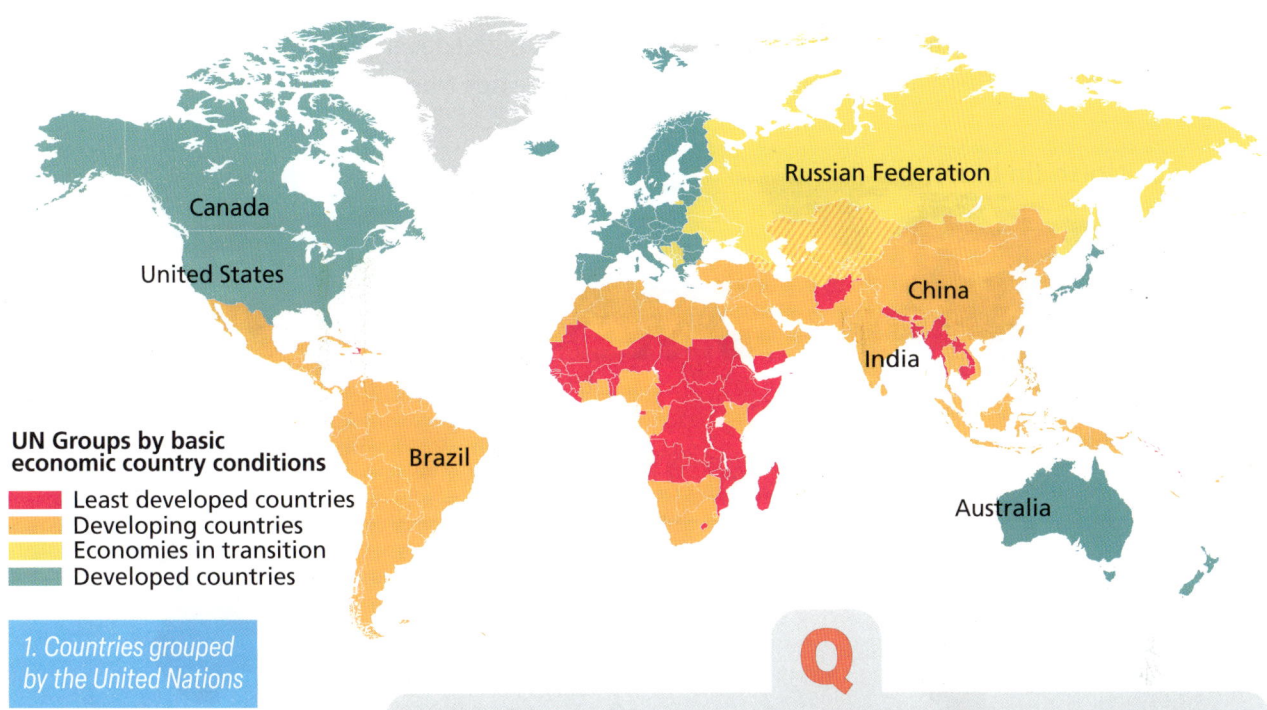

1. Countries grouped by the United Nations

Q

1. Examine figure 1. Which continent has the most developed countries?
2. Which continent has the most countries defined as 'least developed'?

What Do These Categories Mean?

Category of Country	Level of Economic Development
Developed country	A country that has a high quality of life, a well-developed economy and advanced technological infrastructure compared to other less industrialised nations
Developing country	A country that has not reached a high level of industrialisation given its growing population. In most cases, it has a medium-to-low standard of living
Least developed country	A country that has a low level of income and faces major barriers to development, such as its population having a low quality of life, no access to healthcare and little to no income
An economy in transition (changing)	A country that is changing its economy from being under government control to a market economy. This means companies are not controlled by the government

Rio de Janeiro in Brazil

Junior Cycle Geography CYCLONE

Some countries are very rich and some countries are very poor. This has divided the world into two economic regions. Most of the least developed countries are in the southern hemisphere. They are known as the poor South. Most of the developed countries are in the northern hemisphere. They are known as the rich North.

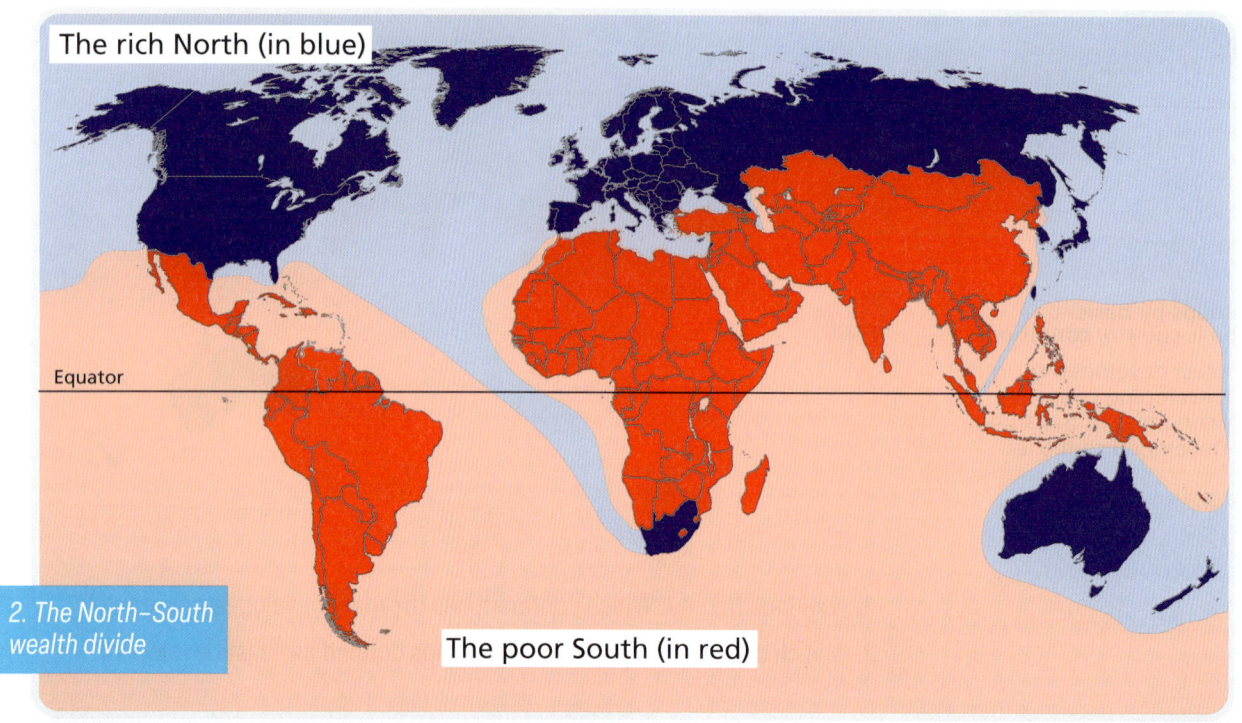

2. The North–South wealth divide

Managing information and thinking

With the person sitting next to you, examine figure 2. Identify two countries/two cities from the rich North and two countries/two cities from the poor South.

Abu Dhabi in the UAE

27.2 How We Measure Economic Development

There are a number of different indicators of economic development. Each can be measured to identify how developed a country's economy is or used to compare the economic development of different countries.

Indicators of Economic Development

Gross national income (GNI) is the total amount of money earned by a nation's people and businesses from making goods and providing services in a financial year. It is measured in US dollars (US$). To calculate its amount per capita (which means per single person), divide GNI by the population of the country.

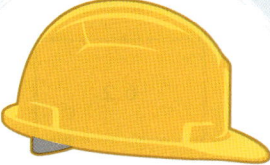

The numbers employed in secondary and tertiary activities

Education rate means the average number of years spent in education in a country.

Indicators of Economic Development

The higher the rates for each of these indicators, the more developed a country's economy is considered to be.

Average life expectancy is the average number of years a person can expect to live from time of birth.

The adult literacy rate is the percentage of adults in a country who can read and write.

Junior Cycle Geography CYCLONE

Country		GNI per person in US$	Years in education (Average)	Adult literacy rates (%)
Ireland		79,450	9.4	99%
Brazil		15,550	4.9	92.9%
Australia		55,290	10.9	99%
Nigeria		2,100	6.2	59.6%
India		2,170	5.1	74%
Iceland		55,920	8.8	99%
China		11,800	6.4	96%

The table on the left gives the GNI, average years in education and adult literacy rates in a range of countries in 2021.

1. Referring back to figure 1 if you need to, group the countries listed in the table as developed, least developed, developing countries and economies in transition.
2. Choose one developed, one least developed and one developing country and explain two characteristics used to describe their economies.
3. Draw a bar chart comparing any two of the countries listed, using an indicator of your choice (e.g. GNI and years in education).

There are two rows left blank in the table above. Choose one country in the rich North and another in the poor South and complete the details for each in your copy. Search online for:
- 'World Population Review – GNI by country'
- 'World by Map – Literacy Rates'
- 'Nationmaster Education average years of schooling of adults: Countries compared'

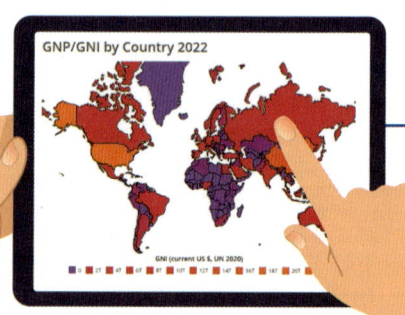

Question Time

1. List the four terms used to categorise levels of development for countries.
2. Explain each of these four terms.

Employment as an Indicator of Economic Development

As we learned on p. 403, the economic development of a country can be measured by examining where and how people are employed. In developed countries, most people are skilled workers involved in the tertiary sector, e.g. as doctors and educators. In developing countries, more people find employment in the primary sector, e.g. in farming, or in the secondary sector, e.g. in manufacturing.

Employment by Sector: Ireland vs Nigeria

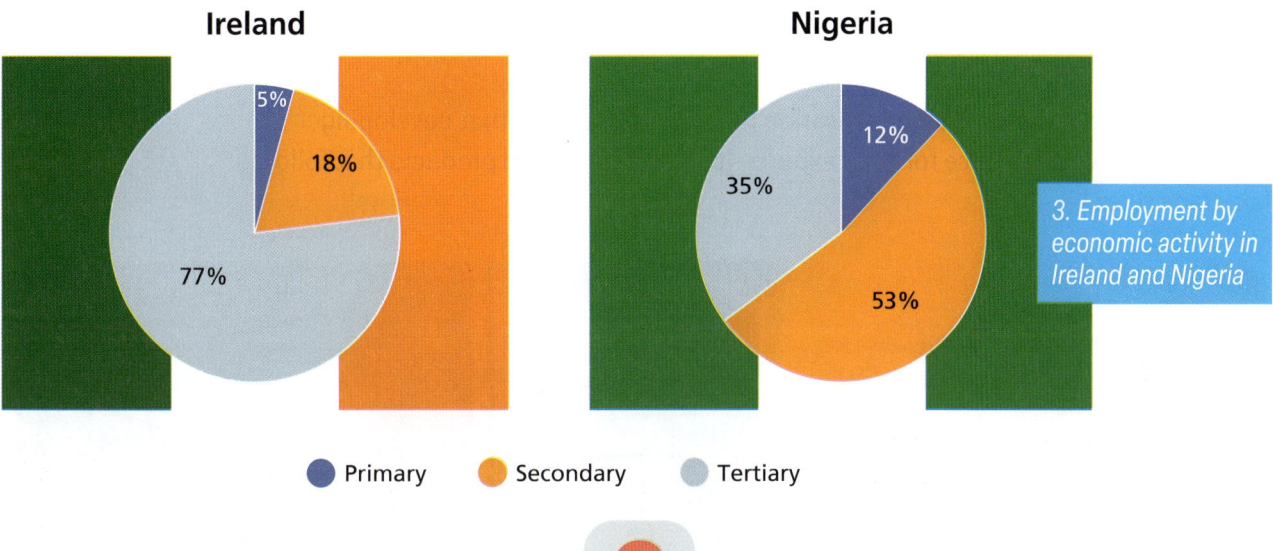

3. Employment by economic activity in Ireland and Nigeria

Q

1. According to the pie charts, which country is more economically developed? Give evidence to support your answer.
2. Thinking back to what you learned in Chapters 18 and 23–25, explain why employment in the primary sector remains high in Nigeria.

27.3 Causes of Unequal Economic Development

There are many causes of unequal economic development. We will look at **imperialism**, **unfair trade**, **globalisation** and the impact of the Covid-19 pandemic.

1. Imperialism

Imperialism is when one country takes over territory that belonged to another country. This is done by creating colonies. A colony is a country or area under the full or partial control of another country. It is occupied by settlers from the more powerful country. There are many examples throughout history, e.g. the Ulster plantation.

In the past, European countries such as Portugal, Spain and England took control of large areas of the world. Imperialism slowed the development of the countries that had been colonised. They were forced to supply the powerful countries with food and minerals. This caused the world to be economically divided.

2. Unfair Trade

International trade is the exchange of capital (money), goods and services across international borders. International trade, including **imports** and **exports**, is vital for the growth of a country's economy.

> **Imports**: Goods or services brought into a country
> **Exports**: Goods or services shipped out of a country

Developed countries try to pay as little as possible for their imports. These imports are often provided by developing countries. When developing countries do not receive a fair price for their exports, it is called unfair trade. Unfair trade prevents the economies of countries from developing.

Tackling Unfair Trade

The Fairtrade Foundation is an organisation that works to ensure that developing countries receive a fair price for their exports. It gives its stamp to products that follow fair-trade rules.

> **Managing myself**
>
> When next in your local supermarket, note how many products you see that display the Fairtrade logo. Are they products you buy regularly?

Case Study: Ben & Jerry's Fairtrade Ice Cream

In 2005, the Ben & Jerry's Ice Cream company made a commitment that its five main ingredients
- Coffee
- Sugar
- Vanilla
- Cocoa
- Bananas

would be Fairtrade-certified.

27. Economic Development and Inequality

The coffee beans
- Sourced from the Huatusco Coffee Cooperative in Mexico since 2010
- Fairtrade premium paid in 2017; payment used to build a medical centre, a library and schools. A fairtrade premium is an extra payment given to workers and farmers, which goes into a shared fund to improve social, economic and environmental conditions of the community

The sugar
- Sourced from Belize in Central America, one of the world's leading producers and exporters of sugar
- Fairtrade premium paid in 2017; payment used to train farmers in new farming methods, which makes production more efficient and lessens use of harmful pesticides

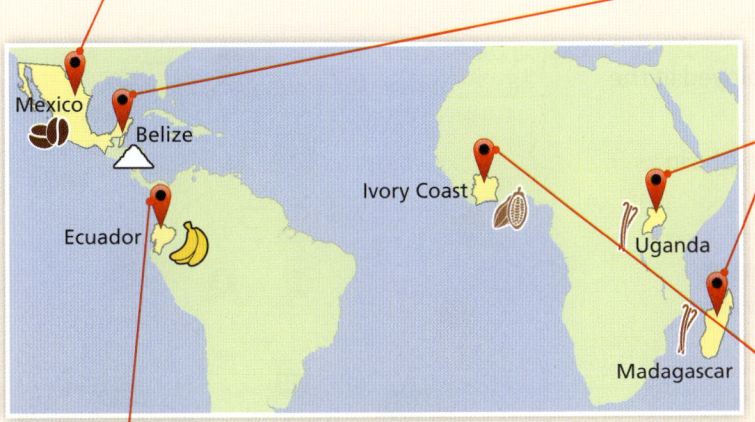

The vanilla
- Sourced from Madagascar and Uganda
- Unfair trade had left farmers struggling to survive in both countries
- Malagasy farmers (farmers from Madagascar) were paid a Fairtrade premium in 2017; payment used to repair a health centre and a school and to build an office

The bananas
- Sourced from Ecuador
- More than 50 tonnes of fairtrade bananas purchased in 2014
- Fairtrade premium paid in 2017; payment used to supported projects, such as providing scholarships for local children, helping children in need with language and physical therapy, and making healthcare more readily available to farmers, workers and their families

The cocoa
- Sourced from the Ivory Coast, which produces almost 40% of the world's cocoa
- Fairtrade premium paid in 2017; payment was used to build a new medical clinic, hire a nurse, install solar panels and purchase a water pump

 Go to YouTube and search for 'Ben and Jerry's – Fairly Nuts' (0:29) to see a short video on the company's commitment to fair trade – and to see the real Ben and Jerry!

Fair trade might not be enough to significantly improve the economy of a country. Some need more assistance from developed countries. This will be explored in Chapter 28.

Question Time

1. Explain the difference between fair trade and unfair trade.
2. In your copy, create a table with the headings 'Health', 'Education' and 'Other' to list how Ben & Jerry's suppliers have been using their Fairtrade premium payments. How do you think these extra payments help these countries grow their economy? Think back to the five indicators of economic development.

3. Globalisation

We learned in Chapters 18 and 19 that globalisation is the process whereby the world has become more interconnected. Globalisation has had many positive impacts on the economic development of developed and developing countries. For example, **multinational companies (MNCs)** setting up in countries around the world bring employment to local communities.

- In 2021, 275,000 people were directly employed in the multinational sector in Ireland.
- The same year, MNCs created 29,000 new jobs in Ireland.

FUN FACT! Ireland has over 1,000 leading MNCs! Eight of the world's top 10 video gaming companies have their European headquarters here.

4. Location of MNCs in Ireland

The number of MNCs per town/city/county

Managing information and thinking

1. Is there an MNC located near your school, home or town?
2. Find out how many people are employed in the MNC.
3. Are there any people from your local area employed there?
4. Discuss the effects the MNC leaving might have on the area.

Globalisation has also had a negative effect on the economic development of developing countries.

Brain Drain	Over-reliance	Dominance of MNCs
• Economic migrants are people who move to a country with a flourishing economy. Well-educated individuals and families leave countries with poor economic development because of push and pull factors. • The countries they leave suffer from a shortage of highly skilled workers, e.g. Eastern European migrants moving to Ireland during the 'Celtic Tiger' (a period when Ireland's economy grew rapidly).	• Countries can be so well-connected that they become too reliant on each other's economies. If the economy of a developed country goes into recession, it has a knock-on effect for a developing country that exports to it. • Trade could be slowed or halted altogether, further increasing the economic divide.	• MNCs that set up in developing countries can endanger local industries. Some may be forced to close as they are unable to compete. This can cause unemployment. • MNCs can also use up the natural resources of an area, e.g. water, minerals and forests. This means that local people and companies cannot use these resources. This can also damage the environment of the developing country, where there may not be strict laws governing the actions of MNCs. • MNCs can also leave a country with very little warning, when it is no longer profitable to operate there.

In order to be fair, globalisation must be properly managed and monitored by the governments of both developed and developing countries.

27. Economic Development and Inequality

4. Covid-19 and Economic Development

The Covid-19 pandemic was not only a major health crisis, it also had a huge impact on world economies. The effects of the pandemic varied from country to country. It increased poverty and inequalities in developing and less developed countries.

Governments around the world gave money or subsidies to people made unemployed due to the pandemic. Developed countries could afford to give more than developing or less developed countries.

US$695 per person per month in wealthy countries

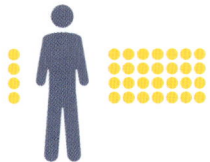

US$4 to US$28 per person per month in low- and middle-income countries

5. Average amount of support (US$) governments gave to people per month in wealthy and in low- and middle-income countries during the Covid-19 pandemic

Question Time

1. List one positive and one negative effect of globalisation on economic development.
2. Briefly explain your chosen positive effect and negative effect.

27.4 Developing an Economy

It is possible for countries to change from being classified as 'developing' to 'developed'. An increase in a country's level of economic development can be explained by gradual improvements made to each of the five economic indicators.

1. **Increased GNI**: If people are healthier, enjoy a higher standard of living and work in a wider variety of skilled jobs, the GNI will grow. Products and services will be both imported and exported.
2. **Type of employment**: Better-educated people are prepared for highly skilled jobs. These jobs also tend to be better paid, which means people have more money to spend in the economy and enjoy a higher standard of living.
3. **Education rate**: More years spent in education means that people are likely to work in tertiary activities, which indicates a growing economy.
4. **Adult literacy rate**: If more adults can read and write, it increases their chances of working in more skilled secondary or tertiary activities. This indicates a growing economy.
5. **Average life expectancy**: Improved access to healthcare and medication in developing countries may increase the average life expectancy of their populations. Healthier people can mean more people working and contributing to the economy.

Reviewing Development

The UN reviews the status of countries every three years. Using many factors, including GNI, the UN examines and decides whether countries can be classified as more or less economically developed.

Slowly Developing and Quickly Developing Countries

This table of data from the UN shows the status of a selection of developing countries. The years indicate when the country was/will be considered as 'quickly developing', or when it was deemed to be 'slowly developing'.

- **Countries with green arrows**: These economies are considered to be quickly developing (or are on track to be considered as such). They have improved (or will improve) to such an extent that they are now considered more economically developed than previously.
- **Countries with blue arrows**: These economies are considered to be slowly developing, or, worse still, to have stopped developing altogether.

Year	Direction	Country
2021	→	Angola
2020	→	Vanuatu
2017	→	Equatorial Guinea
2014	→	Samoa
2012	←	South Sudan
2011	→	Maldives
2007	→	Cabo Verde
2003	←	Timor-Leste
2000	←	Senegal

Question Time

1. Using an atlas or Google Maps, find the locations of these countries and determine whether they are found in the rich North or the poor South.
2. Can you identify a common trend in the locations of the majority of these countries?
3. All of the countries listed here have one thing in common. Can you guess what that might be? (Hint: think of the empires of Britain, France, Portugal and others.)

Sample questions

1. **Examine** the infographic and answer each of the following questions:

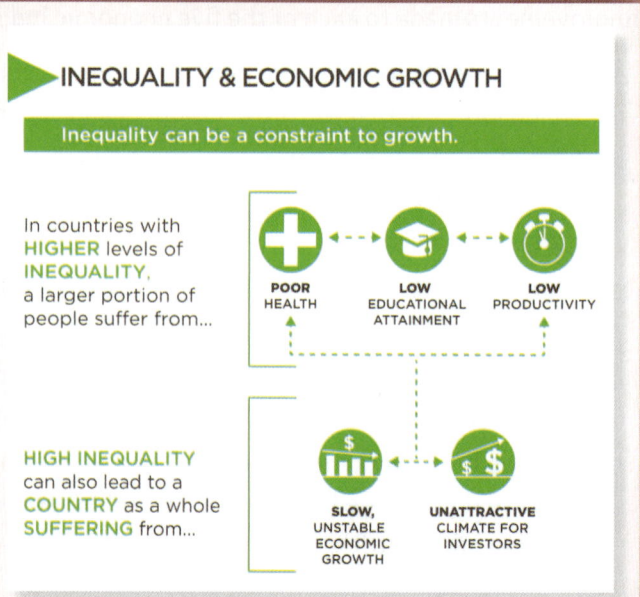

INEQUALITY & ECONOMIC GROWTH

Inequality can be a constraint to growth.

In countries with **HIGHER** levels of **INEQUALITY**, a larger portion of people suffer from... POOR HEALTH, LOW EDUCATIONAL ATTAINMENT, LOW PRODUCTIVITY

HIGH INEQUALITY can also lead to a **COUNTRY** as a whole **SUFFERING** from... SLOW, UNSTABLE ECONOMIC GROWTH, UNATTRACTIVE CLIMATE FOR INVESTORS

27. Economic Development and Inequality

(a) **List** three indicators of economic development that you can see mentioned on this infographic.

(b) **Explain** how each indicator you listed can be measured to identify the economic development of a country.

(c) **Explain** why equality in a country is needed for development.

> **Exam hint!**
> Explaining the indicators of economic development is not enough to answer part (c). You also need to suggest how improving these indicators will help the development of a country as a whole. For example, improving literacy rates in adults and keeping children in school for as long as possible may lead to better job opportunities in their future.

2. (a) **Explain** the terms 'international trade', 'imports' and 'exports'.

   (b) Read the article below and answer each of the following questions:

   > Fairtrade works on behalf of small-scale farmers and also for workers employed on Fairtrade certified plantations. Investment of the Fairtrade monies into education, better housing, better schools and medical facilities is highly valued. These investments of the Fairtrade Premium by companies in community development projects is improving the quality of lives of rural communities. For workers, Fairtrade sets living wage benchmarks (standards) and is taking concrete steps towards decent wages for workers on Fairtrade plantations.

   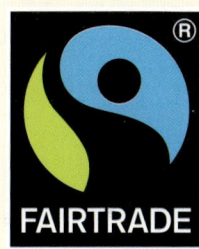

   (i) **Name** a product you have studied that operates under the Fairtrade Foundation.

   (ii) **Explain** how the product you have named benefits its producers because of Fairtrade using the following headings:

    • Economically

    • Socially

    • Environmentally

   > **Exam hint!**
   > To answer part (ii), you can use examples you have studied for developing AND developed countries. Describe at least two indicators of economic development. Describe how money can improve the levels of the indicators you have described (you can use the information from the Ben & Jerry's Fairtrade ice cream case study on pp. 408–409).

Learning Outcomes: 3.8, 3.9

28 Human Development and Development Assistance

Learning Intentions

You will be able to:
- Describe what human development is
- Explain the indicators used to measure human development
- Explain where aid comes from and its different types
- Examine the role Irish Aid plays in helping human development
- Evaluate the advantages and disadvantages of aid on human development
- Outline how technology aids in human development.

key words

- Well-being
- Human development
- Human Development Index (HDI)
- Aid
- Non-governmental organisation (NGO)
- Non-government aid
- Bilateral aid
- Multilateral aid
- Emergency aid
- Conditional aid
- Untied aid
- Long-term development aid
- Intermediate technology

28.1 What Is Human Development?

We learned about economic development in Chapter 27. But economic development does not always reflect the **well-being** of people. Human well-being depends on **human development**.

Human development is dependent on all people being able to access their basic human needs and rights. It also provides the opportunity for all people to improve their own standard of living and personal well-being, without interfering with the legitimate rights of others.

28. Human Development and Development Assistance

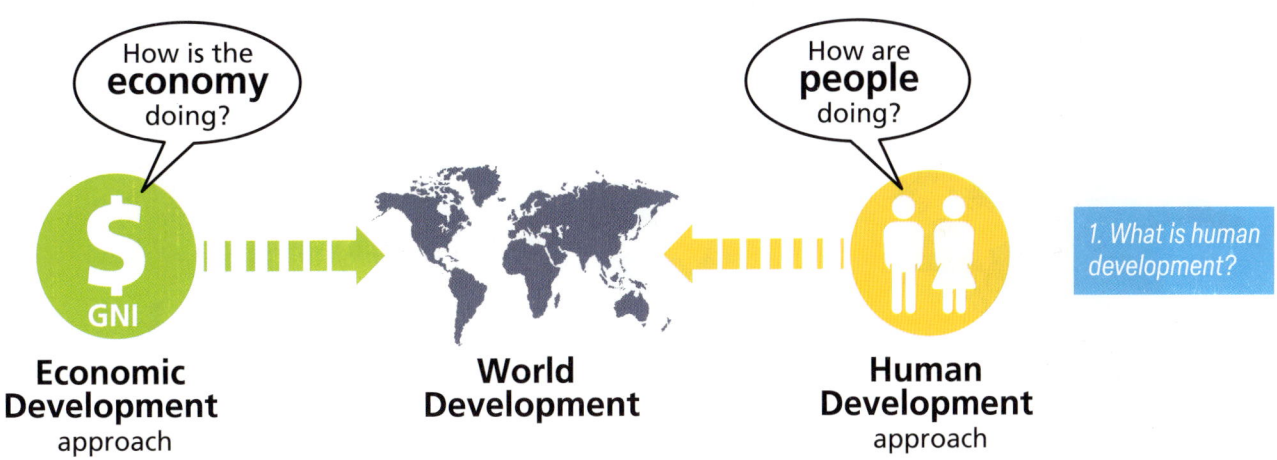

1. What is human development?

In Chapter 27, we also learned that there is a divide between the rates of economic development of different countries. The same is true for the rates of human development.

Measuring Human Development

Human development is as important as a country's economic development.

The UN decides the level of human development of a country by gathering and recording information on three main dimensions:

1. Long and healthy life, indicated by life expectancy at birth
2. Knowledge, indicated by number of years in school
3. Decent standard of living, indicated by gross national income (GNI) per capita.

Using this information, the UN has developed the **Human Development Index (HDI)** to measure the human development of a country.

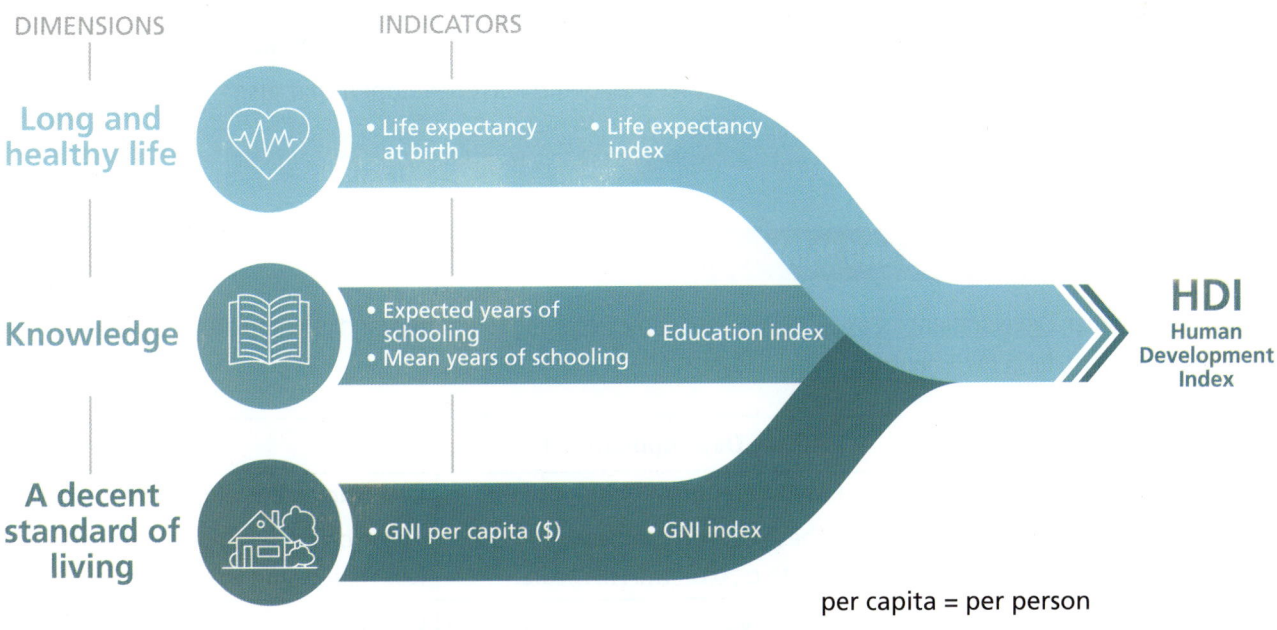

per capita = per person

2. Indicators of human development

415

Junior Cycle Geography CYCLONE

The HDI goes from 0 to 1. Countries with a rank of 0 have the lowest levels of human development in the world, while those at 1 are considered to have very high levels of human development.

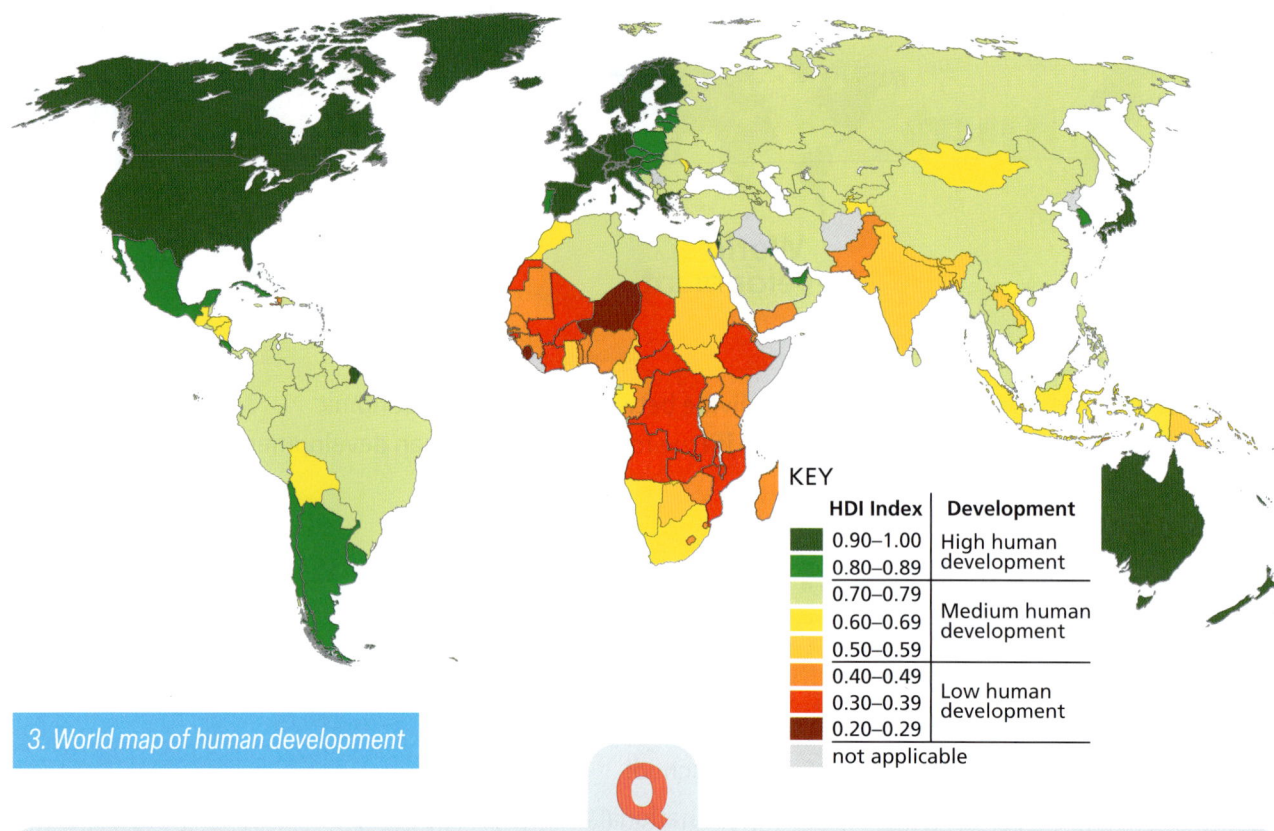

3. World map of human development

Q

1. Examine figure 3 alongside the map showing countries grouped by the UN on p. 403 of Chapter 27. Identify:
   (a) A least developed country, with low economic and low human development
   (b) A developed country (other than Ireland), with high economic and high human development.
2. Which continent has the most countries classed as being least developed that also have a low HDI score?

📁 Managing information and thinking

Copy the Human Development Index table into your copy. Referring to figure 3 (and using an atlas/ Google Maps if needed), fill in two countries for each level.

Human Development Index Table		
High Human Development	**Medium Human Development**	**Low Human Development**
These countries have a score of 0.8 to 1 on the HDI scale.	These countries have a score of 0.5 to 0.79 on the HDI scale.	These countries have a score of less than 0.5 on the HDI scale.
Country 1: _____	Country 1: _____	Country 1: _____
Country 2: _____	Country 2: _____	Country 2: _____

28. Human Development and Development Assistance

Question Time

1. Explain the difference between economic and human development.
2. Name the three main dimensions used to measure human development. How do you think each contributes to high levels of human development?

Being numerate

Search online for 'UNDP composite HDI scale'.
1. Identify what rank Ireland has on the HDI scale.
2. Calculate the difference between 'Expected years in education' and 'Mean (average) years in school' in Ireland.
3. What is the gross national income per person in Ireland?
4. Compare Ireland's gross national income with that of any country with low human development. What is the difference in figures?

28.2 Human Development Aid

What Is Aid?

Aid refers to a donation of resources by one government or **non-governmental organisation (NGO)** to a country experiencing short- or long-term need. These resources can be in the form of:

- Emergency supplies (medicines, clothing, water)
- Food
- People with skills (doctors, engineers)
- Technology (medical equipment)
- Money (grants or loans).

Grant: money that does not have to be paid back
Loan: money that has to be paid back

Sources of Aid

1. **Non-government aid** is assistance provided by NGOs, such as Trócaire. The money for this aid is mainly provided by public donations from individuals and businesses.
2. **Bilateral aid** is assistance given by one government directly to the government of another country. For example in 2020, the Irish government spent over €860 million on Ireland's bilateral aid programme.
3. **Multilateral aid**, also called official development assistance (ODA), is assistance provided by multiple governments to international organisations. These organisations then decide where and when aid is distributed in developing countries. For example, the World Bank and the United Nations have worked together to reduce debt in the developing world.

Types of Aid

There are four different types of aid.

1. Emergency Aid or Short-term Aid	2. Conditional Aid or Tied Aid	3. Untied Aid	4. Long-term Development Aid
Emergency aid is the immediate provision of food, water and emergency shelter. For example, in 2017 Tropical Storm Harvey caused severe damage in Texas (US), and the state's population needed immediate aid.	**Conditional aid** is when the government of one country donates money or resources to the government of another country (bilateral aid), but with conditions attached. These conditions will benefit the more developed country, e.g. goods and services bought with the donated money must be from the donor country.	**Untied aid** is assistance (in money form) given to developing countries by governments (bilateral aid) without conditions attached. This money can be used to purchase goods and services in virtually all countries.	**Long-term development aid** involves providing local communities with education and skills for sustainable development, e.g. the building of schools and hospitals. Irish Aid, the Irish government's official agency for international development, provides long-term aid.

The UN's 17 Sustainable Development Goals (SDGs) are the plan to achieve a better and more sustainable future for all. They tackle the global challenges faced by all countries, including poverty, inequality, climate change, environmental degradation, peace and justice.

The UN provides funding for projects for each SDG. This is another vital source of aid for countries around the world.

Question Time

1. What is aid? What forms can it take?
2. Name and give a brief description of the three sources of aid.
3. Name and give a brief description of the four types of aid.

28. Human Development and Development Assistance

28.3 Ireland's Bilateral Aid

Geography in the News

500,000 Covid-19 vaccines donated by Ireland arrive in Nigeria

29 November 2021

A consignment of 500,000 Covid-19 vaccines donated by Ireland arrived in Nigeria in May of 2021. This is the first consignment of vaccines donated by Ireland. The donation of 500,000 Janssen vaccines form part of Ireland's commitment to donate 1.3 million vaccine doses as part of a major scale up of the global vaccination campaign.

In addition to the 1.3 million doses of vaccine Ireland has agreed to sharing, Ireland has committed €7 million in financial support for the global sharing of vaccines worldwide.

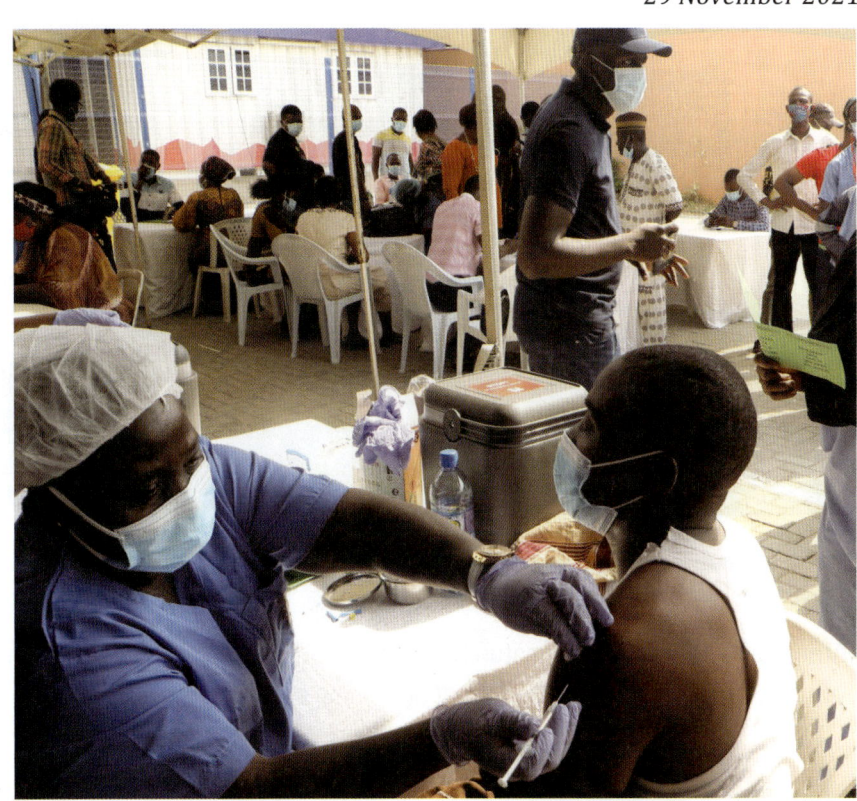

1. **What** two types of aid is Ireland contributing to by donating vaccines and money?
2. **How** would a virus like Covid-19 weaken the human development of a country like Nigeria?

Being creative

Imagine you are an RTÉ news reporter. Create a TikTok video on Irish aid being donated and used for human development around the world. You can use the information provided in this chapter or visit the Irish Aid agency website: www.irishaid.ie.

Untied Aid and Irish Aid

Untied aid is provided by a number of Irish NGOs. We will focus on the untied aid provided by the Irish government agency Irish Aid. This agency supports work in a large number of countries throughout the developing world. Its main focus is on sub-Saharan Africa, Palestine and Vietnam.

4. The countries which received the most from Irish Aid in 2020

Question Time

1. Looking at figure 4, which African country received the most funding from Irish Aid in 2020?
2. Name a country not found in Africa that Irish Aid supports.
3. What do the letters NGO stand for?

Managing information and thinking

Africa has a huge number of natural resources. Yet, it has many countries in need of development aid. Why do you think this might be? Have a look at figure 5 to come up with an answer.

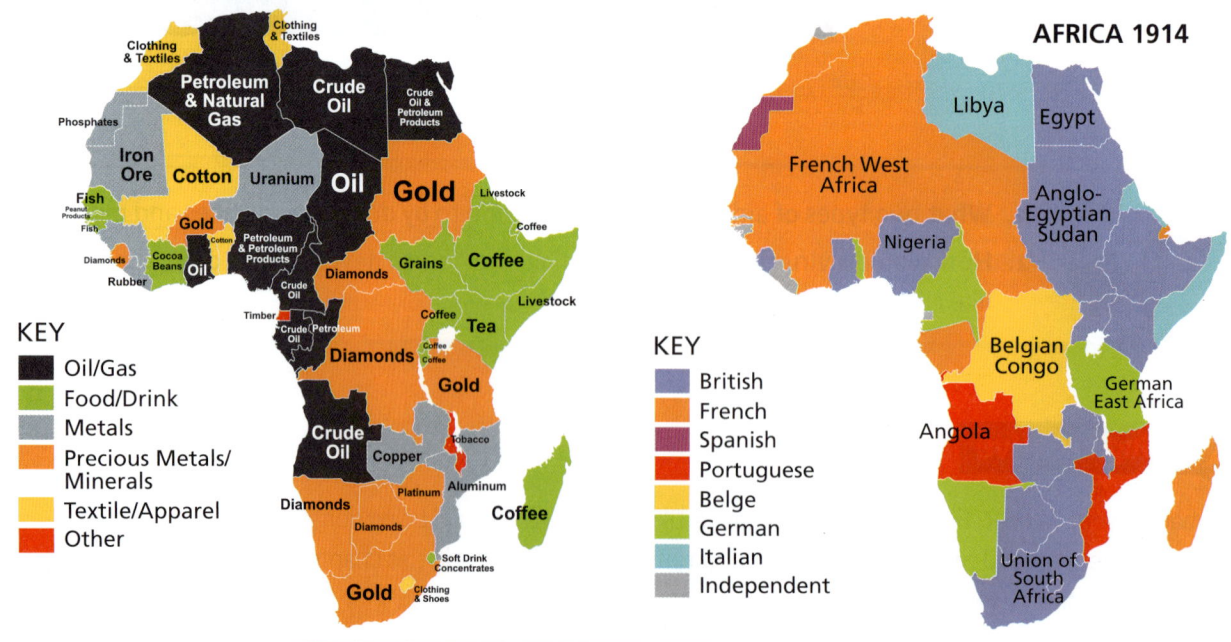

5. Africa's natural resources and the colonies in Africa in 1914

28. Human Development and Development Assistance

How Does Irish Aid Help Human Development?

The Irish government is helping efforts towards achieving the United Nations' Sustainable Development Goals. As part of this, Irish Aid's main areas for ODA include:

Eliminating poverty:
Irish Aid is committed to eradicating poverty in its partner countries through long-term aid.

Eliminating hunger:
Irish Aid responds to emergency situations by providing short-term assistance in the form of food. It is also committed long term to eradicating global hunger.

Promoting gender equality:
Irish Aid is committed to focusing on women's and girls' rights in developing countries, e.g. by providing and promoting access to education, or campaigning against gender-based discrimination and violence.

Sustainability:
'A sustainable world requires the protection of our natural world.' That means Irish foreign assistance is committed to being sustainable and supporting countries to achieve climate resilience and sustainable resource management.

Irish Aid – Helping Human Development

Coping with climate change:
Irish Aid supports poor communities to better cope with the impacts of climate change.

Supporting education:
Irish Aid works to improve access to and quality of education for all, especially girls.

Improving health:
Irish Aid works to ensure that all poor people, especially women and children, have access to health facilities and are enabled to live healthier lives free from preventable diseases.

421

Junior Cycle Geography **CYCLONE**

Case Study: Irish Aid and Young Scientist Kenya (YSK)

In July 2017, Kenya launched its own version of the Young Scientist Exhibition, called Young Scientist Kenya (YSK). Funding from Irish Aid and from the Kenyan government has supported the development of YSK. It is hoped that YSK will:

- Encourage interest in and development of the subject of Science and Technology in secondary schools
- Assist in the development of Science teachers
- Provide links and contacts between secondary schools and third-level institutes specialising in Science and Technology throughout the country.

The competition has gone from strength to strength. The 2020 exhibition featured:

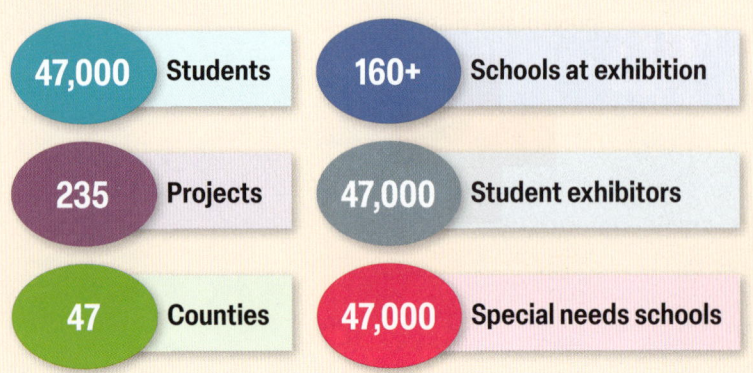

Q In what ways do you think Science and Technology are important in the economic and human development of a country? Why is it important to promote and develop these areas within the young population of a developing country?

28.4 Advantages and Disadvantages of Aid

There are advantages and disadvantages of aid.

Advantages

- Emergency aid in times of disaster saves lives.
- Aid helps to rebuild housing and industry after a disaster.
- Provision of medical training, medicines and equipment can improve health and standards of living. It also enables people to become self-sufficient.
- Aid in the form of resources helps people to help themselves, e.g. by increasing years spent in school and improving literacy levels.

Disadvantages

- Aid can increase the dependency of developing countries on donor countries. Also, sometimes aid is tied and recipients are forced to buy goods and services from the 'donor' country.
- Aid may not reach the people who need it most. Corruption may lead to local politicians using aid for their own ends or for political gain.
- Aid can be used to put political or economic pressure on the receiving country. The country may end up owing a donor country or organisation a favour.
- Aid in the form of defence, i.e. military aid, can help countries defend their borders. This aid often promotes the military or political ambitions of the powerful donor country. It can also be used to promote anti-democratic dictators in the countries receiving this type of aid.

Being literate

Your teacher will read aloud the statements below. If you agree with the statement, thumbs up (or put your hand up), and if you do not agree, thumbs down (or keep your hand down). Count and note for each statement how many students' thumbs/hands go up and how many thumbs/hands stay down. You must then explain in writing why you agreed or disagreed with each statement.

- **Statement 1**: Giving money is the best way to fight world poverty.
- **Statement 2**: The Irish government should give less money to foreign countries.
- **Statement 3**: We should only give aid to countries with strong democratic governments.

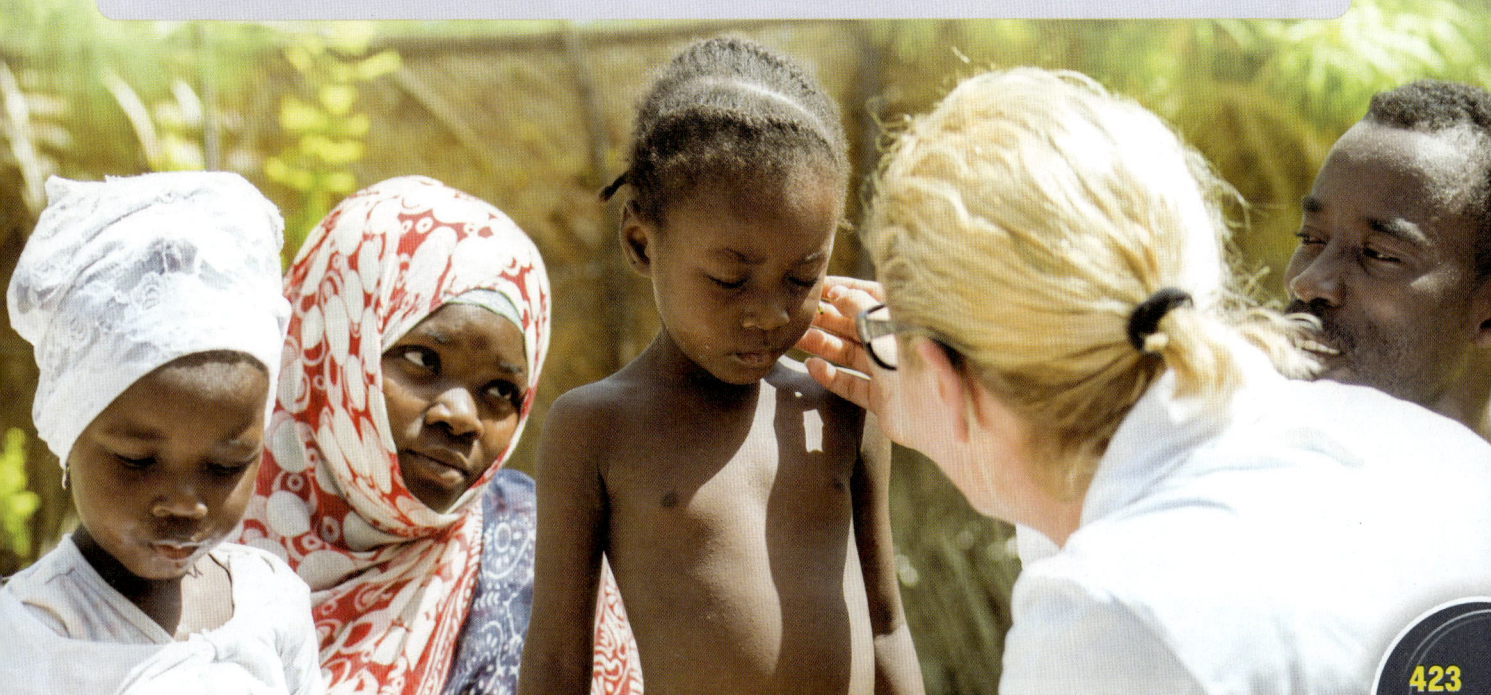

Junior Cycle Geography CYCLONE

28.5 Technology and Human Development

 Give a man a fish and you feed him for a day, teach him how to fish and you feed him for life.

Providing a developing country with '**intermediate technology**', or appropriate technology, can help its human development. Intermediate technology is the middle way between high technology, which might not be suitable for a developing economy, and low technology. It can be used to solve complicated problems for developing countries using cheap and readily available materials. The aim is to help people move away from a dependency on aid to being self-sufficient. Intermediate technology can be used to improve health services, farming techniques and water supplies for a developing country.

The photograph shows an example of intermediate technology. The people of this community were assisted in using materials readily available to them to make a water pump. This pump brings water from a nearby river and is used to irrigate crops. This provides them with a sustainable food source as well as a product to sell.

Question Time

1. List four ways in which Irish Aid is helping human development in Africa.
2. Explain two of these ways in detail.
3. Explain two advantages and two disadvantages of aid.
4. Examine the photo above and name the intermediate technology that has been created. Identify the ready-made materials used to make this piece of technology.

28. Human Development and Development Assistance

Sample questions

1. Complete each of the following sentences by using the correct term from the box below.

   Emergency aid   Tied aid   Development aid

   (i) Aid given with conditions attached is known as …
   (ii) Training local teachers and nurses is an example of …
   (iii) Aid given in times of crisis is known as …

2. Read the article about aid in South Sudan below and answer each of the following questions.

   ## Aid Project in South Sudan

   When the aid project started in 2016, there were no schools and there were 1.3 million primary school age children out of school. Since then, the community has built new school buildings and parents are strongly involved in their children's education through Parent-Teacher Associations. A total of 130 teachers received training with the project, and more than 7,000 pupils can access learning spaces and materials.

   (i) **How** many teachers were trained as part of this aid project?
   (ii) **What** is the type of aid described in the article most commonly known as: emergency aid or development aid?

3. Read the article about aid in Mozambique below and answer the question below.

   ## Humanitarian Crisis in Mozambique

   During his visit to Mozambique in May 2022, Minister for Overseas Development, Colm Brophy announced an Irish Aid investment of €8 million, which will save lives through improving health services for mothers and children. This funding will support the government of Mozambique to help reduce the infant mortality rate in the country. With the support of Irish Aid funding, infant mortality in Mozambique has reduced by 50% in certain areas.

   Which type of assistance do you think is most appropriate to help the people of Mozambique: tied aid, emergency aid or development aid? **Justify** your choice.

   Exam hint!

   The action verb 'Justify' requires you to give valid reasons or evidence to support an answer or conclusion. Explain the type of aid you chose in the first part and then use evidence from the article above to support your choice.

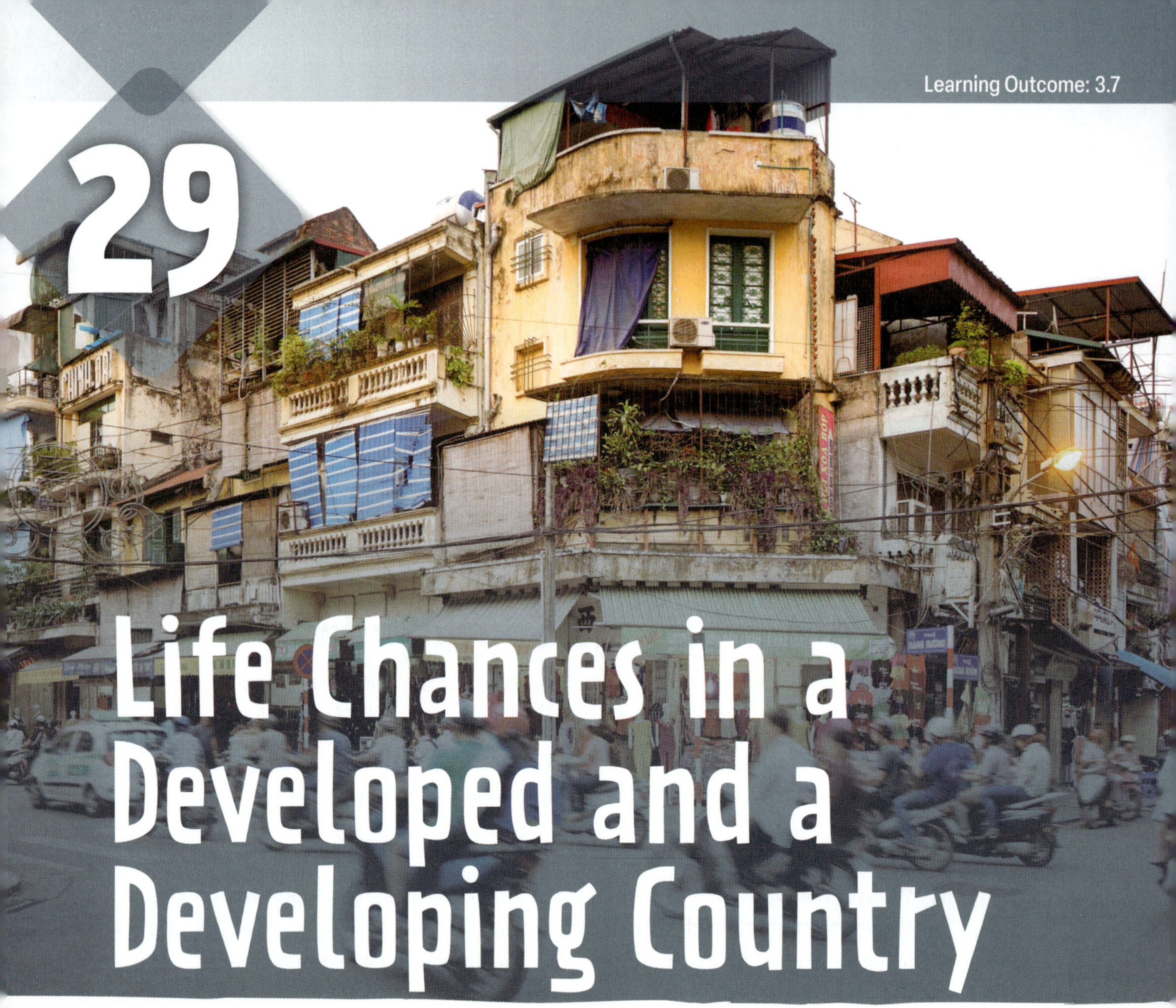

Learning Outcome: 3.7

29 Life Chances in a Developed and a Developing Country

Learning Intentions

You will be able to:
- Describe a country as developed, developing or least developed
- Explain how access to healthcare can impact on the life chances of a young person in a developed and in a developing country
- Examine how educational opportunities can impact on the life chances of a young person in a developed and in a developing country
- Describe how gender can impact on a young person's life chances in a developed and in a developing country
- Compare the employment opportunities for a young person in a developed and in a developing country.

Developed
Quickly developing
Slowly developing
Least developed

29. Life Chances in a Developed and a Developing Country

29.1 Developed and Developing Countries

In the previous chapters, we examined how the economic and human development of a country can be used to describe its level of development. We also looked at population characteristics, how they change and how they can be studied to identify the development of a country.

Managing information and thinking

Examine these photos. Identify whether each image shows a developed or a developing country. Give reasons for your answers.

Countries can be grouped into three categories:

Developed	Developing		Least Developed
	Quickly Developing	**Slowly Developing**	
• High-to-very-high levels of human development on the Human Development Index • Considered more economically developed • Most people work in the tertiary economic sector • Stages 4 and 5 of the demographic transition model	• Medium levels of human development on the Human Development Index • Considered economically developed • Most people work in the secondary economic sector • Stage 3 of the demographic transition model	• Low levels of human development on the Human Development Index • Considered less economically developed • Most people work in the primary economic sector • Stages 1 and 2 of the demographic transition model	• Very low to no levels of human development • Economy severely underdeveloped • Huge unemployment levels and low income • Stage 1 of the demographic transition model

Junior Cycle Geography CYCLONE

Development and Life Chances

A country's level of development can affect the life chances of its young people. The figures used in this chapter are for those aged 15–24 years. Countries use this age bracket to define 'young people' when carrying out large studies.

In this chapter, we compare the life chances of a young person in Ireland (a developed country) with the life chances of a young person in Nigeria (a slowly developing country) under the following headings:

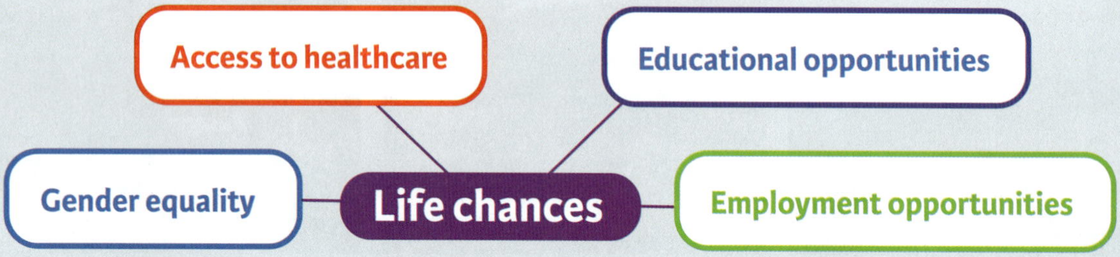

- Access to healthcare
- Educational opportunities
- Gender equality
- Life chances
- Employment opportunities

1. The distance between Ireland and Nigeria

Distance: 7,202 km

Ireland 2022
- Population: 5,045,330
- Size: 84,421 km²
- Population density: 72 per km²
- 0.9 on HDI scale (high human development)
- Stage 4 of demographic transition model

Nigeria 2022
- Population: 216,099,479
- Size: 923,768 km²
- Population density: 226 per km²
- 0.5 on HDI scale (low human development)
- Stage 2 of demographic transition model

Examine the country fact files and answer the questions.

1. What is the difference in population between the two countries?
2. What is the difference in size between the two countries?
3. How might the difference in population density put pressure on resources such as water and soil in Ireland and in Nigeria?
4. The fact files tell us that Ireland is at stage 4 of the DTM and Nigeria is at stage 2. Outline what this means for the level of economic development of each country. (Look back to Chapter 18: Population to help.)
5. Do you agree or disagree that the indicators used to explain levels of development are accurate enough? Explain your choice.

29. Life Chances in a Developed and a Developing Country

29.2 Healthcare in Ireland and Nigeria

The availability of healthcare considerably affects the life chances of a young person in every country around the world. The following table compares access to healthcare, mortality rates and life expectancy rates in Ireland and Nigeria.

Indicators (2021)	Ireland	Nigeria
Population	5,045,330	216,099,479
Number of hospitals	48 Public 18 Private	39,914 operational hospitals – a mix of public and private Cost of private healthcare makes access difficult for most Nigerians
Number of doctors per 1,000 people	3.47	0.7
Infant mortality per 1,000 (Male)	3	72
Infant mortality per 1,000 (Female)	2	66
Life expectancy (Male)	80	54
Life expectancy (Female)	84	56

Cancer and circulatory and respiratory illnesses are the most common causes of death from disease for young people in developed countries, including Ireland. The survival rates for these diseases are higher in developed countries because of access to healthcare and better health facilities in general.

In Nigeria, malaria continues to be a leading cause of death of young people, though rates have decreased in recent years. National data shows that 4.2% (or over 790,000 young people aged 15–24) were living with HIV in 2016. In Ireland, the figure was 512 (or 0.08% of young people aged 15–24).

FUN FACT! Nigeria climbed from 163rd in the world healthcare ranking in 2020 to 144th in 2021. This is out of 195 countries in the world.

Question Time

1. Name one barrier that prevents Nigerian people from accessing healthcare.
2. How has the availability of hospitals and doctors affected life expectancy in both Nigeria and Ireland?
3. (a) Name a type of aid which could help the development of healthcare in Nigeria.
   (b) Briefly explain why you chose this type of aid.
4. Name a leading cause of death in 15–24-year-olds in Nigeria.

29.3 Education Opportunities in Ireland and Nigeria

Access to education for young people in developed and developing countries can differ significantly. The infographic shows the key differences between educational opportunities in Ireland and Nigeria.

Education in Ireland

School attendance is compulsory for all children aged 6–16yrs

Years in Education (2021)
- 13.9 Expected Years in School
- 12.4 Average Years in School

Literacy level (2021)
- 99% literate
- 78% of Ireland's second-level students progress to third-level education
- More than half (53%) of Irish 25 to 64 year olds have a third level education according to CSO

Education in Nigeria

School attendance is compulsory; **however, 40%** of Nigerian children aged 6–11 do not attend school

Years in Education (2021)
- 10.2 Expected Years in School
- 7.9 (an increase from 6 in 2017) Average Years in School

Literacy level (2020)
- Literacy rates in males over 15 yrs old: 55%
- Literacy rates in females over 15 yrs old: 44%
- 86% Urban, 70% Rural areas (males)
- 74% Urban, 53% Rural areas (females)
- Less than 33% of students progress to third-level education

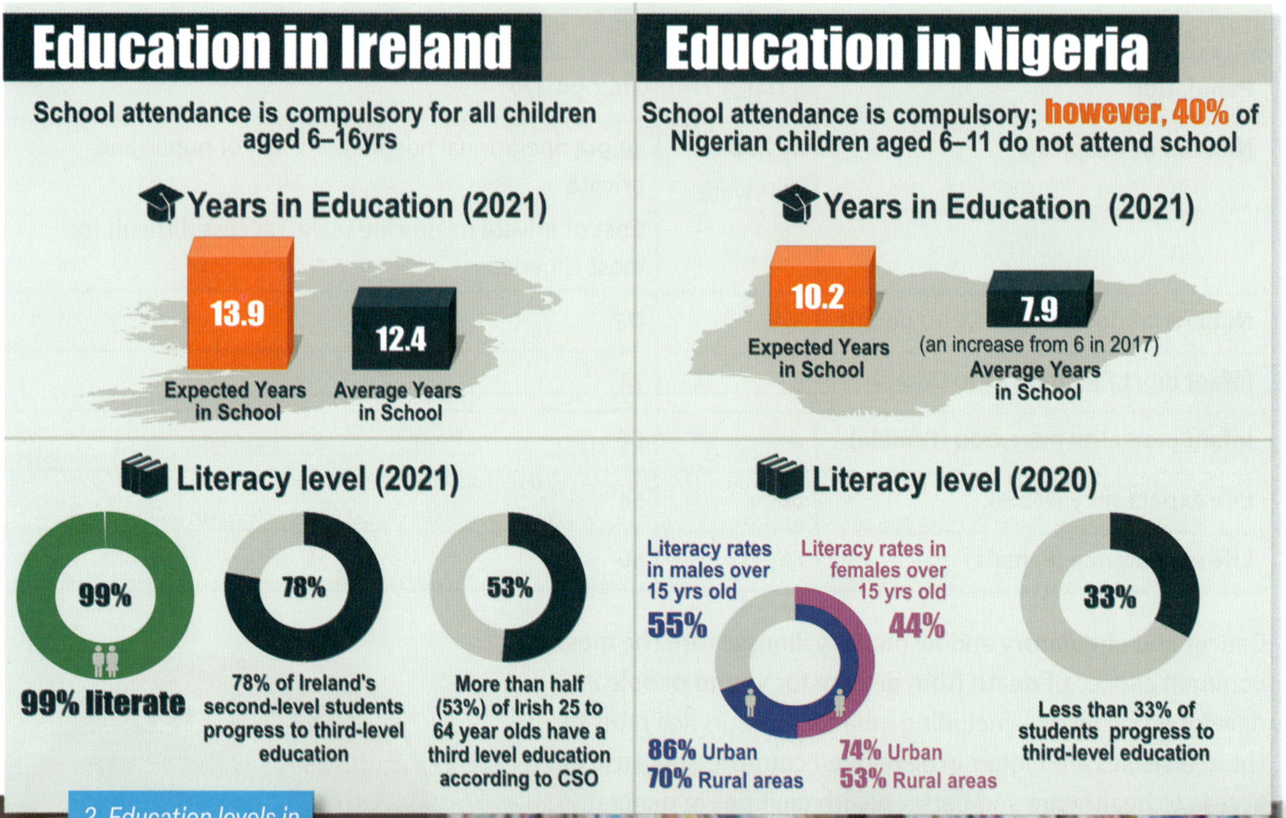

2. Education levels in Ireland and Nigeria

Working with others

Examine figure 2 and, with the person sitting next to you, make a list of the key differences in education for young people in Ireland and Nigeria.

Barriers to Education

The UN's SDG 4 targets education.

By the year 2030, the UN wants primary and secondary education to be provided free of charge till the age of 12 for all children across the world. It also aims to set the number of years that must be spent in school at 9.

There are practical and social barriers to education in Nigeria and Ireland.

Barriers to education in Nigeria	
Practical	**Social**
• Population growth means there are not enough schools to serve the entire population. Many classes contain more than 100 students. • Nigeria has a teacher shortage. There is less than 50% of the required number of secondary school teachers. • There is a lack of basic school resources and equipment within schools, e.g. science labs. This causes many students to leave school early.	• Child labour: Children in poorer areas of the country usually work to earn money to help support their families. • Early marriage is common in parts of Nigeria. Boys can continue their education after marriage, but girls usually have to stop their education and take care of the home. • Gender: According to the United Nations as few as 20% of young women in the north of Nigeria have access to education.

Barriers to education in Ireland	
Practical	**Social**
• Schools have limited capacity. Some students may have difficulty obtaining a place in their local secondary school and may need to travel further from their home to find a suitable school place. • Ireland is experiencing a teacher shortage. In a 2022 survey, 55% of post-primary principals reported that that they had unfilled vacancies. This leads to classes or year groups being sent home or teachers still in training being rushed into the classroom.	• In Ireland, the main barrier to education is cost. Many young people cannot afford college fees, and so education ends after secondary school. • Students from areas perceived as disadvantaged are less likely to go to college. For example, in some wealthy areas of Dublin, 99% of students go on to third level, compared to just 15% from the less well-off areas in Dublin.

Question Time

1. Explain how two social barriers to education have affected education in Nigeria.
2. Compare and contrast the practical barriers to education in Ireland and Nigeria.
3. In your opinion, which are the bigger barriers to education in Nigeria: social or practical? Give one reason explaining your choice.

Junior Cycle Geography CYCLONE

29.4 Gender Equality for Young People in Ireland and Nigeria

The United Nations defines gender equality as:

> *When women and men enjoy the same rights and opportunities across all parts of society, including job opportunities and decision-making, and when the different behaviours, aspirations and needs of women and men are equally valued and favoured.*

 Research online the gender equality ranking of the following countries. Make sure the website you choose is reliable. Identify the gender equality ranking of the following countries:

1. The United Kingdom
2. Brazil
3. The United States

In 2022, 18.5 million children in Nigeria did not attend school. Over 10 million of these children were girls.

Child marriage is illegal in Nigeria but remains widespread because the government has not enforced laws to stop it. Nigeria has one of the highest rates of child marriage in the world. These marriages are usually between young girls and older men. According to the 2020 UN Development Program, 43% of women aged 20 to 24 in Nigeria had been married by the age of 18.

According to the 2021 Global Gender Gap Report, Ireland ranked ninth out of 156 countries for gender equality. This was a fall from sixth in 2016. Nigeria fell from 118th to 139th between 2016 and 2021.

Gender Inequality and Income Levels in Ireland and Nigeria

The average salary earned by males and females is an indicator of a country's level of development. The greater the pay gap between the genders, the lower the level of development of the country. If we compare the income levels between the genders in both Nigeria and Ireland, clear discrimination (unfair/unequal treatment) can be seen in both countries.

The Just a Girl campaign was launched in 2022 to highlight the issue of child brides

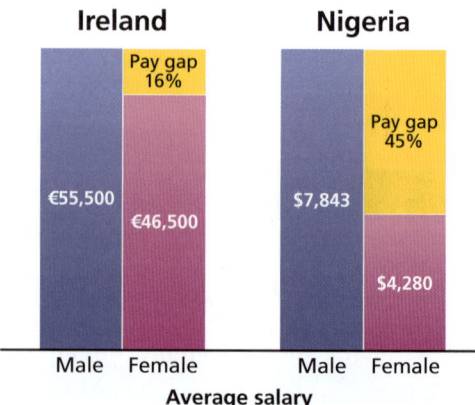

3. The gender pay gap in Ireland and Nigeria

29. Life Chances in a Developed and a Developing Country

29.5 Employment Opportunities in Ireland and Nigeria

There is a strong link between level of education and employment opportunities. The longer a young person remains in education, the more attractive that person is to employers.

Taking the average number of years spent in school (12.4 years in Ireland, 7.9 years in Nigeria; see figure 2), we can see that young people in Ireland may have better employment opportunities than young people in Nigeria.

Youth Unemployment in Ireland and Nigeria

Figures 4 and 5 show the difference between young people not in employment, education or training, e.g. apprenticeships, in Nigeria and Ireland. This rate is known as NEET. In Nigeria, the number of young unemployed people remains consistently higher between 2010 and 2019. In a developed country such as Ireland, a far greater percentage of this age group are in second- and third-level education. This reduces the numbers of unemployed people in this age bracket.

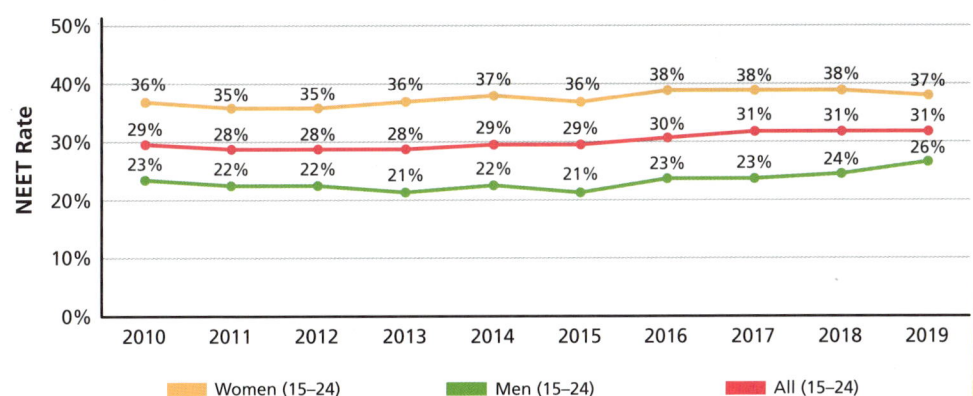

4. Percentage of 15–24-year-olds not in employment, education or training in Nigeria, 2010–2019

Exam hint!
When studying these graphs, keep in mind Nigeria's larger population of young people. Ireland's table also includes people aged 15–29, rather than 15–24.

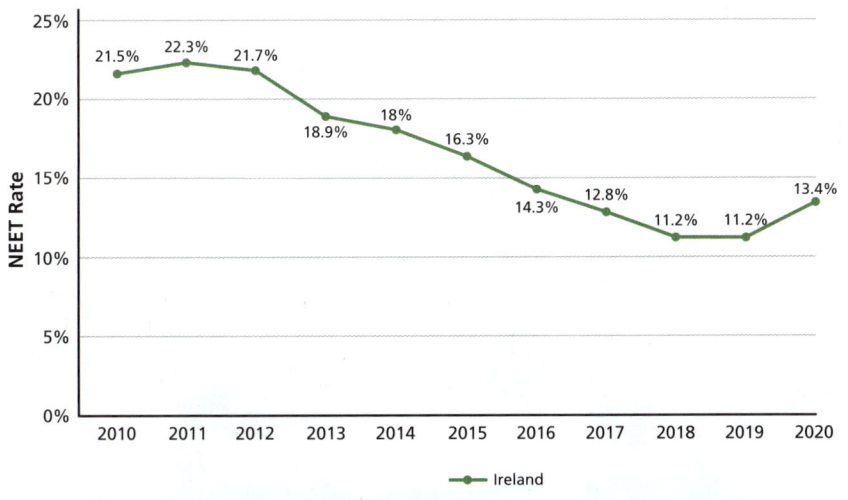

5. Percentage of 15–29-year-olds not in employment, education or training in Ireland, 2010–2019

Q

1. Examine figure 4 showing the NEET rate for young people in Nigeria. Describe the link between Nigeria's low gender equality ranking and the difference between the NEET rate for males and females.

2. What unexpected event may have caused the NEET rate to increase for young people in Ireland in 2020?

Youth Employment: Ireland vs Nigeria

Sector	Ireland	Nigeria
Primary	• Farming has become a less attractive source of employment for young people in recent years. • Young people who once inherited family farms prefer to work in different jobs, which requires further education.	• In Nigeria, 70% of the young people live in rural areas and depend on agriculture as their main means of survival. • A programme called Nigeria's Young Farmer Network has almost 1 million young farmers registered nationwide.
Secondary	• In Ireland, there is a negative perception of certain manufacturing jobs as a source of employment for young people. • As a result, there is a shortage of people working in certain trades, such as tool makers and sheet metal workers. • There are signs of improvement. In 2019, 11% of third-level students enrolled in engineering, manufacturing and construction courses.	• The Nigerian Youth Action Plan 2021–2024 identified manufacturing as a priority area to reduce unemployment among young people. • It hopes to promote short- and long-term skills development in manufacturing to reduce the 75% unemployment rate for young people.

29. Life Chances in a Developed and a Developing Country

Sector	Ireland	Nigeria
Tertiary	• In Ireland, 17-year-olds can get a part-time job, e.g. in the catering industry, the fast-food sector or behind the counter in a shop. • The Covid-19 pandemic saw a huge demand for students to take up jobs in seasonal work such a fruit picking, once carried out by migrant workers.	• The Nigerian Youth Action Plan 2021–2024 discovered that the services sector appears to be the preferred option of employment for young people. • The same plan revealed that staying in education remained a priority for young people. • Many 15–18-year-olds are not available for part-time work as they need to work in the household or on the family farm, which is unpaid.

Youth Employment Laws: Ireland vs Nigeria

Age	Ireland	Age	Nigeria
14–15	In Ireland, 14-year-olds cannot work during school time. From the age of 15, young people can do eight hours of light work a week during the school year	Under 12	A child (under 12 years) cannot be employed in any type of work except in a family enterprise, e.g. farming
		14	Children ages 13–14 can be employed on daily wage and day-to-day-basis
16–17	Cannot legally be asked to work more than eight hours a day or 40 hours a week	Under-18	Employees who are under 18 are not allowed to work for four hours consecutively and more than 8 hours a day. Young persons are also not allowed to work at night except under circumstances specified by the law
Minimum wage for under-18s, 2022	€7.35 per hour	Minimum wage for all age groups in Nigeria	30,000 Nigerian naira per month = €67.92 in 2022. It is €6.50 an hour for adults but it is often not enforced.

Question Time

1. Explain how youth employment in the secondary sector is different between Ireland and Nigeria.
2. Why is part-time work for young people in the service sector less accessible in Nigeria?
3. How do labour laws protect young people in Nigeria and Ireland?
4. How can salary be used as an indicator of gender inequality between Nigeria and Ireland?

Junior Cycle Geography CYCLONE

Sample questions

1. (i) Countries can be grouped into three categories of development. **List** each category.
   (ii) **Describe** a country found in each category using the following indicators: stage in the demographic transition model and level of human development.
   (iii) **Explain** how access to healthcare for young people can affect life expectancy.

Exam hint!

To answer part (iii), you could compare access to healthcare between a developed and a developing country you have studied. Next, list the life expectancy of your two chosen countries. Explain how the availability of healthcare is different, e.g. number of hospitals or doctors available. Lastly compare the leading causes of death in each country.

2. Read the article below and answer each of the following questions.

GENERATION UNLIMITED LAUNCHES IN NIGERIA

A joint programme between the UN, the private sector of Nigeria and the Nigerian government aims to help 20 million young people in the country. Extending across 12 of the 36 states in the country, the public-private-youth programme will reach 20 million with skills training, employment, entrepreneurship, and social impact opportunities by 2030. This initiative will bring together the government, business and financial institutions, e.g. banks, and young people to work together to to solve major challenges and invest in young people as they prepare to transition to adulthood.

(i) **How** many young people will benefit from this new programme?
(ii) **What** groups are involved in starting this programme?
(iii) **How** many states will benefit initially from this programme?
(iv) The article mentions 'major challenges' that must be overcome to help young people transition to adulthood.

Gender inequality, health service availability and education are some of the major challenges faced by young people in Nigeria. **Explain** two of these challenges and **describe** how they could be overcome.

29. Life Chances in a Developed and a Developing Country

Let's get started! Sample starter

Gender inequality is a major issue facing young people in Nigeria. 18.5 million children are out of education — over 10 million of these are girls. Unemployment has been higher for young females for the past 10 years.

Using development aid offered by countries like Ireland to fund better access to education for young females will help more to attend school. Gender equality and education are two of the UN's Sustainable Development Goals. The UN outlines how governments can accomplish these goals ...

... now complete the answer in your copy.

Exam hint!

To complete this answer, choose another indicator and give facts you have examined in this chapter to explain the situation in Nigeria. Use knowledge gained from previous chapters to explain a simple idea to help improve your chosen indicator.

Learning Outcome: 3.9

30 Globalisation

Learning Intentions

You will be able to:
- Explain what is meant by globalisation
- Describe the three types of globalisation
- Discuss the effects that globalisation has on population, settlement and human development.

key words

Cultural globalisation
Economic globalisation
Political globalisation

30.1 What Is Globalisation?

In the past, most regions were more economically isolated and self-sufficient than they are today. This meant they produced food, fuel and raw materials locally and often for their own needs. Trade, travel and communications between different regions were much more limited than now. Most businesses survived by using local raw materials and by selling their products to local markets.

30. Globalisation

Today, people around the world are much more connected to each other. Modern communication systems allow money and information to move almost instantly throughout the world. Improved regional and international transport allows many people to travel widely and quickly. Goods and services that are produced in one part of the world can be available to purchase in another part of the world within hours.

These changes have given rise to rapidly growing globalisation. We can define globalisation as the increasing cultural/social, economic and political interconnection of people and countries around the world.

> **Being literate**
>
> Write your own definition of 'globalisation' as a tweet (no more than 280 characters!). #Globalisation

Q The Aalsmeer Flower Auction in the Netherlands (photo A) is the largest flower auction in the world. Flowers from all over the world are traded, and around 20 million flowers are sold daily. Describe in three sentences how communication systems, transport and goods and services are linked to make sure flowers can be sold in Moore Street, Dublin (photo B) on the same day.

30.2 Types of Globalisation

1. Cultural Globalisation

The ever-increasing movement of people and ideas has resulted in the spread throughout much of the world of different languages, religions, beliefs, attitudes, customs, entertainments and eating habits. This can be referred to as cultural (or social) globalisation. You can see it in the spread of the English language, of Instagram or TikTok, of fast-food restaurants like McDonald's or Nando's or of American or British popular music. These are all symptoms of **cultural globalisation**.

Managing myself

Make a list of all the social media networks for which you have an account. Then note how many of your followers/contacts on these sites live: (a) outside of your home county; and (b) outside of Ireland.

2. Economic Globalisation

The economies of most countries are now becoming more interconnected. This process is called **economic globalisation**. There are several reasons for this:

- Import–export taxes and other trade 'barriers' between countries have decreased.
- International free-trade areas, such as the EU, allow the free movement of goods, services and citizens between their member states.
- MNCs control more and more trade and manufacturing operations. Most of these companies have their headquarters in the US or Europe, with branches in many parts of the world. These powerful companies thrive in free-trade conditions. They can put enormous pressure on individual governments to adopt free-trade laws and policies. Smaller companies tend to decline as MNCs grow.
- Many MNCs will set up factories in developing regions where they can get cheap labour and other resources. This lowers the cost of manufacturing their products and increases their profits.

Managing information and thinking

1. List the MNCs that you see depicted on this map. Name three other MNCs that are not depicted on the map.
2. What message do you think the map is giving?

3. Political Globalisation

In the past, almost all the major policies or laws of independent countries were made by the individual governments of those countries. Now, many policies and laws are made or greatly influenced by international political bodies. The European Union, for example, has made laws on environmental protection and trade that all EU countries must follow. This is an example of **political globalisation**.

The table below shows the 2021 KOF Globalisation Index. This index measures how globalised a country is and gives it a ranking. The higher the ranking, the more globalised the country is. Many small countries can have a high score. This is because they need to reach out globally for their economies to thrive and to access all of the goods and services that their populations need. Large countries such as the United States can have a low score because they are capable of trading to a high level within their own borders. Many developing countries can also be placed low on the index because they have lower levels of economic, social and political development.

2021 KOF Globalisation Index		
Rank	**Country**	**Globalisation Index, overall**
1	Netherlands	90.91
2	Switzerland	90.45
3	Belgium	90.33
4	Sweden	89.44
5	United Kingdom	89.31
6	Germany	88.73
12	Ireland	85.75
204	West Bank and Gaza	39.54
205	Afghanistan	38.39
206	Central African Rep.	38.32
207	Eritrea	30.88
208	Somalia	30.49

Managing information and thinking

1. (a) What is Ireland's overall position on the KOF Globalisation Index?
   (b) Give two reasons why you think Ireland appears above the United States on the KOF index?
2. Look at the countries in the bottom of the list. Why do you think they have such a low score on the index?

30.3 Some Consequences of Globalisation

We will now examine the effects of globalisation on:
- Population
- Settlement
- Human development.

Globalisation and Population in Ireland

How Globalisation Has Affected Ireland's Population

- Globalised communication by means of smart phones, satellite TV, email, internet, Instagram, TikTok, etc. allows people to become aware of the attractions of other countries.
- Cheaper and more accessible global transport allows more people to travel between countries.
- Some international bodies allow people to migrate freely between countries. The EU, for example, allows EU citizens to migrate between one EU state and another.

International migration increases, especially from economically poorer countries to countries that seem to offer better employment opportunities. Between the 1990s and 2008, for example, many immigrants arrived in Ireland. Most of these were from other EU states and came in search of employment during a period of economic boom in Ireland.

During those years of economic boom:
- Immigration led to an increase in Ireland's population.
- Ireland's population profile became 'younger', because most immigrants were young adults.
- Ireland's population became more diverse (varied), as people from many different countries and cultures entered the country.

Globalisation and Settlement in Ireland

Many global MNCs have set up large factories and other businesses in Ireland. Most of those plants are located in or close to large cities, where skilled employees and facilities such as airports and motorways are close at hand.

These companies offer well-paid employment to large numbers of people. They also stimulate local economies, which creates more employment.

Many people migrate to cities to work for MNCs. Some of these employees bring their families with them.

This contributes to the growth of cities, which is called urbanisation.

Q List some multinational companies in the city closest to you. (Do research online if you need to.)

Case Study: Globalisation, Population and Urbanisation in Rio de Janeiro

Rio de Janeiro is the second-largest city in Brazil and has a population of 6.5 million people. Rio has grown rapidly over the last 50 years to become a major industrial, administrative, tourist and commercial centre. The staging of the 2014 World Cup and the 2016 Olympic Games – major international events – brought global attention to the city.

Economic growth and globalisation have attracted many internal and international migrants to the city. This has resulted in population growth and this has made Rio an ethnically mixed city, with migrants coming from many different places.

1. The location of Rio de Janeiro

- Many have migrated from rural areas of Brazil to Rio.
- Many have migrated from Argentina and other South American countries.
- Migrants from South Korea and China have come seeking new business opportunities.
- Skilled workers have come from the UK and the US and other developed countries.

The contrast in Rio – favelas alongside high-rise buildings

Q

1. Describe one positive impact that globalisation has had on population change in Rio.
2. Explain one negative consequence of globalisation for Rio.

These migrants have brought with them many skills, cultures and ideas, which have helped the economic development of the city.

Globalisation has led to changes in the way people live. One change has been the urbanisation of the world's population. Rural populations have slowly moved into cities. Cities have expanded into what was once good farmland. Growing urban populations have put pressure on the supply of housing, employment, education and sanitation. In Rio, this is seen in the makeshift settlements known as 'favelas'.

There are up to 1,000 favelas in the greater Rio area. Unemployment in the favelas is as high as 20%, and many of those who have jobs are on low salaries with little job protection. Some favelas were demolished to clear areas needed for new developments before the Olympic Games in 2016. However, favelas close to the Olympic park were redeveloped and the inhabitants were forced from their homes. Many people found employment during the construction of the Olympic facilities.

Globalisation and Human Development

Globalisation affects numerous aspects of our everyday lives. It can present benefits and difficulties for human development (the economic, social and cultural development of people).

Question Time

Imagine that two foreign-based, multinational companies propose to set up a large factory and a large 'superstore' in an urban area near you. The statements labelled A–J on the next page outline some effects of these proposals on the human development of local people.

1. Indicate whether each statement refers to a 'benefit' or a 'difficulty' for human development. (Make two separate lists and refer to each statement by its letter label. You do not need to rewrite the statements.)

2. Write out the two statements that you think are the two most important 'benefits' and the two statements that you think are the two most serious 'difficulties' of the arrival of the MNCs for local human development.

3. Come up with one benefit and one difficulty not referred to in the statements A–J.

30. Globalisation

A The proposed new factory will provide many well-paid jobs for highly trained workers.

B Local, family-owned shops cannot compete with a large, foreign-owned superstore. If they are forced to close, many locals lose their livelihoods.

C Some MNCs do not want to negotiate with workers' trade unions. This denies workers their rights.

D The new superstore will increase competition with other superstores and can sell goods more cheaply than small, local shops. The prices of things might fall.

E Workers must pay up to 40% tax on their incomes, while MNCs pay just 12.5 –15% corporation tax on their profits. Some multinationals have even managed to pay no corporation taxes at all.

F The multinational store will probably provide a greater choice of goods for people to buy.

G The profits of foreign-owned MNCs will be moved abroad. They will not be used for the human development of people in Ireland.

H Some big employers want to hire people on 'zero-hour contracts'. These contracts give workers few or no guaranteed working hours, which means little or no pay some weeks. This creates great financial uncertainty and stress for workers and their families.

I Employees at the new plants will pay income tax that will contribute to the upkeep of hospitals, schools and other important public services.

J A new factory and superstore might provide spin-off business and employment for local cafés, cleaners, electricians and other service industries.

Junior Cycle Geography CYCLONE

Sample questions

1. **Examine** the article below about a multinational company moving the location of one of its factories and answer each of the following questions.

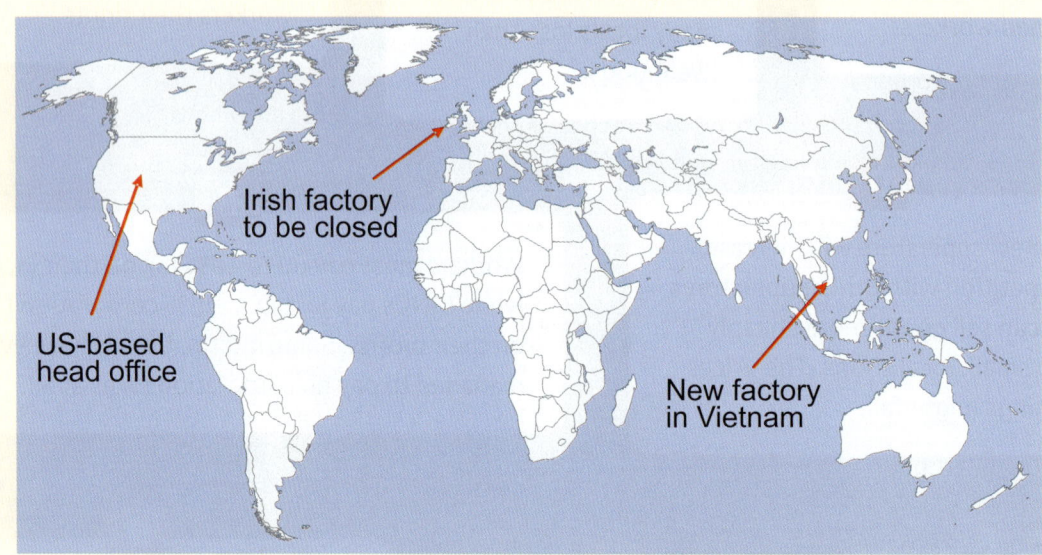

A multinational company, whose head office is in the US, has decided to relocate one of its factories from Ireland to Vietnam in Asia. This will result in the loss of 400 jobs in the large Irish town where the factory is currently located.

(i) **Name** one impact of the relocation mentioned in the article above.

(ii) **State** one reason why a multinational company would decide to move a factory from a developed country, like Ireland, to a developing country, like Vietnam.

(iii) **How** does this decision by the multinational company to move one of its factories to another country show that we live in a globalised and interconnected world?

(2022 SEC exam paper, Q2 (c) (i)–(iii))

Exam hint!

To answer part (ii), you should consider the financial costs and savings that the company may benefit from.

30. Globalisation

2. Globalisation is a process that is leading to the world becoming more connected due to the exchange of goods, people, ideas and information. **Read** the article below about a rural Irish town becoming more connected to the rest of the world and answer each of the following questions.

High-speed internet puts Skibbereen at the centre of the world for connectivity

In 2016, Skibbereen became Ireland's first one gigabit town. Before this, Skibbereen was like many rural towns, with very low levels of broadband internet. The access to high speed broadband has attracted businesses to the town who can now have a global presence. Fifteen families have moved or relocated to Skibbereen since May 2018, attracted by the high-quality jobs, as well as the quality of life. They tell of escaping the daily commuting nightmare, lower housing costs and lower childcare costs. High-speed broadband is clearly having a positive effect on economic development in the town and is contributing to a sustainable future for Skibbereen.

(i) **How many** families have moved or relocated to Skibbereen since May 2018?

(ii) **Name** two pull factors mentioned in the article that attract families to live in the area.

(iii) Imagine you are setting up a business in Skibbereen. **How** will having access to high-speed broadband be of benefit to your business?

Exam hint!

Think back to the push and pull factors for migration that we studied in Chapter 19.

Acknowledgements

For permission to reproduce images, the authors and publisher gratefully acknowledge the following:

© Achill Tourism: 171T; © Adobe Stock: v, 9B, 42c, 42d, 45T, 53T, 127L, 151T, 160, 162, 169T, 172R, 240L, 240B, 243L, 317BR, 342B, 351, 352CR, 365, 369, 375CL, 375CR, 384Cc, 384BL, 389, 438T, 443; © akg-images: 327BL; © Alamy: 13, 24TR, 25, 28, 35, 59L, 100, 121, 127R, 131B, 134, 135C, 135R, 136R, 139, 140T, 141, 142R, 145B, 149, 151B, 156, 163, 164CL, 164CR, 164TR, 165B, 166B, 168, 169B, 171B, 182B, 184, 185, 186, 196T, 202C, 205CL, 205CR, 209C, 209R, 217, 240CT, 246, 247, 254TR, 258, 271, 274, 277, 285, 291L, 296B, 298, 323, 327T, 334B, 335TL, 336, 339BL, 339BR, 358R, 374, 380, 384Ca, 384BR, 390, 393, 394T, 402, 408CL, 408CR, 408B, 419, 424, 439; © Andrew Downes: 220BR; © BahamasLocal.com: 218; © Banagher Precast Concrete: 379B, 381B; © Barrow Coakley: 104, 145T, 290, 291R, 292T, 293, 294, 295, 296T, 300, 301, 302, 303T, 304, 309, 311; © Central Statistics Office: 352B; © Cork County Council: 147; Image used with permission of the Cornwall and Scilly Historic Environment Record © Cornwall Council 2022: 297; © Eddie and Mary Hilda Cavanagh: 57; © Erich Osterberg: 236C; © Eric Jones (CC BY-SA 2.0): 127C; © Evan White: 90; © Fairtrade Ireland: 413; © Freepik: 2, 3, 9T, 33, 55, 361, 363, 379T; © Getty Images: 17T, 24BR, 30, 32R, 61, 64B, 117, 119, 126T, 131T, 132, 164B, 220T; © Gilles Montnach: 76; © Irish Image Collection/Rex Roberts: 183T; © Islandroads.com: 115B; © iStock/Getty Premium: 1, 16, 22, 27L, 39, 41, 42a, 42b, 42e, 45B, 49T, 46, 49, 50, 51, 52, 53B, 54, 59C, 64T, 69, 72, 73, 78, 86, 87, 89, 94, 95, 97, 98, 101, 102, 108, 111, 115T, 120, 122, 125, 136L, 136C, 140B, 148, 151C, 152, 154, 155, 157, 164TL, 165T, 165C, 170, 172C, 173, 175, 177, 182T, 187, 189, 192TL, 192B, 198, 200, 202T, 202B, 203, 204, 205T, 208, 209L, 210, 211, 212, 213B, 223, 226, 232, 235, 239T, 239C, 240CB, 248, 249, 252R, 254TL, 254BR, 268, 269, 273, 278B, 283R, 287T, 287B, 292B, 312, 322, 327BR, 341, 342T, 342C, 345L, 345TR, 345CR, 352CL, 354, 355, 357, 362, 364, 372, 373, 375C, 375BL, 384Cb, 384Cd, 387, 392, 394B, 399T, 403, 404, 423, 427TL, 426, 427BL, 427BR, 430, 434TL, 435L, 437, 438B, 440T, 441, 442, 444, 445; © Joe O'Shaughnessy, Connacht Tribune, Galway: 130; © John Herriott: 317BL, 319; © Jonathan Billinger (CC BY-SA 2.0): 142L; © Kerry Group: 383; © Lisney Estate Agents: 338; © Liz White: 38; © Łukasz Pawlik: 128L; © mike-page.co.uk: 166T; © NASA: 213T; © National Oceanic and Atmospheric Administration: 216B; © NSF, USGS, National Ice Core Lab: 236B; © Ordnance Survey Ireland: 305, 307, 310; © Paul Naessens/Western Aerial Survey: 297R; © Peter Bijsterveld: 179; © Pixabay: 264, 396, 399B; © ProVision Photography: 146, 220BL; © Reuters: 23, 32L, 281; © RIA: 330; © Richard Dorrell (CC BY-SA 2.0): 159; © Rolling News: 335TR, 335B; © RTÉ Archives: 334C; © Ruth Medjber: 280; © Science Photo Library: 252L; © Shutterstock: 11R, 17C, 17B, 24C, 26, 59R, 126B, 129, 174, 183B, 243R, 282, 283L, 283C, 286, 316, 317T, 343, 345BR, 358L, 427, 434TR, 434BL, 434BR, 435R; © Sikkim NOW!: 24BL; © Image courtesy Southern California Earthquake Center, University of Southern California: 34; © Swords Celtic FC: 278T; © Take Off Photography: 400; © Teagasc: 254BL; © Transport Infrastructure Ireland: 398B; © Photo courtesy U.S. Army Corps of Engineers, Seattle District: 144; © UNHCR/Gordon Welters: 288; © United Nations Sustainable Development Goals, https://www.un.org/sustainabledevelopment/. The content of this publication has not been approved by the United Nations and does not reflect the views of the United Nations or its officials or Member States: 370, 418; © Wikimedia Commons: 135L, 192TR, 192TC, 216T, 331; by Andrew Tatlow (CC BY-SA 2.0): 128R; by Jonathan McIntosh (CC BY 2.0): 427TR; by Khalili Collections (CC-BY-SA 3.0): 64C; by Simon Ledingham (CC BY-SA 2.0): 143; © Young Scientists Kenya: 414, 422.

The authors and publisher have made every effort to trace all copyright holders. If, however, any have been inadvertently overlooked, we would be pleased to make the necessary arrangement at the first opportunity.

2ND EDITION

CYCLONE
COMPLETE JUNIOR CYCLE GEOGRAPHY

SKILLS BOOK

CHARLES HAYES

Gill Education
Hume Avenue
Park West
Dublin 12

www.gilleducation.ie

Gill Education is an imprint of M.H. Gill & Co.
© Charles Hayes 2023

ISBN: 978 07171 95299

All rights reserved. No part of this publication may be copied, reproduced or transmitted in any form or by any means without written permission of the publishers or else under the terms of any licence permitting limited copying issued by the Irish Copyright Licensing Agency.

Design and layout: Liz White Designs

Illustrations: Keith Barrett, Derry Dillon and Andrii Yankovskyi

For permission to reproduce images, the author and publisher gratefully acknowledge the following:

© Adobe Stock: 16TR, 16B, 17T, 18TR, 18BR, 37, 40, 49, 63TC, 63BC, 158T; © Alamy: 13, 18BL, 38CL, 38CR, 43BL, 43BR, 47T, 53, 54, 56, 57, 58, 79BL, 93T, 112B, 116T, 120BR, 142T, 150; © Barrow Coakley: 51, 100, 102, 105, 120TL, 120TR; © brickbats.co.uk: 119B, 152T, 153T; © Carbouval/Dreamstime.com: 79BR; © Freepik: 4T; © Getty Images: 43T; © iStock/Getty Premium: 1T, 16TL, 16C, 18C, 20, 21, 38T, 38C, 43BC, 45, 52T, 58T, 63T, 63TL, 63BL, 63BR, 74, 77T, 79T, 81T, 111, 121T, 122, 123, 125T, 127, 130T, 135T, 136T, 136B, 137T, 140, 152T, 155B, 156; © Jeff Danziger: 161B; © Kevin O'Dwyer: 133; © Liz White: 14T; © NASA: 69T; © Ordnance Survey Ireland: 27T, 37, 50; © Pixabay: 73T, 85T; © Reuters: 9T; © Shutterstock: 18TL, 112T, 136C, 137B, 155T, 155C; © Tom Mulligan/The Cobblestone: 120BL; © Young Scientists Kenya: 149T.

The author and publisher have made every effort to trace all copyright holders. If, however, any have been inadvertently overlooked, we would be pleased to make the necessary arrangement at the first opportunity.

Ordnance Survey Ireland Permit No. 9278
© Ordnance Survey Ireland/Government of Ireland

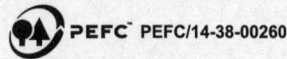 PEFC/14-38-00260

Contents

About this Skills Book iv

1. Dynamic Earth 1
2. Volcanic Activity 4
3. Earthquakes 9
4. Fold Mountains 14
5. Rocks 17
6. Natural Energy Resources 22
7. Geographical Skills: Ordnance Survey Maps 27
8. Weathering 38
9. Mass Movement 43
10. Rivers 47
11. The Sea 52
12. Glaciation: The Work of Moving Ice 58
13. Measuring and Forecasting Weather 63
14. Severe Weather 69
15. Global Climates 73
16. Climate Change 77
17. Soils 81
18. Population 85
19. Migration 93
20. Geographical Skills: Aerial Photographs, Charts, Graphs and Infographics 100
21. Rural and Urban Settlement in Ireland 112
22. Urban Change in Dublin 116
23. Primary Economic Activities 121
24. Exploitation of Natural Resources 125
25. Secondary Economic Activities 130
26. Tertiary Economic Activities 135
27. Economic Development and Inequality 142
28. Human Development and Development Assistance 149
29. Life Chances in a Developed and a Developing Country 154
30. Globalisation 158
31. **Welcome to the Junior Cycle CBA Section** 163

CBA 1: Geography in the News 165
CBA 2: My Geography 180
The Written Assessment Task 197

Junior Cycle Geography CYCLONE SKILLS BOOK

About this Skills Book

Cyclone Skills Book is the ideal homework, revision and exam-preparation book. Combined with *Cyclone* textbook, it ensures you derive maximum benefit from the Junior Cycle Geography programme.

- ✓ It **reinforces**, **examines** and **enlivens** what students learn in class.

- ✓ It ensures that students understand and repeatedly practise all the key **geographical skills** that are a vital element of Junior Cycle Geography.

- ✓ Its repeated use of **examination-type questions** ensures that students learn to answer questions fully and effectively in the Junior Cycle examination.

- ✓ It exposes students to **active learning techniques** that make learning effective and enjoyable.

- ✓ It contains a comprehensive 42-page section on **Classroom-Based Assessments (CBAs)**. This step-by-step section guides students clearly and carefully through the processes involved in preparing their best possible assessment **projects**. It will also prepare students to answer the **Written Assessment Task** associated with the second CBA project.

Dynamic Earth 1

1. In the grid provided below, link each of the items labelled **A–F** in Figure 1 with its matching term (name) from **List 1** and its matching description from **List 2**. Two matches have been made for you.

List 1 (terms)
- Crust
- Inner core
- Mantle
- Outer core
- Plates
- Plate boundary

List 2 (descriptions)
- Large slabs of the earth's crust
- Made of molten nickel and iron
- Where plates meet
- Thick layer of molten and semi-molten rock beneath the earth's crust
- Thin, solid layer all over the surface of the earth
- At the centre of the earth

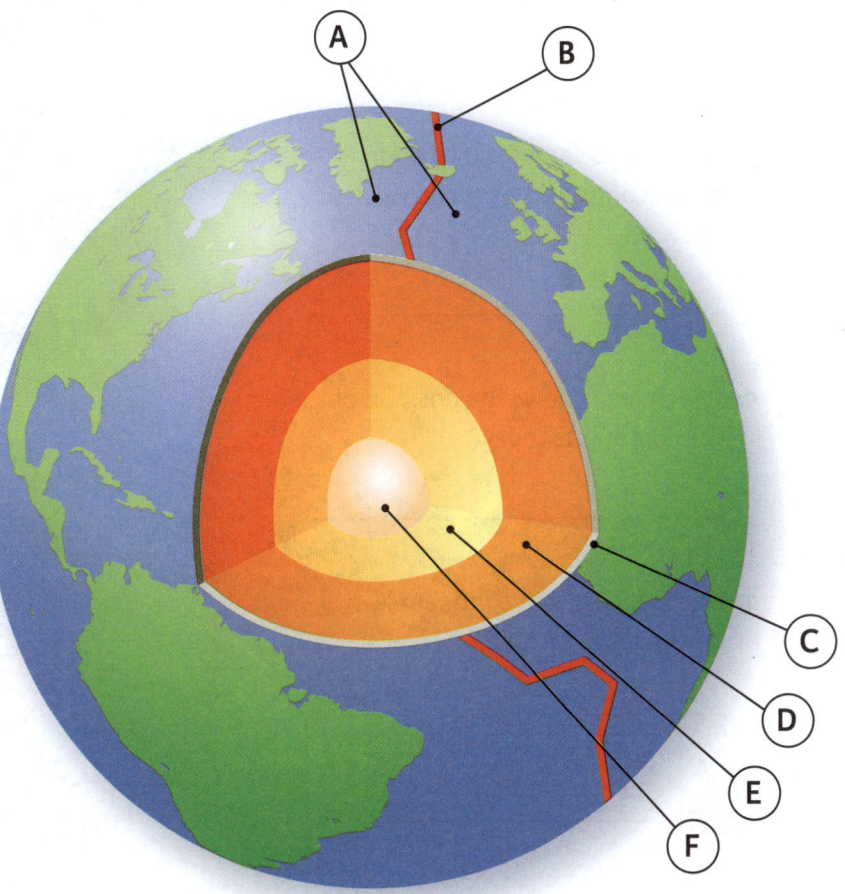

Figure 1: The earth's layers

Label	Term (from List 1)	Description (from List 2)
A	Plates	Large slabs of the earth's crust
B		
C		
D		
E	Outer core	Made of molten nickel and iron
F		

2. Over the past 250 million years, the world's **continents have been slowly moving apart** from each other. **Maps A, B and C** in Figure 2 show the approximate positions of these continents at three different times.

   a. Which map shows the position of the continents 250 million years ago?

   _____

   b. Which map shows the position of the continents 65 million years ago?

   _____

   c. Which map shows the position of the continents today?

   _____

   d. Name each of the continents labelled **A–E** on Map B.

A	
B	
C	
D	
E	

   e. What term is used to describe the movement of continents?

   _____

   _____

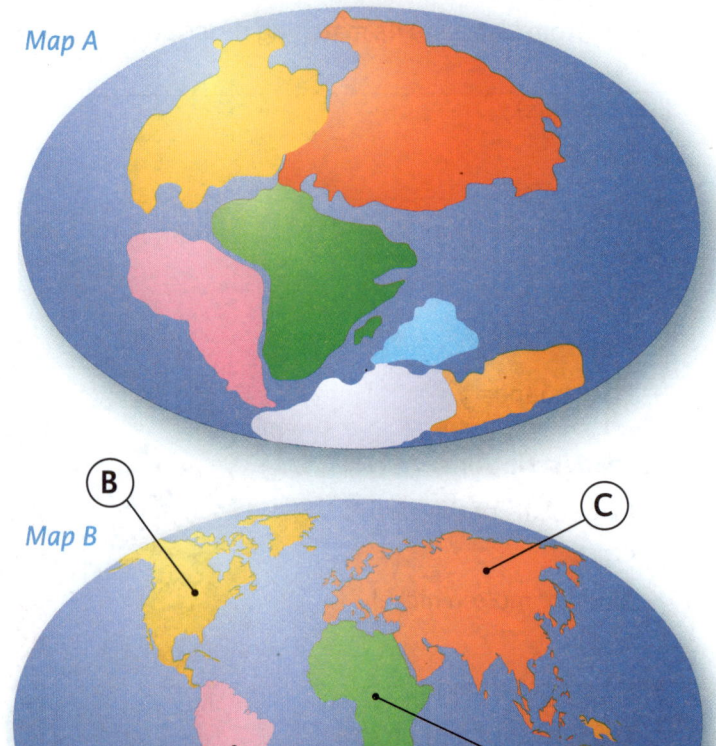

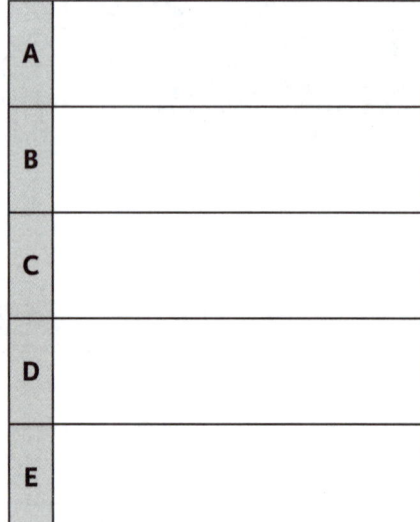

Figure 2

3. The map in Figure 3 shows **plates and plate boundaries**. Study the map and answer the following questions.

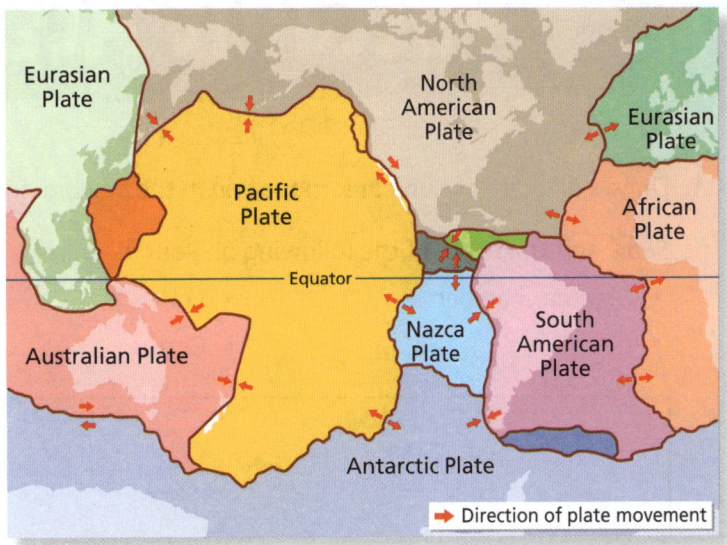

Figure 3

a. On which of the following plates is Ireland located? Tick (✓) the correct answer.

   Nazca Plate ☐   Eurasian Plate ☐
   Pacific Plate ☐   African Plate ☐

b. Name two plates that collide with each other.

   Plate 1: _____

   Plate 2: _____

c. Place an **X** on the map where the Eurasian and North American Plates are shown to separate from each other.

d. Figure 4 shows an example of a plate boundary. Describe what is happening at this plate boundary.

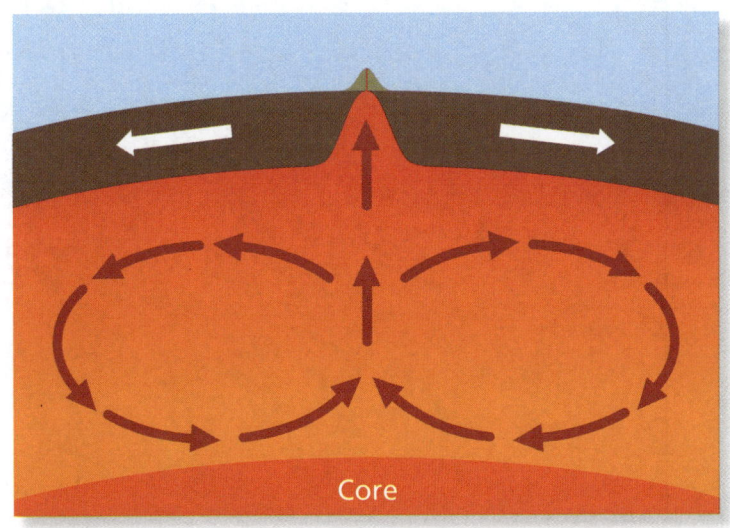

Figure 4

_____
_____
_____
_____
_____
_____

2 Volcanic Activity

1. Draw a diagram of a **volcanic mountain** in the box provided below.

   Show and *label* each of the following on your diagram:
   - Magma chamber
   - Cone
   - Vent
   - Crater
   - Lava
   - Ash cloud

2. Volcanic Activity

2. Many volcanic mountains exist at or near **destructive plate boundaries**.

   a. What are destructive plate boundaries?

   _____

   _____

   _____

   _____

   b. The diagram in Figure 1 shows the **development of a volcano at a destructive plate boundary**. Describe the **processes** (things that happen) that cause a volcanic mountain to develop at a destructive plate boundary. The labels A–D on Figure 1 might help you.

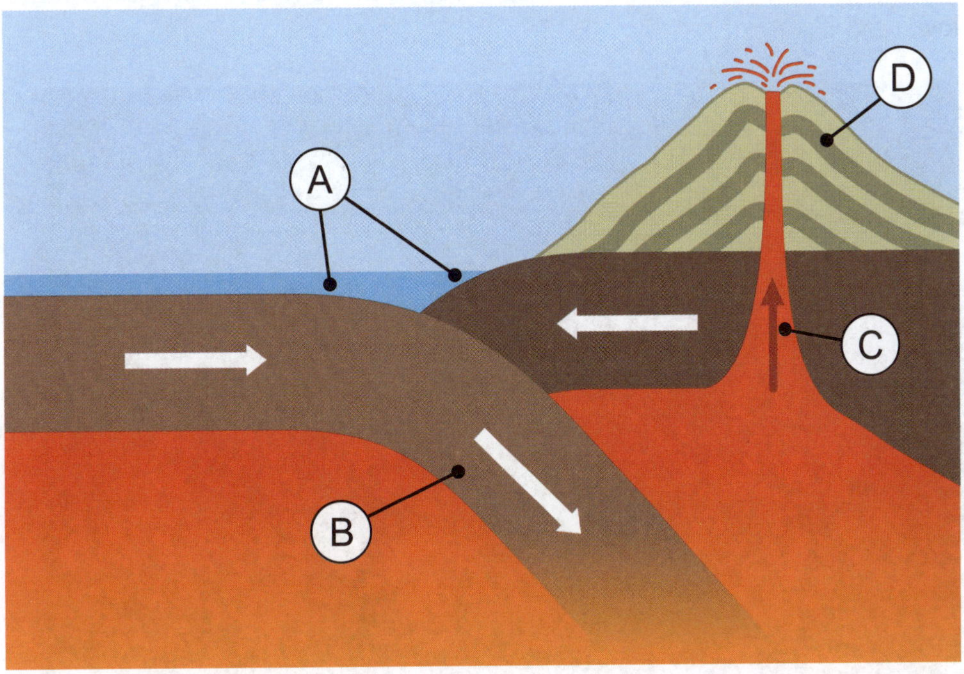

   Figure 1

   _____

   _____

   _____

   _____

   _____

   _____

3. Select the correct terms in the box to complete the table below. The table refers to **three stages in the life cycle of a volcanic mountain**.

> Extinct   Has not erupted for a long time, but is expected to erupt again
> Mount Etna   Slemish   Erupts regularly   Dormant

Stage	Description	Example
1. Active		
2.		Kilimanjaro
3.	Is not expected to erupt again	

4. The map in Figure 2 shows the location of the ash cloud that resulted from a volcanic eruption of Eyjafjallajökull in Iceland in 2010. Examine the map and *circle the correct option* in each of the statements below.

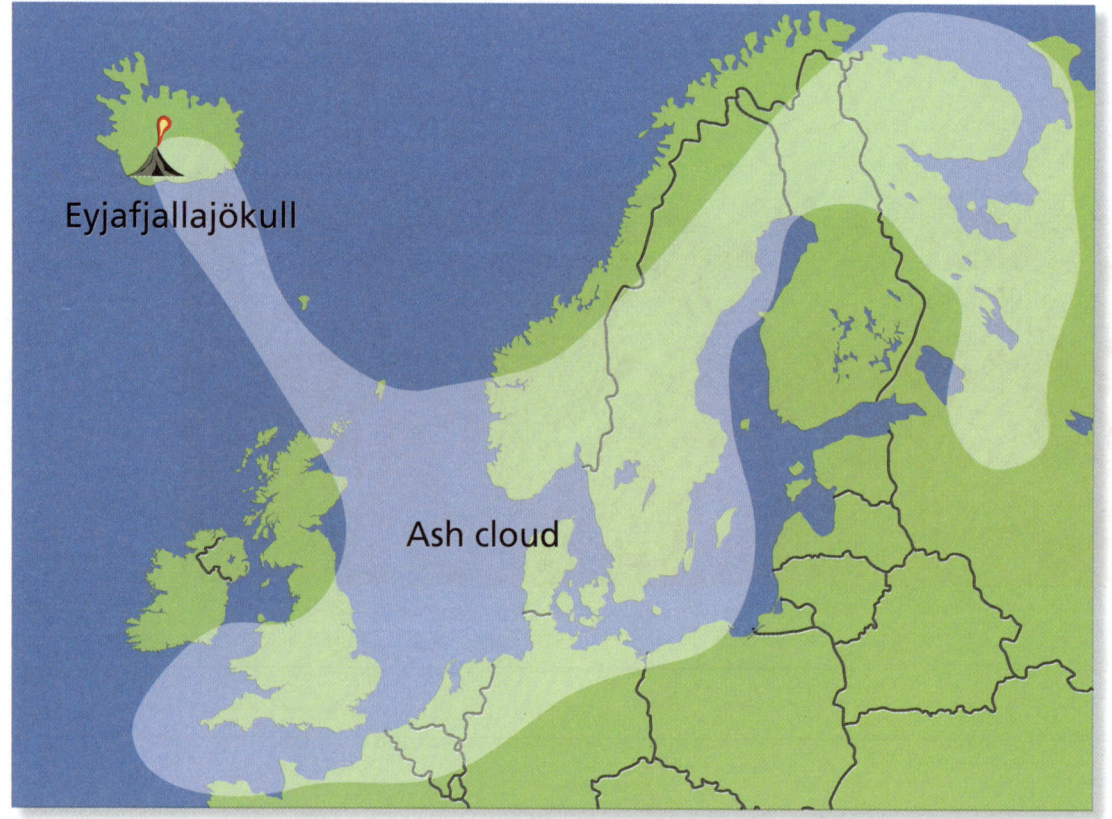

Figure 2

a. The ash cloud originated **over Iceland/south of Iceland/east of Iceland**.

b. The cloud settled mainly to the **northwest/southeast** of Iceland.

c. The cloud affected **a total of four countries/more than six countries/all EU countries**.

d. The cloud covered the whole of **Norway/Denmark/the UK**.

e. The cloud **did /did not** reach Ireland.

2. Volcanic Activity

5. Describe briefly two **negative (bad) impacts** of volcanic activity.

Impact 1: _____

_____

Impact 2: _____

_____

6. Read the news article below, which refers to two **positive (good) impacts** of volcanic activity in Iceland.

   a. Name the type of energy referred to in the article.

   _____

   b. How is that energy created?

   _____

   _____

   _____

   _____

   c. What uses are made of that energy in Iceland?

   _____

   _____

   _____

   d. How did volcanic eruptions benefit tourism in the Reykjanes Peninsula?

   _____

   _____

   _____

Volcanic activity not all bad news for Iceland

Volcanic activity is usually thought of as 'bad news'. But natives of Iceland talk also of the advantages that volcanic activity brings to that northerly island.

One Icelander reminded me of the very cheap and ecologically 'clean' energy that volcanic activity provides. Volcanic activity 'superheats' underground water to above boiling point. *Geothermal energy* is then created by passing circuits of piped water through these superheated underground areas. This energy is used to generate electricity and to provide abundant hot water and central heating.

Another Icelander pointed to the boost in tourism that prolonged 'low-risk' eruptions in 2021 brought to the Reykjanes Peninsula in the southwest of the country. Up to 12,000 sightseeing visitors began to pour into the area each day. 'After-dark' sight-seeing became particularly popular, because it offered extra-spectacular views of the glowing eruptions.

7. Complete the word game below.

It's FUN! and it will help you to revise Chapters 1 and 2.

Clues Across:

3. This is the top of a volcanic mountain.
4. Between crust and core.
7. The home country of Etna.
8. The shape of a volcanic mountain (it might remind you of ice cream).
10. The pipe through which magma flows.
11. A piece of the earth's crust (not the kind from which you eat dinner).
12. Ireland is on this crustal plate.
13. The name of a crustal plate and of a great ocean.

Clues Down:

1. The outside layer of the earth (or of a loaf of bread).
2. The gradual movement of continents over time.
5. Comes out of a volcano – hot stuff!
6. The centre of the earth (or of an apple).
9. Nigeria and Sudan are on this plate.

Earthquakes 3

1. Read the passage below, which describes **how earthquakes happen**, and circle the correct options.

   Most severe earthquakes happen near the **boundaries/centres** of plates. They happen especially at places where plates **collide/separate** or where plates move past each other.

   Where the edges of plates **push up/slide gently** against each other, they create friction and huge stress builds up. Eventually this stress becomes so great that the plates slip. When this happens, massive energy is released in the form of **sea waves/shock waves** that spread out in all directions. These waves are called **breakers/tremors** and they produce earthquakes when they cause the ground to **shake/explode**.

2. Examine the diagram in Figure 1.

   Write each of the following terms in the correct box on the diagram.

   | Fault | Epicentre | Focus | Shock waves | Plate movement |

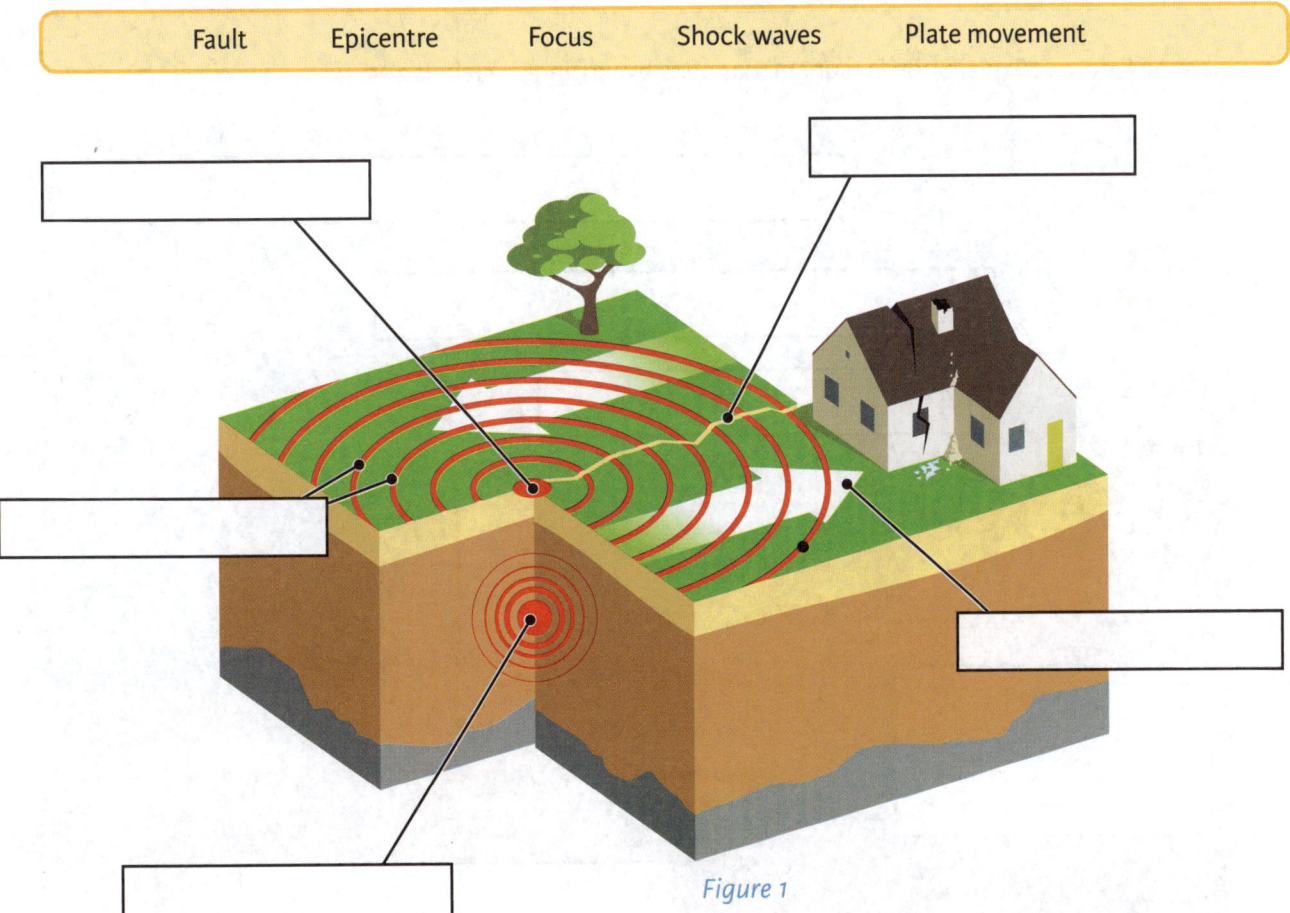

Figure 1

3. Match each of the **terms** in the list with the matching **description** in the grid below.

Terms
- Tsunami
- San Andreas
- Aftershock
- Moment magnitude scale
- Modified Mercalli scale
- Seismologist

Term	Description
	The name of a fault line in California, USA
	A person who studies earthquakes
	A scale that records the effects of an earthquake
	A giant wave that can result from an offshore earthquake
	A scale that records the strength of an earthquake
	A tremor that sometimes happens after the main earthquake tremor

4. Examine Figure 2. Explain how a tsunami happens by writing the labels **1–5**.

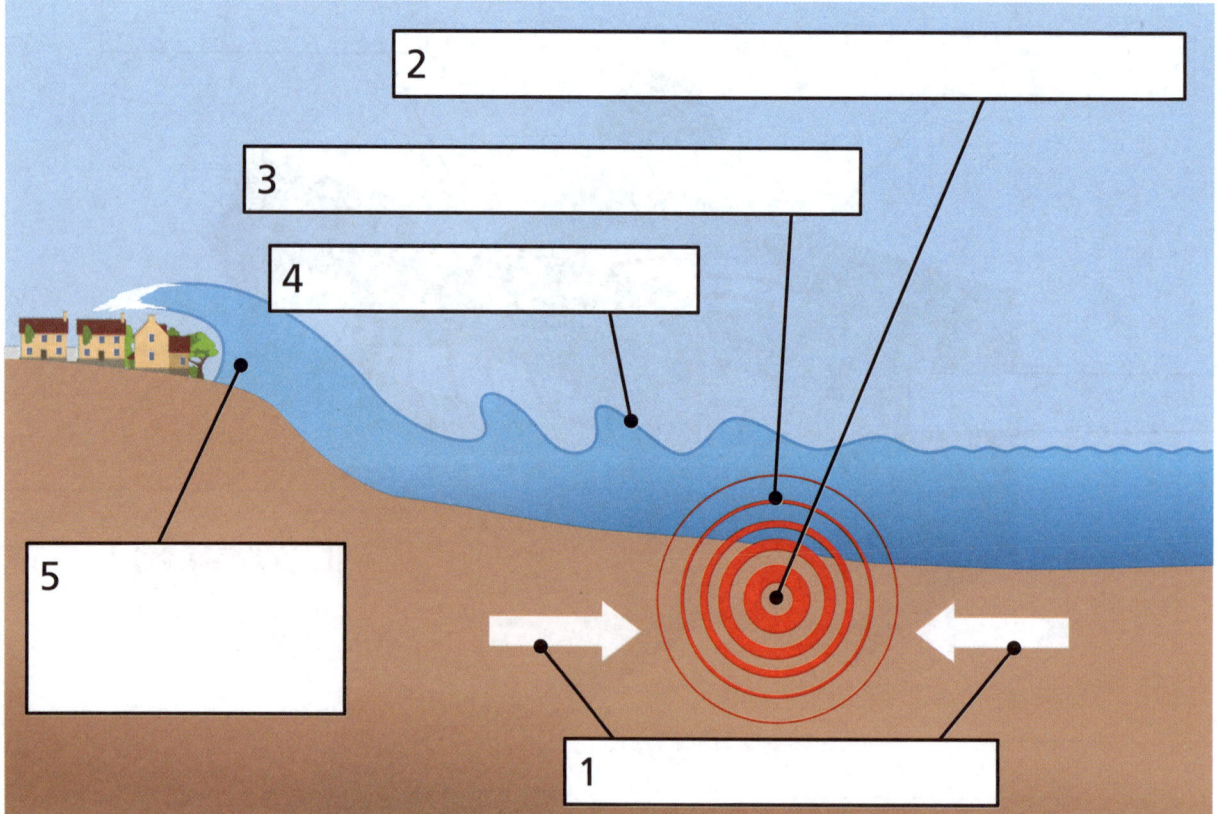

Figure 2

3. Earthquakes

5. Study the table and partially completed chart in Figure 3, which give information on the eight strongest earthquakes ever recorded. Then answer the questions.

Location and year of the eight strongest earthquakes on record (by magnitude)								
Location	Chile	Alaska, USA	Indian Ocean	Japan	Russia	Chile	Ecuador/Colombia	Alaska, USA
Year	1960	1964	2004	2011	1952	2010	1906	1965
Magnitude	9.5	9.2	9.1	9.1	9.0	8.8	8.8	8.7

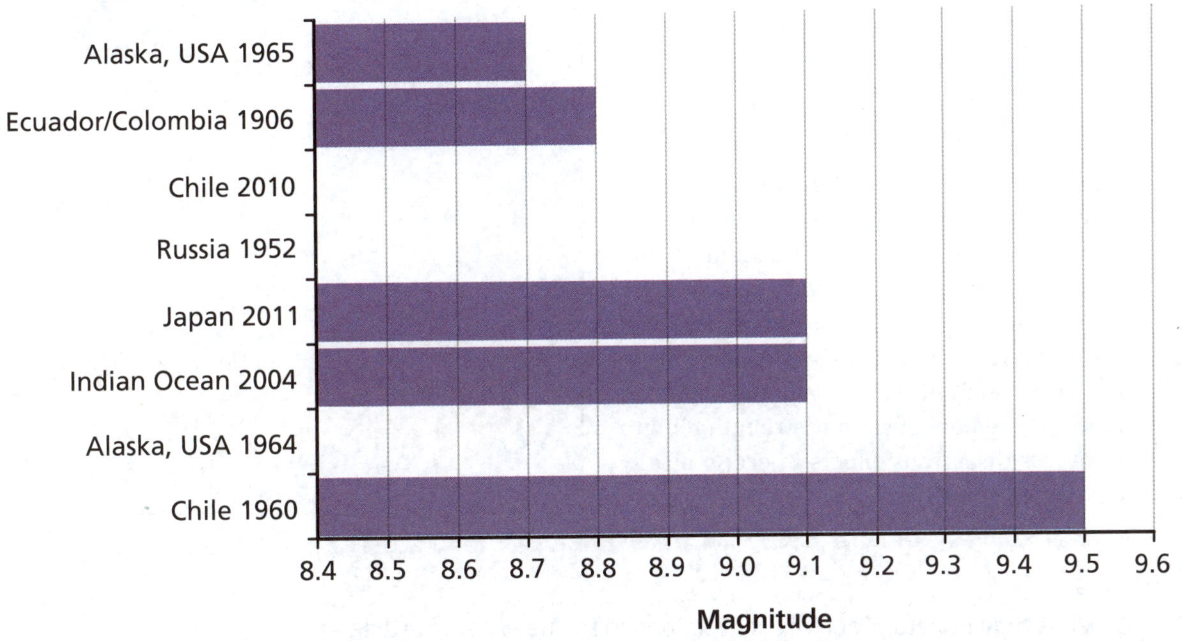

Figure 3

a. Use the information from the table to insert the **three** missing bars in the chart.

b. What was the magnitude of the earthquake in Japan in 2011?
_____

c. Name **one** instrument used to measure the strength of an earthquake.
_____

6. Geography in the News

Natural disaster in low-income Caribbean country

At 8.30 a.m. on 14 August 2021, the island nation of Haiti was struck by an earthquake of 7.2 magnitude. The epicentre was in the southwest of the country, about 125 kilometres from the capital city, Port-au-Prince. The earthquake resulted from the Caribbean plate and the North Atlantic plate pushing against and scraping by each other at an oblique angle.

The effects of the earthquake were very considerable. By 6 September, 2,248 people were confirmed dead, 12,763 injured and 329 still missing. These severe results were caused in part by the high magnitude of the earthquake. But they were also due to the fact that Haiti is an economically poor country. Tremors of similar strength tend to have less disastrous effects in economically developed countries.

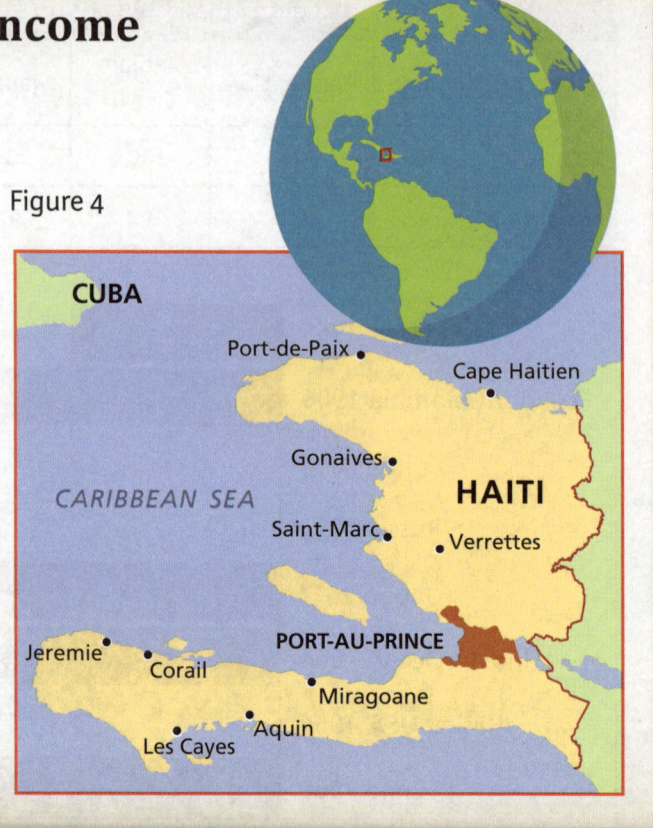

Figure 4

a. **What type** of natural activity is described in the newspaper article above? _____

b. **When** did it occur? _____

c. **Where** did it occur? _____

d. **Why** did it occur? _____

_____

_____

e. According to the article, what were the **effects** of the earthquake on local people?

_____

_____

_____

3. Earthquakes

f. What do you think might have happened to the 329 people who were still missing?

_____
_____
_____

g. The effects referred to in questions **e** and **f** were **immediate** effects of the earthquake. Name some possible **long-term** effects of the earthquake. (The pictures on this page might help you.)

_____
_____
_____
_____

h. 'Tremors of similar strength tend to have less disastrous effects in economically developed countries.' Why, in your opinion, is this the case?

_____
_____
_____
_____
_____
_____
_____
_____

4 Fold Mountains

1. Study the map in Figure 1 and complete the passage below by *circling the correct options* and *filling in the blanks*. You might need to consult an atlas or map to fill in some of the answers.

   The map in Figure 1 shows crustal plates in the region of the **Atlantic/Pacific** Ocean. The map also shows the locations of two great fold mountain ranges. The North American mountains labelled **A** are called the _____. The South American mountains labelled **B** are the _____.

   The map confirms that fold mountains tend to form **far from/on** plate boundaries, where plates **collide with/move apart from** each other.

   The 'Pacific Ring of Fire' is a wide horseshoe-shaped band that encircles the **northern/southern** boundary of the Pacific Ocean. Strong e_____ and active v_____, as well as fold mountains, are common along this 'Ring'. The area is called a 'Ring of Fire' because of the presence there of **earthquakes/volcanoes**.

Key
- major plate boundary
- strong earthquakes
- active volcanoes
- fold mountains
- Pacific Ring of Fire

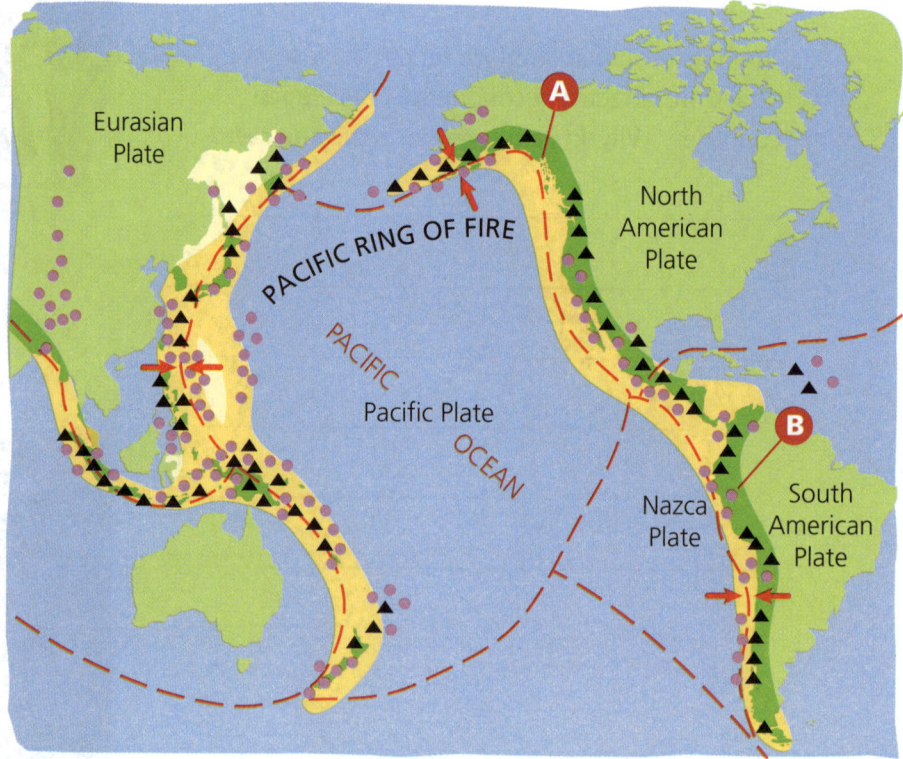

Figure 1

4. Fold Mountains

2. *Draw a labelled diagram* in the space provided, showing the **formation of fold mountains** at a **colliding plate boundary**. Label each of the following on your diagram:

- Mantle
- Two colliding plates
- Convection currents
- Fold mountains

3. Complete the table below by writing each of these terms in the correct place.

- Alpine period
- 400 mya
- Galtee Mountains
- 250 mya
- Caledonian period
- Alps

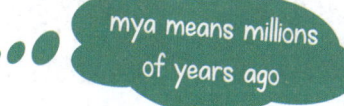

mya means millions of years ago

TIME

Fold mountain period	Millions of years ago	Example	
	35 mya		
Armorican period			
		Dublin–Wicklow Mountains	

15

Junior Cycle Geography CYCLONE SKILLS BOOK

A

B

C

D

E

4. Referring to each of the pictures labelled **A–E**, write an account of some **social and economic impacts of fold mountains**.

_____
_____
_____
_____
_____
_____
_____
_____
_____
_____
_____
_____
_____
_____
_____
_____
_____
_____
_____
_____
_____
_____
_____
_____

Rocks

1. Fill in the empty spaces in the grid below, which **classifies rocks into three groups**.

Formation	Magma **cools** beneath or on the earth's surface	Weathered/eroded rocks (or animal bones or vegetation) are **compacted and cemented**	Great **heat or pressure** causes rocks to change
Rock group			
Examples of rock type	1.	1.	1.
	2.	2.	2.

2. **True or false?**

   Indicate whether each of the statements **(a)** to **(h)** is true or false by ticking (✓) the appropriate box.

	True	False
a. Sandstone, limestone and quartzite are all sedimentary rocks.	☐	☐
b. Limestone is Ireland's most common rock type.	☐	☐
c. Marble is a metamorphic rock that was once limestone.	☐	☐
d. The world's first rocks must have been igneous rocks.	☐	☐
e. Igneous rock is found at the Giant's Causeway, Co. Antrim.	☐	☐
f. Granite and basalt are both igneous rocks.	☐	☐
g. Fossils can be found in all rock types.	☐	☐
h. Limestone is the main rock found in the Burren, Co. Clare.	☐	☐

Junior Cycle Geography CYCLONE SKILLS BOOK

3. **Recognising rocks:** Examine the rock types in the pictures labelled **A–E**. In the grid provided, name each rock, say which group it belongs to, suggest a possible use for it and name a location in Ireland where it is commonly found. (*Note:* one of the rocks shown is not studied in Chapter 5 of your *Cyclone* textbook. You might have to do some research to find the answers for this one.)

A — This rock was once limestone

B

C — Ireland's most common rock / A grainy rock

D — Hot stuff!

E — Rock with visible crystals that formed inside the earth's surface

	Name of rock	Rock group	Possible use	Where found
A				
B				
C				
D				
E				

4. Examine the map of Co. Mayo showing the rock types in the county. Then answer the questions that follow.

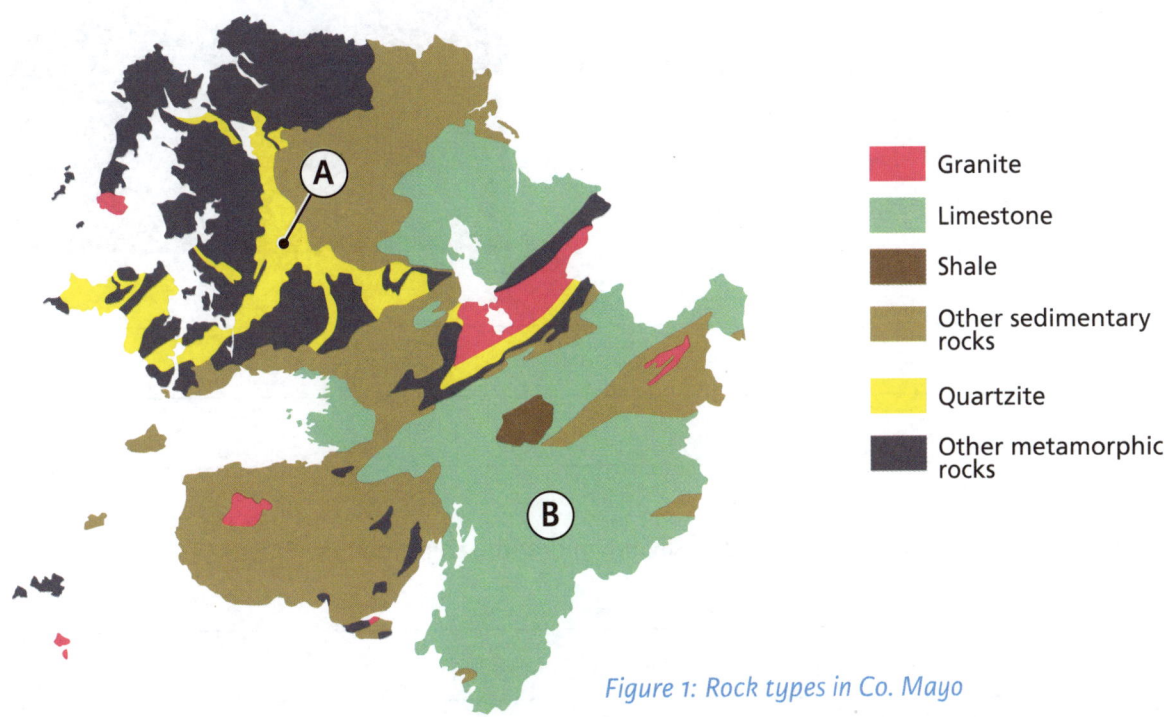

Figure 1: Rock types in Co. Mayo

a. Which of the following rock types can be found at the location labelled **A**? Tick (✓) the correct box.

   Granite ☐   Limestone ☐   Quartzite ☐

b. Name the sedimentary rock found in the area marked **B**. _____

c. Name **one** type of igneous rock shown on the map above. _____

d. Describe in detail how the sedimentary rock labelled **B** on the map was formed.

   _____
   _____
   _____
   _____

e. Identify two uses that can made of the sedimentary rock labelled **B**.

   Use 1: _____

   Use 2: _____

5. People interact with rocks by mining or quarrying them. The picture on the right shows a large **quarry**.

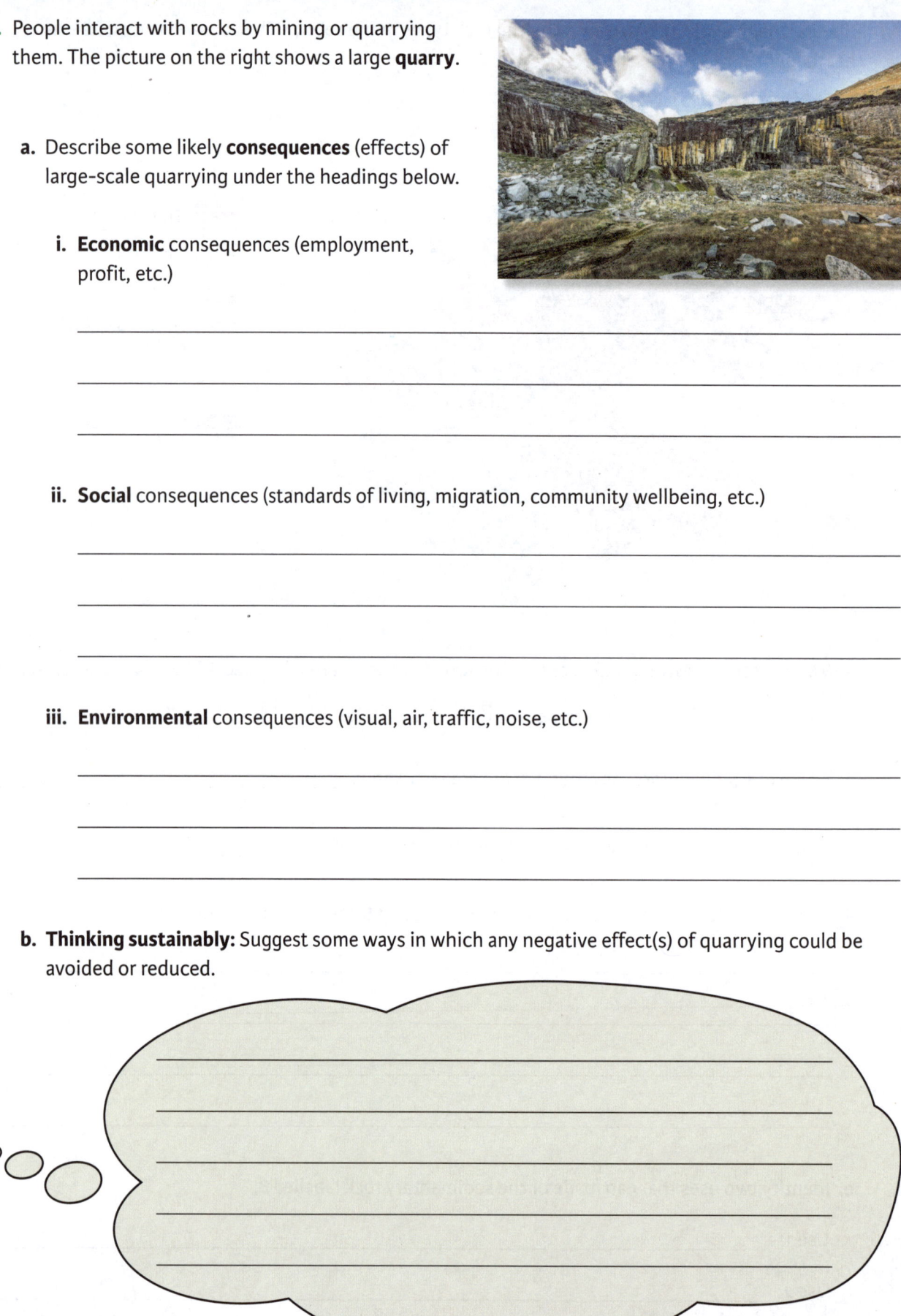

   a. Describe some likely **consequences** (effects) of large-scale quarrying under the headings below.

      i. **Economic** consequences (employment, profit, etc.)

      _____
      _____
      _____

      ii. **Social** consequences (standards of living, migration, community wellbeing, etc.)

      _____
      _____
      _____

      iii. **Environmental** consequences (visual, air, traffic, noise, etc.)

      _____
      _____
      _____

   b. **Thinking sustainably:** Suggest some ways in which any negative effect(s) of quarrying could be avoided or reduced.

5. Rocks

6. Complete this word puzzle, which will help you to draw together the information you have learned in this chapter.

Clues Across:

2. Great heat and pressure form this rock group.
3. Quartzite is found on _ _ _ _ _ _ Patrick, Co. Mayo.
4. Ireland's most common rock.
5. An intrusive igneous rock.
6. Picture B (below) shows one of these.
7. When magma reaches the earth's surface it is referred to as _ _ _ _.
8. One of three rock groups.
9. Sedimentary rocks often form in layers called _ _ _ _ _ _.
10. _ _ _ _ _ mining involves drilling vertically into the ground.
11. The rock _ _ _ _ _ is the process by which rocks constantly change.
12. Hot, molten material beneath the earth's surface – the source of igneous rock.
13. A famous limestone area in Co. Clare.
14. An extrusive igneous rock.
15. Sedimentary rocks dominate MacGillycuddy's _ _ _ _ _ in Co. Kerry.
16. Granite is found in this eastern county of Ireland.
17. An often brown or reddish sedimentary rock.
18. A place where rock is extracted from the earth's surface.

Clue Down:

1. The location of the basalt columns shown in Photograph A above (3, 6, 8).

Photograph A

Photograph B

6 Natural Energy Resources

1. Define the following terms:

   Renewable energy: _____

   _____

   _____

   Non-renewable energy: _____

   _____

   _____

2. a. Match each **energy source** in the yellow box below with the correct **statement** in the table. One match has been made for you.

   b. In the column on the right, write a tick for each energy source that is **renewable**.

   > hydroelectricity   gas   wind energy   tidal energy   fossil fuels
   > solar energy   biomass   geothermal energy   crude oil

	Statement	Energy source	Renewable?
1	Saudi Arabia is a big producer of this energy source.		
2	Extracted from the Corrib Field off Co. Mayo.		
3	Harnessed using the natural rise and fall of the tides.		
4	Generated from the sun's heat and light.		
5	Produced from waste wood or vegetation products.		
6	Generated from falling water.		
7	Produced by extracting heat from the ground.		
8	A common term used for oil, gas, coal, timber and peat.		
9	Produced by large turbines, often in exposed areas.	wind energy	✓

6. Natural Energy Resources

3. a. Explain fully how **acid rain** is formed. The diagram in Figure 1 will help you.

   _____

   _____

   _____

   _____

   _____

   _____

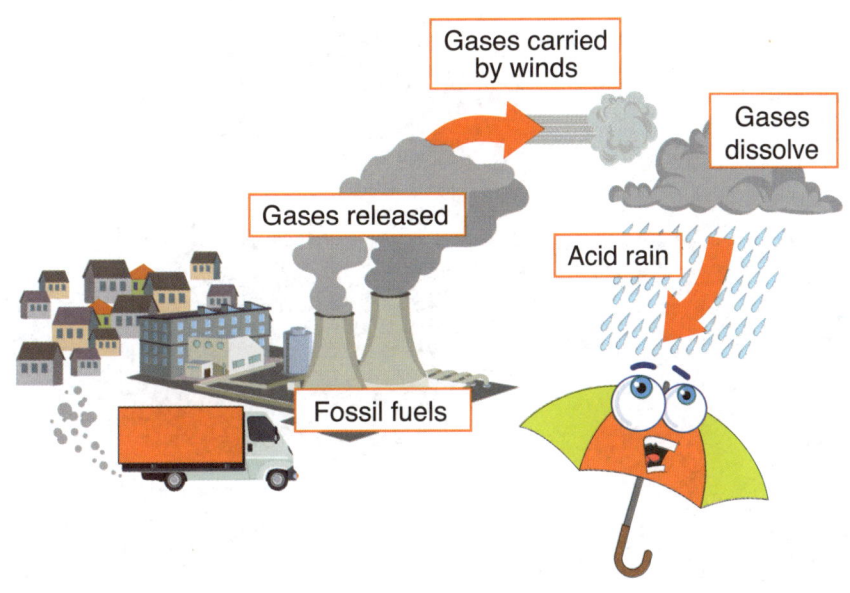

Figure 1: How acid rain is formed

b. Describe **two** effects of acid rain.

   Effect 1: _____

   _____

   Effect 2: _____

   _____

c. Describe **two** ways in which acid rain has been or could be reduced.

   Method 1: _____

   _____

   Method 2: _____

   _____

23

4. Examine the Ordnance Survey map extract and information below, showing the **village of Templetuohy** and **Bruckana Wind Farm**.

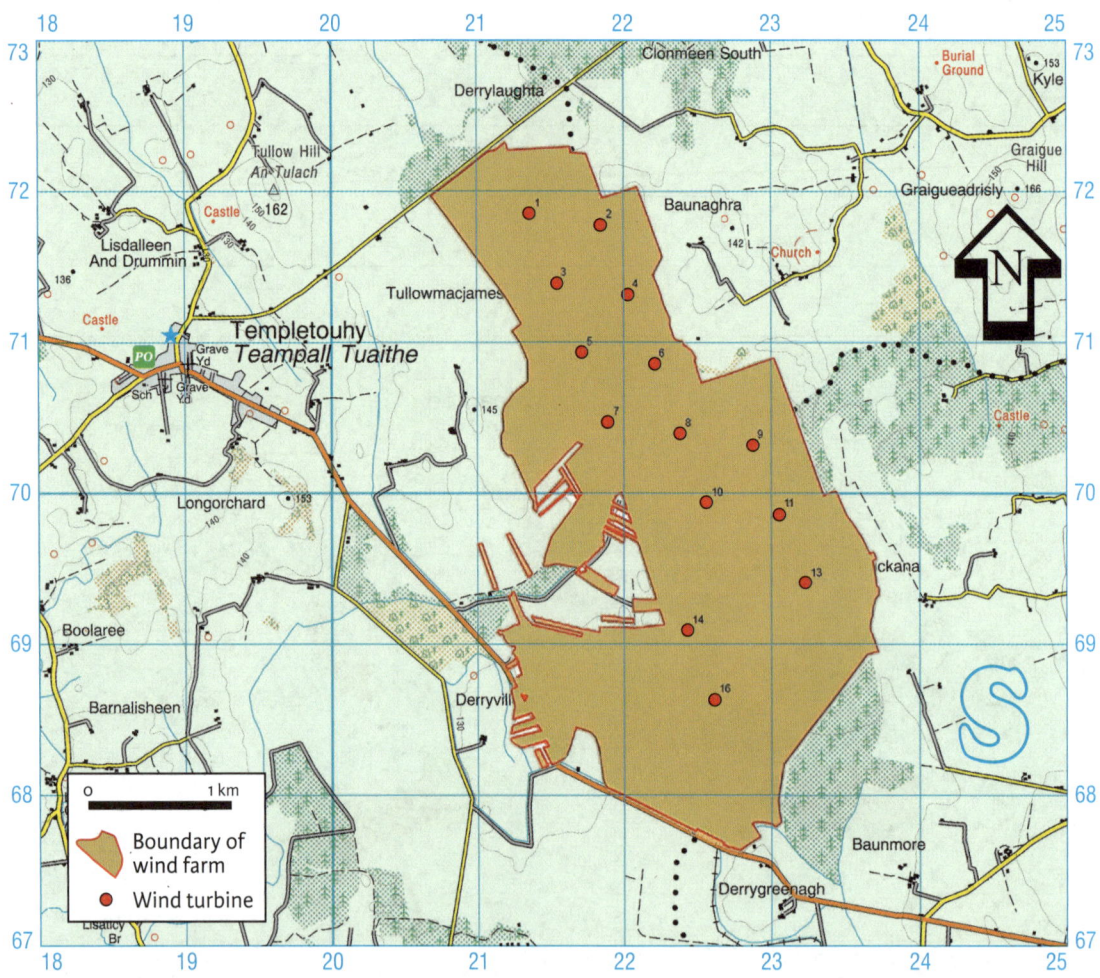

Bord na Móna started to develop Bruckana Wind Farm in 2014. It has the capacity to power 25,000 homes per year.

Figure 2

a. How many homes can be powered by Bruckana Wind Farm per year?

_____

b. How many wind turbines are there in the wind farm?

_____

c. What is the approximate area of the wind farm? Tick (✓) the correct answer.

4 km² ☐   7 km² ☐   15 km² ☐

Hint for doing Question (c).
- This map is divided into grid squares.
- Each grid square is 1 km² in size.
- Count the grids that are at least half-filled by the wind farm. That will indicate the approximate size in km² of the wind farm.

6. Natural Energy Resources

d. The approximate location of the wind farm is (tick (✓) the correct box):

two and a half kilometres east of Templetuohy ☐

seven kilometres east of Templetuohy ☐

three kilometres west of Templetuohy ☐

e. When Bruckana Wind Farm was in the planning stage the people living in the nearby village of Templetuohy had arguments for and against it. Explain in detail **two arguments for** and **two arguments against** the development of the wind farm that the people of Templetuohy might have raised.

Arguments for

1. _____

_____

_____

_____

_____

_____

_____

2. _____

_____

_____

_____

_____

_____

Arguments against

1. _____

_____

_____

_____

_____

_____

_____

2. _____

_____

_____

_____

_____

_____

25

Junior Cycle Geography CYCLONE SKILLS BOOK

5. Examine the statistics in the table. They show the average electricity usage per person in a selection of countries in 2019. Then answer the questions that follow.

Yearly per capita (per person) usage to nearest 1,000 kilowatts (2019)	
Germany	6.3
India	1.2
Ireland	5.5
Nepal	0.2
Cuba	1.4
USA	12.1

Source: Energy Information Administration (EIA)

a. List the six countries in the table in *rank order*, with the country with the highest usage at number 1 and the country with the lowest usage at number 6.

1. _____   4. _____

2. _____   5. _____

3. _____   6. _____

b. Three of the countries in the list are **economically developed** countries. *Draw a circle* around each of these countries in your list.

c. Based on what you have learned from the table, indicate whether each of the two statements below is true or false. Tick (✓) the correct box.

Statement	True	False
People in economically developed countries use much more electricity per person than people in economically poorer countries.		
All economically developed countries use the same amount of electricity per capita.		

d. Many Irish families need to reduce the amount of energy that they use in their homes. Explain three **practical** ways by which this could be achieved.

Consider what could be done in your own home.

1. _____

_____

2. _____

_____

3. _____

_____

26

Geographical Skills: Ordnance Survey Maps

7

1. To answer the following questions, you will need to examine the **OS (Ordnance Survey) map** of the **Rosslare area** on page 29 of this book. Tick (✓) the correct box to answer each question.

   You might need to take a look at the OS symbols on page 85 of your Cyclone textbook.

   a. The scale of the map on page 29 is:

   1:1 000 ☐   1:10 000 ☐   1:50 000 ☐

   b. The **shortest distance** between the railway station at Rosslare and the railway station at Rosslare Harbour is:

   3.5 km ☐   4.5 km ☐   6.5 km ☐

   c. The **distance by rail** between the railway station at Rosslare and the railway station at Rosslare Harbour is:

   3.5 km ☐   4.5 km ☐   6.5 km ☐

   d. If you travelled by road from the village of Killinick (shown on the west of the map) to the village of Lady's Island (shown on the south of the map), which **type of road** would you use?

   National primary road ☐   Regional road ☐   Third class road ☐

   e. The **general direction** from Killinick to Lady's Island is:

   Southwest ☐   Northwest ☐   Southeast ☐

   f. The **distance by road** between Killinick and Lady's Island is:

   5.5 km ☐   6.0 km ☐   7.5 km ☐

   g. The **total area** represented by the map is:

   50 km² ☐   90 km² ☐   99 km² ☐

h. The approximate area of Rosslare Bay is:

15 km² ☐   21 km² ☐   30 km² ☐

i. At grid reference T 06 14, there is:

A coniferous plantation ☐   A hill ☐   A national primary road ☐

j. The location of the village of Lady's Island is identified by the following four-figure grid reference:

T 11 08 ☐   T 10 07 ☐   T 07 10 ☐

k. At grid reference T 057 107, height above sea level is:

Between 10 and 20 m ☐   23 m ☐   30 m ☐   100 m ☐

l. The six-figure grid reference for the church at Lady's Island village is:

T 077 107 ☐   T 107 077 ☐   T 083 117 ☐

m. The main settlement pattern at T 07 10 is:

Nucleated ☐   Linear ☐   Dispersed ☐

7. Geographical Skills: Ordnance Survey Maps

Rosslare area

Figure 1

2. Examine the OS map fragment of the Carlow area in Figure 2. Altitude (height above sea level) is shown in four different ways on this map. Match each **number** in Column A with its corresponding **letter** in Column B in the grid below. One match has been done for you.

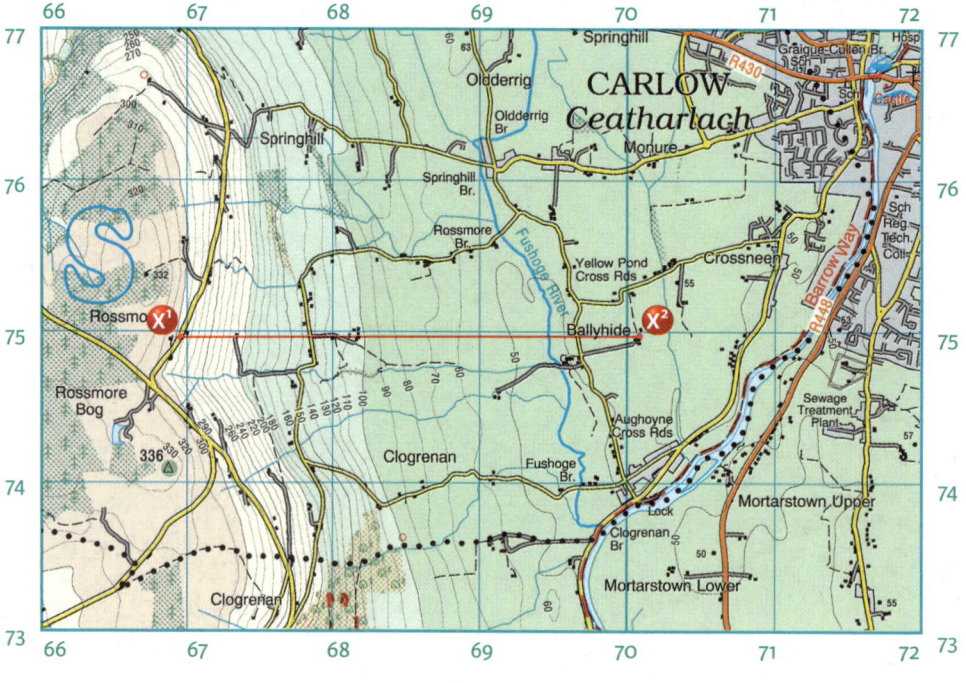

Figure 2

	Column A (location)
1	S 669 741
2	S 683 752
3	S 700 758
4	S 706 735

	Column B (how altitude is shown)
A	Contour
B	Spot height
C	Triangulation station
D	Colour coding

	Grid
1	C
2	
3	
4	

3. For each of these locations on the map in Figure 2, indicate using the letters A, B or C whether the land is (A) mostly very steep; (B) mostly gently sloping; or (C) mostly flat.

   a. S 70 75 _____

   b. S 67 74 _____

   c. S 68 76 _____

7. Geographical Skills: Ordnance Survey Maps

4. Examine the map (Figure 2) on the previous page. On the grid below, match the number of each **location** in Column C with its **height** in Column D. One match has been done for you.

Column C (location)	
1	S 690 750
2	S 685 740
3	S 669 741
4	S 706 735

Column D (altitude in metres)	
A	50
B	56
C	90
D	336

The grid	
1	B
2	
3	
4	

5. A red 'section line' labelled **X¹–X²** appears on the map in Figure 2. Study the positions of the contours that this line crosses.

   a. Which of the diagrams **A–D** in Figure 3 below best represents the **slope** shown by these contours?

   _____

   b. Name the **type of slope** shown in diagrams A–D. Write your answer in the space below each diagram.

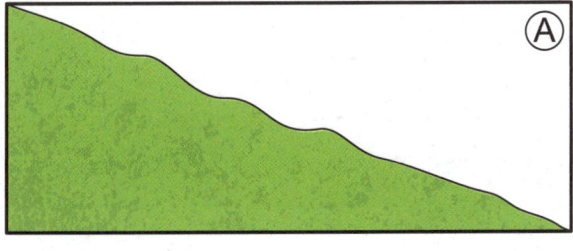

_____

_____

_____

_____

Figure 3

Waterford

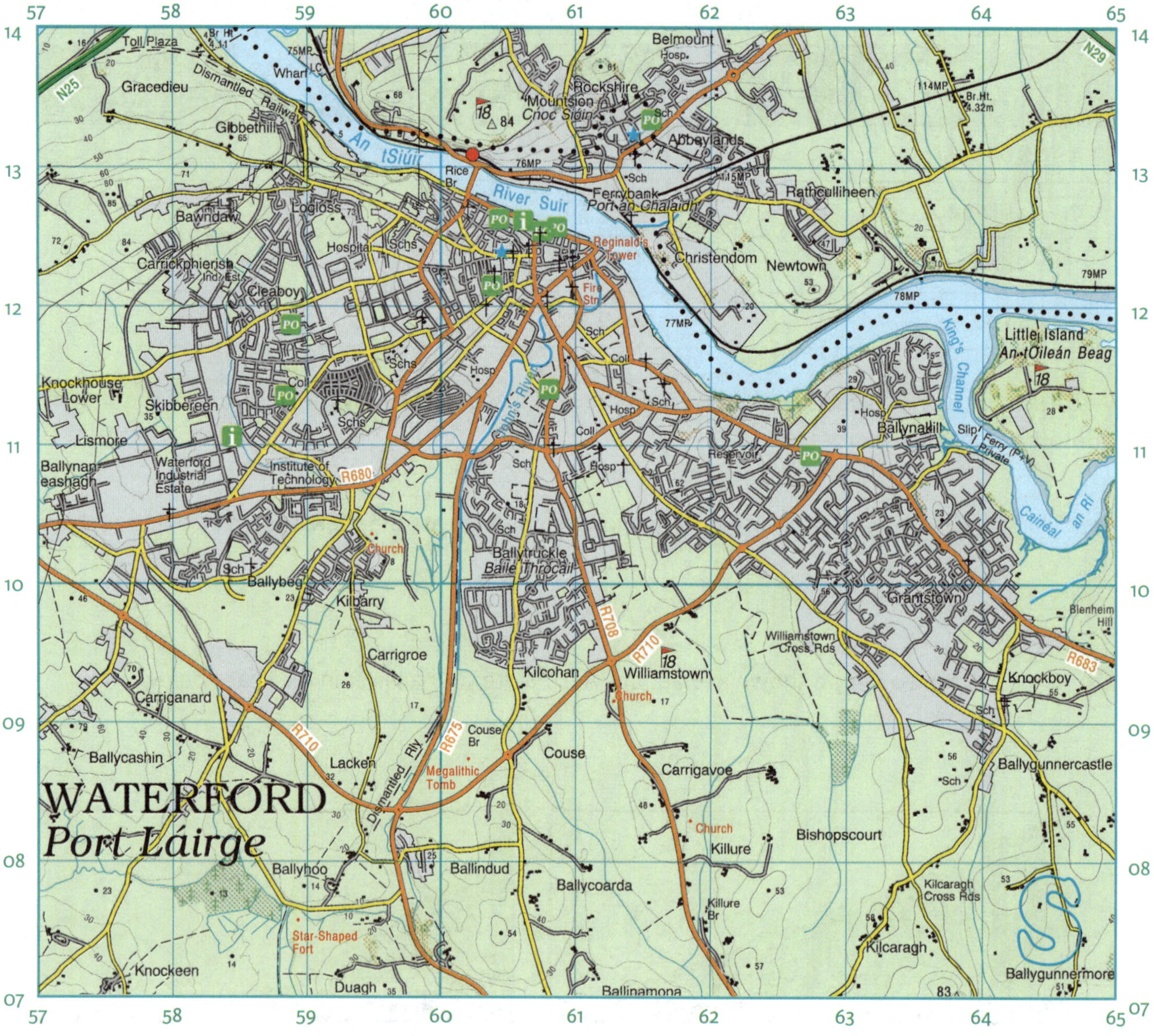

Figure 4

6. Examine the 1:50 000 OS map of Waterford in Figure 4.

   a. Complete the sketch map of Waterford in Figure 5 by *showing* and *labelling* each of the following:

   - A county boundary
   - The area of the urban (built-up) area north of the River Suir
   - An old fort south of northing 08
   - The most easterly golf course on the map.

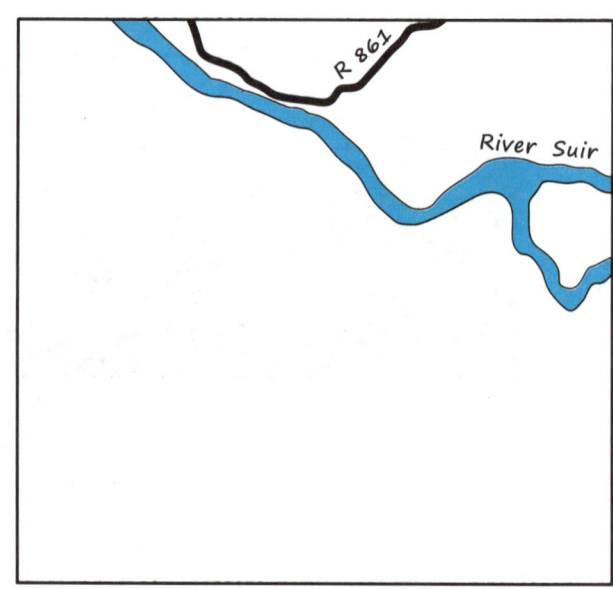

Figure 5: Sketch map of Waterford

b. Explain *three* factors that have led to the **development** over time of Waterford City. Support your answer with evidence from the map in Figure 4.

_____
_____
_____
_____
_____
_____
_____
_____
_____
_____
_____
_____

c. Describe *three* **past and/or present functions** of Waterford City. Refer to Figure 4 in your answer.

_____
_____
_____
_____
_____
_____
_____
_____
_____
_____
_____
_____

Junior Cycle Geography CYCLONE SKILLS BOOK

Clonakilty

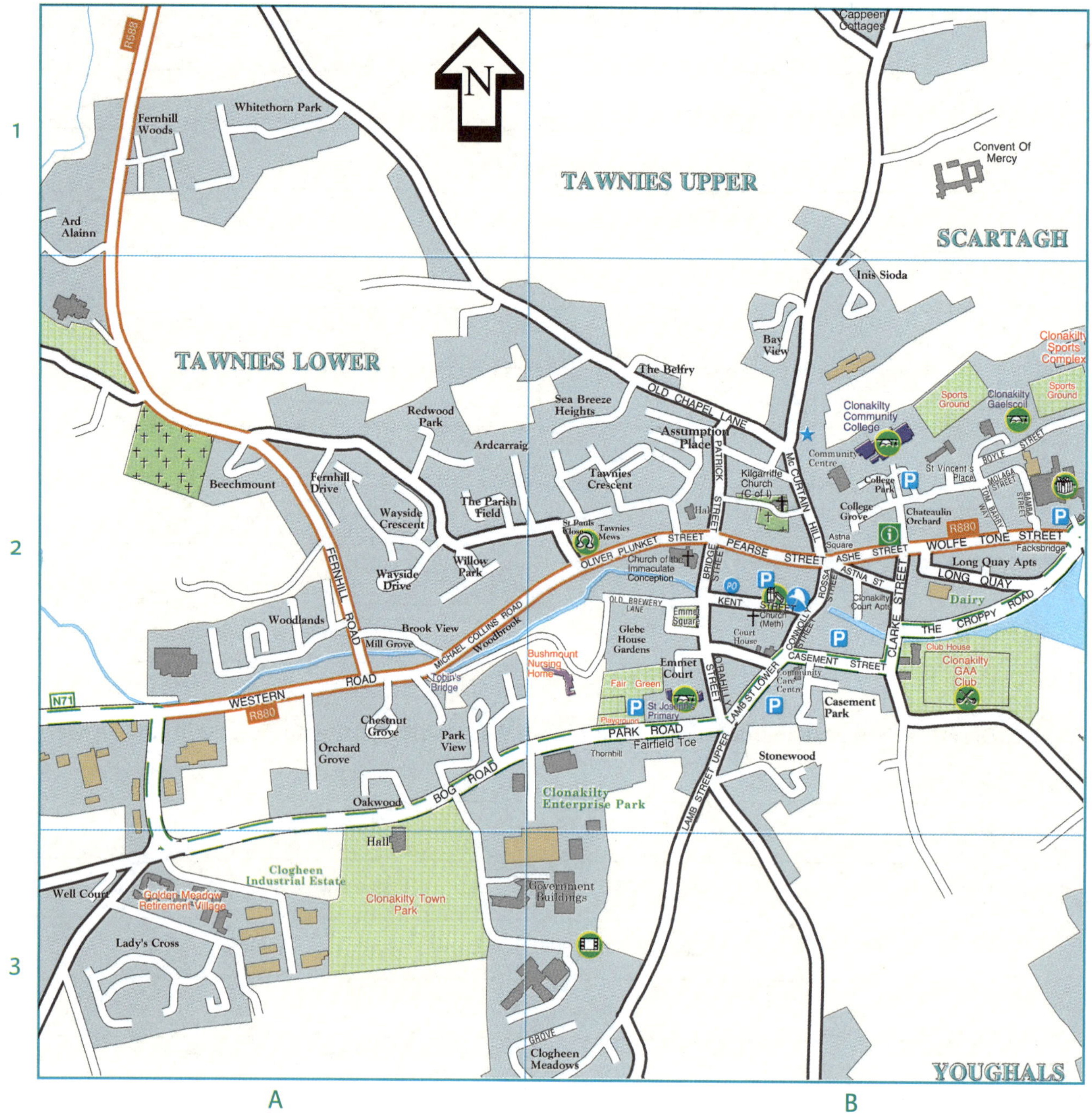

Figure 6

7. Examine the 1:10 000 map of Clonakilty in Figure 6.

   a. Calculate, in kilometres, the distance by road from the roundabout at the east of Wolfe Tone Street (B2) to the junction between Fernhill Road and Western Road (A2).

   _____

b. Complete the sketch map of Clonakilty by *showing* and *labelling* each of the following:

- The entire routeway that connects Cappeen Cottages in the north of the map with Clogheen Meadows in the south
- Golden Meadows Retirement Village
- A museum
- A cemetery
- A tourist office.

You might need to examine the symbols shown on the 1:10 000 maps on page 108 of your *Cyclone* textbook.

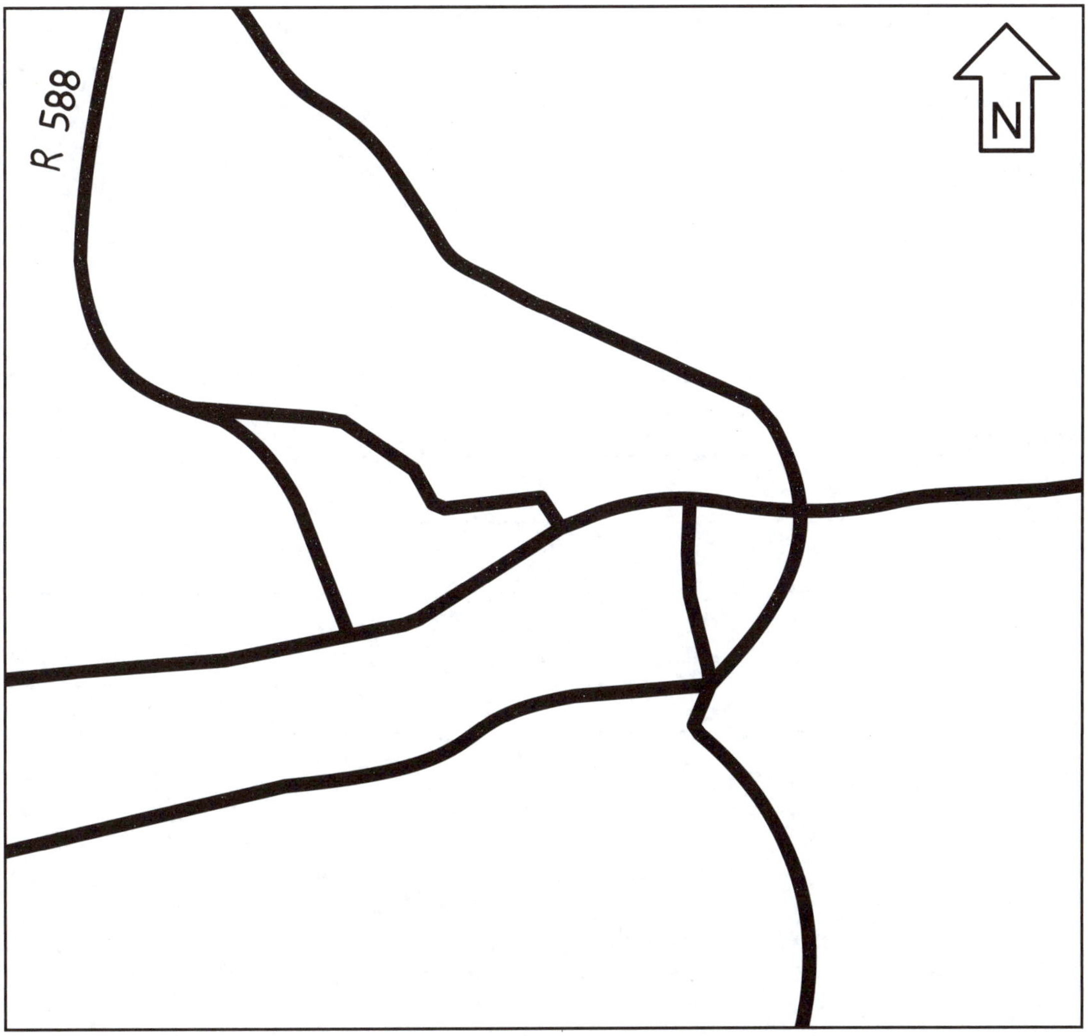

Figure 7: Sketch map of Clonakilty

c. The table below lists evidence of four **functions** in the town of Clonakilty. Use the map of Clonakilty on page 34 to complete the table by:

- naming the function most associated with each piece of evidence
- stating the grid square(s) where each piece of evidence can be found on the map.

One row has been completed for you.

Function	Evidence of function	Grid square on the map
Residential	Wayside Crescent	A2
	Clonakilty Community College	
	Convent of Mercy	
	Government Buildings	

d. 'Irish towns take steps to reduce traffic congestion.' Using evidence from the map, describe two ways in which Clonakilty strives to reduce or control traffic congestion.

Method 1: _____

_____

_____

_____

_____

Method 2: _____

_____

_____

_____

_____

7. Geographical Skills: Ordnance Survey Maps

Before you finish this chapter, revise your symbols!

8. The symbols printed below appear frequently in 1:50 000 Ordnance Survey maps. Write the meaning of each symbol alongside it.

The symbols shown on this page have been enlarged. A full list of symbols appears on page 85 of your Cyclone textbook.

a.  PO _____
b. Motorway _____
c. N11 _____
d. R574 _____
e. ✝ _____ 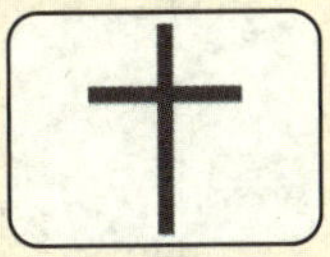
f. P _____
g. _____ 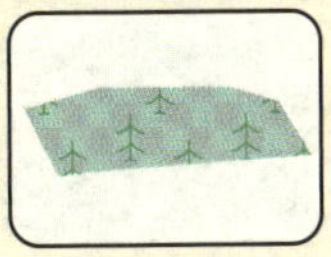
h. _____ 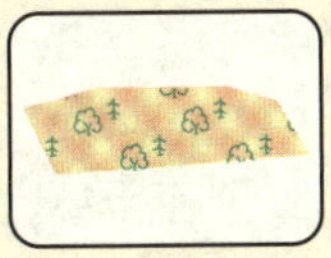
i. _____ 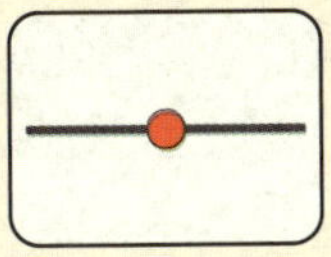
j. _____ 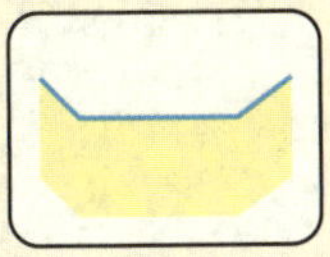
k. _____

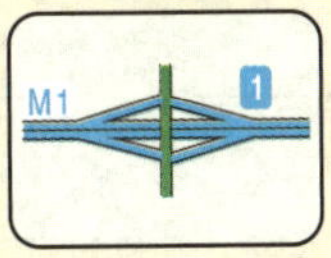

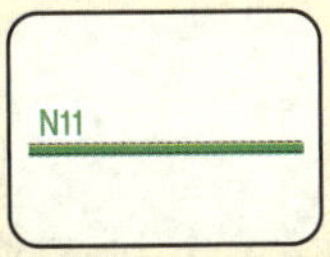

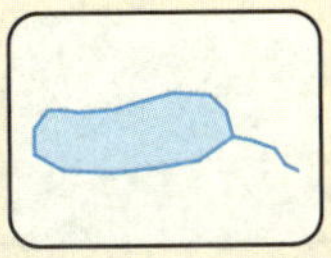

l. _____ 
m. _____ 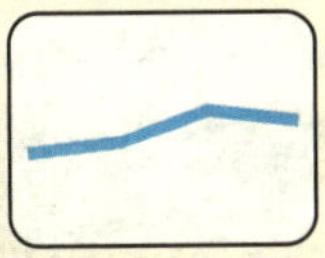
n. _____ 
o. _____ 
p. _____ 
q. _____ 
r. _____ 
s. _____ 
t. _____ 
u. _____

8 Weathering

1. Briefly define each of the following terms:

   Weathering: _____
   _____
   _____

   Erosion: _____
   _____
   _____

2.
   A  B  C

   In the grid provided, match each of the pictures labelled **A**, **B** and **C** with the correct **type of weathering** (**1**, **2** and **3**).

1	Freeze-thaw action
2	Biological weathering
3	Carbonation

A	
B	
C	

38

3. Add labels to the diagrams in Figure 1 to explain how freeze-thaw action takes place in upland areas in Ireland during winter. Include the phrases below in your labels, but **not** in the order given.

> water freezes and expands by 10%   rocks shatter   water collects in cracks in the rock
> scree collects at the base of slopes   this forces cracks to expand
> gravity causes shattered rocks to fall downslope

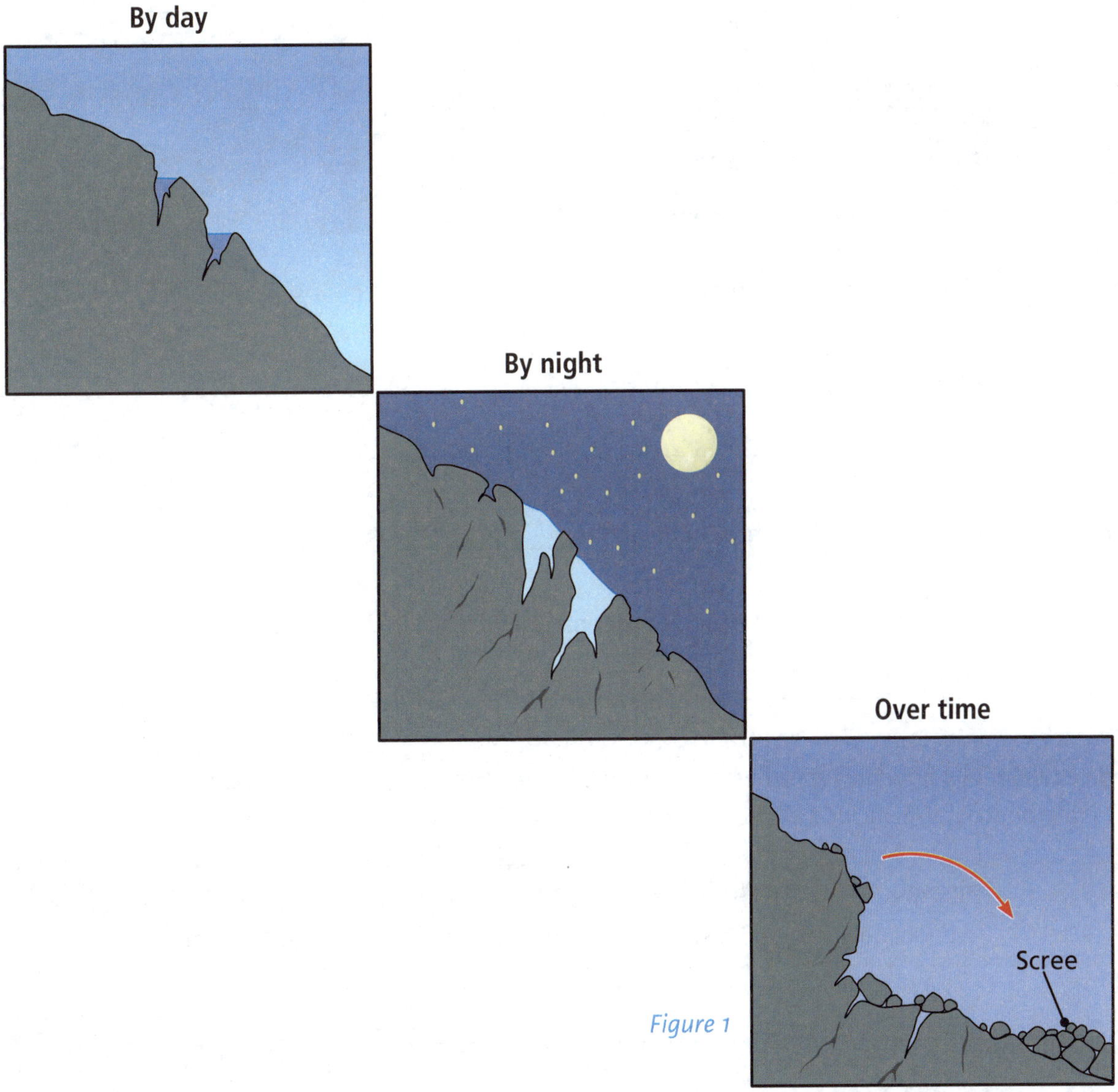

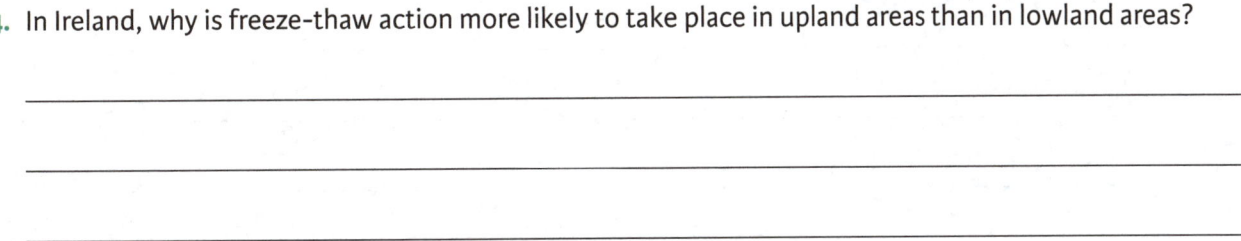

Figure 1

4. In Ireland, why is freeze-thaw action more likely to take place in upland areas than in lowland areas?

_____
_____
_____
_____

5. The picture above shows a karst area.

   a. What is a karst area? _____
   _____

   b. Name one well-known karst area in Ireland. _____

   c. Name the type of chemical weathering that has taken place in the photograph.
   _____

   d. Explain fully how the type of weathering you named in question (c) takes place in the area shown in the photograph. Include all the following words and phrases in your answer:

   > rainwater   carbon dioxide   carbonic acid   calcium carbonate
   > carbonation   permeable limestone   joints   bedding planes

   _____
   _____
   _____
   _____
   _____
   _____
   _____

8. Weathering

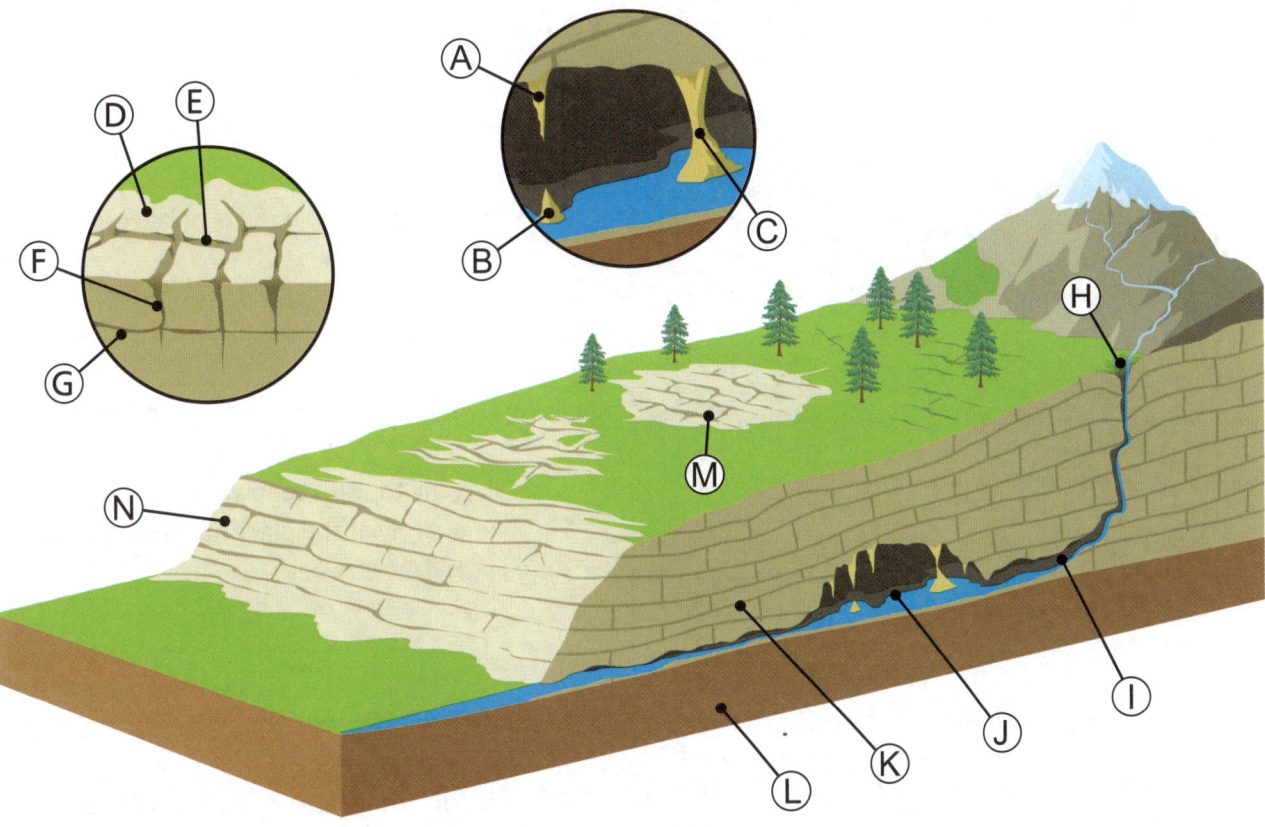

Figure 2: Features of a karst region

6. a. Study Figure 2 and the grid on the right. Pair each feature named in the grid with the matching label (**A–N**) in Figure 2. Two matches have been done for you.

   b. Explain how the feature labelled **C** in Figure 2 was formed.

   _____

   _____

   _____

   _____

   _____

Feature	Letter
cave	
clint	D
impermeable rock*	
pillar	
limestone pavement	
bedding plane	G
grike	
stalactite	
swallow hole	
stalagmite	
limestone (rock)	
passage	
limestone cliff	
joint	

*Impermeable rock is rock through which water cannot pass.

7. Examine the 1:50 000 Ordnance Survey map in Figure 3.

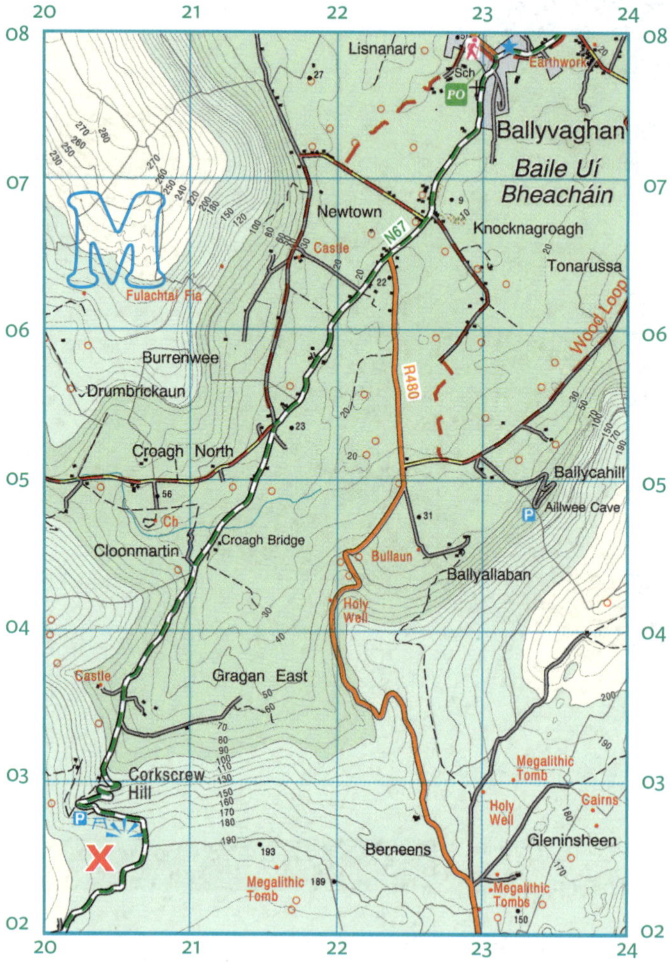

Figure 3

a. Suggest briefly:

   i. Why almost no rivers or lakes appear on this map

   _____

   _____

   ii. Why the N67 roadway winds very sharply at the place named Corkscrew Hill (M 204 029)

   _____

   _____

   iii. Why there are no houses at M 20 07.

   _____

   _____

   _____

b. Imagine that a local businessperson proposes to erect a large hotel in the area labelled **X** on the map. Explain **two** reasons why you think this proposal should **or** should not be supported. Refer to the map in your answer.

   Reason 1: _____

   _____

   _____

   _____

   Reason 2: _____

   _____

   _____

   _____

Mass Movement   9

1. **a.** Complete the word puzzle by naming **four** factors that influence mass movement.

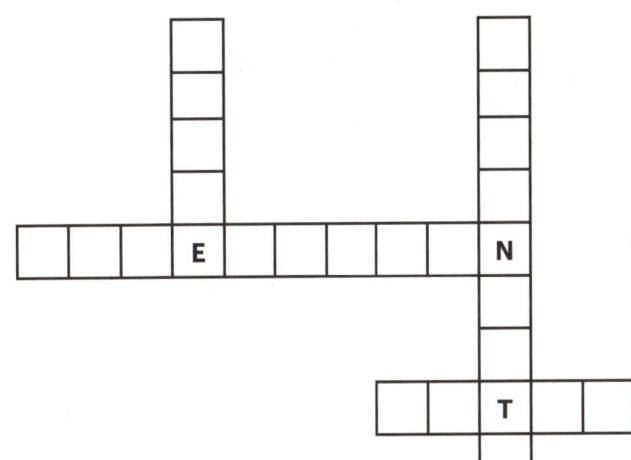

   b. Complete the grid below by naming each factor.

Factor	How factor influences mass movement
	Helps to cause mass movement by lubricating and adding weight to soil.
	Absorbs water and roots bind soil particles together. This helps to prevent mass movement.
	Steeper slopes cause faster mass movement.
	Helps to cause mass movement by removing vegetation or by cutting into hillsides.

2. Identify the **three** types of mass movement shown in the photographs below. In the circles, rank each type according to its speed: 1 (fastest), 2 or 3 (slowest).

_____   _____   _____

43

3. **Geography in the News: Case Study**

Look at page 131 of your *Cyclone* textbook, which refers to a mass movement disaster in Ecuador, South America. Then answer the following questions.

You could use this case study in the exam if you are asked about a mass movement event.

a. What type of mass movement occurred? _____

b. Where and when did it occur? _____

c. Explain briefly what **triggered** the event and **what happened**.

_____

_____

_____

_____

d. List **three consequences** of the event.

i. _____

ii. _____

iii. _____

e. List **three** relatively fast **human responses** to the event.

i. _____

ii. _____

iii. _____

f. How does **climate change** contribute to the risk of events such as this?

_____

_____

_____

9. Mass Movement

g. Suggest some long-term responses that governments and individuals can make to **reduce climate change**.

You could refer to Chapter 16, page 244, of your *Cyclone* textbook for some ideas.

_____

_____

_____

_____

_____

_____

_____

_____

_____

_____

4. Describe how this photo shows that human activities can lead to mass movement.

Cliff excavated to make room for road

Road built on steep slope

_____

_____

_____

_____

_____

_____

_____

_____

45

5. **a.** The unfinished diagram in Figure 1 shows some of the **effects of soil creep**.

   i. Fill in the *three blank labels*.

   ii. Complete the diagram by *drawing and labelling* three effects of soil creep other than those already shown.

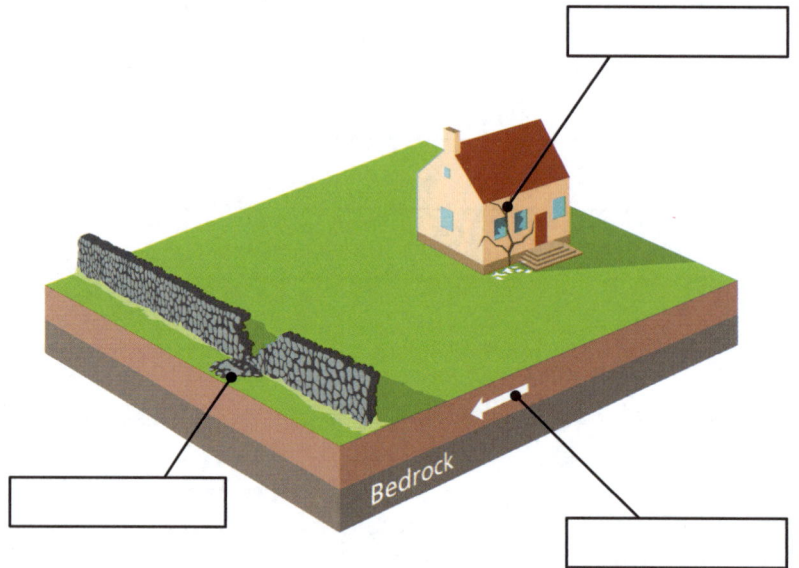

Figure 1

b. Examine the Ordnance Survey map extract in Figure 2.

Figure 2

   i. At which of the places marked **X**, **Y** and **Z** would soil creep be most likely to occur? Tick (✓) the correct answer.

   X ☐   Y ☐   Z ☐

   ii. Explain briefly why you chose the option you ticked.

   _____

   _____

   _____

   _____

   _____

   _____

   _____

Rivers

10

1. Complete the account on **river processes** below. Fill in the blank spaces using words from the yellow box and circle the correct options.

   Rivers **erode, transport and deposit** materials as they flow **from higher to lower/from lower to higher** ground.

   Rivers use four different processes to **erode**:
   - The power of moving water erodes material by a process known as _____ _____.
   - Eroded material (called the river's **load**) is hurled against the beds and banks of a river and wears them down by a process called _____.
   - The load is itself broken down by _____ as particles and pebbles rub and bounce off each other.
   - River water can also dissolve rock such as limestone. This process is called _____.

   Rivers **transport** their loads in different ways:
   - Heavier stones and large pebbles are rolled and dragged along the river bed in a process called _____.
   - Small pebbles are bounced along the river bed in a process called _____.
   - Extremely light and tiny particles can be carried along in _____ with in the river water.
   - Dissolved minerals are carried along in _____.

   Rivers **deposit** (drop) their loads where they **slow down/speed up**, which causes them to **lose/gain** energy. This happens, for example, when a river reaches **steep/flat** ground, or when it **enters/leaves** a lake or sea. Most deposition tends to take place in the **lower/upper** courses of rivers.

attrition	saltation	abrasion
solution	hydraulic action	
suspension	traction	solution

47

2. Complete this word puzzle on **river features** (landforms) by identifying the features labelled **1–12** in Figures 1, 2, 3 and 4.

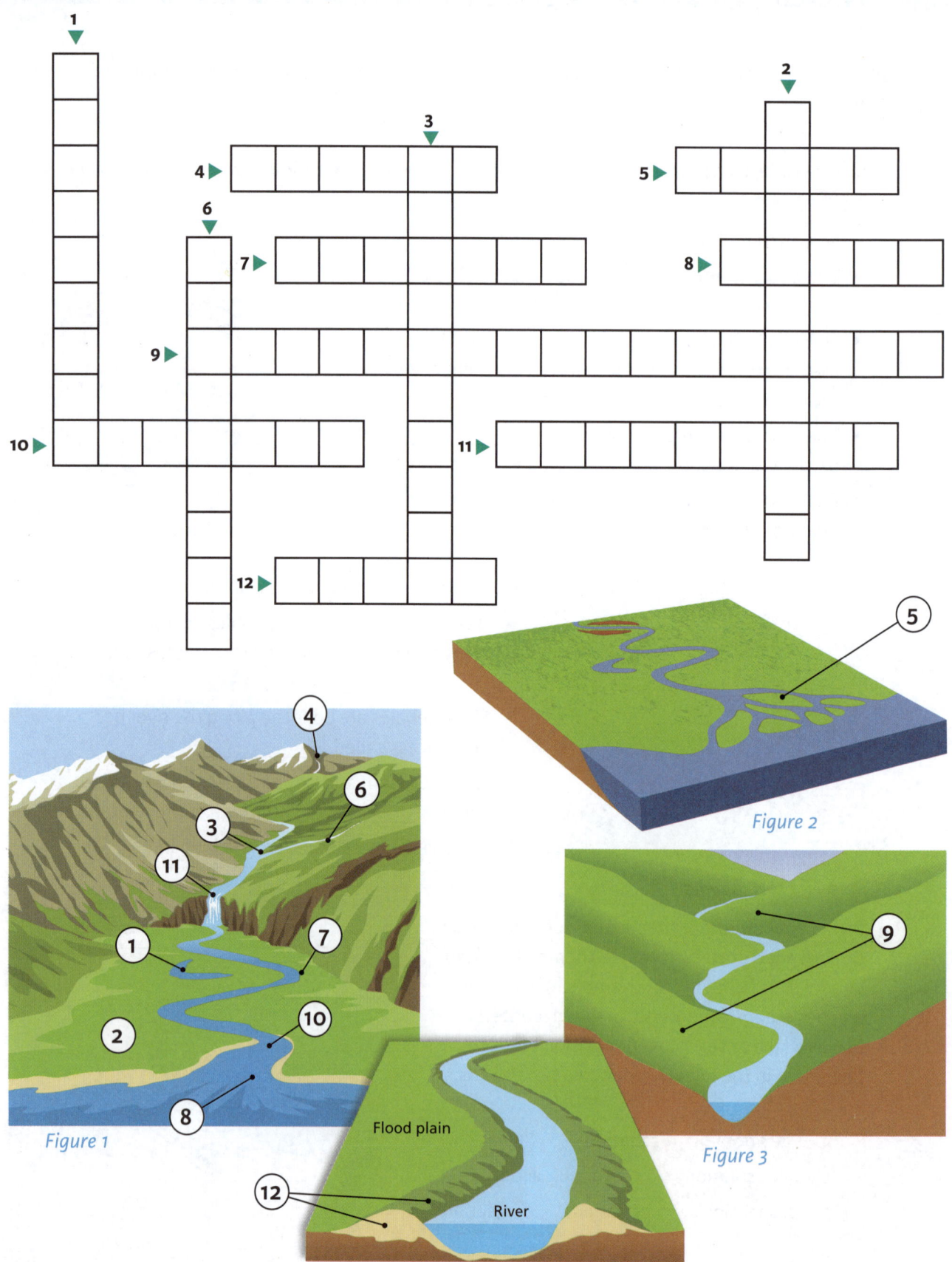

Figure 1

Figure 2

Figure 3

Figure 4

3. **a.** What type of river feature is shown in this photograph?

   _____

   b. Name one Irish example of this type of feature.

   _____

   c. Is this feature one of river **erosion** or **deposition**? Is it normally found on the **upper** or **lower** course of a river? (Tick (✓) the correct option in each case.)

   d. With the help of a labelled diagram, explain how this type of feature is formed. Draw your diagram in the box provided.

   Erosion ☐   Upper course ☐
   Deposition ☐   Lower course ☐

   _____
   _____
   _____
   _____
   _____

Junior Cycle Geography CYCLONE SKILLS BOOK

Rivers and human settlement: In questions on OS maps or aerial photographs, you may be asked to suggest ways in which a **river** may have influenced the **location or development of a town**.

Here are some handy hints for answering this type of question.

HANDY HINTS

Use evidence to support your answer. In the case of OS maps, name and locate rivers, roads, etc. You could use grid references to locate features.

✓ The river will almost certainly be crossed by a **bridge** in or very close to the town. **Roads converge** (come together) to cross the bridge. Where roads converge, people converge and trade may develop. Where trade develops, towns grow.

✓ A river might provide an **amenity value** that would help a town to develop. Signs of a riverside park or a walking route or of boating activity (in a photograph) would support this point. The presence of a boating activities symbol (⛵) on a 1:50 000 map would do likewise.

✓ **Long ago** (before there was piped water), the river might have been used to provide **water supplies** for drinking, cooking and washing. This might have influenced the location of a new settlement. (*Note:* this does **not** apply to a river estuary or mouth, where there is usually tidal salt water.)

✓ **Long ago**, castles were built close to rivers, which could provide moats as a means of **defence**. People sometimes settled close to these castles in the hope that the castle owners might provide employment, or protection in times of war.

✓ If the river is **wide** (double blue lines on a 1:50 000 OS map), it might have been used as a means of **water transport** in the past, when roads were few and of poor quality. If the river is **very wide**, it might even now be used for transport. The presence of docks, quays or wharfs could confirm this.

50

10. Rivers

4. a. Examine the aerial photograph, which shows the area around a town in Co. Cork. Answer the questions that follow.

i. Is the river shown in the photograph in the upper or lower course? Tick (✓) the correct box.

Upper course ☐   Lower course ☐

ii. Name **one** river feature that can be seen in the photograph. _____

b. Give **two** reasons why towns often develop near rivers.

Reason 1: _____

_____

_____

_____

Reason 2: _____

_____

_____

_____

11 The Sea

1. In the grid provided, match the letter naming a **process of sea erosion** in **Column X** with the number of its definition in **Column Y**. One match has been done for you.

	Column X		Column Y		Grid
1	Hydraulic action	A	Some rock dissolves in sea water	1	E
2	Abrasion	B	Stones carried by waves erode each other	2	
3	Attrition	C	Air becomes trapped by incoming waves	3	
4	Compressed air	D	Stones carried by the waves erode the coast	4	
5	Solution	E	The physical erosive power of the water	5	

2. In the grid below, name each of the **coastal features** numbered **A–N** in Figure 1 (the diagram and photographs on the opposite page). Indicate whether each feature is formed by erosion or deposition by writing the letter **E** or **D** in the column beside it.

	Name of feature	E or D
A		
B		
C		
D		
E		
F		
G		

	Name of feature	E or D
H		
I		
J		
K		
L		
M		
N		

11. The Sea

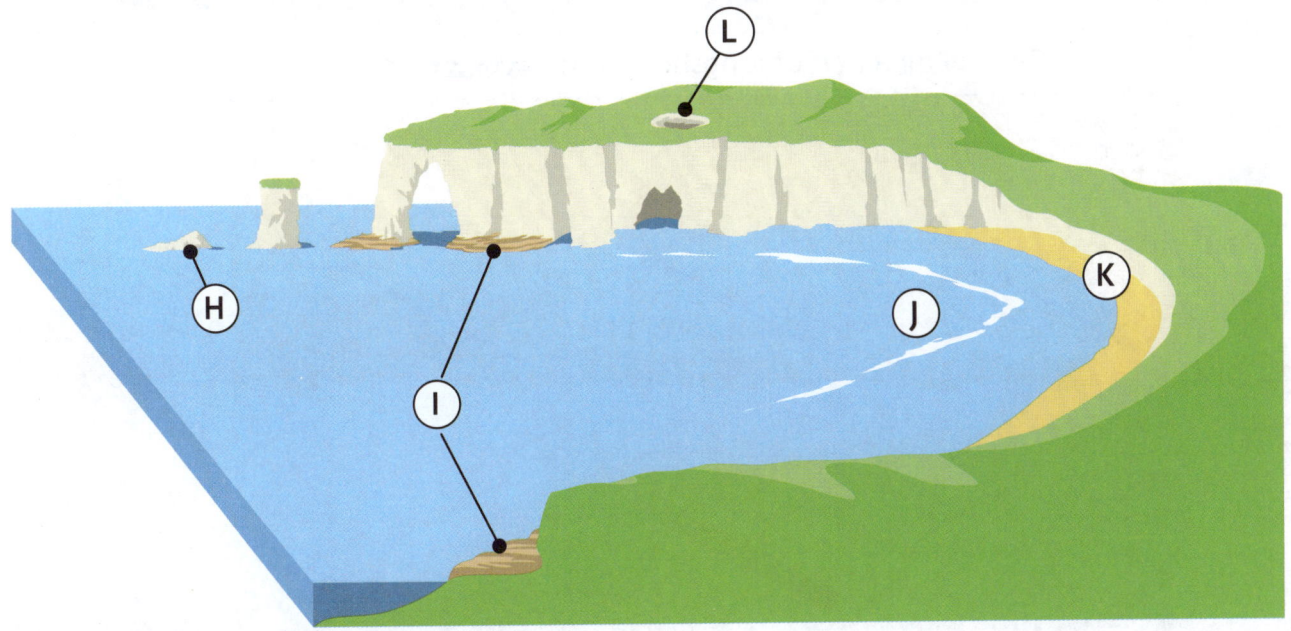

Figure 1: Coastal features

3. a. Explain how **longshore drift** transports material along a coast. Use the diagram in Figure 2 to help you. Label each of the following on the diagram: *wave direction, swash, backwash*.

___

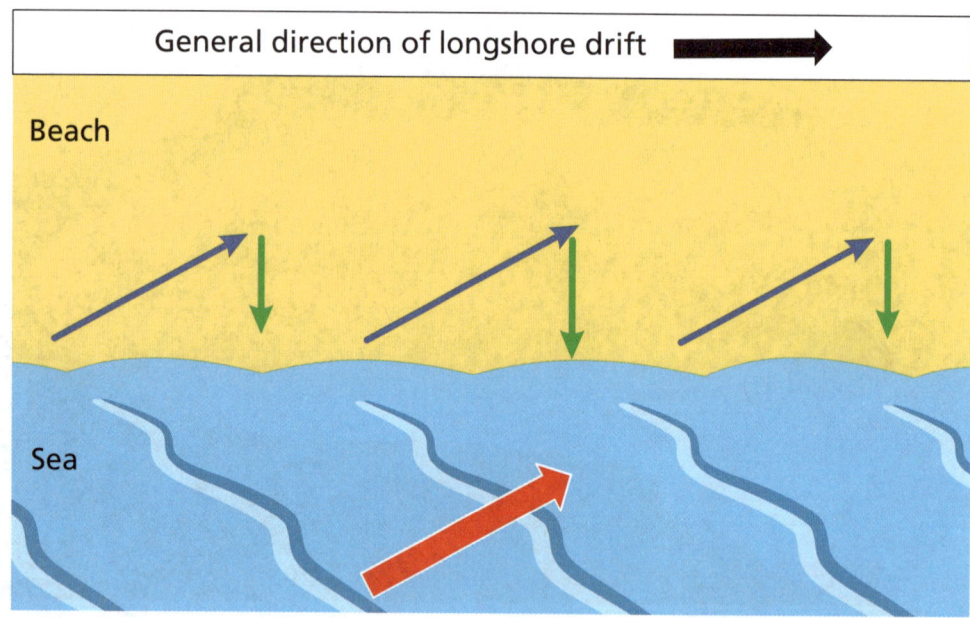

Figure 2

b. What is the name and purpose of the human-made features shown in the photograph?

Name: _____

Purpose: _____

11. The Sea

4. Answer the following questions on *either* a **sea stack** *or* a **beach**:

   a. State what the feature is. _____

   b. State the location of one Irish example of the feature. _____

   _____

   c. Describe in detail how the feature is formed. Draw a labelled diagram in the box below to support your answer.

   _____

   _____

   _____

   _____

   _____

   _____

   _____

   _____

Junior Cycle Geography CYCLONE SKILLS BOOK

5. Examine the photograph above.

   a. State in one sentence how you know this coastal area is being eroded.

   _____

   b. State in one sentence why this particular coastline might experience rapid erosion.

   _____

   c. Name and describe measures that could protect this area from coastal erosion. Explain why these measures might be successful.

   _____
   _____
   _____
   _____
   _____
   _____
   _____
   _____

6. Figures 3 and 4 show *An Daingean/Dingle*, a coastal town in Co. Kerry.

   a. Referring to the map in Figure 3, explain how the shape of the coast influenced the location of Dingle.

   _____

   _____

   _____

   _____

   _____

   b. What evidence can you find in Figures 3 and 4 that **tourism/recreation** is a function of Dingle?

   _____

   _____

   _____

   _____

   _____

   _____

   _____

   _____

   _____

   _____

   _____

   _____

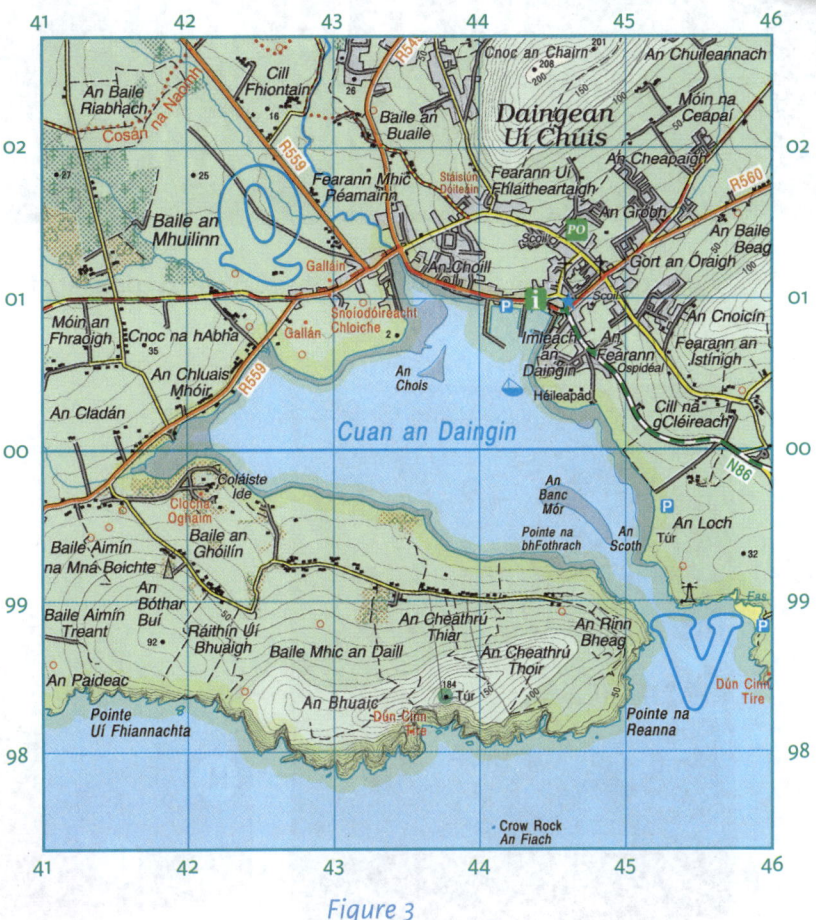

Figure 3

Figure 4

12 Glaciation: The Work of Moving Ice

1. Use the clues on this and the opposite page to complete the word puzzle.

Photograph A

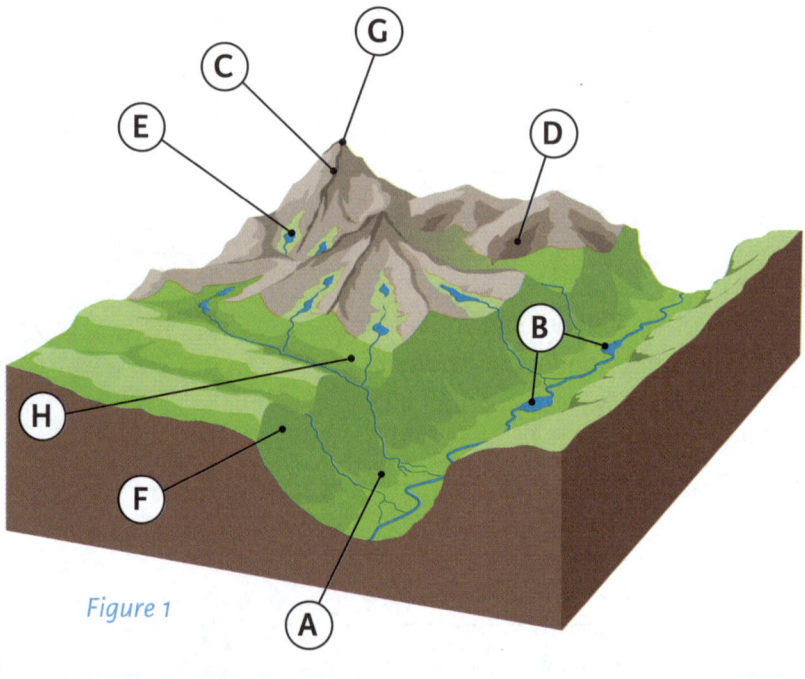

Figure 1

Photograph B

Photograph C

12. Glaciation: The Work of Moving Ice

Clues Across:

3. Low, fairly flat area of sand and gravel deposits.
4. Materials transported **along the surface of** a moving glacier.
5. The feature labelled **C** in Figure 1.
6. The process that drops material on glacial lowlands.
7. The feature labelled **D** in Figure 1 (also shown in Photograph C).
8. The feature shown in Photograph A.
9. The 'out of place' boulder shown in Photograph B.
10. Oval-shaped hill made of boulder clay.
11. The feature labelled **E** in Figure 1 (also shown in Photograph C).
12. Materials transported **within** a moving glacier.
13. The feature labelled **F** in Figure 1 – a truncated _ _ _ _.
14. The general process by which glacial material is moved from uplands to lowlands.
15. The feature labelled **G** in Figure 1.
16. Glacial erosion that scrapes and smooths.
17. The feature labelled **H** in Figure 1.
18. A long, narrow lake on the floor of a glaciated valley.
19. The moraine that marks the end point reached by a glacier.
20. Glacial erosion that pulls rock from the ground.
21. The general glacial process that is most common in highlands.

Clues Down:

1. The type of valley labelled **A** in Figure 1.
2. The linked lakes labelled **B** in Figure 1.

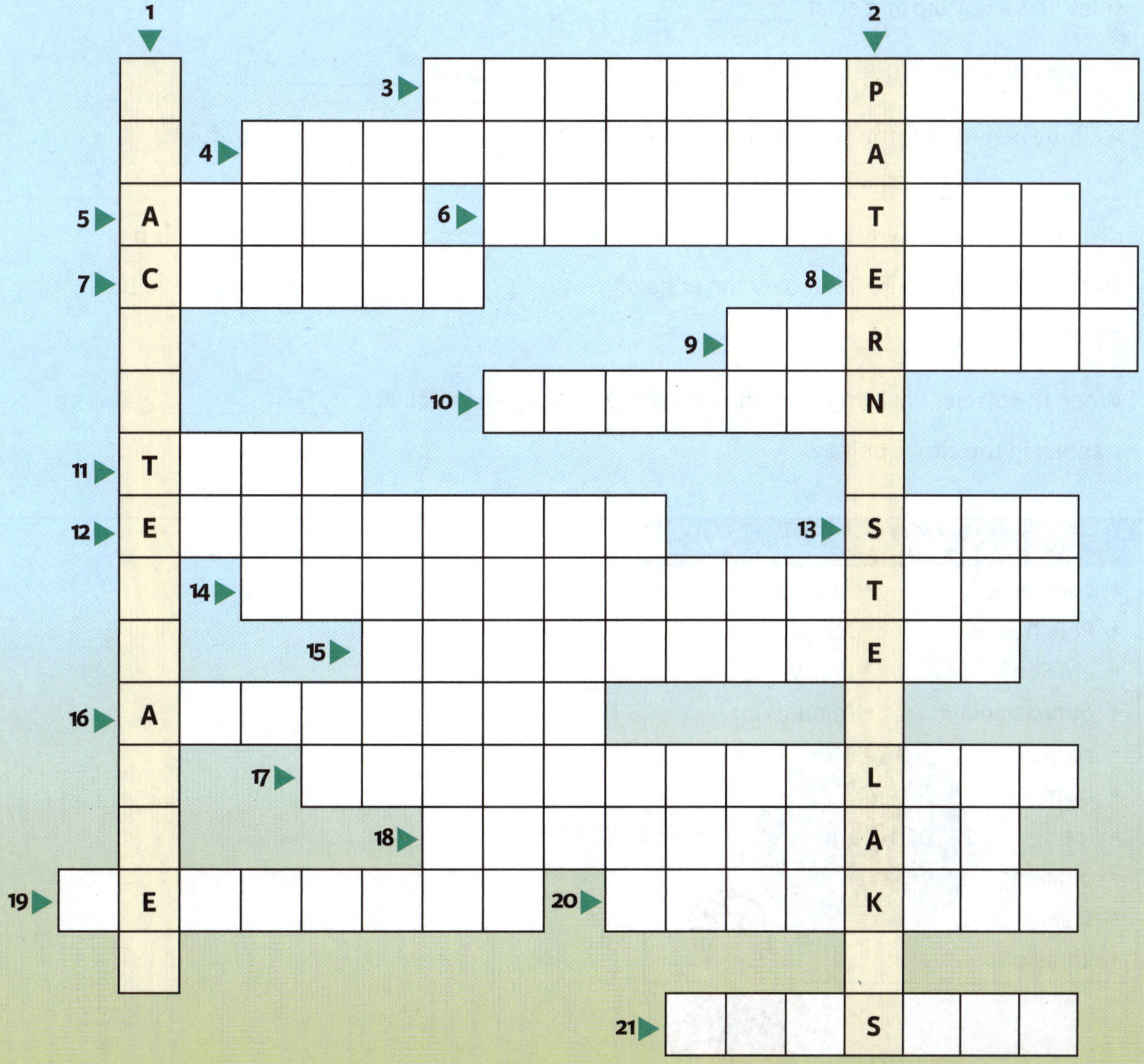

59

2. Describe the erosional processes of **plucking** and **abrasion**.

Plucking: _____
_____
_____
_____

3. The account below describes and explains the formation of a **cirque**. Select words from the Word Box to fill in the gaps. Then draw a labelled diagram of a cirque in the space provided below.

A cirque, also called a _____, is a deep mountain hollow with very _____ sides. It is a feature of glacial _____.

One example of a cirque is the Devil's _____ in Co. _____.

A cirque begins to form when snow in a mountain hollow is _____ to form _____. The erosional glacial processes of _____ and _____ deepen the hollow, which then becomes the 'birthplace' of a glacier. When the glacier is big enough, it begins to flow out over the edge of the _____ and starts to move slowly down the _____.

When the glacier finally melts at the end of the ice age, a lake called a _____ may be trapped in the cirque hollow.

Word Box
- compacted
- Punch Bowl
- erosion
- outwash plain
- corrie
- tarn
- ice
- plucking
- steep
- abrasion
- Wexford
- cirque
- Kerry
- mountain

Ignore the two terms that do not belong here.

(for diagram)

12. Glaciation: The Work of Moving Ice

Abrasion: _____

_____

_____

_____

4. The account below describes and explains the formation of **drumlins**. Choose words from the Word Box to fill in the gaps. Draw a labelled diagram of a drumlin in the space provided below.

Drumlins are _____-_____ hills that often occur in large numbers called _____. They are made up of boulder _____ (a mixture of sand and clay) and are features of glacial _____.

A swarm of drumlins can be found in _____ Bay, Co. _____.

Drumlins are formed in three phases. First, an ice sheet advances and lays down boulder clay in _____ heaps. Then the ice retreats. Finally, the ice _____ again and _____ the heaps of boulder clay into rounded, oval-shaped hills.

The steep slope of a drumlin points to the direction from which the ice _____. The _____ slope points to the direction in which the ice was travelling.

Word Box
- deposition
- smooths
- Clew
- gentle
- plucking
- advances
- oval-shaped
- eskers
- came
- swarms
- Mayo
- irregular
- clay

Ignore the two terms that do not belong here.

(for diagram)

5. Read the extract below and answer the questions that follow.

People and melting ice

The little things in life really matter. The long and 'just-for-the-craic' car drives. The quarter-filled washing machine churning merrily. The kitchen lights left on all night. Little things like these help to make us part of a quiet but deadly global-warming process that hastens the end of the last active remnants of the ice age.

The great ice sheets of Antarctica and Greenland are shedding gigantic shelves of melting ice into the oceans. Sea-ice in the Arctic Ocean is diminishing. Those glaciers that still grace many of the world's highest mountains are melting back more and more quickly.

Melting ice will have a massive impact on Planet Earth as we know it. Sea levels are rising and will continue to do so ever more rapidly. What will this mean for low-lying countries? Countries such as Bangladesh, with a population of 165 million people, many of whom live at current sea level. What will rising sea levels mean for low-lying urban areas closer to home? Parts of Dublin, Cork and Limerick, for example.

As glacial ice melts, rising sea temperatures will bring us more intense and frequent coastal storms. These, in turn, will increase the sea erosion of our coasts. Melting ice might even halt the flow of the North Atlantic Drift, which helps to keep Ireland's harbours ice free. The results of that could be considerable. Imagine Irish harbours being ice-bound in winter!

a. Summarise in one sentence the message of the opening paragraph of the passage above.

_____

_____

b. Name three consequences of melting ice that are mentioned in the extract.

1. _____

2. _____

3. _____

Think and discuss

a. What would happen if large, heavily populated areas of the world were to become flooded by rising sea levels? What might the consequences be in both **flooded areas** and **areas that were not flooded**?

b. 'The little things in life really matter.' What can we do personally to help reduce global warming?

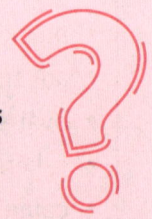

Measuring and Forecasting Weather

13

1. Examine the **weather instruments** labelled **A–G** in Figure 1. Then fill in the grid below. Some answers have been filled in for you.

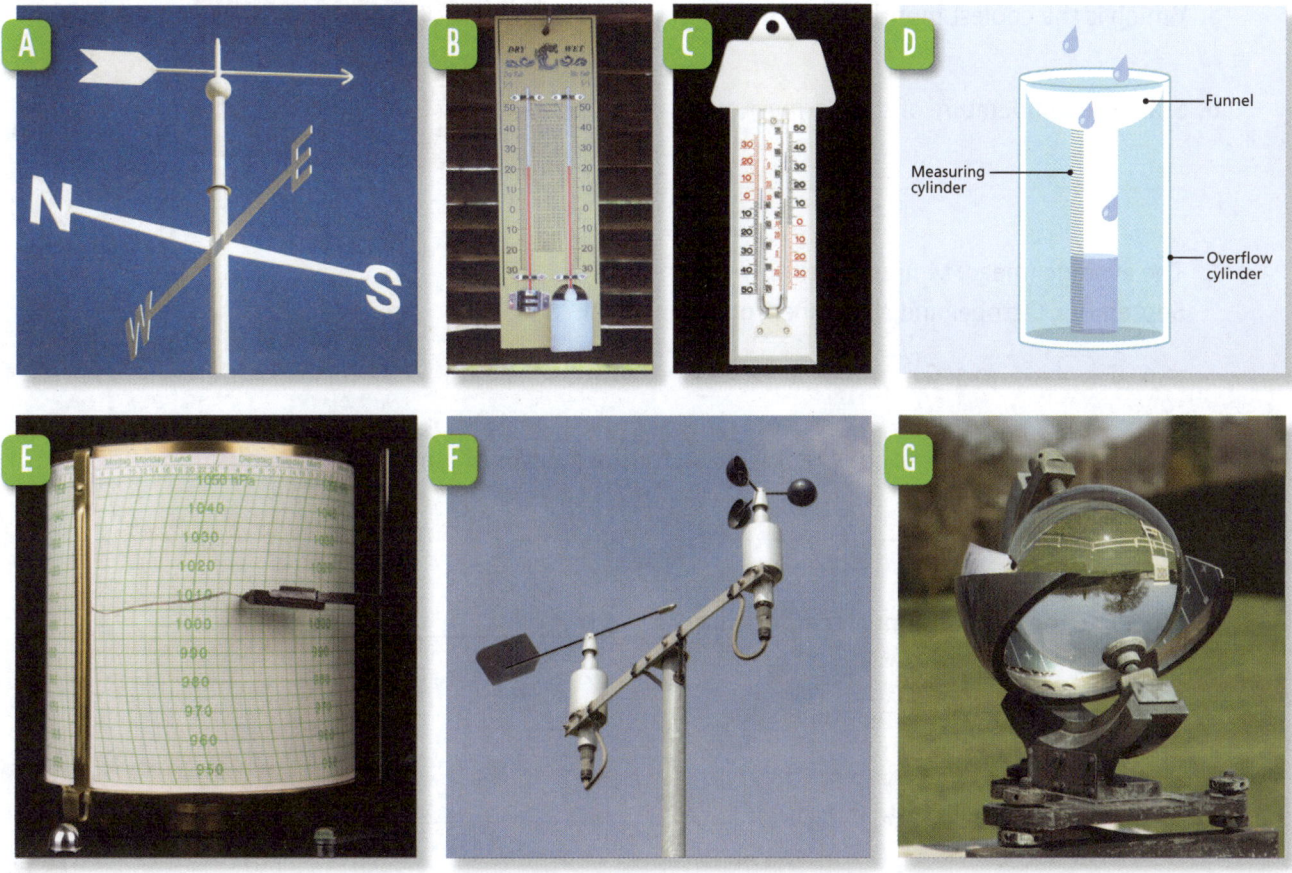

Figure 1

	Name of instrument	What it measures	Unit of measurement used (where relevant)
A			
B			
C	Maximum and minimum thermometer		
D			
E		Atmospheric pressure	
F			
G			Hours per day

63

2. Examine the table below, which shows **average temperature and precipitation*** at Salthill, Co. Galway. Temperatures are given in degrees Celsius (°C) and precipitation is given in millimetres (mm).

	Salthill, Co. Galway											
Month	Jan	Feb	Mar	Apr	May	Jun	Jul	Aug	Sept	Oct	Nov	Dec
Temperature (°C)	5	6	7	12	11	14	15	16	13	9	9	6
Precipitation (mm)	50	111	51	17	39	30	23	20	40	76	98	102

a. Which is the coolest month? _____

* Precipitation includes rain, snow, sleet and hail – any form of water (liquid or solid) that falls from the sky.

b. State the temperature of the warmest month.

_____

c. The average (mean) temperature for the months of September, October and November together is (✓):

12.6 °C ☐   10.3 °C ☐

d. Calculate the average monthly temperature for the months of June, July and August together.

_____

e. Calculate the average annual temperature. _____

f. Calculate the annual temperature range.* _____

* To calculate the annual temperature range, subtract the temperature of the coldest month from the temperature of the hottest month.

g. Name the wettest month. _____

h. Name the month with the least precipitation. _____

i. State the precipitation for the wettest month. _____

j. Calculate the total annual precipitation.

_____

k. Calculate the mean precipitation for the months of October, November and December together.

_____

3. Figure 2 shows a **climate graph**. It records **average temperature and rainfall** figures over a 12-month period.

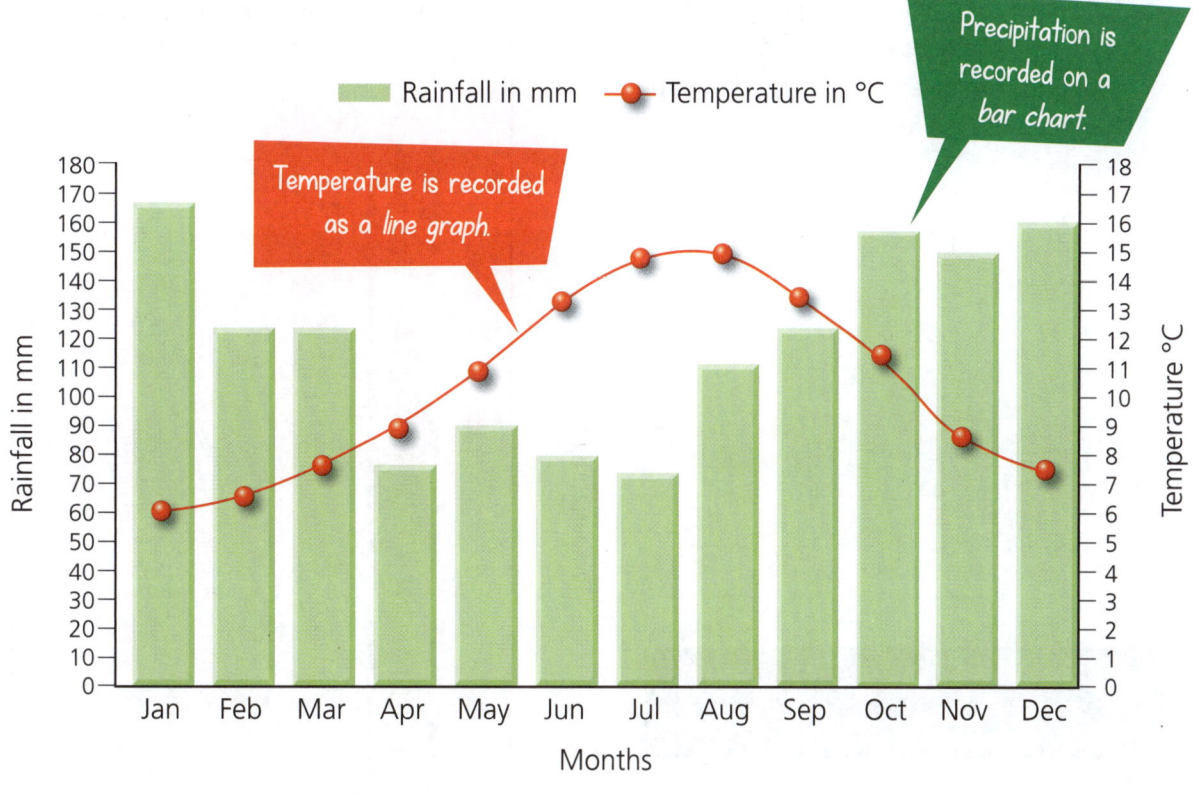

Figure 2

a. According to the weather chart, which month is the coolest? _____

b. Which is the warmest month, and what is its average temperature? _____

c. Calculate the annual temperature range. _____

d. Calculate the average temperature for April and May together.

_____

e. Which month gets most precipitation? _____

f. Which month is the driest? _____

g. State the precipitation in millimetres for the month of November. _____

Junior Cycle Geography CYCLONE SKILLS BOOK

4. The grid below shows **temperature, precipitation and wind direction** at Dublin Airport over a seven-day period.

   a. On Figure 3, complete a climate graph recording the **temperature and precipitation** given in the grid. The graph has been started for you.

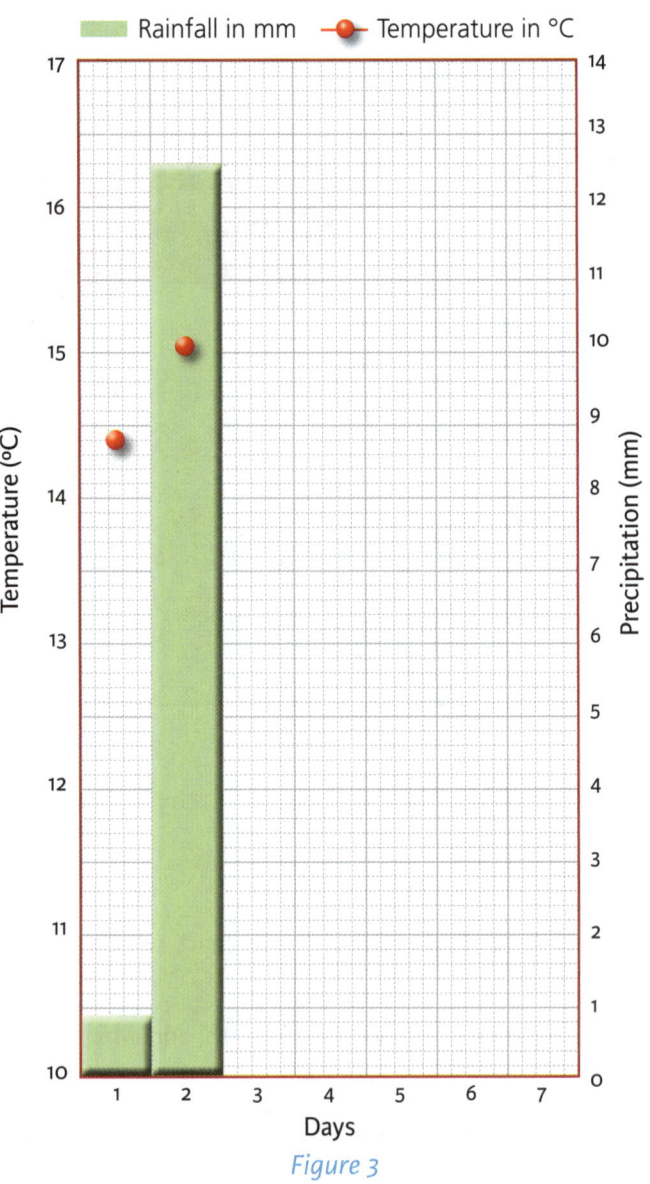

Figure 3

Day	Average temperature (°C)	Precipitation (mm)	Wind direction
1	14.4	0.9	W
2	15.0	12.6	S
3	16.8	11.3	S
4	14.5	10.4	SW
5	14.7	0.0	SW
6	13.5	1.6	SW
7	15.1	0.4	W

   b. Examine the **wind direction** data shown in the grid.

   i. From which direction was the wind most commonly blowing?

   _____

   ii. Draw an arrow in the box to show that wind direction.

   iii. What was the wind direction on Day 7? _____

66

5. The **rainfall map** in Figure 4 shows that areas along the west coast of Ireland receive large amounts of rainfall. Explain fully, with the aid of a labelled diagram, why this is so.

_____

_____

_____

_____

_____

_____

_____

_____

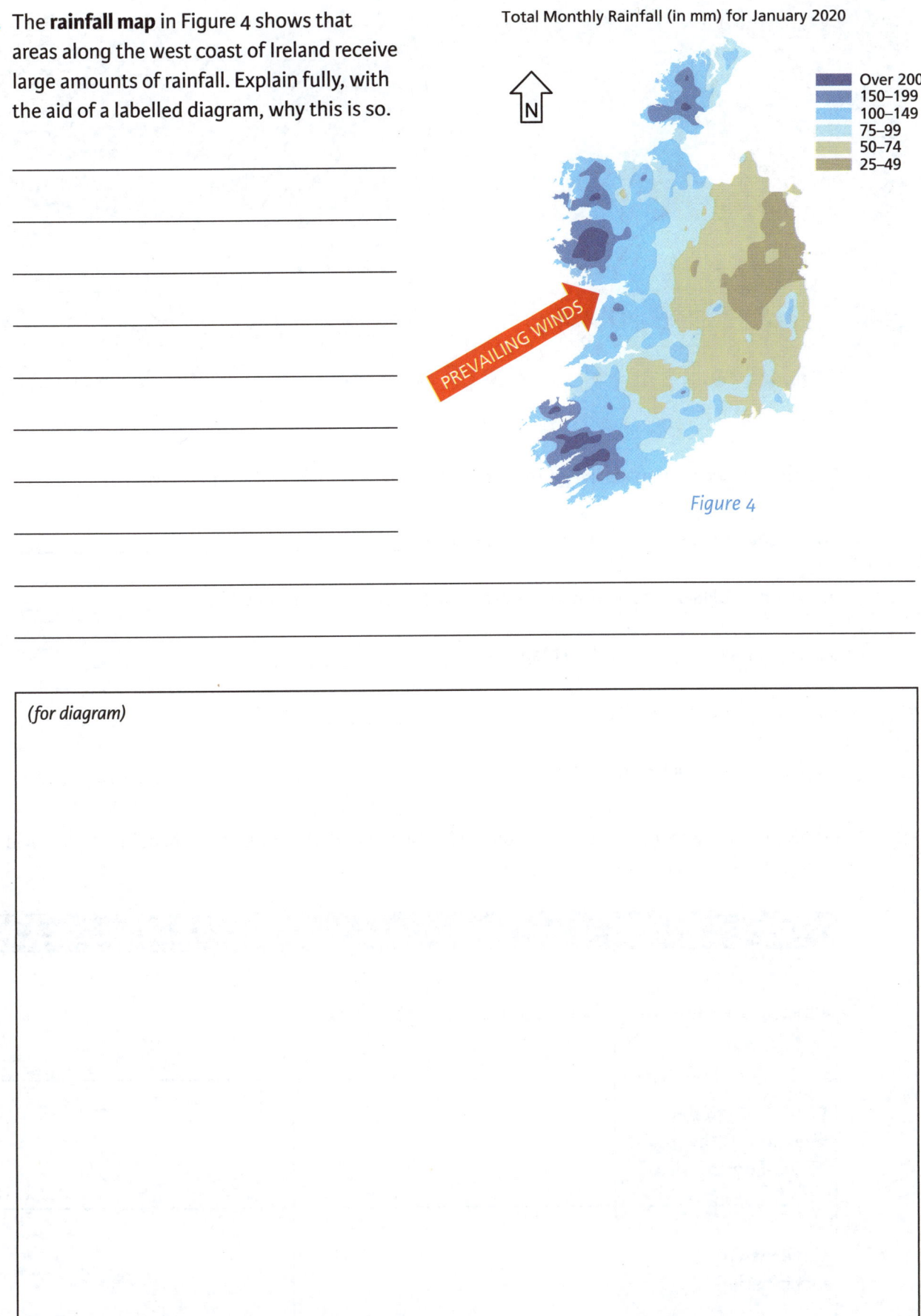

Figure 4

_____

_____

(for diagram)

Junior Cycle Geography CYCLONE SKILLS BOOK

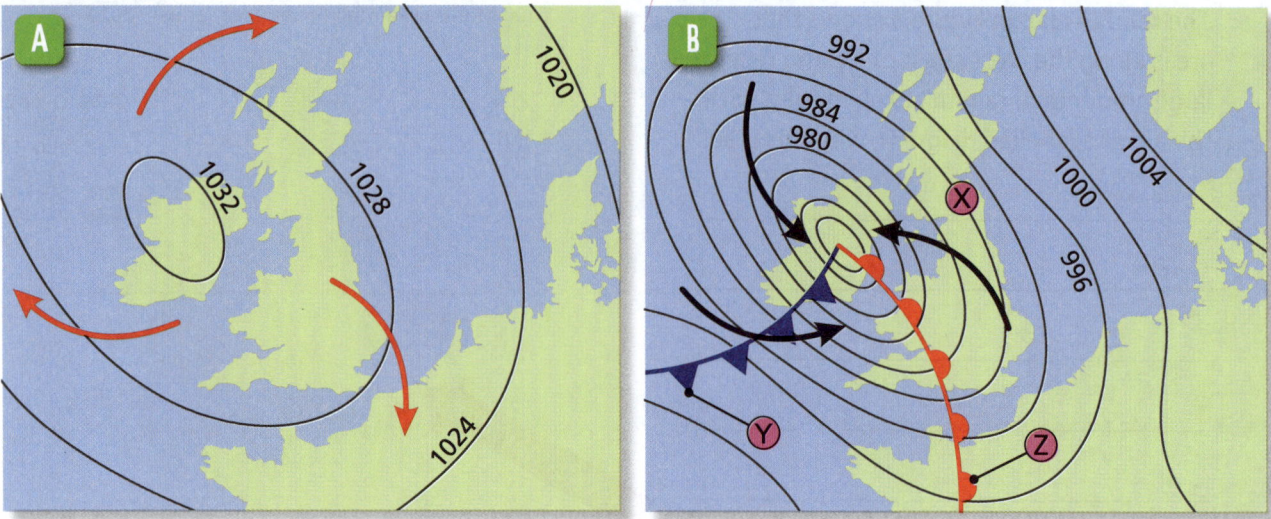

Figure 5

6. **a.** Examine the weather maps in Figure 5 above and complete the sentences below.

   i. On these maps, the numbered lines are called _____.

   ii. The numbers on the lines indicate atmospheric pressure measured in _____.

   iii. The missing number at **X** on Map B is _____.

   iv. The feature labelled **Y** is a _____.

   v. The feature labelled **Z** is a _____.

   b. Briefly contrast weather conditions indicated in Map A and Map B using the headings in the grid below. One entry has been made for you.

	Map A	Map B
Atmospheric pressure	High pressure (up to 1032 millibars)	
Wind strength and direction off the south coast of Ireland		
Likelihood of precipitation		

Severe Weather

14

1. What is meant by a **'significant weather event'**?

   _____
   _____
   _____
   _____
   _____
   _____

2. Study the map below, which shows the path of Cyclone Kenneth in 2019. Answer the questions that follow.

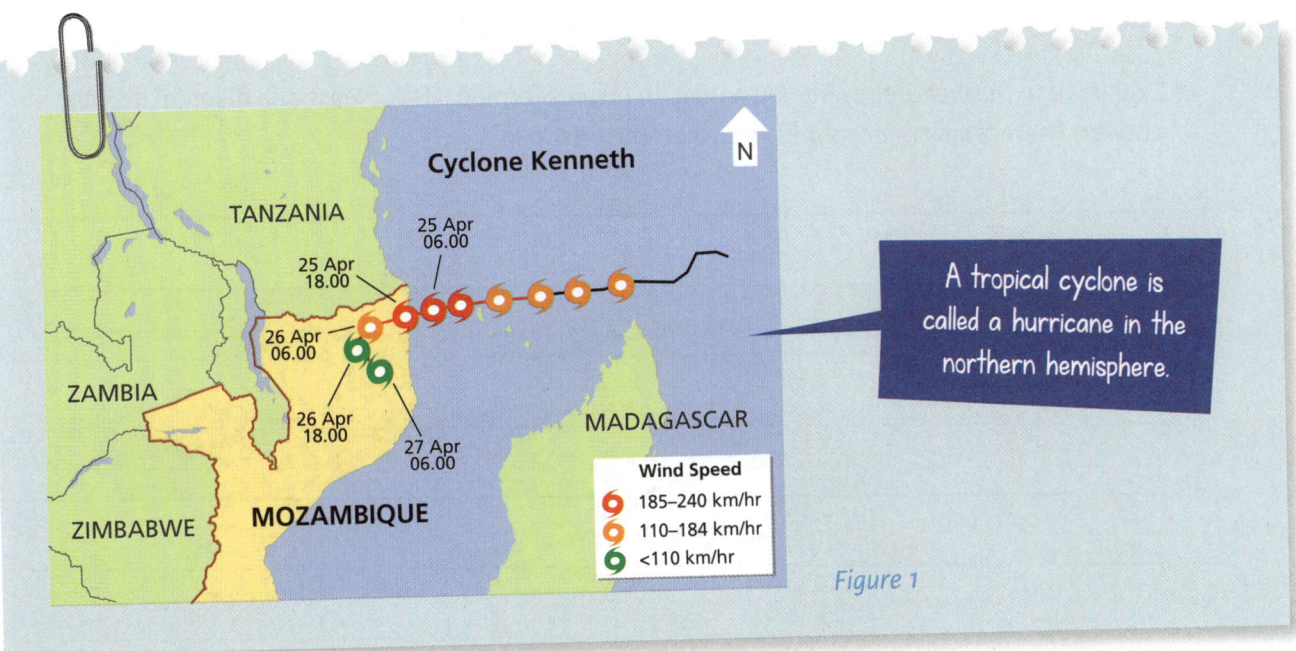

Figure 1

   a. What was the wind speed of Cyclone Kenneth at 06.00 on 26 April?

   _____

   b. Did the wind speed of the hurricane increase or decrease after it reached land?

   _____

3. **a.** Name **one** example of a **significant weather event** (not Cyclone Kenneth) that you have studied.

_____

_____

> Hmmm! A topic such as this might be suitably adapted to use as a 'Geography in the News' Classroom-Based Assessment.

b. When did this weather event happen?

_____

c. Name a location that was affected by this weather event.

_____

d. Name a weather instrument that could have been used during this weather event and say what it measures.

Weather instrument: _____

What it measures: _____

e. Explain how the weather event you named in (a) was formed. Use at least one diagram in your answer. There is space for your diagrams on the next page.

_____
_____
_____
_____
_____
_____
_____
_____
_____
_____
_____
_____
_____

f. Describe **three** significant consequences (results) of the weather event.

g. Outline **two short-term responses** and **two long-term responses** that governments or aid agencies might make to help people affected by the weather event.

Short-term responses

1. _____
_____
_____
_____
_____

2. _____
_____
_____
_____
_____
_____

Long-term responses

1. _____
_____
_____
_____
_____
_____

2. _____
_____
_____
_____
_____
_____

h. Describe **one** practical measure that a government might take to help prepare its citizens to cope with similar weather events in the future.

_____
_____
_____
_____
_____

Global Climates 15

1. The columns below relate to **factors affecting climate**. In the grid provided, match the *number* of each factor in Column A with the correct *letter* of its description in Column B. One match has been done for you.

	Column A		Column B		Grid
1	Prevailing winds	A	Inland areas may have colder winters and warmer summers than coastal areas.	1	C
2	Altitude	B	Temperatures generally decrease as distance from the equator increases.	2	
3	Latitude	C	Southwesterly air masses (winds) bring temperate conditions and rain to Ireland.	3	
4	Distance from the sea	D	Temperatures decrease as height above sea level increases.	4	
5	Aspect	E	In the northern hemisphere, slopes facing south tend to be warmer and sunnier than slopes facing north.	5	

2. Examine the diagram in Figure 1. Then *circle* the correct **options** in the passage.

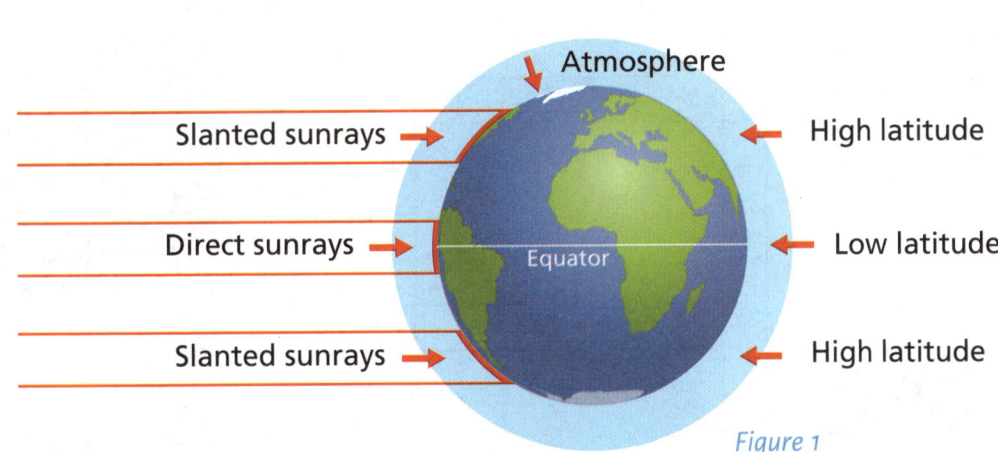

Figure 1

Sunrays shine almost directly over **low/high** latitudes. Direct rays are concentrated on smaller areas of the earth's surface and so give **little/great** heat. Places in high latitudes receive **slanted/direct** sunrays. These are spread out over wide areas and give **more/less** heat.

The earth's atmosphere absorbs and so **increases/reduces** the sun's heat. Slanted sunrays have to travel through **less/more** atmosphere than direct sunrays. That is another reason why places in high latitudes are **colder/hotter** than places close to the equator.

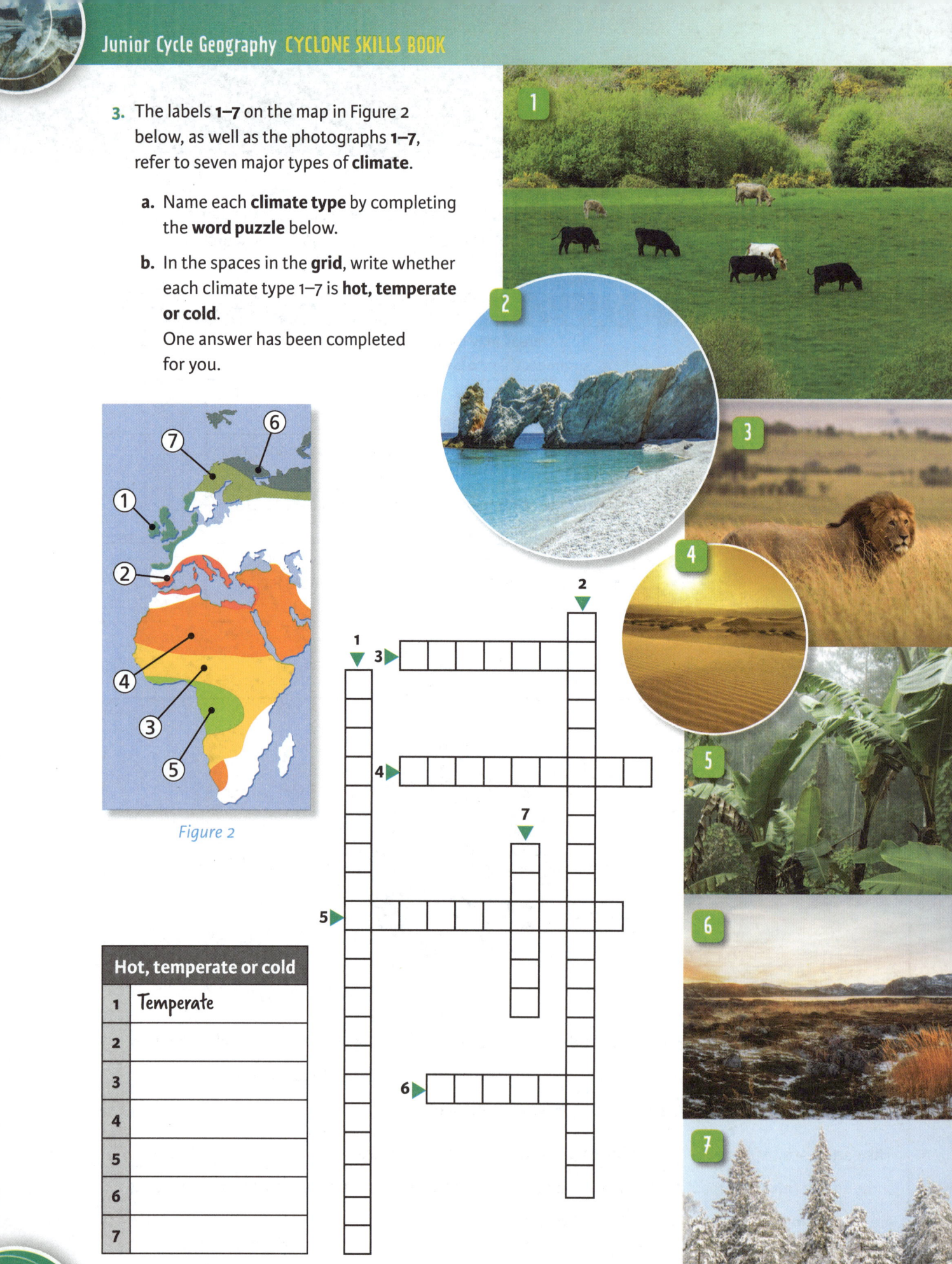

4. The map in Figure 3 shows three air masses (winds) that affect Ireland's climate. For each air mass, refer briefly to (a) where it comes from and (b) the type of weather it usually brings.

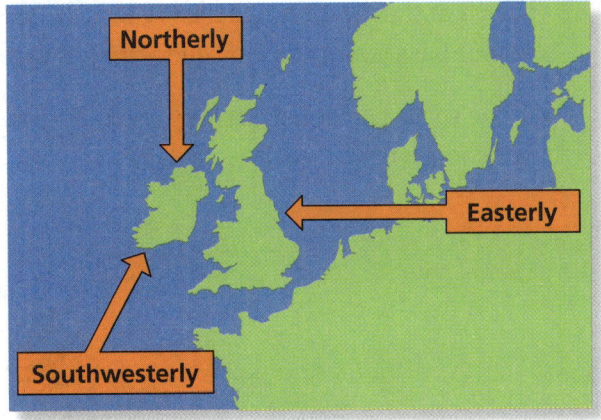

Figure 3

Southwesterly
(a)
(b)

Northerly
(a)
(b)

Easterly
(a)
(b)

5. The map in Figure 4 shows mean January (coolest month) and mean July (warmest month) temperatures in Ireland.

   a. Calculate the **annual temperature range** at the area labelled **A** and at the area labelled **B**.

   A: _____

   B: _____

   b. Explain briefly why the annual temperature range is greater at **B** than it is at **A**.

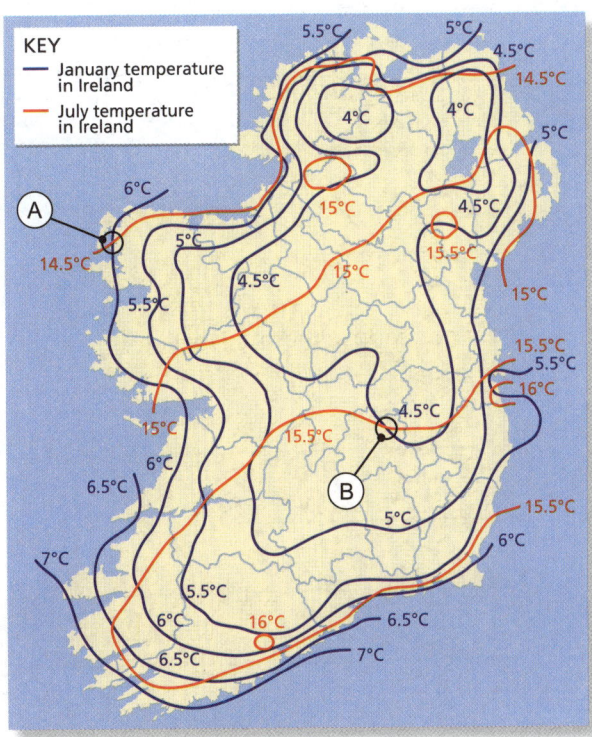

Figure 4

6. Which of the locations labelled **A**, **B** and **C** in Figure 5 would you consider the best site for a family home? Base your choice on geographical factors such as aspect, slope, drainage and accessibility.

In the spaces provided, identify and give reasons for your choice of location. Explain also why you did not choose the other sites available.

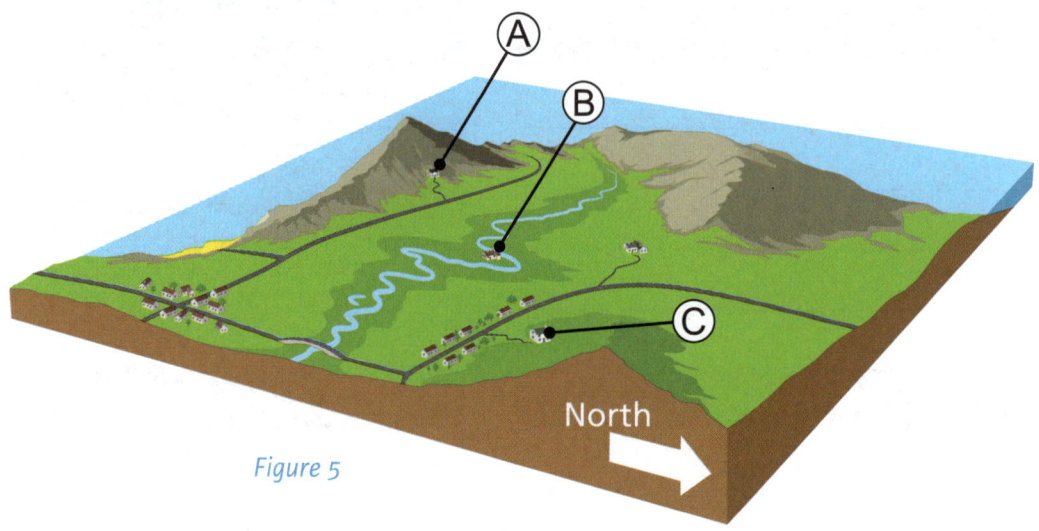

Figure 5

I chose the site labelled ◯ because:
_____
_____
_____
_____

I did not choose the site labelled ◯ because:
_____
_____
_____
_____

I did not choose the site labelled ◯ because:
_____
_____
_____
_____

Climate Change 16

1. The diagram in Figure 1 shows **how global warming happens**. Complete the diagram by selecting labels from the Selection Box and writing them into the correct boxes labelled **A–D**.

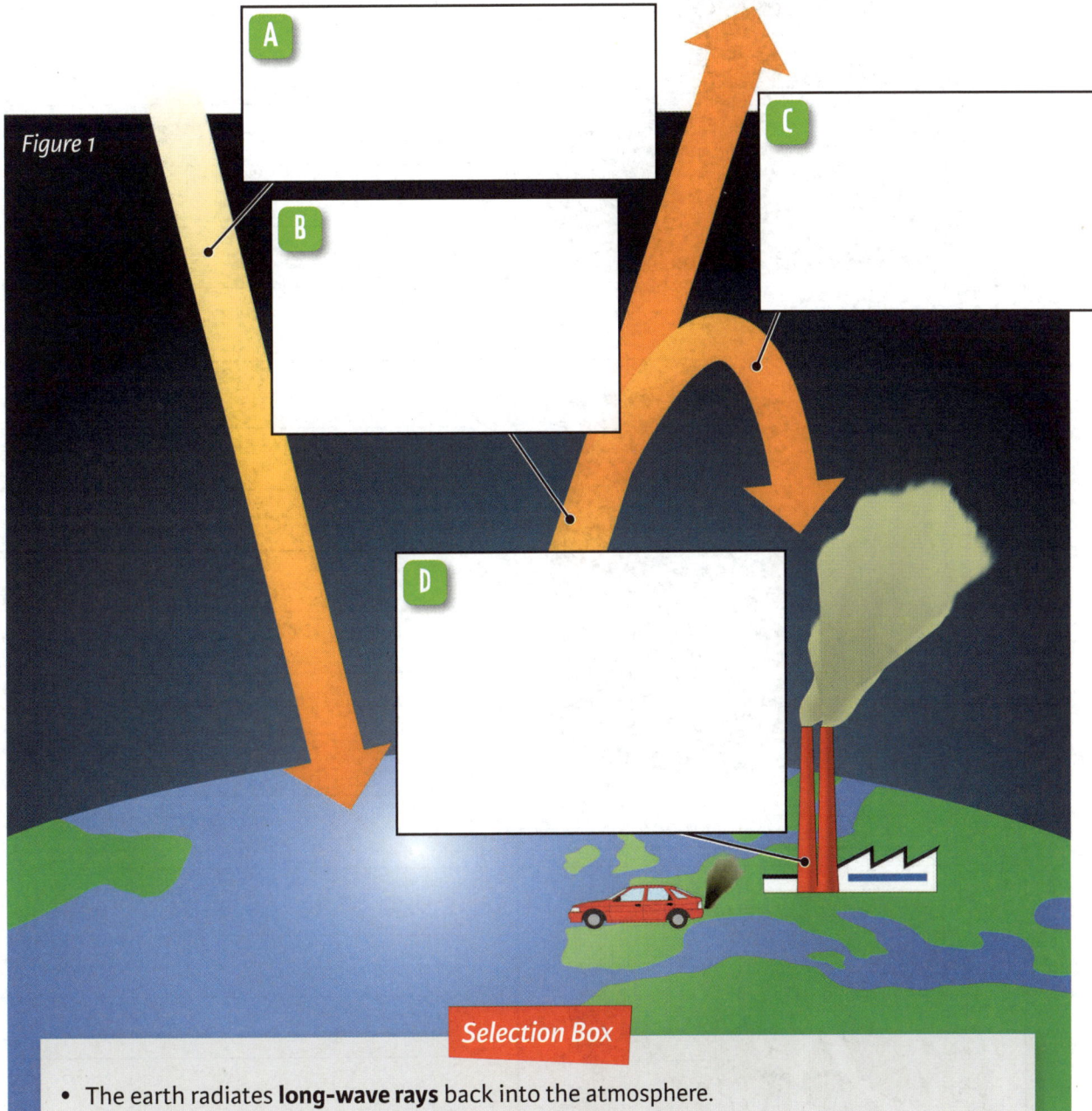

Figure 1

Selection Box

- The earth radiates **long-wave rays** back into the atmosphere.
- **Short-wave sunrays** pass through the atmosphere and heat the earth.
- Human activities create huge amounts of CO_2, methane and other greenhouse gases. These increase the 'greenhouse effect' so much that they cause **global warming**.
- Some long-wave rays are trapped in the atmosphere because they cannot pass through the atmosphere's 'greenhouse gases'. This natural process is called the '**greenhouse effect**'. It prevents the earth from freezing.

2. In the boxes provided, describe **three** greenhouse gases that cause **global warming**. These graphics might help you.

C _ _ b _ _ di _ _ _ _ _

1. _____

M _ _ _ _ _ _ e

2. _____

N _ _ _ _ _ _ _ o _ _ _ _

3. _____

16. Climate Change

3. In the spaces below, describe **four results of global warming**.

Desertification	Extreme weather events

Melting ice and rising sea levels	Threats to wildlife

4.

Figure 2: How rising sea levels might affect Dublin by 2100

The map in Figure 2 shows the **possible** effects of very severe global warming on Dublin by the year 2100. According to the map, which **three** of the following statements are correct?

- A. Ballsbridge, Ringsend and Sutton may all be submerged by the sea.
- B. Coolock, Clonskeagh and Ranelagh would all survive the rising sea levels.
- C. The north Dublin suburbs of Beaumount, Baldoyle and Donnycarney would all definitely survive rising sea levels.
- D. The shores of Dublin Bay would retreat eastward.
- E. Howth may become a new island.

The correct statements are (tick [✓] the correct box):

A, C, E ☐   A, B, E ☐   B, C, D ☐   A, C, D ☐

Discuss

Some scientists expect sea levels to rise by 30 cm by 2050 and (if global warming is not controlled) by 1 metre by 2100.

Might any coastal area near you be in danger from rising sea levels?

Soils

17

1. The divided bar in Figure 1 shows the **composition of soil**.

   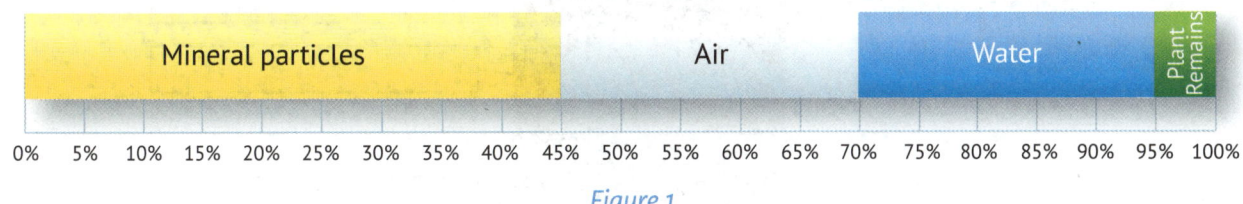

   Figure 1

   a. What percentage of the soil is composed of each of the following?

   Mineral particles ☐   Air ☐   Water ☐   Plant remains ☐

   b. Label the pie chart in Figure 2 to illustrate your answer to (a).

   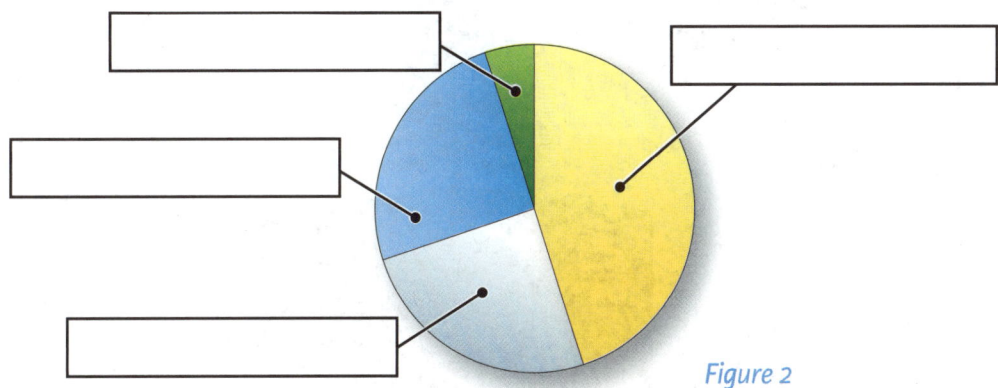

   Figure 2

2. The statements **A–E** below refer to **processes involved in the formation and development of soil**. A key word is missing from each statement. Discover each of these words in the puzzle. Then insert each word into the correct statement.

   A. _____ provides water, frost and other things vital to soil formation.

   B. _____ and erosion break down rock into mineral particles.

   C. Living _____ change plant litter into fertile humus.

   D. _____ is needed for slow soil-making processes to take place.

   E. _____ may improve or damage the development of soil.

81

3. Use the terms in the box to label aspects of **soil profile** shown in Figure 3.

Bedrock

Topsoil

Subsoil

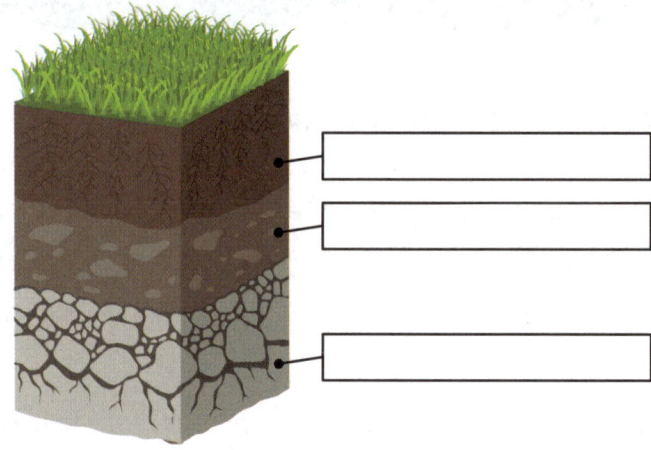

Figure 3

4. Examine the soil profile in Figure 4.

   a. On the labels provided on the diagram, identify the **A**, **B** and **C** horizons.

   b. Name the feature labelled **X**.

   _____

   c. What does the presence of the feature labelled **X** suggest about climatic conditions in this area?

   _____

   _____

   _____

   d. Is this soil likely to be fertile? Explain your answer in one sentence.

   _____

   _____

   _____

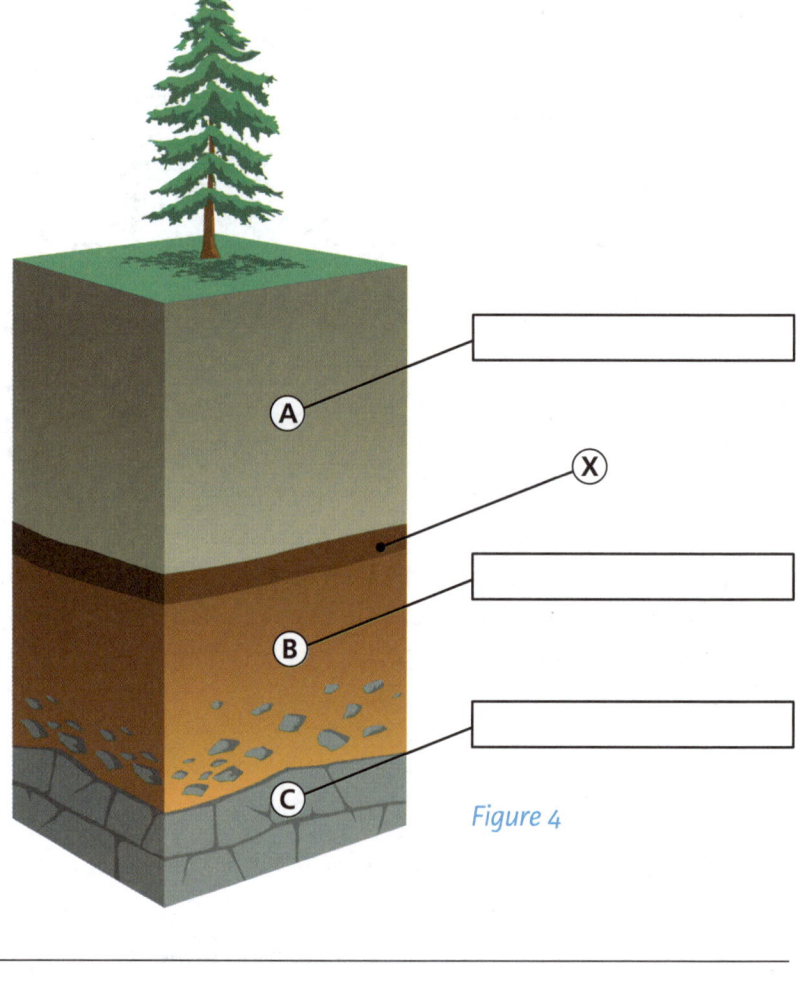

Figure 4

5. 'In nature, the health of vegetation and soil are interconnected.' With the help of Figure 5, explain what this statement means.

Figure 5: The soil–vegetation nutrient cycle

_____
_____
_____
_____
_____
_____
_____

6. 'Human activities can be good or bad for the soil.' Choose **any two** of the activities listed below. In the spaces provided, *name* and *briefly describe* each activity and explain why it is likely to be **good or bad** for the soil.

| Deforestation | Irrigation | Overgrazing | Terracing |

You might need to find out what some of these activities are.

Activity 1: _____
_____
_____
_____
_____
_____
_____
_____
_____
_____

Activity 2: _____
_____
_____
_____
_____
_____
_____
_____
_____
_____

7. Complete the word puzzle.

Clues Across:
5. Horizontal layers within a soil profile.
6. This influences frost, rain, vegetation, etc., which all affect the formation of soil.
7. _ _ _ _ _ _ _ is found in the B horizon.
8. The colour of Ireland's most common soil type.
9. You cannot see or touch this, but soil needs lots of it to form.
10. The process that turns plant remains into humus.
11. A process that helps to break down rock into mineral particles.
12. Slender, brownish animals that burrow in the earth.
13. Waterlogged soil, often blue-grey in colour.
14. Life-giving liquid that makes up about 25% of soil.
15. Micro-organisms mentioned in 18 across.
16. Soil of the A horizon.
17. A type of soil often found in areas of coniferous forest.
18. Bacteria can break down plant litter into this black, jelly-like substance.
19. What we find in the C horizon.
20. Substance that makes up about 25% of soil.
21. Brown soils are usually found in areas that that were once covered in this kind of forest.
22. The breakdown of rock by frost action, temperature changes, etc.

Clues Down:
1. A thin, crusty layer of leached soil.
2. A thing of nature useful to people – soil is an example.
3. Leaves, etc. that have fallen on the ground.
4. The percolation of nutrients and minerals down through the soil.

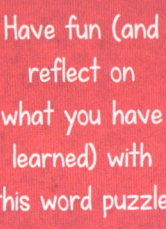

Have fun (and reflect on what you have learned) with this word puzzle.

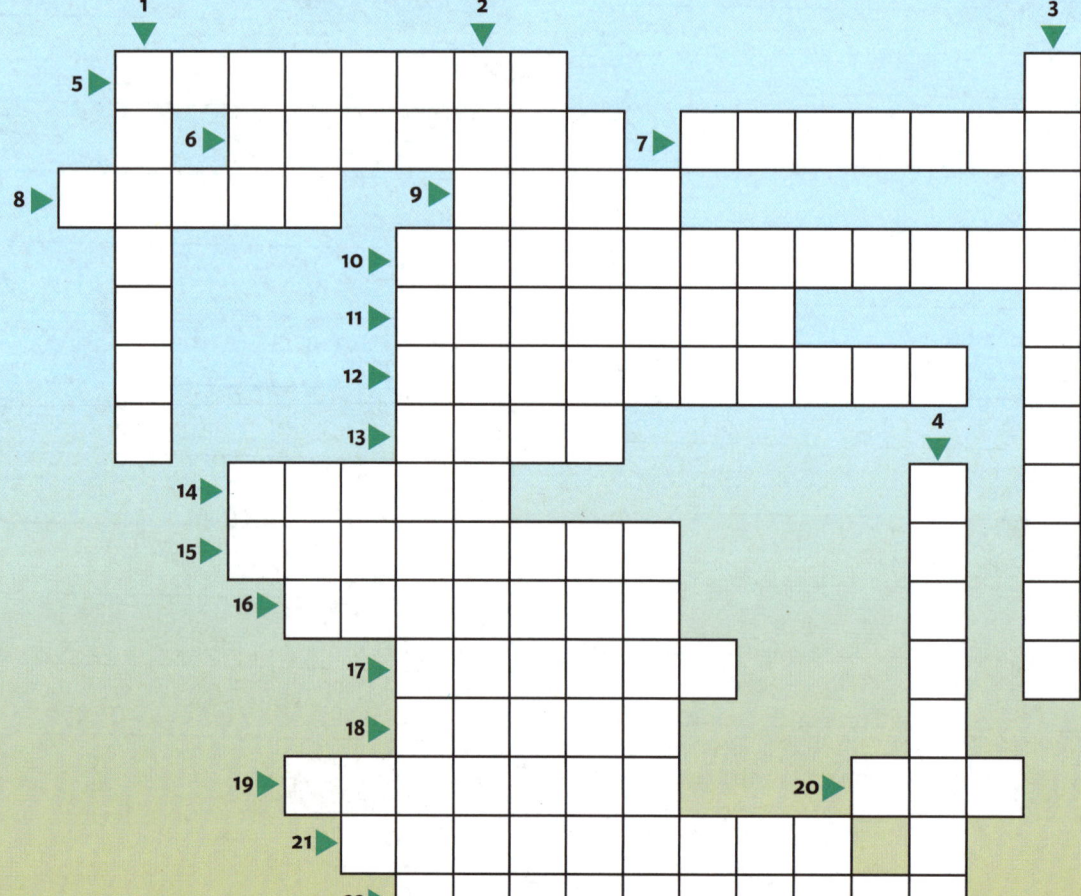

Population 18

1. In the grid, write the correct term for each of the definitions **A–E**. One has been completed for you. The rest can be found in Chapter 18 of your *Cyclone* textbook.

	Definition	Term
A	The number of live births per 1,000 population in one year	
B	This happens when birth rate is higher than death rate	
C	The average number of people per square kilometre	Population density
D	The number of deaths per 1,000 population per year	
E	This happens when death rate is higher than birth rate	

2. The table below provides data on the **birth rates**, **death rates**, **natural increase** and **natural decrease** in a selection of countries. Some statistics have been omitted. Calculate each of the omitted figures and enter in the correct place in the table.

\<Figures given below are per 1,000 of population\>				
Country	Birth rate	Death rate	Natural increase	Natural decrease
Brazil	16.82	6.80	10.02	None
China	7.52	7.67	None	0.15
Ireland	11.68	6.41		None
Japan	7.11	11.34	None	
Nigeria	36.40	11.42		None
USA	12.01		2.93	None

> Remember:
> Natural increase is birth rate minus death rate.
> Natural decrease is death rate minus birth rate.

85

3. **Population change**

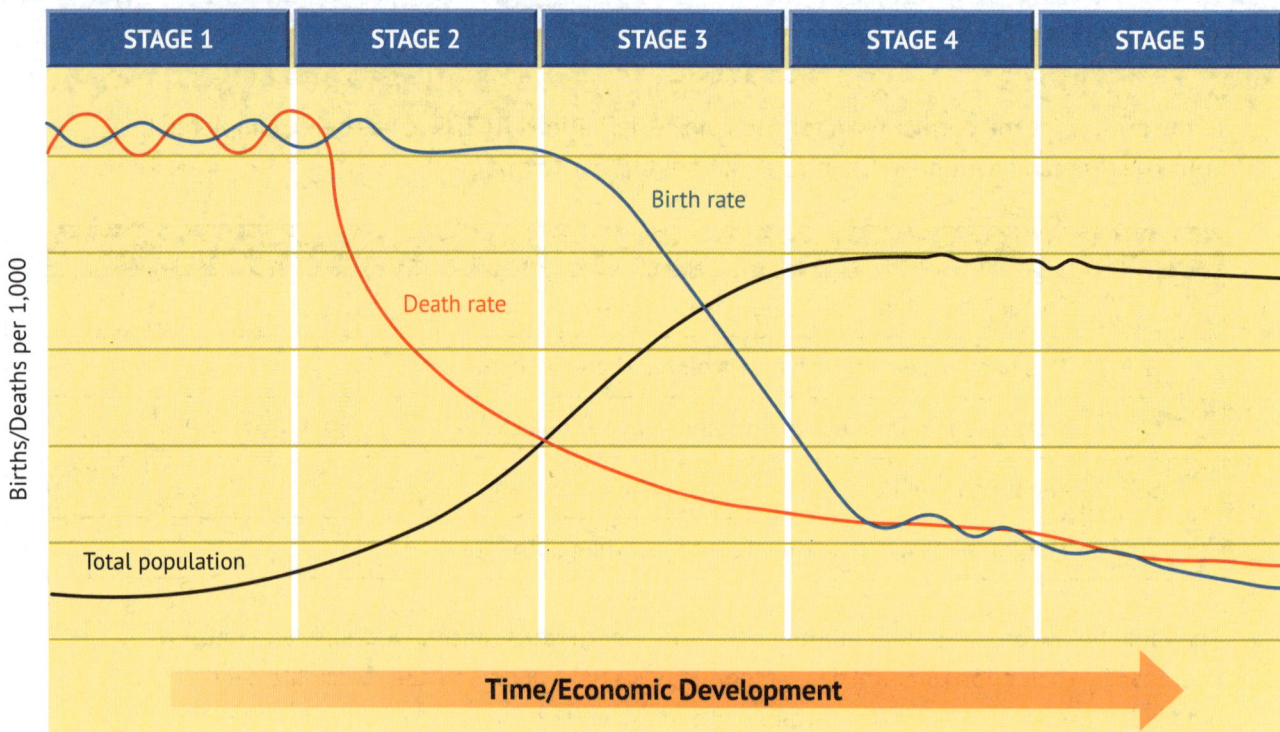

Figure 1

a. Examine Figure 1 and answer the following questions.

   i. What name is given to the geographical 'model' shown in this graph?

   _____

   ii. At which stage of the model is the death rate at its highest? _____

   iii. At which stage does the population grow most rapidly? _____

   iv. At which stage is the population at its lowest? _____

   v. Which stage shows a decrease in total population? _____

   vi. What names are given to Stage 2 and to Stage 4?

   Stage 2: _____

   Stage 4: _____

18. Population

b. The statements below relate to information given in Figure 1 on the previous page. Some of the statements are true and some are false. Write **True** or **False** beside each statement.

   i. Birth rate is always higher than death rate. _____

   ii. Total population growth is very slow at Stage 1. _____

   iii. The death rate declines evenly throughout Stages 2 and 3. _____

   iv. There is a rapid fall in the birth rate at Stage 3. _____

   v. The total population decreases at Stage 4. _____

   vi. Both birth rate and death rate are at their lowest at Stage 5. _____

   vii. The populations of most Western European countries are at Stages 4 or 5. _____

   viii. Rapid population growth comes to an end as an area starts to develop economically.

   _____

   ix. The model in Figure 1 does not give a full picture of population change because it does not take factors such as human migration into account.

   _____

c. Why, in Figure 1, is there very rapid population growth at Stages 2 and 3?

   _____
   _____
   _____
   _____
   _____
   _____

4. **Population pyramids of two countries**

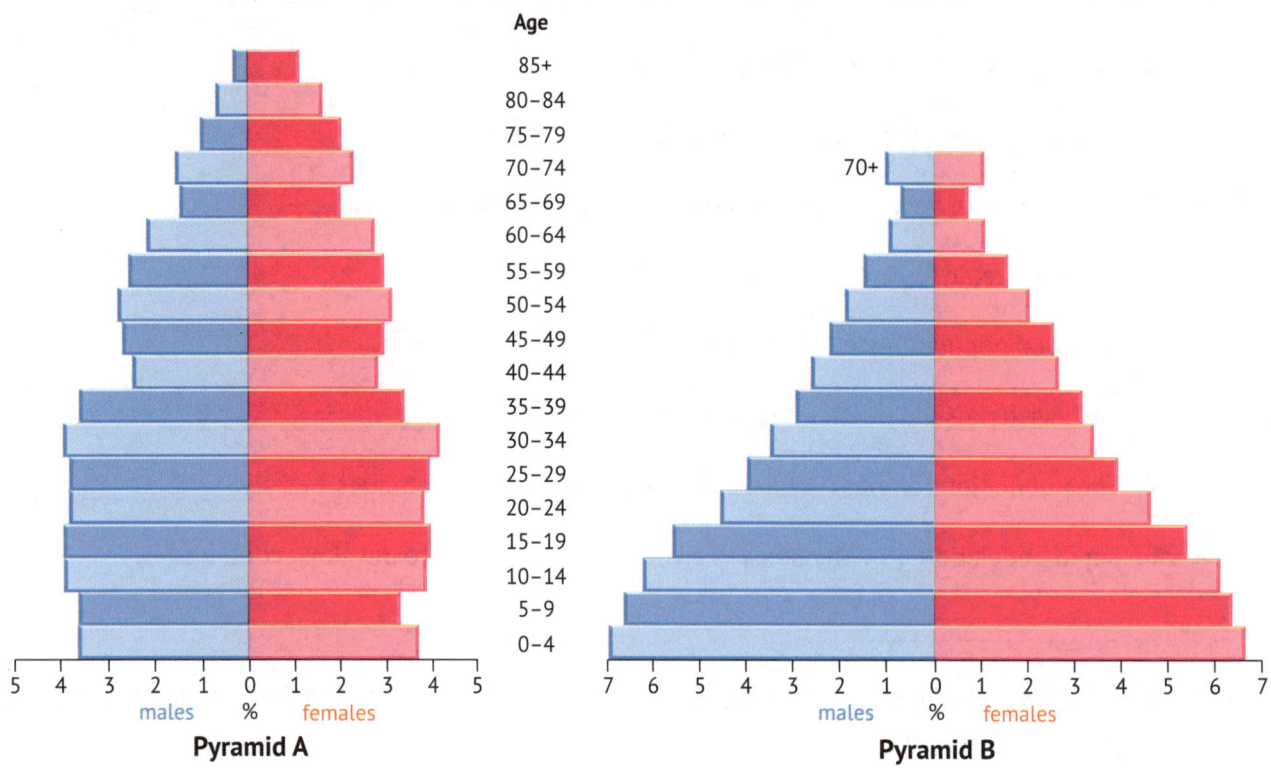

Figure 2

a. Which of the two pyramids, **A** or **B**, presents the population of a developing country?

_____

b. Name one country with a population structure similar to that represented by Pyramid A.

_____

c. Explain one reason why Pyramid B has a wider base than Pyramid A.

_____

_____

d. What percentage of the population shown in Pyramid B is under 5 years old?

_____

e. What percentage of the population shown in Pyramid A consists of women over the age of 85 years?

_____

f. Indicate whether Pyramid A and Pyramid B represent an **expansive**, **stationary** or **constrictive** population. Tick (✓) the correct box for each pyramid below.

	Expansive	Stationary	Constrictive
Pyramid A	☐	☐	☐
Pyramid B	☐	☐	☐

g. Suggest **three** uses that central or local governments might make of information provided by population pyramids.

> Hint: Think about how a government might respond if the population of its country included:
> - a large population of children
> - a large proportion of adults aged between 20 and 60
> - a large proportion of elderly people over the age of 60.

1. _____

2. _____

3. _____

Junior Cycle Geography CYCLONE SKILLS BOOK

5. Class project on population structure

Learning objectives:
- For each student to create a population pyramid that illustrates the population structure of all class members' families.
- For each student to comment on the population structure illustrated.

Carry out the steps below for your class project on population structure.

In class:

1. The project is explained to students using the steps outlined here.

2. Each student finds out the age of each person in their family.* The student uses this knowledge to draw an 'unticked' copy of a population pyramid skeleton similar to that shown in Figure 4.

   Include in the family the following living people: parents or guardians, brothers, sisters, stepbrothers, stepsisters, foster brothers, foster sisters, grandparents. Count all the above even if they are not living in the home. Also count any other person who is living permanently in the home.

Figure 3: Sample of one female student's research on family age structure. These results are ticked on the population pyramid skeleton in Figure 4.

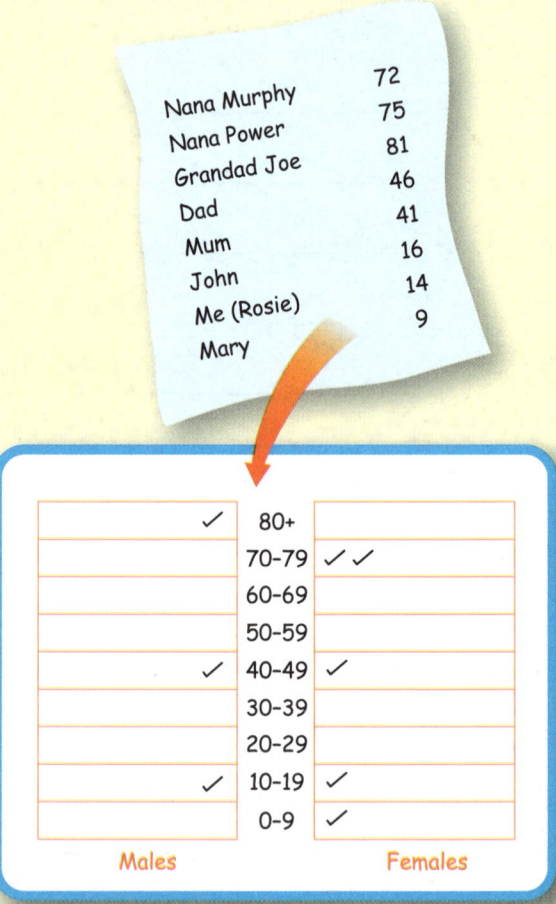

At home:

3. Each student places ticks on the appropriate age-group sections of their pyramid skeleton (see example in Figure 4).

In class:

4. A large copy of the pyramid skeleton is placed on the whiteboard. Each student neatly enters their ticks on this large copy.

5. The total number of ticks for each gender total in each age group is then calculated and written on the appropriate age-group area (see the circled examples in Figure 5 on the opposite page). Each student copies these **totals** on the appropriate age group areas of their own pyramid skeleton.

6. Each student is given a sheet of graph paper.

Figure 4: Population pyramid 'skeleton' (a sample filled in by one student)

18. Population

At home:

7. Each student uses the graph paper and a pencil to create a population pyramid based on the age group totals calculated in class (see example in Figure 6).

Each population pyramid should have:
- A suitable horizontal scale along the base
- Clearly labelled age group sectors up the middle
- Neat and accurate age-bars for both males and females in all age groups
- A title. (See Figure 6.)

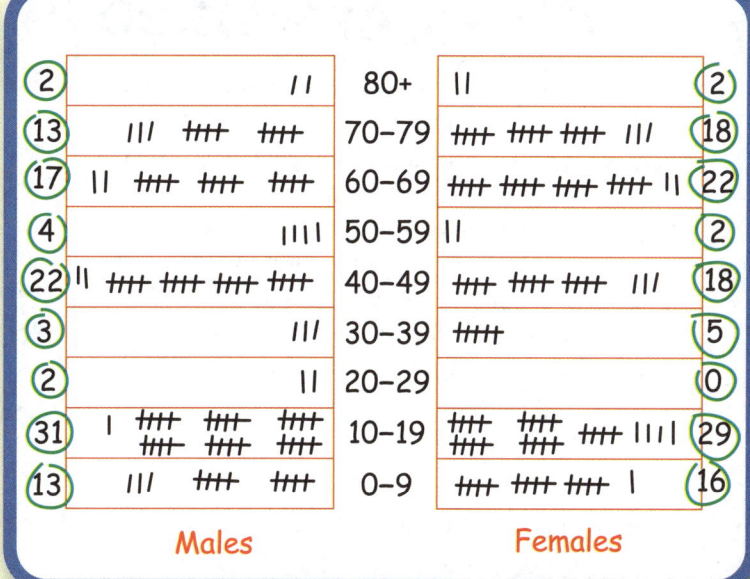

Figure 5: Sample population pyramid 'skeleton' (filled in by all students in class)

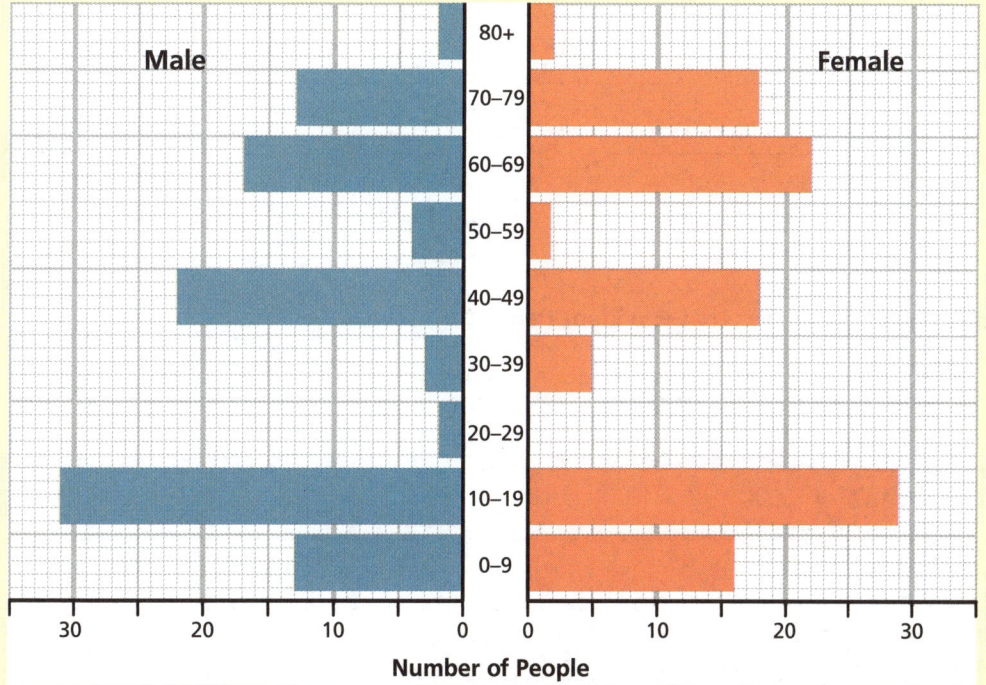

Figure 6: Sample of a complete population pyramid based on information in Figure 5 above

8. Each student writes on their graph paper two interesting comments or observations about the population structure illustrated in their population pyramid. Later, students discuss these comments in class.

6. Read the extract below and answer the questions that follow.

Birth rates and the place of women in society

Birth rates are affected by **gender/racial** issues such as the place of women in society. In places where women enjoy gender equality with men, women are empowered to make **more/fewer** decisions that relate to their own lives. It is clear that the empowerment of women results in **rising/falling** birth rates. In economically developed countries such as **Ireland/Uganda**, an overwhelming majority of women can choose between becoming full-time mothers and taking employment outside the home. This choice leads to **more/fewer** children being born. Prolonged female education also leads to **rising/falling** birth rates. Women who stay longer in education are **more/less** likely to marry young and **more/less** likely to be aware of family planning.

a. *Circle* the correct option from each pair in bold in the text above.

b. *Name* four factors, other than the place of women in society, that affect birth rates and/or death rates.

1. _____
2. _____
3. _____
4. _____

c. *Explain* one of the factors that you identified in (b) above.

_____
_____
_____
_____
_____
_____
_____

Migration

1. Each entry in **Column Y** in the grid describes one of the terms in the Selection Box below. In each space in **Column X**, fill in the term that matches the description in Column Y. One match has been made for you.

	Column X	Column Y
1	Migrants	People who leave one region or country to live elsewhere
2		People who leave their country of origin to live in another country
3		People who come into a country to live
4		People who are forced to leave their homes but remain in their own country
5		The country or region from which people migrate
6		The country or region to which people migrate
7		Migrants who receive special permission to live in another country because of war, persecution, famine or other dangers in their own country
8		Immigrants who seek to be classified as refugees
9		When people are compelled to migrate from their home regions or countries
10		When people migrate of their own free will

Selection Box

Immigrants
Donor country or region
~~Migrants~~
Voluntary migration
Refugees
Emigrants
Asylum seekers
Host country or region
Forced migration
Internally displaced people

2. The graphs in Figure 1 represent **immigration**, **emigration** and **net migration** in Ireland between 2011 and 2021. Study the graphs and answer the questions that follow.

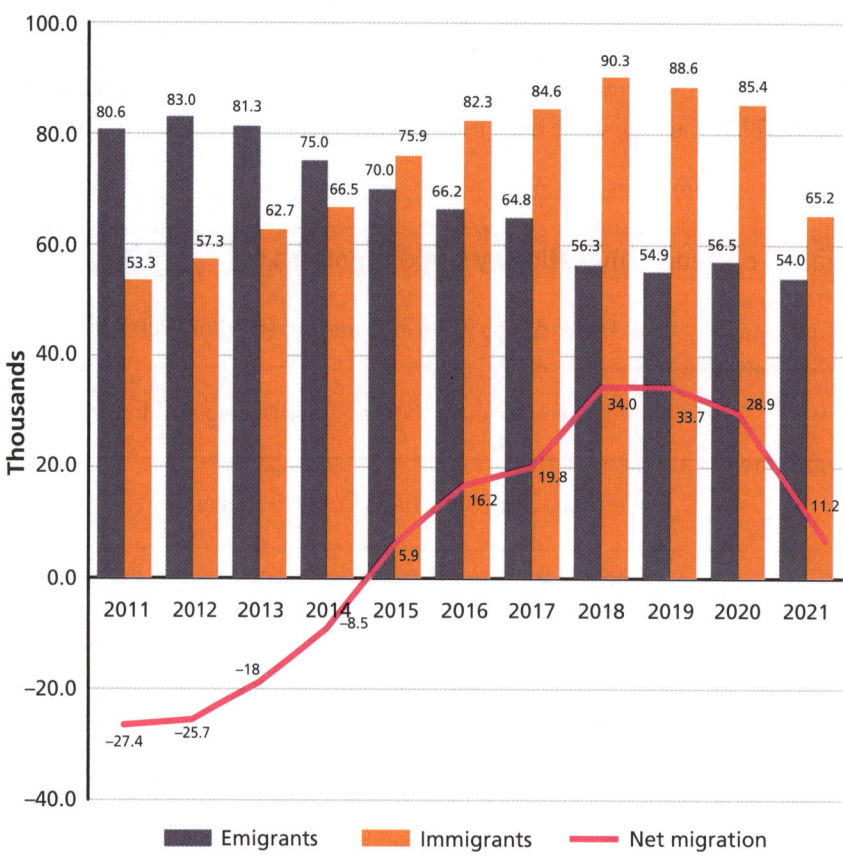

Figure 1: Immigration, emigration and net migration in Ireland between 2011 and 2021

a. i. In which year was immigration at its highest? _____

ii. How many people immigrated to Ireland in that year? (Don't forget that the figures are in thousands.)

_____

iii. In which year was emigration at its lowest? _____

iv. How many people emigrated from Ireland in that year? _____

v. In which year did immigration first exceed emigration? _____

vi. In which years was Ireland a country of net emigration (when emigrants exceeded immigrants in number)?

_____

vii. By how much (how many people) did immigration exceed emigration in 2018?

_____

b. Indicate whether each of the following statements is **true** or **false** by ticking (✓) the correct box.

	True	False
i. Ireland was a country of net immigration for most of the period shown.	☐	☐
ii. Emigration was at its highest in 2012.	☐	☐
iii. Immigration and emigration figures increased in 2013.	☐	☐
iv. Immigration exceeded emigration by 5,900 people in 2015.	☐	☐
v. The figures suggest that Ireland may have enjoyed a period of increasing prosperity between 2015 and 2018.	☐	☐

c. Describe **two benefits** and **two challenges** (problems) that large-scale immigration could create in Ireland.

Benefits:

1. _____

2. _____

Challenges:

1. _____

2. _____

3. Explain each of these factors that affect migration.

- Push factor:

    _____
    _____
    _____

- Pull factor:

    _____
    _____
    _____

- Barrier to migration:

    _____
    _____
    _____

4. The grid below and on the next page shows a list of factors relating to migration. In the empty spaces provided, indicate and explain whether each factor is a **push factor**, a **pull factor** or a **barrier to migration**. One factor has been completed for you.

Factor	Push factor/pull factor/ barrier to migration	Explanation
Better educational facilities	Pull factor	Might lead to prosperity and attract families with young children
Lack of jobs		
Leaving family and friends		

19. Migration

Factor	Push factor/pull factor/ barrier to migration	Explanation
War		
High cost of travel		
Severe drought		
Employment prospects		
Government restrictions on immigration		
Famine		
Overcrowding		
Inability to speak a foreign language		
Positive TV images of city life		

Junior Cycle Geography CYCLONE SKILLS BOOK

5. Most of the world's migrants are young adults who migrate from rural areas or small towns to larger urban centres or cities.

   a. Learn more about **the consequences of rural-to-urban migration** by using words from the box to complete the news item below.

   | workforce | birth rates | home | Dublin | west | rising |
   | urban sprawl | growth | education | services | shortages | decline |

   Young adults have for many decades migrated from rural or small-town communities in the _____ of Ireland to large urban areas such as _____.

   Out-migration resulted in population _____ and in falling _____ in many western areas. This resulted in less demand for services, which in turn caused the closure of shops, cafés, post offices, small town cinemas and some sports clubs. As these _____ declined, life became more difficult for those young people who remained. So, having completed their _____, they too began to leave, creating a spiral of out-migration.

   Rural-to-urban migration affected cities such as Dublin very differently. In-migration contributed to population _____, while young, well-educated migrants provided a versatile _____ that assisted rapid development.

   But in-migration also contributed to severe housing _____ in Dublin. This resulted in rapidly _____ house prices, which by 2022 made it almost impossible for most young people to buy a _____ of their own. In the meantime, Dublin City grew outwards as well as upwards, creating a great _____ _____ across much of Co. Dublin.

   b. Now see if you can recall the consequences of rural-to-urban migration:
   - For rural areas
   - For large urban areas.

19. Migration

Summarising a Case Study on Migration

6. Select **any one** of the three case studies on migration that appear in Chapter 19 of your *Cyclone* textbook. Write the **key ideas** of that case study in the spaces provided below.

Write key ideas only.

Who migrated? _____

From where? _____

To where? _____

Push and pull factors that influenced this migration:

_____
_____
_____
_____
_____

Barriers to this migration:

_____
_____

Consequences of this migration – for the migrant(s); for the (destination) area to which they came; and for the (donor) area from which they came:

_____
_____
_____
_____
_____
_____
_____

20 Geographical Skills: Aerial Photographs, Charts, Graphs and Infographics

Part 1: Aerial photographs

1. Examine the aerial photograph of Dundalk, Co. Louth in Figure 1.

   a. Is this an oblique or a vertical photograph? _____

   b. Was the photograph taken in winter or in summer? How do you know? _____

   _____

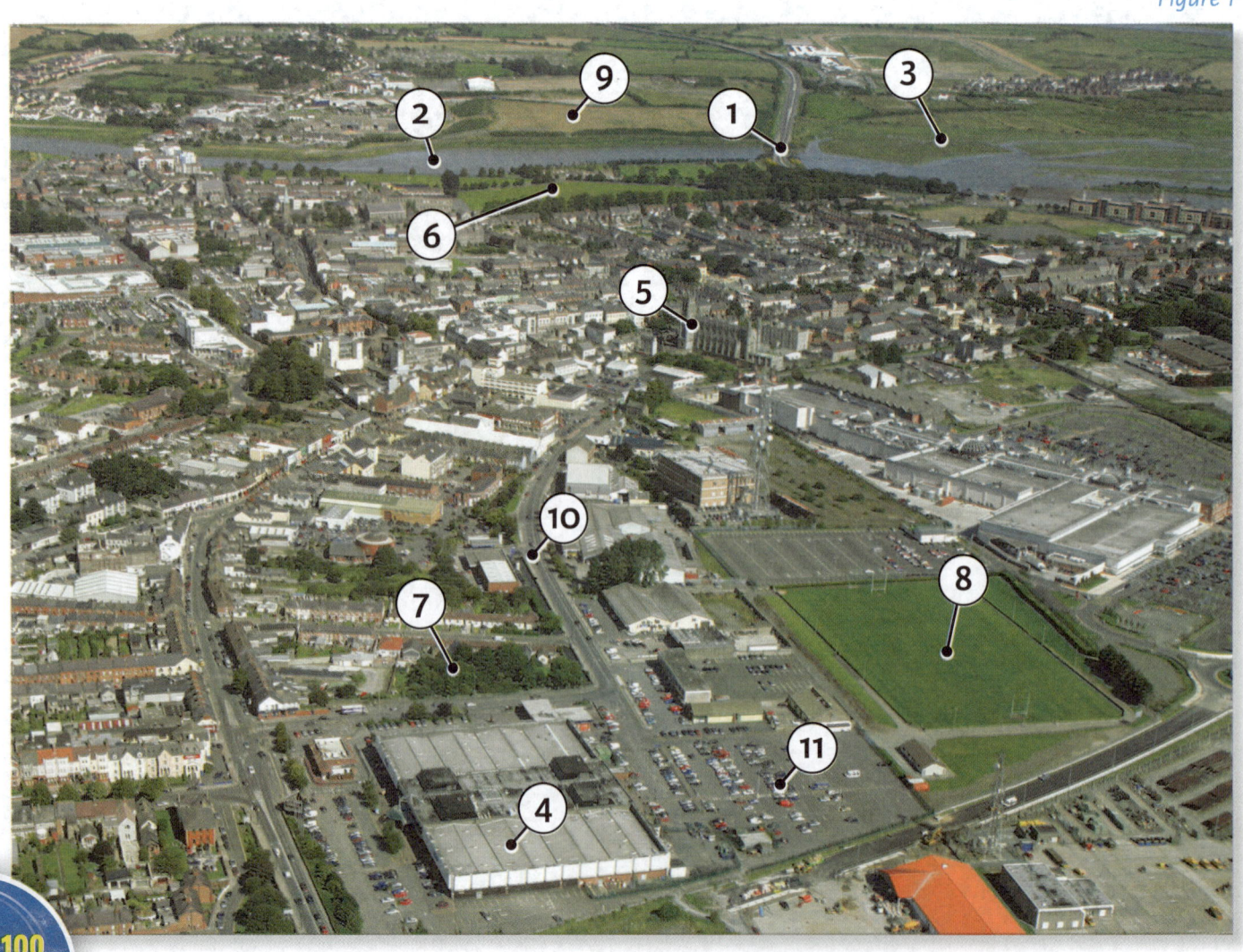

Figure 1

100

c. In the grid below, match each of the features numbered in Figure 1 with the correct landscape feature from the box. One line of the grid has been completed for you.

> Shopping centre   Car park   Sports field   Street   Church
> Deciduous trees   River   ~~Bridge~~   Riverside marshland
> Agricultural land with cereal crop   Pastoral (grazing) agricultural land

Label on photograph	Landscape feature	Location of feature on photograph
1	Bridge	Centre background/Right background
2		
3		
4		
5		
6		
7		
8		
9		
10		
11		

Junior Cycle Geography CYCLONE SKILLS BOOK

Figure 2

2. Examine the aerial photograph of Carrick-on-Shannon in Co. Leitrim in Figure 2.

   a. In the space provided on the opposite page, draw a **sketch map** of the photograph in Figure 2. On your sketch map, *show* and *name* each of the following:
   - The river
   - A bridge over the river
   - The principal street
   - Two facilities that are used for recreation and/or tourism
   - A church in the right foreground of the photograph.

20. Geographical Skills: Aerial Photographs, Charts, Graphs and Infographics

Use this space to draw your map.

b. 'The River Shannon has been important to the early development and to the more recent growth of Carrick-on-Shannon.' To what extent do you think the **river** influenced (i) the foundation and early growth and (ii) the more recent development of Carrick-on-Shannon? In your answer, refer to evidence from Figure 2.

i. Foundation and early growth: _____

_____

_____

ii. More recent development: _____

_____

_____

c. 'Carrick-on-Shannon and its surrounding area engages in primary, secondary and tertiary economic activities.' Referring to the aerial photograph in Figure 2, describe evidence (or possible evidence) to support this statement.

Identify and *locate* features in the photograph that relate to each activity you write about. For example: 'What looks like a race track or racecourse in the right background of the photograph is evidence of recreational tertiary activities.'

Find out the meanings of primary, secondary and tertiary economic activities on page 299 of your Cyclone textbook.

Primary economic activities:
_____
_____
_____
_____
_____

Secondary economic activities:
_____
_____
_____
_____
_____

Tertiary economic activities:
_____
_____
_____
_____
_____

20. Geographical Skills: Aerial Photographs, Charts, Graphs and Infographics

3. Examine the aerial photograph of Galway City in Figure 3.

   a. Do you think the area marked **X** would be a suitable location at which to build a four-storey apartment block? Using evidence from the photograph, give **two** reasons for your answer.

   Is it a suitable location? _____

   Reason 1: _____
   _____
   _____

   Reason 2: _____
   _____
   _____

   b. Do you think the area in the left foreground of the photograph is an area of **planned** urban development? Tick (✓) Yes or No.

   Yes ☐   No ☐

   Briefly explain your answer.

   _____
   _____
   _____

Figure 3

Part 2: Charts, graphs and infographics

The following pages will show you how tables of statistics, graphs, infographics and charts can be used to reveal geographical trends and knowledge. This is an important part of your Geography programme.

Tables of statistics

4. Examine the table of statistics and answer the questions that follow.

| Weight (in thousands of tonnes) of sea fish catches in six European countries in 2008 and 2016 ||
Country	2008	2016
Denmark	691	670
Germany	207	241
Ireland	205	230
Spain	853	860
Italy	232	193
Netherlands	376	368

(*Source:* Eurostat)

a. What does this table show (what is its title)?

_____

b. What was the weight of Ireland's catch in 2008? _____

c. Which country had the largest catch in 2016? What was the weight of that catch?

Country: _____ Weight: _____

d. By how many thousands of tonnes did Ireland's catch increase between 2008 and 2016?

_____

e. Which countries experienced a decrease in catch between 2008 and 2016?

_____

f. In 2016, the combined catches of Ireland and Denmark were greater than Spain's catch.

True or **false**? _____

20. Geographical Skills: Aerial Photographs, Charts, Graphs and Infographics

Line graphs

5. Figure 4 is a **line graph** or **trend graph** showing **change over time**. It shows the number of **immigrants** who came to live in a region between 2010 and 2022.

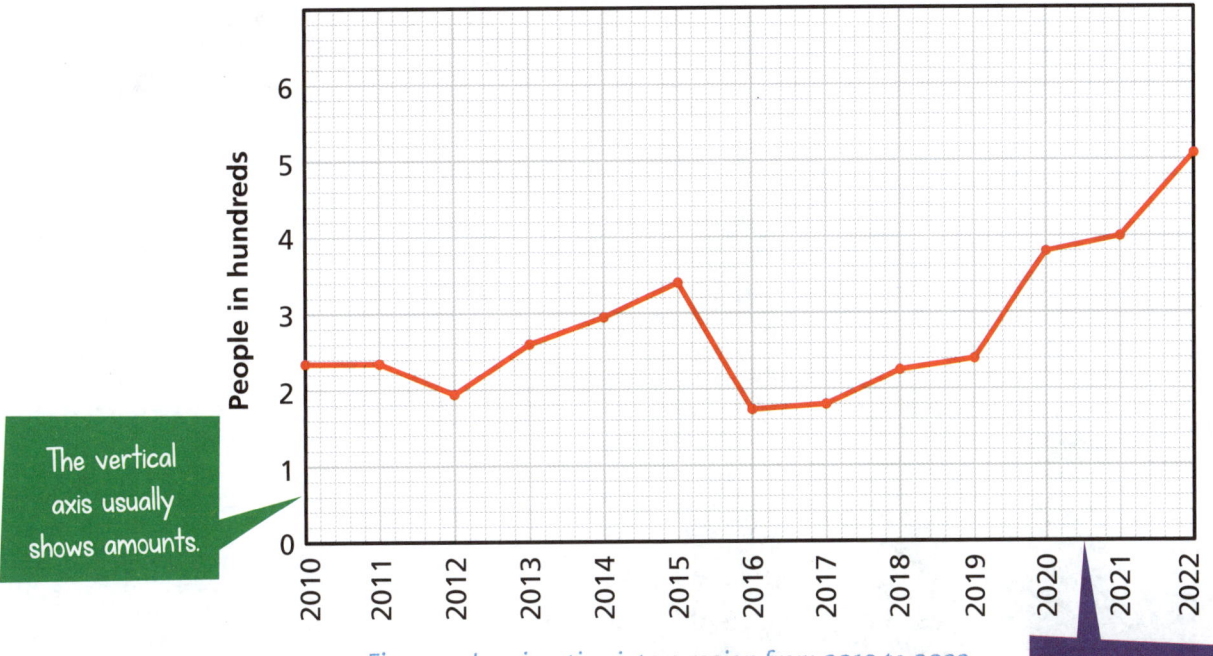

The vertical axis usually shows amounts.

The horizontal axis usually shows time.

Figure 4: Immigration into a region from 2010 to 2022

a. How many immigrants came to this region in 2010? _____

b. In which year did immigration peak (reach its highest point)? _____

c. Between which two years did the greatest **decrease** in immigration occur? _____

d. How many immigrants came to the region in 2021?
_____

e. Examine the statistics in the grid on the right. They show the number of **emigrants** (in hundreds) who left the same region between 2010 and 2022. *Draw another line graph on Figure 4 (in a different colour) to illustrate these statistics.*

f. *Shade that part of the grid* that show **emigration** exceeding the **immigration** figures indicated in Figure 4.

g. By how much did immigration exceed emigration in 2022?
_____

2010	160
2011	200
2012	280
2013	300
2014	290
2015	200
2016	170
2017	140
2018	120
2019	120
2020	140
2021	160
2022	200

107

Bar charts

6. Figure 5 shows a simple bar graph and a grid that shows us **change over time in relation to a single subject** (in this case, membership of a youth club). Study the chart and grid and answer the questions that follow.

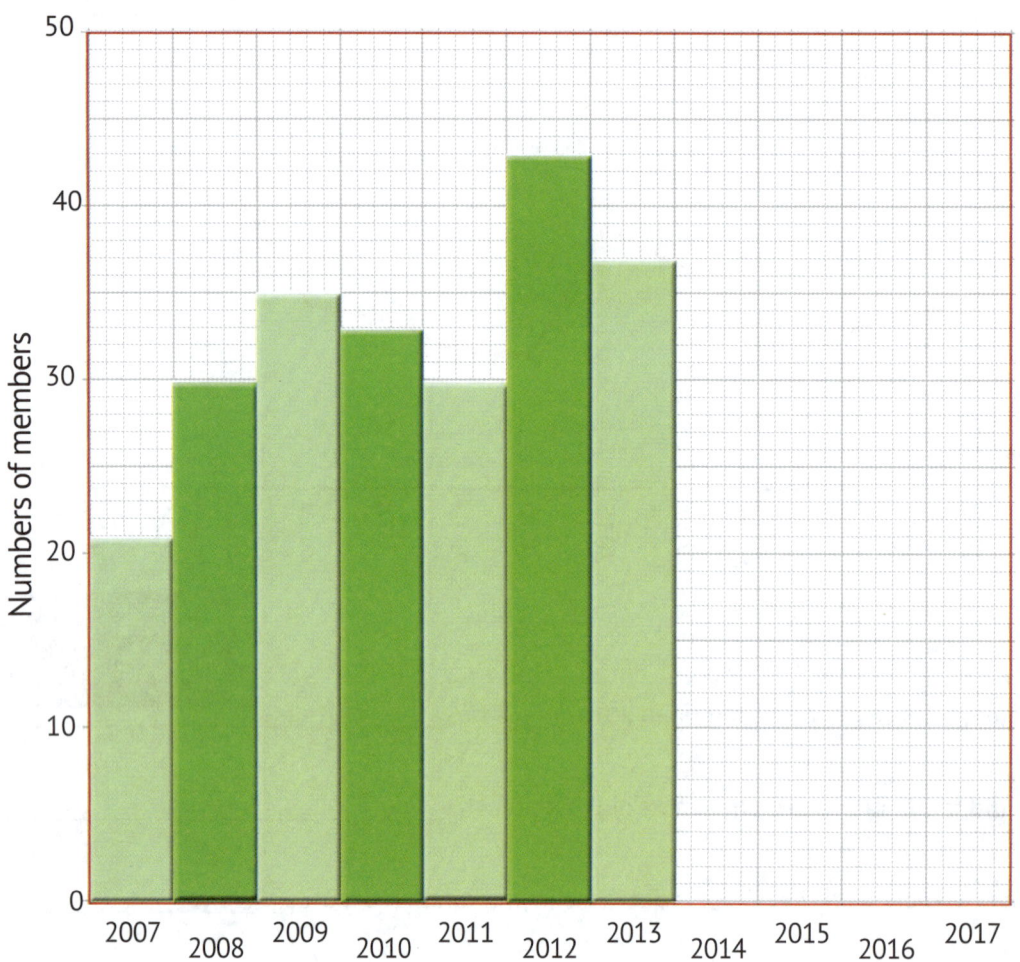

Figure 5: Membership of a youth club in Ireland, 2007 to 2017

a. The statistics in the grid reveal membership of the youth club between 2014 and 2017. Use these statistics to complete the bars in Figure 5.

b. Indicate membership during each of the following years:

2008: ☐   2010: ☐   2013: ☐

c. In which year was the largest increase in membership recorded? _____

d. In how many years was there a reduction in membership? _____

e. What was the difference in membership numbers between 2007 and 2017? _____

20. Geographical Skills: Aerial Photographs, Charts, Graphs and Infographics

Percentage bars (divided bars)

Some bar graphs are called **percentage bars** or **divided bars**. They are divided into segments and show what percentage of the whole is made up by each segment.

Several percentage bars can appear together in the same diagram. These allow us to compare the **percentage make-up** of the various bars.

Note that the bars can be arranged horizontally or vertically. The bars shown in Figure 6 are horizontal bars.

7. Figure 6 shows three percentage bars together. Study Figure 6 and answer the questions that follow.

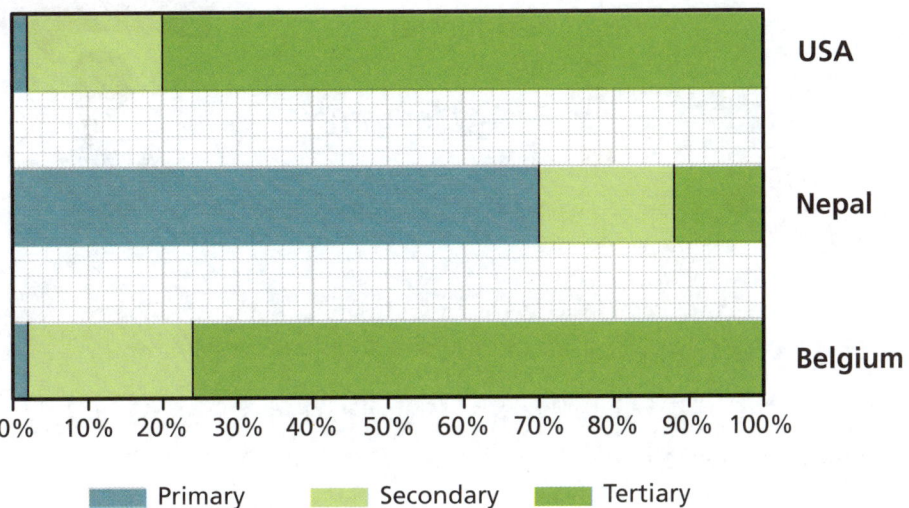

Figure 6: Percentages of people employed in primary, secondary and tertiary economic activities in three countries in 2020

a. Which of the countries shown depended most on tertiary activities in 2020?

_____

b. What percentage of people in Belgium worked in primary activities?

_____

c. What percentage of people in Nepal worked in secondary and tertiary activities combined?

_____

d. What percentage of people in the USA worked in (i) primary, (ii) secondary and (iii) tertiary activities?

Primary: ☐   Secondary: ☐   Tertiary: ☐

Pie charts

Pie charts are circular graphs. They are cut into segments like slices of pizza to show **how something is sub-divided**. Most pie charts show percentages of the whole.

8. Examine the pie chart in Figure 7 and answer the questions that follow.

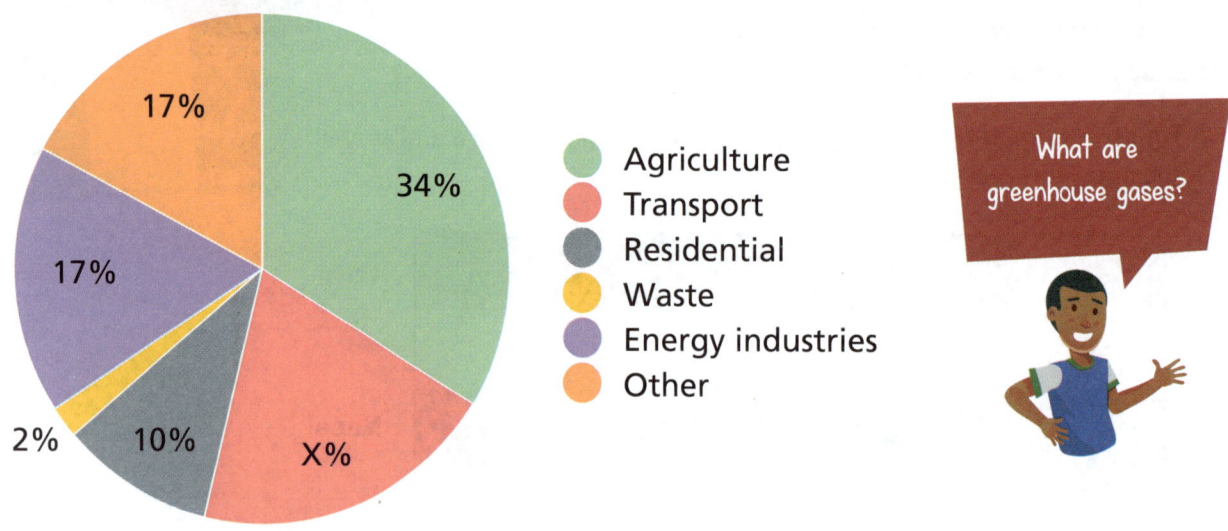

Figure 7: Ireland's greenhouse-gas emissions, percentage share by sector of the economy

a. Fill in the gaps in the following statement:

   The economic sector that produced the greatest amount of greenhouse gas emissions was
   _____. This sector produced _____% of total emissions.

b. 'Agriculture and energy industries combined produced more than half of all emissions.' Is this statement **true** or **false**? _____

c. Calculate **X**, the percentage share of Ireland's greenhouse gas emissions that was generated by transport.

Optional rough work:

Answer: ☐

20. Geographical Skills: Aerial Photographs, Charts, Graphs and Infographics

Infographics

Nigeria and Ireland: how they compare		
Ireland 		**Nigeria** 
84,421 km²	Area	923,768 km²
5.0 m	Population	216.1 m
	Population density	
$74,520	GNI per capita (per person)	$2,200
82	Life expectancy	69
2.5	Infant mortality rate (per 1,000 population)	63.0
3.47	Doctors (per 1,000 population)	0.7
Gaelic football	Most practised team sport	Football
1922	Year of independence	1960

Figure 8

9. Study the infographic above and answer the questions.

   a. How large is Nigeria? _____

   b. Calculate the **population density** of Ireland and Nigeria and write each calculation into the table.

   c. Do you think that Figure 8 is an effective way of showing comparisons between two countries? Explain why/why not.

   _____

   _____

   _____

   _____

Population density is the average number of people per square kilometre. To calculate population density, **divide the population by the area** of the country.

111

21 Rural and Urban Settlement in Ireland

1. Examine the places labelled **A–D** on the Ordnance Survey map extract in Figure 1 on the following page.

   a. In the grid below:
   - Name the **rural settlement pattern** in each of the labelled areas.
   - Explain **why** these settlement patterns occur in the areas identified.
   - Match each of the photographs labelled **1–3** below with the settlement pattern in the grid that it most closely resembles. (One of the boxes in the grid will remain empty.)

Settlement pattern	Why it occurs in that area	Matching photograph
A		
B Absence of settlement		
C		
D		

21. Rural and Urban Settlement in Ireland

Figure 1

b. Using evidence from the map in Figure 1, give **three** reasons for the past and present growth and development of New Ross town.

Reason 1: _____

_____

_____

Reason 2: _____

_____

_____

Reason 3: _____

_____

_____

Junior Cycle Geography CYCLONE SKILLS BOOK

My Geography: My Town

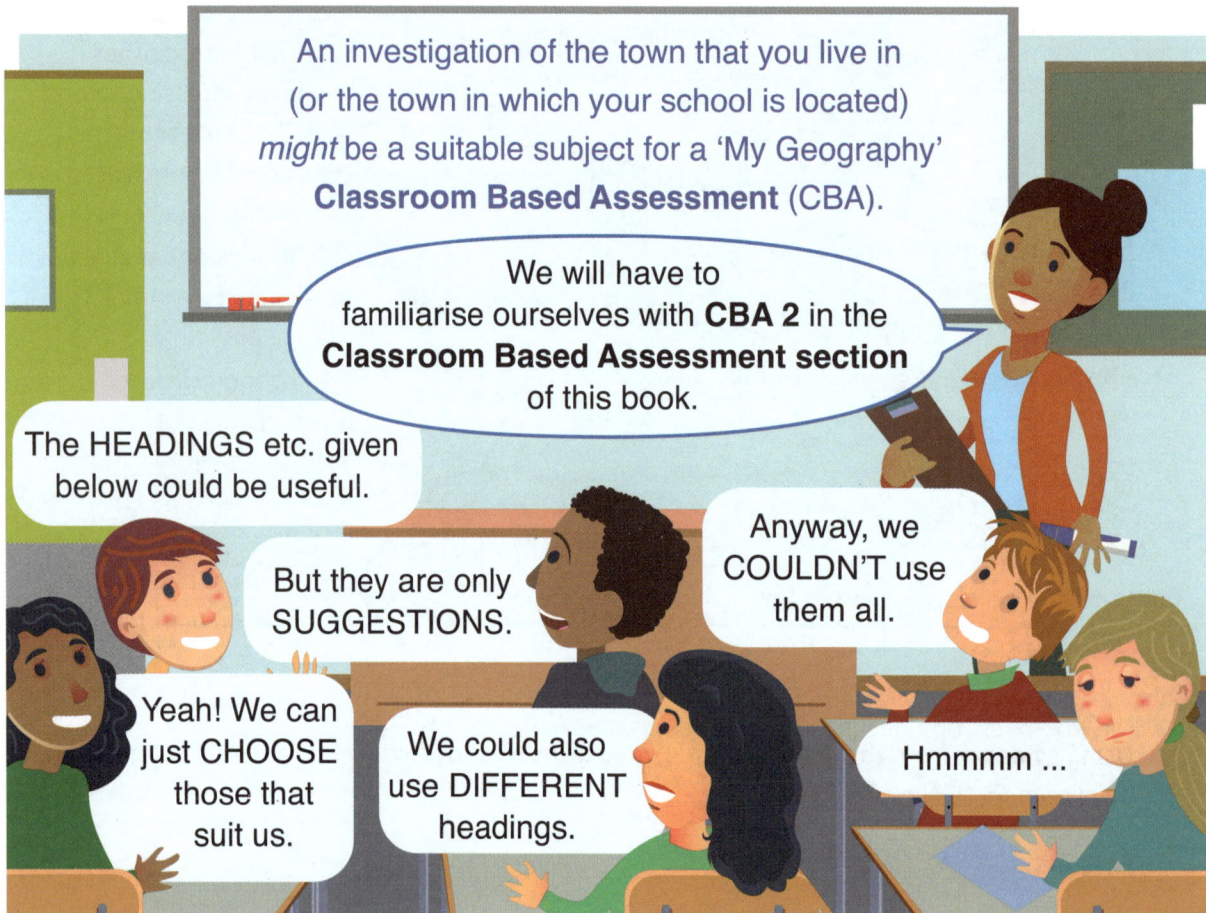

Possible headings

Introduction
- My town's **name** – meaning and origin of name
- Current **population**

Location
- Exact location (**site**) of town
- General location (**situation**) of town in relation to other towns, etc.
- Distances and directions to other places
- Transport connections to other places

Possible sources

- Internet
- Local library
- County council and planning offices
- Local history books, articles or online accounts
- Old and recent maps
- Old and recent photographs
- Past and present population statistics
- Local historians
- Older local people
- Monuments and memorials in your town

21. Rural and Urban Settlement in Ireland

> **Reasons for my town's location:** River (water supply/transport route)? Bridge over river? Meeting point of roads? Sheltered harbour? Port? An easily defended site?

Growth and change over time
- What remains of the original settlement?
- Oldest/most recent streets, buildings, etc.
- Examples of Georgian, nineteenth-century, modern architecture
- Any particularly old or interesting buildings

Causes of growth and change over time
- Past or present **sources of employment**, e.g. factories, mills, hotels, schools, shops, businesses
- **Recreational** facilities or tourist attractions
- In-migration
- Is my town now a **commuter town** of a nearby city? Have improved roads or rail connections facilitated this?
- Did **out-migration** or the closure or decay of any factories, mills or other sources of employment cause my town to **decline**?

Social changes
- How have the lives, living conditions and activities of town-dwellers changed over time?

Current challenges for my town
- Traffic? Flooding? Housing shortages? House prices? Social problems?
- How is my town **responding** to any challenges?

My town's future
- Will my town **grow** in the future? In what direction? Why?
- Can I foresee any **other changes** to any parts of the town? (These predictions should be **evidence-based** – not just guesswork.)

Some suggestions
- A **walkabout survey** could record services (schools, supermarkets, clothes shops, hairdressing salons, cafés, etc.) in each street. Data could be presented in lists and/or bar charts.
- **Line graphs** or **bar charts** could illustrate population change over time.
- **Pre-arranged interviews or school visits** could be conducted with an older townsperson or a local historian. *All such interviews or visits must be arranged or approved by your parents or teacher.* Always prepare in advance the questions you want to ask.
- **Maps and/or aerial photographs** could be photocopied and labelled to show important information.
- **Old and new photographs and/or maps** of the same location could be used to show change over time (and enduring features) in your town.
- **Hand-drawn sketch maps** could be used to present information. Give each sketch a title, a north arrow and (where possible) a line showing scale in metres/kilometres.

115

22 Urban Change in Dublin

1. The **description** column in the grid below describes nine key terms. Each of these terms is listed in the Terms Box. Write each term alongside the correct description in the grid.

   You must know and understand the meaning of key terms relating to urban change.

Term		Description of term
1		The worldwide growth in the size and number of large towns and cities
2		A council region on the northern side of the Dublin Metropolitan Area
3		The spread of an urban area across large areas of former countryside
4		The deterioration of (usually) an inner-city area due to neglect or age
5		When old, run-down houses and buildings are demolished, the original tenants moved to new suburbs and new businesses built on the land
6		The restoration or replacement of run-down buildings and the creation of community centres, parks and other facilities that would encourage existing residents to remain in a formerly decaying area (u _ _ _ _ r _ _ _ _ _ _)
7		A large urban area that is built (usually) outside but close to an existing city; it provides services, employment and homes
8		Another term for urban renewal
9		The dominant city in a country; the population is at least twice as large as the country's next-largest city

Terms Box

Urban renewal	Urbanisation	Urban sprawl
Fingal	Primate city	Urban regeneration
New town	Urban redevelopment	Urban decay

116

22. Urban Change in Dublin

2. Examine the table of statistics below, which shows population changes in Co. Mayo and in Co. Dublin from 1926 to 2022. Co. Dublin is largely urban, while Co. Mayo is mainly rural.

Population (in thousands) of Co. Dublin and Co. Mayo, 1926–2022										
	1926	1936	1946	1956	1966	1976	1986	1996	2006	2022
Co. Dublin	505	586	635	705	795	920	1,012	1,058	1,186	1,430
Co. Mayo	172	161	148	133	115	112	115	110	124	137

a. How do the statistics suggest that **rural-to-urban migration** took place over time in Ireland? Refer to some statistics in your answer.

_____
_____
_____
_____
_____
_____
_____

b. Describe two **causes** of rural-to-urban migration in Ireland.

Cause 1: _____
_____

Cause 2: _____
_____

c. Describe two **effects** on Dublin of rural-to-urban migration.

Effect 1: _____
_____

Effect 2: _____
_____

3. Examine the map in Figure 1, which shows the **growth of Dublin City over time**. Referring to the map, circle the correct options in the passage below.

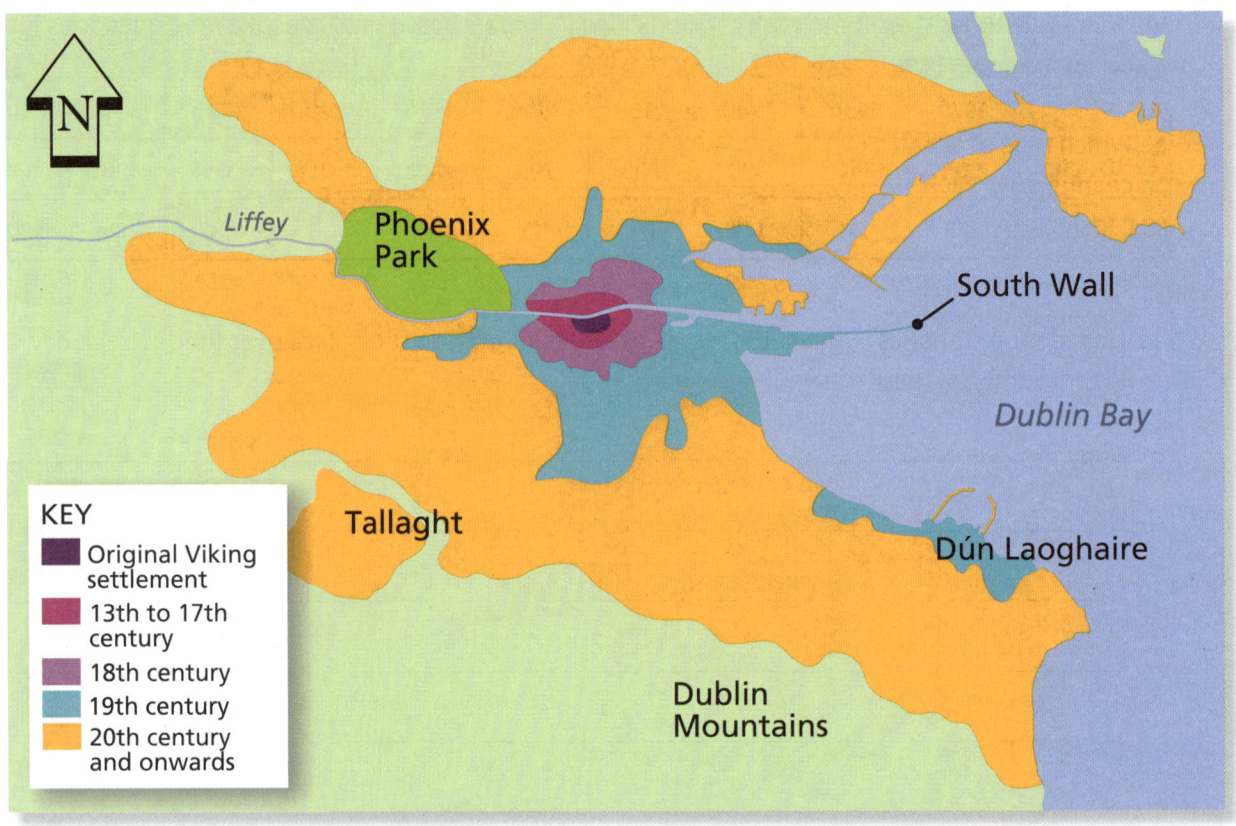

Figure 1: The growth of Dublin over time

Dublin began as a small **Viking/thirteenth-century** settlement on the **north/south** side of the River Liffey. The city grew gradually **from west to east/from the centre outwards**. It grew **slowly/very rapidly** up to the eighteenth century. It then began to grow **more/less** rapidly.

Dublin extended to the South Wall in the **18th/19th** century. At that time, outlying urban areas such as **Dún Laoghaire/Tallaght** also developed to the **southeast/southwest** of the city.

Dublin's most rapid period of urbanisation was in the **19th/20th** century and onwards. Its growth to the east has been limited by the presence of **Phoenix Park/Dublin Bay**. Its southward growth is limited by **Dublin Bay/the Dublin Mountains**, where **local protests/steep slopes and high altitudes** prevent further urban sprawl.

4. The cartoons in Figures 2 and 3 respond to types of urban change, such as inner-city decay, urban redevelopment, urban sprawl and the building of commuter towns.

   With another student, examine each cartoon closely. Prepare answers to the following questions in relation to *each* cartoon.

   a. Which type of **urban change** does each cartoon refer to?

   b. Is the cartoonist **in favour of or critical of** that urban change?

   c. **Why**, do you think, does the cartoonist have that opinion?

   Be prepared to discuss your answers with the rest of your class.

Figure 2

Figure 3

5. a. Look at Figures 4 and 5, which illustrate **urban change**. In each case:
- Name the urban change that is illustrated
- Describe at least one **cause** of that change.

Figure 4

Figure 4

Type of change: _____

_____

Cause(s): _____

_____

_____

_____

_____

Figure 5

Figure 5

Type of change: _____

_____

Cause(s): _____

_____

_____

_____

_____

b. Explain **one negative effect** of urban change in a *named* town or city you have studied.

_____

_____

_____

_____

_____

Primary Economic Activities 23

1. Are the workers listed below involved in **primary, secondary or tertiary economic activities**? Group each profession listed in its correct category in the grid provided.

> musician   baker   accountant   engineer   miner   teacher   video editor
> farmer   garage worker   shop assistant   peat bog worker   printer   hairdresser
> priest   quarry worker   carpenter   weaver   professional footballer   tour guide
> factory worker   forestry worker   refuse collector   surgeon   market gardener
> actor   nurse   oil driller   postman/woman   taxi driver   potter   cement
> manufacturer   bar worker   fisherman/woman   software programmer

Primary activities	Secondary activities	Tertiary activities

2. The Fact File below describes and explains some **connections between physical landscape and four types of agriculture**. With the help of the Fact File, identify the types of farming best suited to each of the areas labelled **A** and **B** in the OS map extract in Figure 1 and to the areas labelled **C** and **D** in the photographs. Account for (give reasons for) your answer in each case.

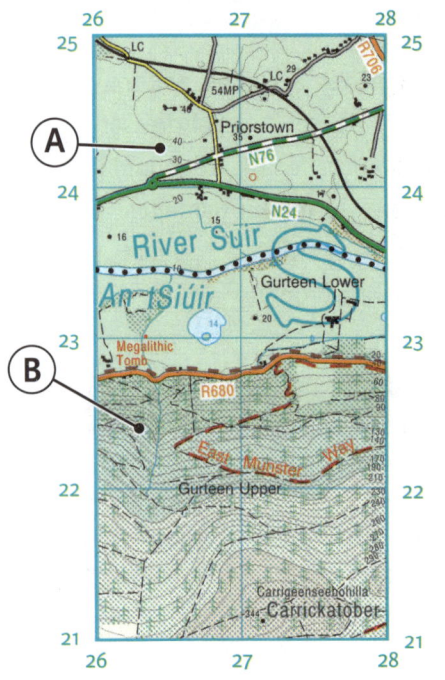

Figure 1

Fact File

Cattle farming

Flat or gently sloping lowlands

- Fertile lowlands provide rich grass
- Not too cold or windy
- Cattle are not sure-footed on steep ground
- River floodplains provide grass even in dry periods

Tillage farming

Flat or gently sloping lowlands

- Likely to have deep, fertile (brown) soil
- Flat or gentle gradients (and big rectangular fields) suit large machinery
- Better if land is well drained by a river

Coniferous woodlands

Infertile ground or lower mountain slopes

- Needle-like foliage can tolerate cold
- Shallow roots can survive in poor, shallow soils
- Roots bind soil together and prevent soil erosion on hill slopes
- Cannot survive in very high, exposed mountains

Highland sheep farming

High, steep-sided mountains

- Sheep are protected from cold, windy conditions by their woolly coats
- They are sure-footed on steep slopes
- They can survive on very rough mountain pasture

23. Primary Economic Activities

A

Type of farming: _____

Reasons: _____
_____
_____
_____
_____

B

Type of farming: _____

Reasons: _____
_____
_____
_____
_____

C

Type of farming: _____

Reasons: _____
_____
_____
_____
_____

D

Type of farming: _____

Reasons: _____
_____
_____
_____
_____

3. Every farm operates as a **system** with **inputs** (things needed for or used in the farm), **processes** (farming activities) and **outputs** (things produced by the farm).

   Examine the items in the Selection Box and write each appropriate item in its correct place – **Inputs**, **Processes** or **Outputs** – in Figure 2.

   Note: Some farm outputs can also be used as inputs in the farm system. For example, cattle manure is an output that later becomes an input when it it is used to fertilise the soil. Write these outputs in the **Feedback loop** in Figure 2.

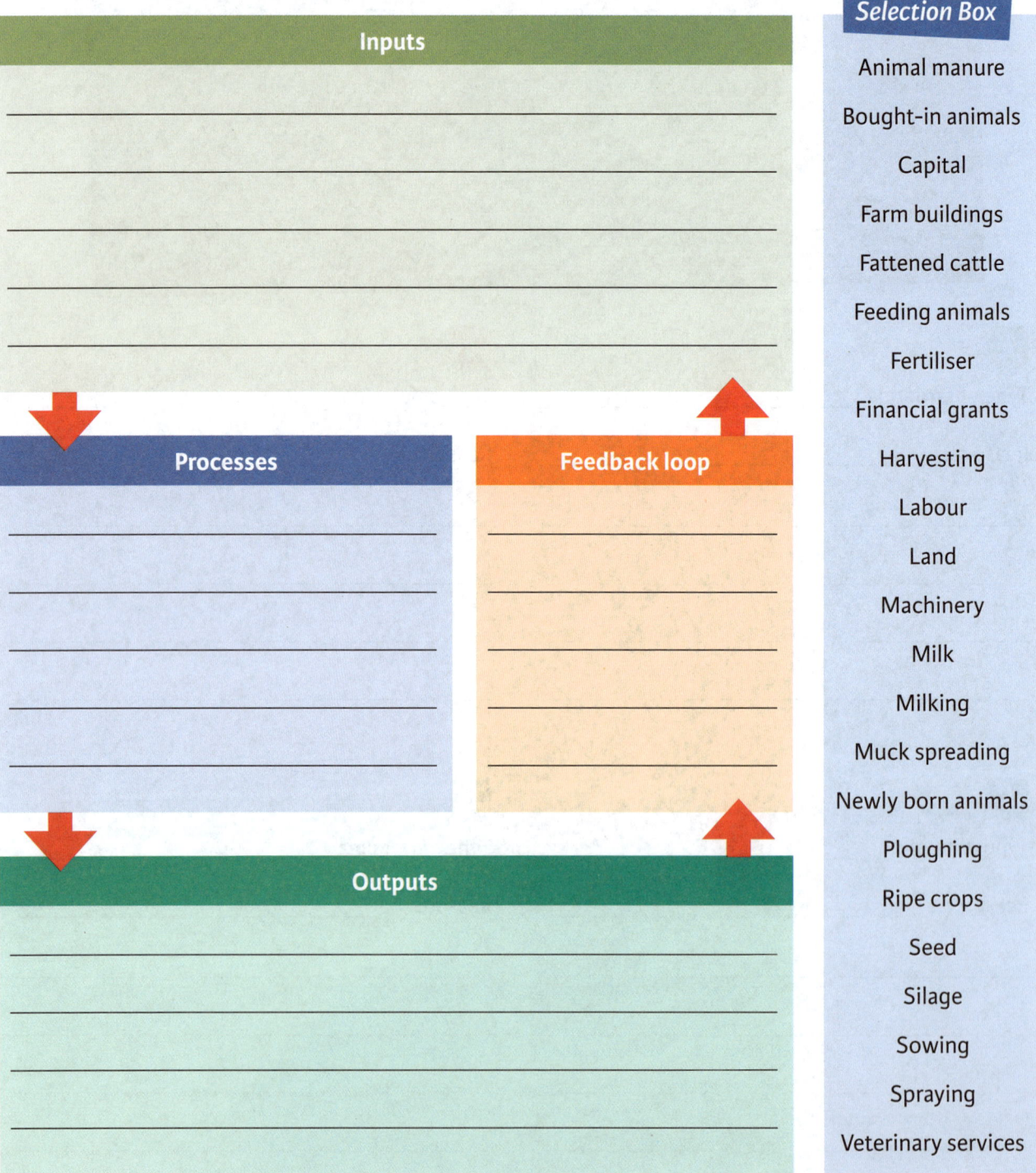

Figure 2: A farm as a system

Exploitation of Natural Resources

24

1. Use this **word game** to identify important terms, names and places.

Clues Across:

1. The type of exploitation that allows a resource to be available and usable in the future.
2. The company responsible for water services in Ireland.
3. This country developed a project called the Great Green Wall to combat desertification.
4. Limits on fish catches set by the European Union.
5. A site, usually in a rural area, that has not previously been built on.
6. The production, processing and selling of agricultural products by large businesses.
7. Serious reduction in quality or quantity of a resource, usually by overexploitation.

Clues Down:

8. The sea labelled **X** on the small map on this page.
9. Watering land by artificial means to promote plant growth.
10. Breeding, rearing and harvesting animals and plants in water. Fish farming is an example.
11. The large-scale removal of forests.

125

Junior Cycle Geography CYCLONE SKILLS BOOK

Special focus: Cod fishing

Sustainability and unsustainability in the North Sea

2. Examine the graph in Figure 1 and answer the questions that follow.

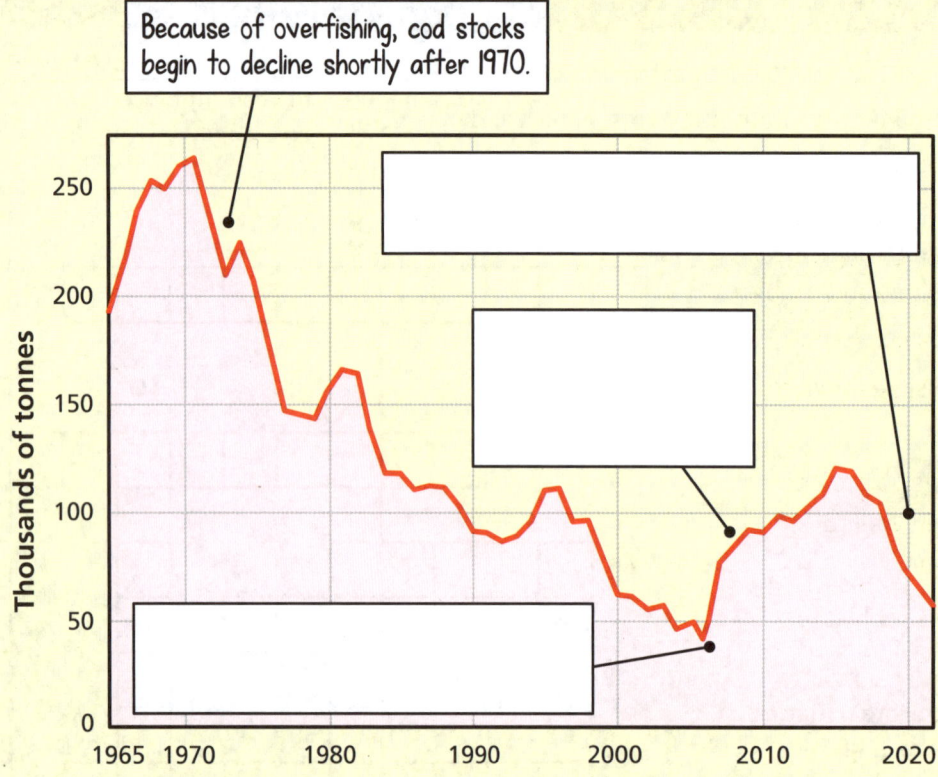

Figure 1: Change over time in North Sea cod stocks, 1965–2020

a. What does the graph show?

_____

b. In which year were cod stocks at their highest? _____

c. What was the approximate size of cod stocks in that year? _____

d. In which year were cod stocks at their lowest? _____

e. What was the approximate size of cod stocks in that year? _____

f. Calculate the approximate range of (difference between) the highest and lowest cod stocks shown. Tick (✓) the correct answer.

  Less than 100,000 tonnes ☐

  Between 100,000 and 200,000 tonnes ☐

  More than 200,000 tonnes ☐

24. Exploitation of Natural Resources

3. Read the news article below and carry out the activities that follow.

The sad story of North Sea cod

For many years the North Sea has been a major source of European cod. Up until the 1970s, North Sea cod stocks seemed to stand firm against the ruthless efficiency of modern fishing methods.

Then, shortly after 1970, cod stocks buckled under the pressures of overfishing. Stock depletion led to rapidly declining catches. By 2006, catches had declined by more than 80% and panic began to strike the fishing industry. Something had to be done before North Sea cod disappeared.

A European 'cod recovery plan' sought to restore stocks to **sustainable** levels, from which they could continue to grow naturally. The numbers of fishing days were limited.

Some boats were taken out of commission. Hunting young fish in 'cod nursery areas' was banned. New nets were made with larger mesh, so that more young fish could escape through them.

The cod recovery plan seemed to work. Both stocks and catches began to recover, and by 2015 some experts began to believe that a secure future was in sight for the cod-fishing industry. But more disaster was brewing.

Global warming had caused water temperatures in the North Sea to rise. This negatively affected the breeding of cold-water fish such as cod.

Fishing trawlers were required, under a quota system, not to bring ashore more than a given weight of fish. Some trawlers kept to their quotas by simply dumping dead (usually young) fish back into the sea.

Then, in 2020, came Brexit – the UK left the EU. This allowed Britain (which controlled more than 40% of the entire North Sea cod quota) to fish its own North Sea waters without being restricted by EU fish-conservation rules.

Since 2017, already depleted cod stocks have been declining rapidly. We do not know what the final outcome will be. But we would do well to be aware of the fact that the continued **overexploitation** of a resource can cause the **depletion** and often the **extinction** of that resource.

a. Using the above news article and the graph in Figure 1 (page 126), write each of the following labels into the correct box in Figure 1. (One label has been entered for you.)

- ~~Because of overfishing, cod stocks begin to decline shortly after 1970.~~
- Stocks reach an all-time low by 2006.
- Stocks recover under the 'cod recovery plan'.
- Stocks decline again after 2015, largely owing to global warming.

127

b. The first paragraph in the news article on page 127 refers to 'the ruthless efficiency of modern fishing methods'. Using information from your textbook, describe some of the fishing methods to which the news article might be referring.

_____
_____
_____
_____
_____
_____
_____
_____

c. Describe the 'cod recovery plan' referred to in the news article.

_____
_____
_____
_____
_____
_____
_____
_____

d. Why did some experts believe that the cod recovery plan was working?

_____
_____
_____
_____

24. Exploitation of Natural Resources

e. Explain how global warming affected stocks of cod in the North Sea.

_____
_____
_____
_____
_____
_____

f. Explain in your own words the meaning of each of the following terms, which are given in bold in the news article:

sustainable (levels of fish stocks)

_____
_____
_____

overexploitation (of a resource)

_____
_____
_____

depletion (of a resource)

_____
_____
_____

extinction (of a resource)

_____
_____
_____

25 Secondary Economic Activities

1. The box at the bottom of the page lists the **terms** that are defined in the table. In the spaces provided, write each term next to the correct **definition**. Then learn the definitions.

Term	Definition
Multinational corporation (MNC)	An industry with its headquarters in one country but with branches in many countries
Heavy industry	Makes large, bulky products from heavy/bulky raw materials
Light industry	Makes relatively small products from small (or small quantities of) raw materials
Footloose industry	An industry not tied to one location – it can operate successfully from a wide range of places
Capital	Money needed to set up and operate an industry
Industrial linkages	Connections that might encourage some industries to locate close to each other. Materials often flow between one factory and another – for example, flour (the output of a mill) becomes an input for a bakery
Market	The place(s) where a finished product is sold
Labour	The workers needed to operate an industry. Many industries are located near large towns or cities that provide many workers with a range of skills and qualifications
Sustainable industry	Industry that does not harm people or the planet. It does not waste energy or other natural resources and does not damage the environment, communities or human rights

Terms Box

Capital
Footloose industry
Heavy industry
Industrial linkages
Labour
Light industry
Market
Multinational corporation (MNC)
Sustainable industry

25. Secondary Economic Activities

2. Examine the Ordnance Survey map extract of part of Cork City in Figure 1 on page 132. Find the **industrial estate** located at **W 67 69** on the map.

   a. What is an industrial estate?

   _____

   _____

   _____

   _____

   _____

   b. Would you be more likely to find **heavy industry** or **light industry** in an industrial estate?

   _____

   c. Industrial estates offer the benefits of **industrial linkages** and **shared services** to individual industries. What is meant by each of these terms?

   Industrial linkages: _____

   _____

   _____

   _____

   _____

   Shared services: _____

   _____

   _____

   _____

   _____

d. Are the factories in the industrial estate at W 67 68 suitably located in relation to each of the factors listed below? (*Refer to the map* in Figure 1 in each of your answers.)

Suitable site

Transport facilities

Labour force

Possible local market

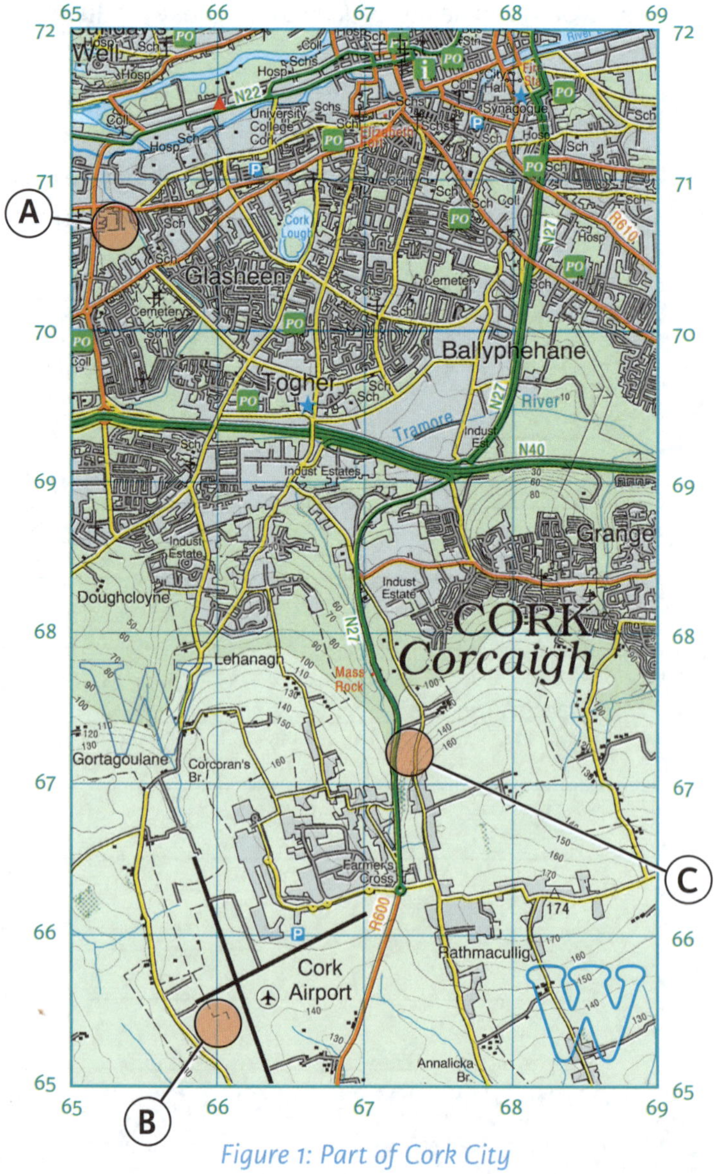

Figure 1: Part of Cork City

e. Imagine that you are a Cork City or Cork County planning engineer and that you refused three separate applications to build a large factory at each of the locations labelled **A**, **B** and **C** on the map in Figure 1. Give one **reason** why you refused each planning application, referring to the map in each of your answers.

Location A: _____

_____

Location B: _____

_____

Location C: _____

_____

3. Examine the aerial photograph of Enniscorthy, Co. Wexford in Figure 2.

Would the large field in the background of the photograph be a suitable location for a new factory manufacturing healthcare products? Give one reason **for** and one reason **against** the proposal. Refer to the photograph in each case.

For
_____
_____
_____
_____

Against
_____
_____
_____
_____

Figure 2

4. **a.** Give the *name* and *function* of a **local factory** that you have studied.

_____

b. Complete the diagram below to show how this local factory **functions as a system** with **inputs**, **processes**, **outputs**, and other factors.

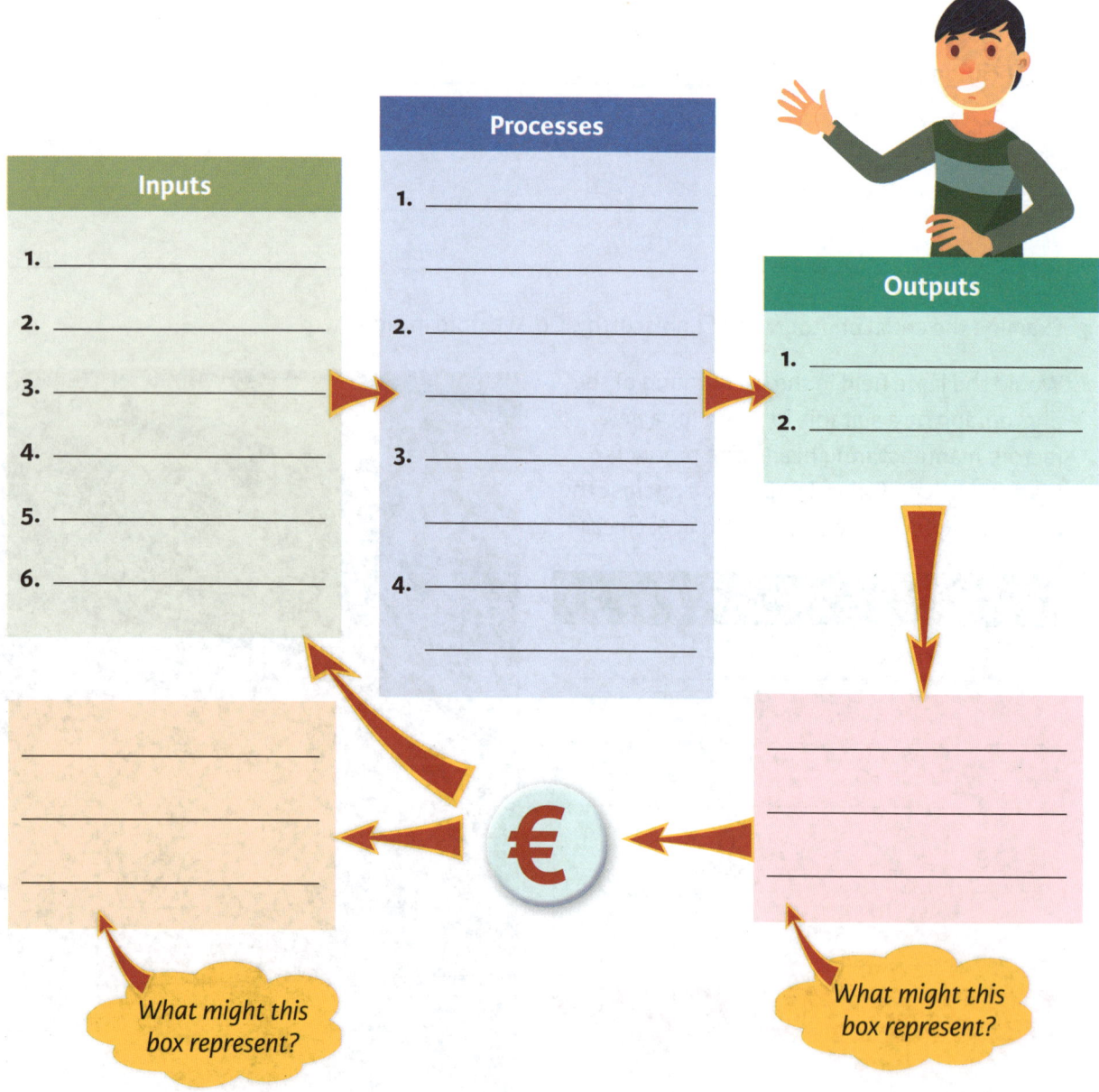

c. Name one **local benefit** and one **benefit to the country** of the factory you named in (a) above.

Local benefit: _____

Benefit to the country: _____

_____

Tertiary Economic Activities   26

1. What is meant by the term **'tertiary economic activities'**?

    _____

    _____

2. Six types of tertiary economic activity are listed in **Column A** of the table below. The yellow box lists six professions that are associated with these activities.

    a. In **Column B**, match each profession with the correct economic activity.

    b. In **Column B** of the **Tourism** section of the table, name three other professions associated with tourism.

Column A	Column B
1. Information technology	
2. Transport	
3. Professional services	
4. Personal services	
5. Retail	
6. Tourism	

Shop worker   Hairdresser   Travel agent

Truck driver   Computer programmer   Lawyer

Bunowen Bay, Connemara, Co. Galway

The Gap of Dunloe, MacGillycuddy's Reeks, Co. Kerry

Cruising on the Shannon

The Burren, Co. Clare

3. 'Ireland's physical landscape attracts tourists to our **coastline**, **mountains** and **inland waterways**, as well as to **unique areas** such as the Burren, Co. Clare.'

Explain how the **physical landscapes** listed above would attract tourists. Refer in your answer to examples of Irish features/places. If you wish, use the photos on this page to help you.

26. Tertiary Economic Activities

4. 'The physical geography of Ireland has stimulated tourism. But tourism also impacts on our country's physical geography.'

   Explain how tourism has **impacted on Ireland's physical landscape**. Refer to specific places in your answer. If you wish, use the news article and the photographs on this page to help you.

   _____
   _____
   _____
   _____
   _____
   _____
   _____
   _____
   _____
   _____
   _____
   _____
   _____
   _____
   _____
   _____
   _____
   _____

How might large caravan parks, holiday villages or hotels change the natural environment of tourist areas? (Consider visual impacts, sewage disposal, noise, road traffic.)

The Wild Atlantic Way

The Wild Atlantic Way is a series of roads that run along the beautiful west coast of Ireland. Increasing tourist numbers have led to many of these roads being upgraded. Wider roads, roundabouts and car parks impact on the physical landscape.

The remains of a beach party in Ballybunion, Co. Kerry. How might the beach, the sea and sea creatures be impacted by this kind of litter?

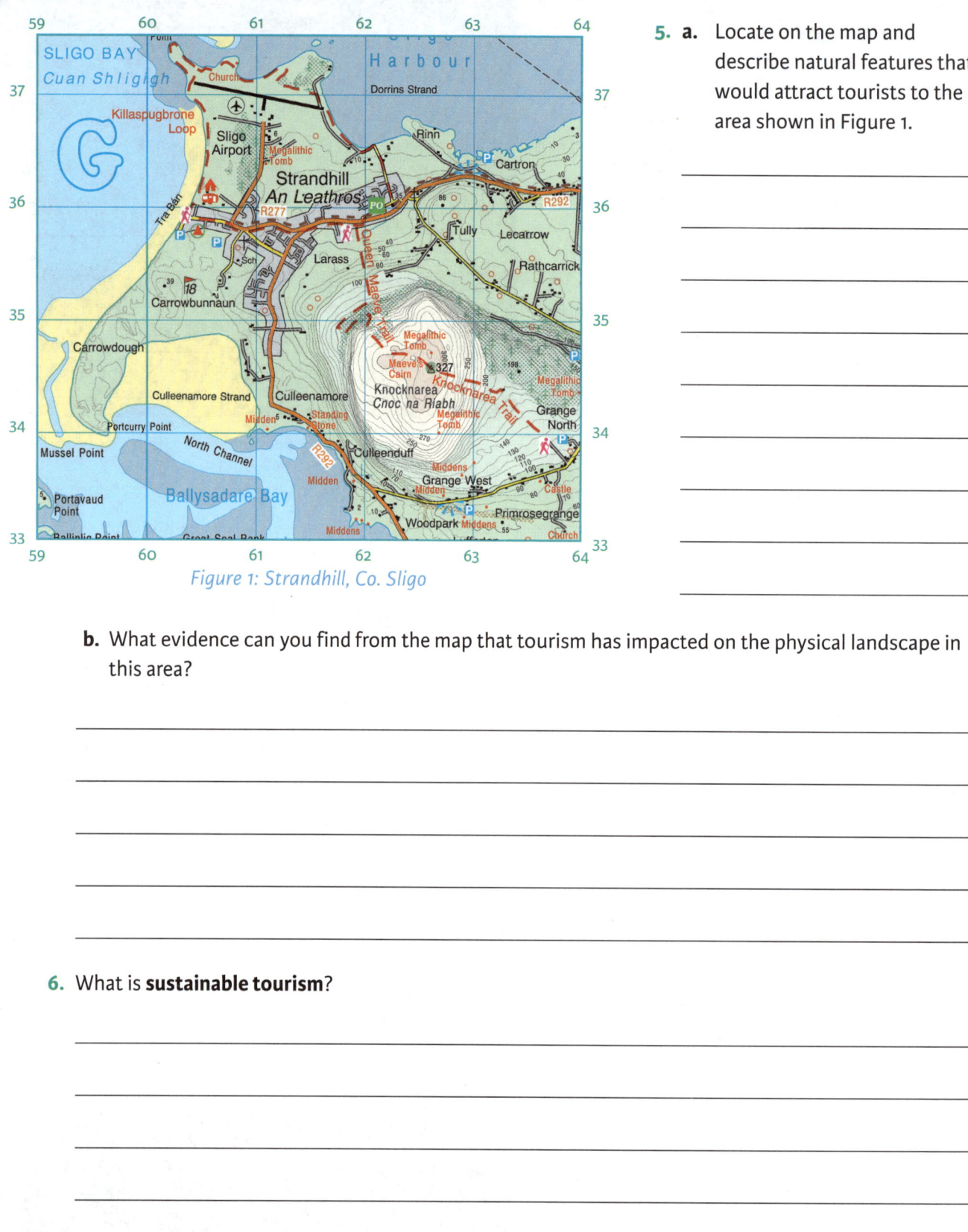

Figure 1: Strandhill, Co. Sligo

5. a. Locate on the map and describe natural features that would attract tourists to the area shown in Figure 1.

___

b. What evidence can you find from the map that tourism has impacted on the physical landscape in this area?

___

6. What is **sustainable tourism**?

___

26. Tertiary Economic Activities

¡Hola!

More than 80 million tourists visit Spain every year. Most of them flock to Spain's Mediterranean coastline in search of 'sun holidays', especially during the hot, dry and sunny summer season.

Climatic conditions in a Spanish Mediterranean coastal tourist resort												
Month	J	F	M	A	M	J	J	A	S	O	N	D
Temperature (°C)	9	11	13	16	21	26	28	28	19	17	15	12
Precipitation (mm)	64	53	60	39	20	5	0	1	14	50	90	70
Daily hours of sunshine	5	6	6	7	9	10	11	11	8	5	5	4

7. Using the information provided in the table above, explain fully, under three separate headings, why many tourists visit Spain's Mediterranean coast during the months of June, July and August.

139

A Spanish coastal resort in summer

8. a. Apart from the climate, what features of **physical geography** attract tourists to the resort shown in the photograph?

   _____

   _____

   b. How has mass tourism altered the physical landscape of the area shown?

   _____

   _____

   c. Do you think that mass tourism has had a positive or a negative impact on the area shown? Explain your answer.

   _____

   _____

   _____

   _____

   _____

26. Tertiary Economic Activities

Figure 2: Mass tourism and transport along Spain's Mediterranean coast

9. **a. Why** and **in what ways** has mass tourism helped to cause the development of major transport facilities along the coastal area shown? Refer to Figure 2 in your answer.

   _____
   _____
   _____
   _____

   b. How might the development of major transport facilities help to encourage the further growth of mass tourism in the area shown? Refer to Figure 2 in your answer.

   _____
   _____
   _____
   _____

27 Economic Development and Inequality

1. Economically, the world is very unevenly divided. Many wealthy countries are located in Europe and North America. Most poor countries are located where the majority of the world's population lives: in Africa, South and Central America and Asia.

   The following terms are used to broadly describe the wealthy and less wealthy parts of the world:

   > the North   the majority world   the developing world
   > the developed world   the minority world   the South

   State which terms are used to refer to:

   - the economically wealthier parts of the world: _____
   _____

   - the economically poorer parts of the world: _____
   _____

2. Examine the map in Figure 1 on the opposite page, which groups the countries of the world into four income categories.

   a. Which **continent** has the largest number of high-income countries? _____

   b. Which **continent** has the largest number of low-income countries? _____

   c. Find out the names of the countries labelled **A–D**. In the grid below, name each country and state its economic category.

	Country	Economic category
A		
B		
C		
D		

27. Economic Development and Inequality

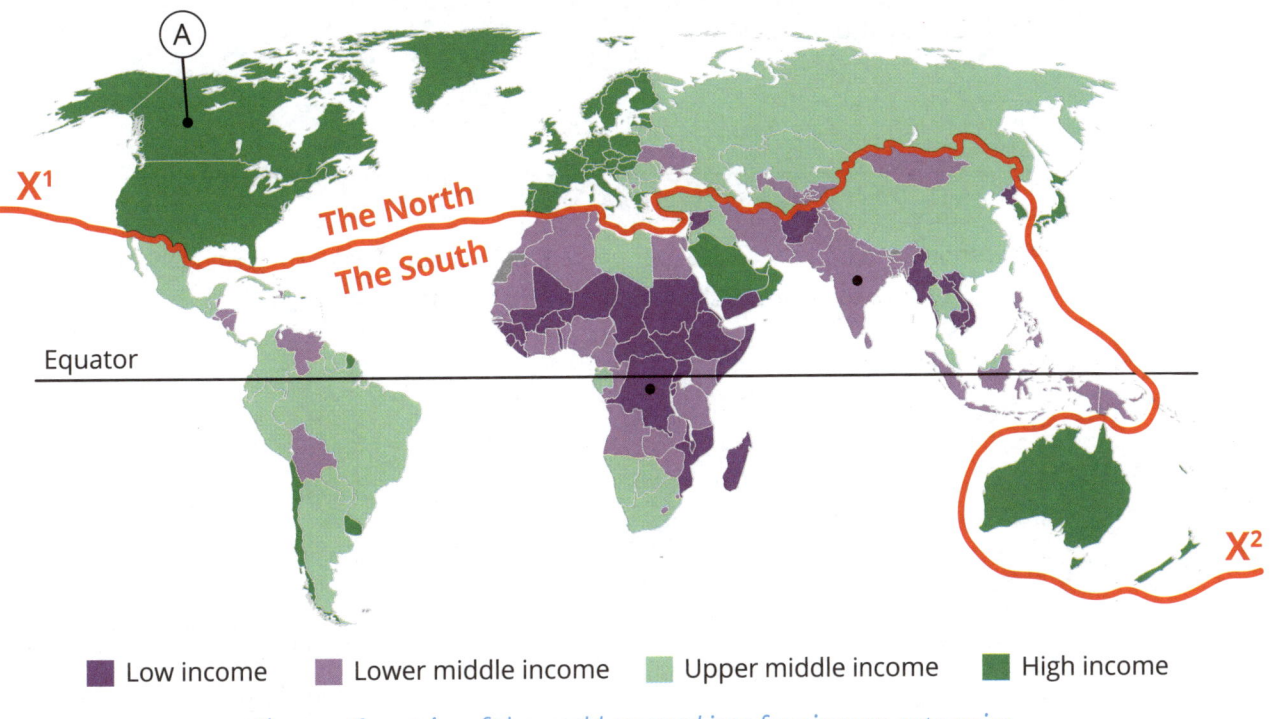

Figure 1: Countries of the world grouped into four income categories

d. What does the line **X¹–X²** represent? _____

e. Explain the term 'majority world'.

_____

_____

_____

f. Is the country labelled **A** part of the majority world or of the minority world? _____

g. The area north of the line **X¹–X²** is called 'the North' because (tick (✓) the correct box):

All the countries in that area are north of the equator. ☐

Most of its countries are located north of the countries in 'the South'. ☐

All of its countries are located north of the countries in 'the South'. ☐

3. **Gross national income** (GNI) is widely used as an **indicator of economic development**.

   a. Define the term 'gross national income'.

   _____

   _____

   b. The grid below shows the GNI per capita (per person) of four countries.* Draw and label a suitable graph or chart in Figure 2 to illustrate the information in the grid.

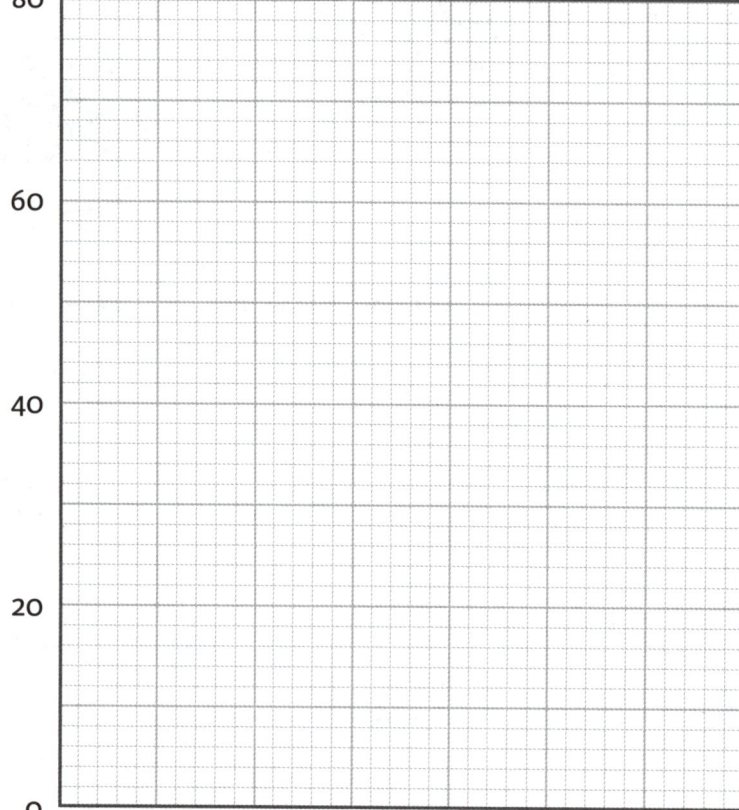

China	$11,890
Ireland	$74,520
Nigeria	$2,100
UK	$45,380

* Figures are given in US dollars.

Figure 2

Think carefully before you answer the next three questions.

   c. Does GNI per capita give an accurate idea of the income of every person in a country? Explain your answer.

   _____

   _____

   _____

   _____

27. Economic Development and Inequality

d. Is GNI alone sufficient to measure the levels of people's well-being and development? Explain your answer.

_____

_____

_____

e. Look at the cartoon in Figure 3. What point is it making about GNI as an indicator of economic development?

_____

_____

_____

_____

_____

_____

_____

_____

_____

_____

_____

_____

Figure 3

f. Define the following terms.

Life expectancy: _____

_____

Adult literacy rate: _____

_____

145

4. The newspaper article and Figure 4 help to explain how **unfair trading patterns** are partly responsible for **economic inequality** around the world. Examine the newspaper article and Figure 4 and answer the questions.

   a. Read the news article below. *Circle* the correct options.

   # EUROPEAN EMPIRES:
   ## A historic cause of global inequality

   One of the historic causes of global inequality is the former existence of European empires.

   Several European countries conquered and controlled numerous colonies throughout the world, particularly in Africa, Asia and other parts of the world that we now call **'the South'/'the North'**. These colonies were robbed of their natural resources and used to provide **expensive machinery/cheap raw materials** and other such products for the benefit of the imperial powers. This trade made the **imperial powers/colonies** rich, but kept the colonies **rich/poor**.

   During the twentieth century, the European powers collapsed and their colonies became **independent countries/new imperial powers**. But many of the old, unfair trading patterns remained. These patterns helped to **do away with/maintain** inequalities between North and South.

   b. Examine the map in Figure 4 on the opposite page and answer the following questions.

      i. What is an imperial power? _____
      _____
      _____
      _____

      ii. What is a colony? _____
      _____
      _____
      _____

27. Economic Development and Inequality

The map shows the extent to which Africa was colonised by European imperial powers in the 19th and early 20th centuries. Some colonies came under the control of different European powers at different times.

You may need to look up maps of European and African countries to answer some of the questions that follow.

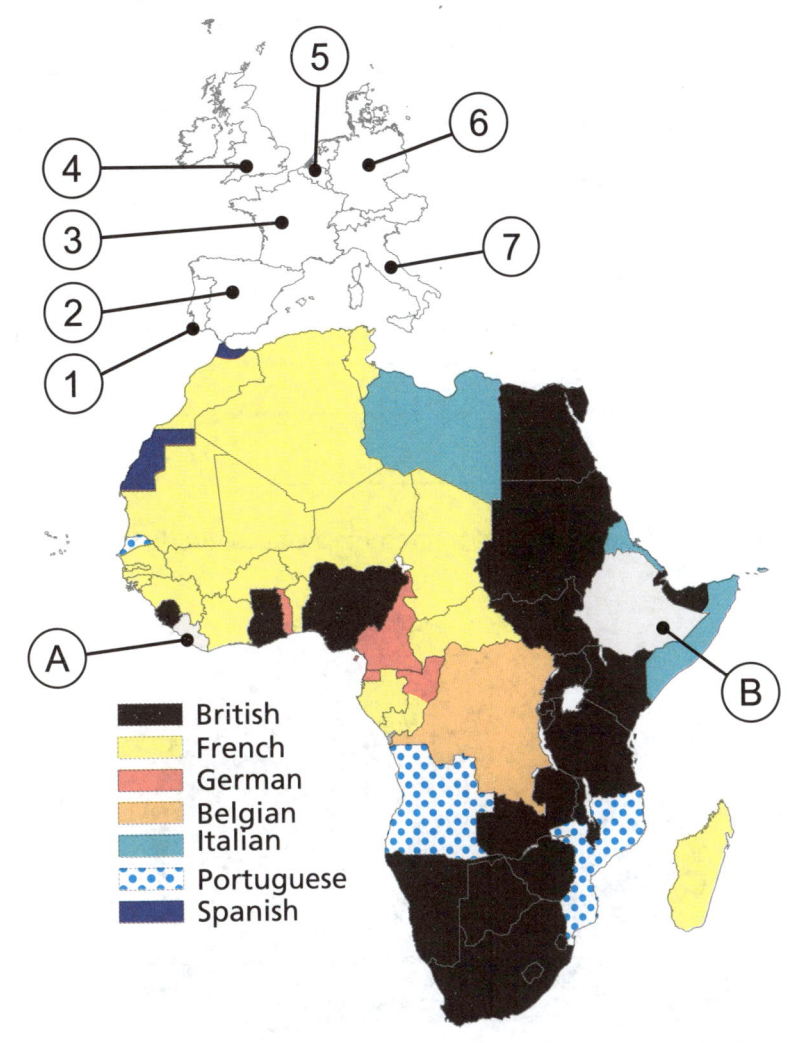

Figure 4: Africa – a conquered continent of European colonies in the 19th and early 20th centuries

c. Judging by the map, what percentage of Africa was colonised by European powers? Tick (✓) the correct box.

Less than 50% ☐

50–90% ☐

More than 90% ☐

d. The European powers that had colonies are labelled 1–7 on the map in Figure 4. Name each of these countries.

1. _____   5. _____

2. _____   6. _____

3. _____   7. _____

4. _____

e. Which two imperial powers had most colonies in Africa?

1. _____   2. _____

147

f. Name the imperial power that colonised each of these present-day countries.

Algeria: _____

Egypt: _____

South Africa: _____

g. Only two African countries, labelled **A** and **B** on the map, were not colonised by European powers in the 19th and early 20th centuries. Name these two countries.

A: _____

B: _____

5. The cartoon in Figure 5 represents a view of **world trade**.

Figure 5: A view of world trade

a. How do we know that the pig in the cartoon represents the commercially developed North?

_____

_____

b. What is the main message of the cartoon?

_____

_____

_____

_____

c. Do you think the cartoon's message is a fair one? Explain your answer.

_____

_____

_____

_____

_____

_____

Human Development and Development Assistance

28

1. What is meant by the term **'human development'**?

   _____

   _____

   _____

   _____

2. Each of the proposals in the Proposals Box presents a viewpoint on how best to promote human development in economically poorer countries of the majority world (the South).

   In the spaces below:
   - Write in rank order the **four** proposals you think would **best promote human development**. (Rank the proposal you like best number 1.)
   - Write two proposals you think would be unhelpful to human development.

   Best proposals:

   1. _____
   2. _____
   3. _____
   4. _____

   Unhelpful proposals:

   _____
   _____

Proposals Box

- Top-class hospitals with top-class equipment
- Basic housing for all
- Very low taxes
- Internet access for all
- Food and clean water for all
- A large army and air force
- Freedom for all to say what they wish
- Education for all, at primary level at least
- Foreign military bases for employment and protection
- Basic healthcare for all
- Western democracy, as in the USA
- Large sports stadiums to promote recreation

Junior Cycle Geography CYCLONE SKILLS BOOK

3. Test your knowledge of key terms and concepts relating to **international aid**. Match the terms in the first table below with the descriptions in the second by writing the number of the description next to the term it matches. One match has been made for you.

Terms	Description number
Emergency aid	7
Tied aid	
Development aid	
Bilateral aid	
Multilateral aid	
Non-governmental aid	
Irish Aid	
Military (or defence) aid	

	Descriptions
1	Aid given with conditions attached that usually serve the interests of the donor country (the country that is providing aid)
2	Aid provided by multiple governments to international organisations
3	Aid provided by voluntary NGOs (non-governmental organisations) such as Trócaire and Concern
4	The provision of arms, military advisers or troops, usually to promote the donor country's military objectives
5	The name of the Irish government's official agency for international development
6	Long-term aid (schools, hospitals, training assistance, etc.) used for the sustainable development of the host country
7	Short-term aid (food, water, emergency shelter, etc.) given in times of crisis
8	Aid given by one government directly to another

What type of aid is shown in this photograph?

_____

150

28. Human Development and Development Assistance

4. What is the **Human Development Index (HDI)** and what does it measure?

   _____

   _____

   _____

   _____

   _____

5. The table below shows the HDI scores of ten countries/regions. A score of 1 is the highest possible HDI score. A score of 0 is the lowest possible score. Examine the table and answer the questions that follow.

The Human Development Index of ten countries		
Country	HDI (2022)	Rank order
Cuba	0.783	
India	0.645	
Ireland	0.955	
Iraq	0.750	
Norway	0.957	
Palestine	0.708	
South Sudan	0.500	
UK	0.930	
Ukraine	0.779	
USA	0.926	

   a. In the table, rank the HDI score of each country/region, starting with 1st for the highest score.

   b. List the countries/regions in the table that are in Europe or North America.

   _____

   _____

   _____

   _____

   _____

   c. The European and North American countries/regions shown in Figure 1 all have higher HDI scores than any other country/region shown. True or false?

   True ☐   False ☐

   d. By how much does the HDI of the highest-ranked country/region exceed the HDI of the lowest-ranked country/region?

   _____

6. Cartoons are often used to comment on important moral issues relating to international aid and trade. Examine the cartoons on this double-page spread and answer the questions.

Cartoon 1

Cartoon 1

a. Who or what do the people giving the money represent?

_____

b. What type of aid is being offered?

_____

c. Name two conditions attached to the aid.

1. _____

2. _____

Cartoon 2

Assume that the man turning the tap represents a wealthy donor (aid-giving) country. Which of the following points does the cartoon make? Tick (✓) the correct box.

Donor countries understand the importance of clean water. ☐

Donor countries do not want to see aid being wasted by receiving countries. ☐

Donor countries sometimes use aid as a means of controlling poor countries. ☐

Cartoon 2

28. Human Development and Development Assistance

Cartoon 3

Cartoon 3

Describe fully the message in Cartoon 3.

_____
_____
_____
_____
_____

Cartoon 4

a. What type of international assistance is referred to in Cartoon 4?

_____

b. Who or what does the hungry person represent?

_____

c. What point is the cartoon making?

_____
_____
_____
_____

Cartoon 4

29 Life Chances in a Developed and a Developing Country

Examine Amina's story on pages 155 and 156 and the useful statistics that compare some aspects of life in Nigeria with those in Ireland. Use Amina's story and the statistics to help you answer the following questions.

On this page and on page 157, briefly compare **your own** life chances with **Amina's** life chances. Use the headings given.

Education opportunities

_____
_____
_____
_____
_____
_____
_____
_____
_____
_____
_____
_____

Healthcare

_____
_____
_____
_____
_____
_____
_____
_____
_____
_____
_____
_____

29. Life Chances in a Developed and a Developing Country

Amina is a 14-year-old schoolgirl from northern Nigeria. She enjoys school and has already spent more years in **education** than the average Nigerian child. Nigeria needs many more schools and trained teachers to educate its growing population. One classroom in Amina's school has more than 100 children. But educational facilities are gradually improving. Sixty per cent of schoolchildren are girls.

Amina's infant brother died last year of a chest infection. Infant mortality rates are high in Nigeria, but they are gradually falling. Life expectancy is increasing. Females fare better than males in both categories. Nigerian **healthcare** would benefit enormously from more doctors. The country currently has an average of less than one (0.9) doctor for every 1,000 people.

Useful statistics

Average years in education:
Ireland 13   Nigeria 7

Number of doctors per 1,000 people:
Ireland 3.4   Nigeria 0.9

Life expectancy:
Ireland   Females 84   Males 80
Nigeria   Females 56   Males 54

Infant mortality rates (per 1,000 live births):
Ireland   Females 2   Males 3
Nigeria   Females 56   Males 74

Amina's story continues on the next page.

Junior Cycle Geography CYCLONE SKILLS BOOK

Amina would like to become a teacher. But she has two brothers at school and her parents will find it difficult to keep her in education for much longer. She is also needed to help her mother with housework and sometimes has to help with farm work. Many rural children leave school early to work in agriculture, which provides **employment** for 35% of Nigerian workers.

Employment (%) in three economic sectors:

	Primary	Secondary	Tertiary
Ireland	4	19	77
Nigeria	35	12	53

Amina's parents and teachers treat her and her brothers with equal respect. But old rural traditions sometimes expect different things from children of **different genders**. Many young men migrate to faraway cities in search of employment; but Amina's father would worry if she went to a strange city to do teacher training. Her mother never migrated further than to a village near where she was born, where she married Amina's father and now cares for her husband and family. Will Amina realise her ambition and become a teacher? Or will she follow her mother's life path?

A rural scene in northern Nigeria

29. Life Chances in a Developed and a Developing Country

Employment opportunities

Gender equality

30 Globalisation

1. In the spaces provided, write the meaning of each of the four terms given below.

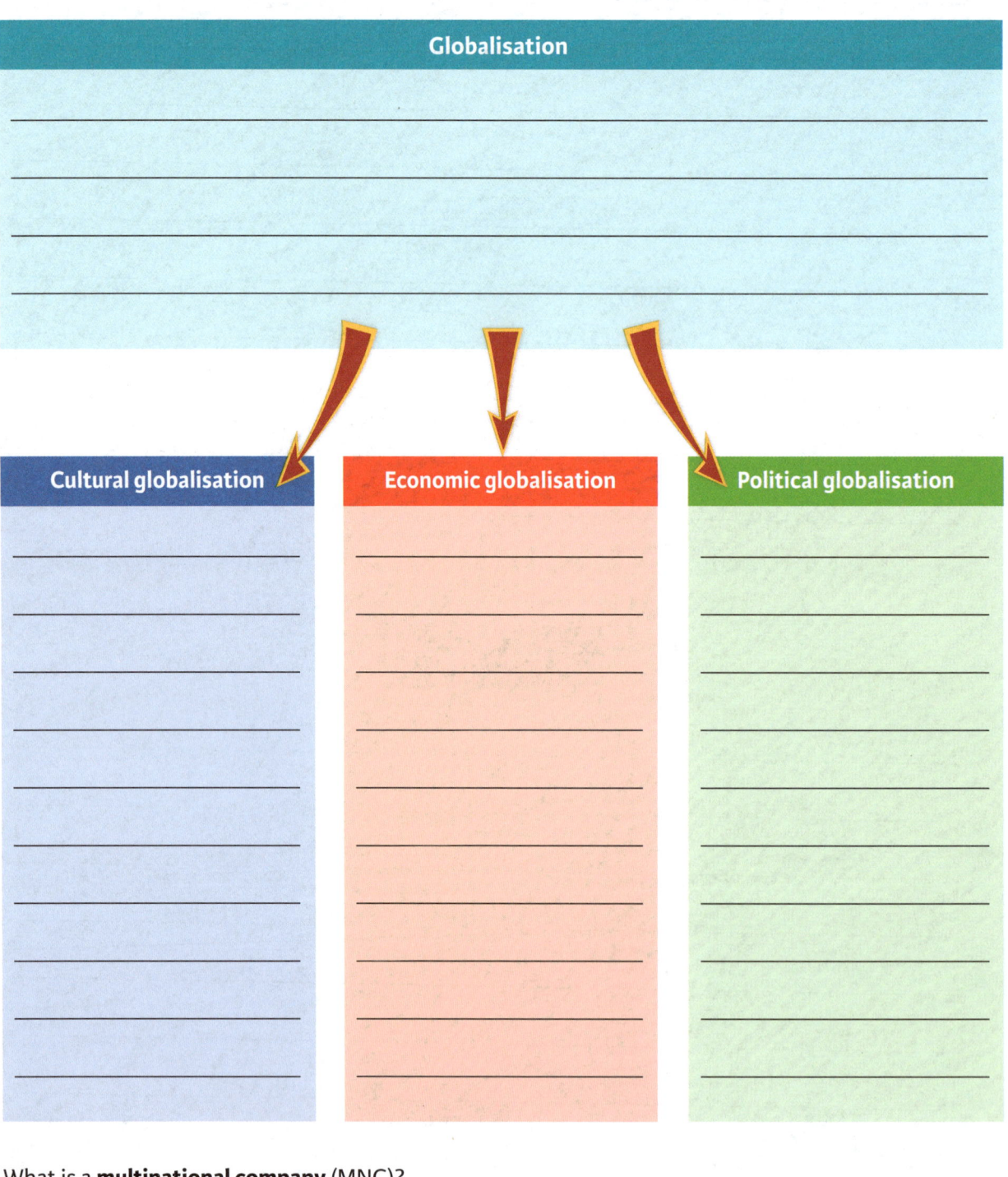

Globalisation
_____

Cultural globalisation	Economic globalisation	Political globalisation

2. What is a **multinational company** (MNC)?

30. Globalisation

3. 'Urbanisation (the growth of cities) encourages globalisation and globalisation encourages urbanisation.'

   Show that this statement is true by writing the four sentences below in their correct order in Figure 1. One box has been filled in for you.
   - Cities grow.
   - MNCs provide employment.
   - ~~MNCs set up bases in growing cities.~~
   - People migrate to cities in search of employment.

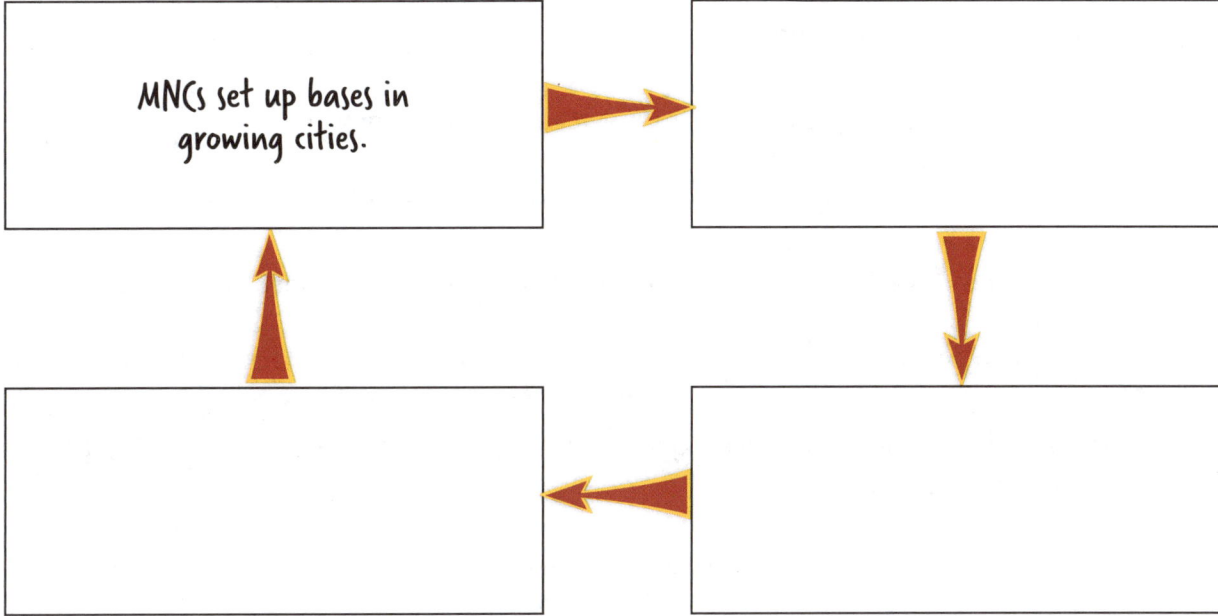

Figure 1: How globalisation and the growth of cities are linked

4. Why do MNCs choose to locate factories and offices in or near growing cities?

   _____
   _____
   _____
   _____
   _____
   _____
   _____
   _____

5. The table shows the number of people (to the nearest thousand) **resident in (living in) but born outside** Ireland between 2001 and 2021. Examine the table and answer the questions that follow.

Year	Number of people
2001	404,000
2006	613,000
2011	767,000
2016	810,000
2021	850,000

   a. Which of the following **trends** do the figures reveal regarding change over time in the number of people living in but born outside Ireland? Tick (✓) the correct box.

   A steady and very even increase in the number ☐

   A constant increase in the number ☐

   Almost no increase in the number ☐

   b. Between which two years did the greatest increase in numbers occur?

   _____

   c. Calculate the increase that occurred between the two years you identified in your answer to (b).

   _____

   d. Do the trends revealed in the table suggest that Ireland became more globalised or less globalised between 2001 and 2021?

   _____

   e. During the years 2001 and 2006, Ireland was experiencing an economic boom (a period of economic prosperity). How and why might this have affected migration in or out of Ireland at that time?

   _____
   _____
   _____
   _____
   _____
   _____
   _____

30. Globalisation

6. The incomplete bar chart in Figure 2 represents the statistics given in the table on page 160.

    a. Complete Figure 2 by inserting a bar for the year 2021.

    b. The bar chart in Figure 2 is a suitable way to represent **change over time** in the number of people resident in but not born in Ireland. Which of the following would also be a suitable way of representing this information? Tick (✓) the correct box.

    A pie chart ☐

    A line graph ☐

    A population pyramid ☐

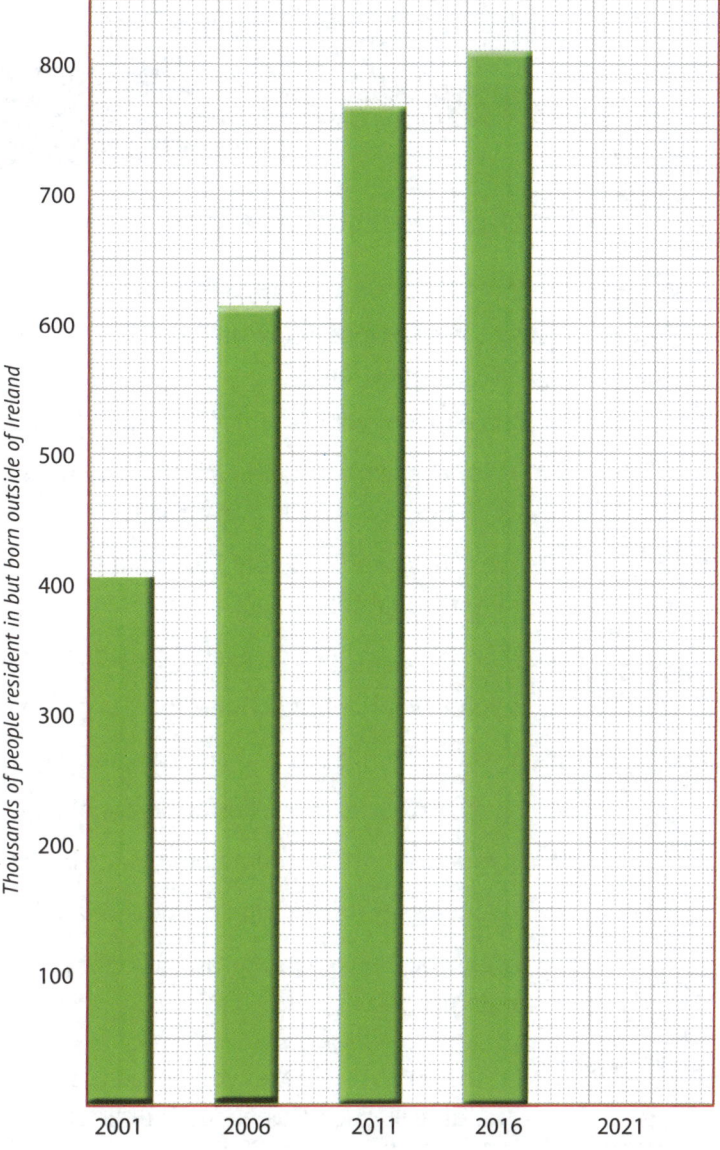

Figure 2

7. The cartoon in Figure 3 is about globalisation. The person in 'USA' represents workers in rich, developed countries. The person in 'Indonesia' represents workers in poorer, developing countries.

    Which **three** of the following points is the cartoon making?
    Tick (✓) the correct boxes.

    Workers in rich countries become unemployed. ☐

    Workers in poor countries are very badly paid. ☐

    Workers in rich countries like to show off expensive shoes. ☐

    People in poor countries hate hard work. ☐

    Fashion items produced cheaply in poor countries are sold at huge profit in rich countries. ☐

Figure 3

How globalised are you and your classmates?

On your own:

Answer the questions below. In each case, if the answer is *yes*, tick (✓) the box opposite the question.

1. Was either of your parents born outside Ireland? ☐
2. Does any member of your immediate family (mother, father, guardian, sister, brother) usually live abroad? ☐
3. Does any member of your immediate family work for a foreign multinational company? ☐
4. Have you ever been on a family holiday outside of Ireland? ☐
5. Is your favourite food originally from another country (e.g. pasta, curry)? ☐
6. Is your favourite sports team from another country? ☐
7. Is your favourite music performer/singer/band from outside Ireland? ☐
8. Is your favourite TV programme made in another country? ☐
9. Is your greatest hero from outside Ireland? ☐
10. Could you hold a basic conversation in a language other than Irish or English? ☐

Count the number of boxes you ticked to see how far globalisation influences your life:

10 boxes ticked	Completely
7–9 boxes ticked	Greatly
4–6 boxes ticked	Moderately
1–3 boxes ticked	Slightly
0 boxes ticked	Not at all

Draw a circle around the category you fit into.

Class activity

Your teacher could ask for a show of hands to find out how many students fit into each of the categories listed above. These results could be listed on the whiteboard.

Have a class discussion on the degree to which your class as a whole is affected by globalisation.

Junior Cycle Geography CYCLONE SKILLS BOOK

What are CBAs?

As part of your Junior Cycle Geography programme, you will complete two CBAs and a Written Assessment Task.

What on earth is a CBA?

CBA is short for **Classroom-Based Assessment**. That is a well-organised geographical project or investigation carried out under your teacher's guidance over a three-week period.

You said there were two CBAs.

Yep! **CBA 1** will be completed during the second term of Year Two of your Junior Cycle course. **CBA 2** will be completed during the first term of Year Three.

Will CBA results be official?

Yes. The results of each CBA will be entered into your Junior Cycle **Profile of Achievement**. That's the official document that records your Junior Cycle results and achievements.

Who will grade my CBAs?

Your **teacher** will judge each CBA to be within one of these categories: (a) exceptional; (b) above expectations; (c) in line with expectations; (d) yet to meet expectations.

You mentioned a Written Assessment Task. What's that?

That's a special **written task** that will be linked to your CBA 2. It will be carried out in Year Three, **after** you have completed your CBA 2. It will be corrected by the Department of Education and will make up 10% of your total Junior Cycle Geography mark.

So CBA 1 in Year Two will be the first of my special tasks?

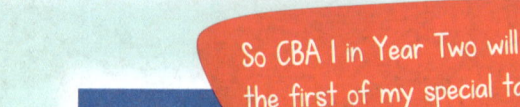

Correct.

What should I do about CBAs before CBA 1?

Do some *Geography in the News* and *My Geography* warm-up exercises from your *Cyclone* textbook and Skills Book. You could also start looking out for a suitable **topic** for your CBA 1.

CBA 1: Geography in the News

(for term two of Year Two)

CBA 1 relates to *Geography in the News*. It requires you to complete **a structured (well-organised) and detailed inquiry into a recent geographical event anywhere in the world**.

Your project can be in written **format** – for example, a well-illustrated booklet or set of posters. It could also be an illustrated oral presentation, a podcast or another suitable format. Your teacher will help you choose the best format for you.

A CBA can be carried out by an **individual**, by a **pair** or by members of a **larger group** working very carefully together. Your teacher can advise you on the best way to carry out your CBA.

Step-by-step guidelines for CBA 1

Choosing an event of geographical significance

The first thing to do is choose a **recent** (real) **event of geographical importance** that has been **reported in the media** and that interests you. Your teacher can help you decide whether the event you have chosen is suitable and recent enough.

Some of the *Geography in the News* events in your *Cyclone* textbook and Skills Book might interest you. News headlines such as the ones below might also provide ideas. Try to choose an event that will provide plenty of interesting material to work with. Draw up a shortlist before making your final choice.

Unusually high temperatures cause concern in …

Violent volcanic eruption in …

Parts of Greenland's massive ice sheet begin to melt at an alarming rate

Massive earthquake in …

Storm Ophelia batters Ireland

MORE PEOPLE SEEK ASYLUM IN IRELAND

Junior Cycle Geography CYCLONE SKILLS BOOK

1. Write a **shortlist of possible topics** here.

   _____
   _____
   _____
   _____
   _____
   _____

   > The step-by-step questions and notes that follow are designed to help guide you towards your **best possible CBA response**. You may not be able to respond fully to everything on these pages. Don't worry! Just do your best. Consult your teacher or your classmates and keep going!

   Discuss your possible *Geography in the News* event with your teacher and, if you're working in a pair or group, with other group members.

2. What is your **chosen** *Geography in the News* **event**? Write the title of your investigation.

   _____
   _____
   _____

Reflection

3. Why did you choose this significant geographical event?

   _____
   _____
   _____
   _____
   _____

4. Why do you think this event is newsworthy or important?

   _____
   _____
   _____
   _____
   _____

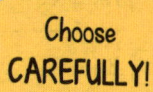

CBA 1: Geography in the News

Choosing a format for your project

You can choose to present your response to your *Geography in the News* event in many different ways. A few examples are:

- In written form – a written/typed report or article with illustrations
- In visual form – a graphic presentation using images or a display such as a set of posters
- In digital form – a blog, web page or slideshow
- In audio form – a podcast, voiceover or oral presentation.

Things to consider when choosing your format:

- Think about your **skills**. What sort of response would you be create best?
- Think about the **resources** available. Do you have the equipment/software needed to create a response in a particular format?
- Think about the **information** you want to communicate. Is it mainly text or mainly visual? This might help determine the most suitable format.

Reminder Box

a. Which **format** have you selected with which to present your project?

_____
_____

b. Why have you chosen this format?

_____
_____
_____
_____

c. On what date did you discuss it with your teacher?

_____

d. What is the deadline for completing your project?

_____

167

Junior Cycle Geography CYCLONE SKILLS BOOK

Sources

Sources are the things, people or places that will provide information for your CBA. It is best to use a **variety of sources**. The table below lists a range of sources. Put a tick (✓) beside the sources you think you might use. There are some blank spaces at the end of the table for you to add any other sources you might think of.

Source		Plan to use? (✓)
Print media	Newspaper	
	Magazine	
	Book	
	Newsletter	
Broadcast media	Radio broadcast	
	Television news	
	Television current-affairs programme	
	Documentary on television or streaming services	
Digital media	Online newspaper	
	Online magazine	
	Geographical website	
	Twitter	
	Blogs, vlogs, podcasts	
Interview		
Visit to site		
Other		

CBA 1: Geography in the News

Sources and data

Reminder Box

Name some sources you used and summarise the data you received from each source. This will help to remind you of things you should include in your project presentation.

A **source** is any person or thing that provides you with information. Sources can be people (through interviews, recordings, etc.), books, newspapers, magazines, questionnaires, maps, pictures, radio, television programmes, podcasts, social-media posts, etc.

Data is the information provided by your sources. Data can be written, photographed, spoken or recorded in any manner or form.

Source 1:

Type of source:

Data (information) received from source:

Source 2:

Type of source:

Data (information) received from source:

Source 3:

Type of source:

Data (information) received from source:

If you wish, you can add more sources and data on a blank sheet of paper and attach it to this page. Or, you could write any extra notes in a special 'CBA notebook', including the relevant page of this Skills Book (in this case, page 169).

Credits

You should **acknowledge** the sources from which you gather information for your CBA. These acknowledgements are called **credits**.

The table below shows one way in which credits might appear at the end of your CBA. The grid can be modified to suit your particular CBA.

Type of source	How to credit this source
Newspaper/magazine/local newsletter	Author, headline/title of article, name of newspaper/magazine, page number(s), date of publication
Book	Author, title, publisher, place and date of publication, page number(s)
Website/online newspaper	Web page title, URL, author (if known), date accessed/published
Other digital	Podcast title, Twitter address (e.g. twitter.com/JCGeography), etc.
Map	Title, scale, publisher, place and date of publication
Interview	Name of interviewee, date of interview
Visit	Location, date of visit

Reflection

1. How easy or difficult was it to find suitable sources for your chosen geographical event?

   _____
   _____
   _____
   _____
   _____
   _____

2. Are you satisfied that all of your sources were reliable and unbiased in the information that they provided? Explain.

   _____
   _____
   _____
   _____
   _____
   _____

'Features of quality' for CBA 1

Your CBA 1 grade will depend largely on how well you incorporate as many as possible of the **'features of quality'** described below.

1. **Detailed**, **informed** and **well-organised** work
2. Use of **geographical questions**
3. Reference to **patterns**
4. Reference to **processes**
5. Reference to **systems**
6. Reference to **sustainability issues**
7. Reference to the **geographical significance** of the event
8. Clear, relevant **conclusions**

Use as many features of quality as possible in your project.

1. Detailed, informed and well-organised work

Your CBA should be well organised and have a clear purpose. It should contain clear, detailed information, taken from a range of sources. Such sources might include news reports, books, magazines, personal accounts, maps, charts, videos, etc. Your CBA must show a clear understanding of the chosen event. It should be suitably illustrated. Illustrations might include relevant sketch maps, charts, labelled diagrams, photographs or video recordings.

2. Geographical questions

The table on the following page lists some **key questions** to keep in mind when carrying out geographical research. They will help to keep your research on track. Several will be relevant to your CBA.

Useful geographical questions

Key questions	Some examples	Other examples
What?	• What was the event? • What were its effects? • What was the response to the event? • What was the importance of the event?	
Where?	• Where did the event happen? • Where were its effects felt?	
When?	• When did the event happen? • When did people experience its effects? • When did people respond?	
How?	• How did the event happen? • How long did it last? • How were people, animals and plants affected? • How many people were affected? • How did people respond?	
Why?	• Why did it happen? • Why did people respond as they did? • Why was the event geographically important?	
Who?	• Who was involved or affected? • Who responded?	

3. Patterns

Patterns occur when **geographical events are distributed or landscape features are arranged in a fairly regular manner**. Many geographical events occur in **set patterns**. Volcanic eruptions, for example, often occur in **linear patterns** (lines) across some of the world's plate boundaries.

Reminder Box

a. Can you identify and describe any **pattern(s)** that occur(s)/occurred in relation to your chosen geographical event?

_____
_____
_____

b. Can you identify any reason for or cause of the patterns you have identified?

_____
_____
_____

4. Processes

A process is a **sequence of natural or human activities that changes our natural, human or economic environment**. Sea and river erosion are examples of **natural processes** that alter the landscape. Human migration is an example of a **human process** that might increase the diversity of our population.

Processes can sometimes be linked to **patterns**. The occurrence of volcanic eruptions in linear patterns results from the **processes** that cause plate edges to sink, melt and expand violently along linear boundaries where plates collide with and separate from each other.

Reminder Box

a. Can you identify and describe any **processes** in your chosen event?

_____
_____
_____
_____

b. Are there any **links** between the processes and patterns you have identified? If so, identify and describe those links.

_____
_____
_____
_____
_____
_____

5. Systems

A system exists when **a number of things or processes work together to make a unified whole**. There are many examples of natural and human systems in our world. For example:

- The circular movement of magma in the earth's mantle is part of a system that causes tectonic plates to move and is why fold mountains, earthquakes and volcanic activity occur near plate boundaries.
- The water cycle is a system in which water is constantly renewed in the atmosphere.
- A factory system uses inputs, processes and outputs to produce manufactured goods and to make a profit.

Reminder Box

a. Can you identify and describe any physical or human **system** connected to the geographical event you are investigating?

_____
_____
_____

b. Are any of the patterns and processes that you identified earlier part of that system? Explain.

_____
_____
_____

CBA 1: Geography in the News

6. Sustainability issues

Something is **sustainable** if it can continue without harming people or the environment now or in the future. Figure 1 shows how something can be environmentally, socially and economically sustainable.

Figure 1: Environmental, social and economic sustainability

- Examine the **impacts** (effects/results) of the event you have chosen.
- Examine how people have **responded** or how they could **respond** to these impacts.
- Explain how these impacts and responses are **sustainable** or **unsustainable**.

Reminder Box

a. Write some key phrases or headings on **sustainability issues** that you have discovered in your CBA 1 investigation.

_____
_____
_____
_____
_____
_____

7. Geographical significance (importance) of the event

Reminder Box

- The event itself might be significant.
- So might the **effects** of the event.
- So might the **responses** to the event.

a. Write key ideas on why *you consider the event that you have examined to be of* **geographical significance**.

_____
_____
_____
_____
_____

8. Clear, relevant conclusions

Summarise the main conclusion(s) you have drawn about the event and its consequences. Begin by naming the event, its location and (if appropriate) the time of its occurrence. Your conclusions should be clear, precise and to the point.

Your conclusions should be at the end of your assignment, followed only by the list of credits for the sources you used.

Event
_____

Where
_____
_____

When
_____
_____

Conclusions
_____
_____
_____
_____

Organising your CBA project

Your CBA 1 project must be **organised** clearly. You must plan this carefully, with your teacher's guidance.

Figure 2 shows one **possible** plan for organising a CBA 1 project. Some of its elements may be useful to you.

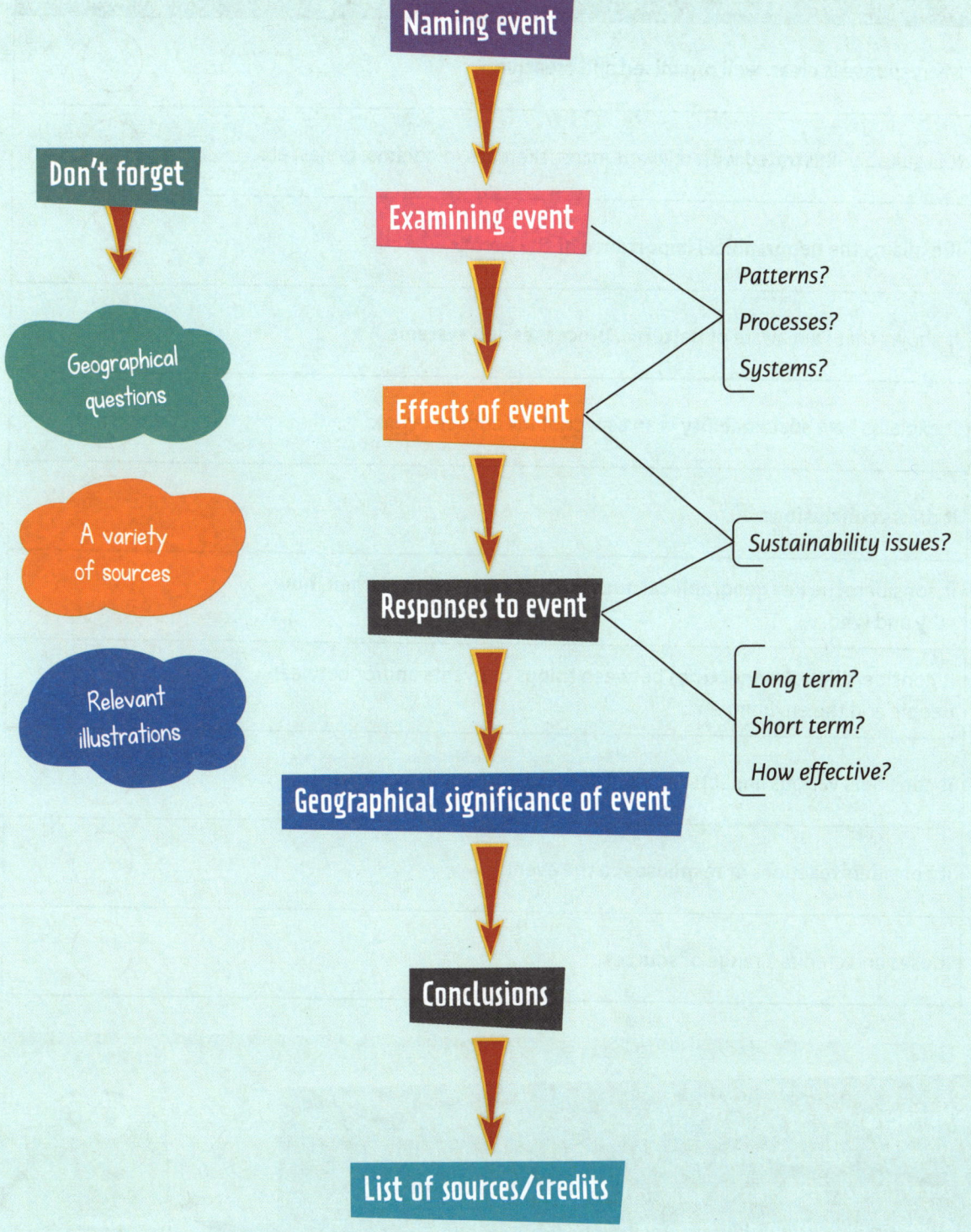

Figure 2: A possible step-by-step plan for a CBA 1 project

Success-criteria checklist for your project

Not all criteria will be relevant to your project.

Success criteria	Tick (✓) when complete
My response is **clear**, **well organised** and **creative**.	◯
It is suitably **illustrated** with relevant maps, sketches, diagrams, tables, etc.	◯
It explains the **geographical importance** of the event.	◯
It shows that I am aware of **patterns**, **processes** and **systems**.	◯
It explains how **sustainability** is an aspect of my chosen event.	◯
It draws **conclusions**.	◯
It considers the **key geographical questions** of **what**, **where**, **when**, **how**, **why** and **who**.	◯
It considers **links/connections** between things or events and/or between people and the environment.	◯
It considers various **impacts** (effects) of the event.	◯
It considers reactions or **responses** to the event.	◯
It uses and credits a **range of sources**.	◯

Reflection

When you have completed your CBA, reflect on (think about) the benefits and difficulties of your task. Make and fill in a copy of the student evaluation and reflection template on the next page. There are no right or wrong answers here. Just answer honestly and frankly. That will help you to learn and develop even further.

CBA 1: Geography in the News

Student evaluation and reflection template for CBA 1

CBA 1: *Geography in the News*	Event chosen:

Write key words only.

What did I enjoy most about completing this CBA? *Give details.*
_____
_____
_____
_____

What were the main challenges/difficulties I faced?
_____
_____
_____
_____

How could I overcome these in the future?
_____
_____
_____
_____

What new geographical skills did I develop when completing this CBA? How could I use these in the future?
_____
_____
_____
_____

Who or what organisation might be interested in my research/the data I collected? Why?
_____
_____
_____
_____

CBA 2: My Geography

(for term one of Year Three)

CBA 2 is called *My Geography* and will be **a structured inquiry into one or more aspects of your local area**.

Remember CBA 1, which you carried out in Year Two?

You won't be able to use **all** of the material that appears in this CBA 2 section. *Choose and use* what you can and what suits your particular investigation.

CBA 2 will be **similar to CBA 1** in the following ways:

1. Your **teacher** will supervise and assess (grade) your CBA.
2. Your **result** will be judged on the quality and detail of your work and will depend largely on the **features of quality** in your CBA.
3. Your result will be entered into your **Junior Cycle Profile of Achievement** and will be rated within one of the following categories:
   - Exceptional
   - Above expectations
   - In line with expectations
   - Yet to meet expectations

But CBA 2 will be **different from CBA 1** in the following ways:

1. The **topic** will be different. (This will affect the way you organise your investigation.)
2. Some **features of quality** will be different.

You will also be required to complete a Written Assessment Task in Year 3. This task will be separate from but based directly on your CBA 2. The assessment will be worth 10% of your Junior Cycle examination marks. (More about that later (page 197).)

CBA 2: My Geography

Before getting started ...

... it is best to **read carefully** through all of this section on CBA 2: My Geography. This could be done in class. You will not understand everything on your first reading. Don't worry. Your teacher and classmates will help to explain things that you don't understand. After some readings and explanations, you will understand better what you have to do. This section might seem to contain a huge amount of information. Again, don't worry. The information is there to help you, but you will use only some of it in your investigation.

Your CBA 2 grade will depend largely on certain **'features of quality'** in your work. It is best to find out about those features of quality before planning the format and outline of your project.

'Features of quality' for CBA 2

Incorporate as many as possible of the features of quality listed and described on the pages that follow. Some features of quality are likely to lend themselves especially to different parts of the project. Other features (such as point 1 opposite) should be evident throughout the entire work.

Features of quality for CBA 2

1. Well-organised and detailed work
2. Use of **geographical questions**
3. Gathering, analysis and presentation of **data**
4. Awareness of **sustainability** concerns
5. Awareness of **processes**
6. Awareness of **patterns**
7. Awareness of **systems**
8. **Conclusions**

1. Well-organised and detailed work

Your CBA should be **well organised** and have a **clear purpose**. This CBA section will show you how to organise your work on a step-by-step basis. Your work should contain **clear and detailed information**, preferably taken from a **variety of sources**, such as interviews, books, newspapers, magazines, personal accounts, questionnaires, maps, photographs, videos, podcasts and fieldwork. Your CBA should be presented clearly and (apart from quotations) in your own words. Relevant **illustrations** (not mere decorations) are very useful. They might include labelled sketch maps, diagrams, charts, graphs, pictures or video recordings.

2. Geographical questions

You learned on page 171 (in relation to CBA 1) that the key geographical questions to keep in mind are **what**, **where**, **when**, **how**, **why** and **who**. When you relate these questions to your particular investigation, they will help to keep your investigation on track.

The table below contains random examples of questions, some of which might be helpful. Empty spaces have been left in the box for you to add examples of your own.

Useful geographical questions

Key questions	Some examples	Other examples
What?	• What data needs to be gathered? • What might be the causes and/or effects of the topic that you are investigating? • What importance might this topic have?	
Where?	• Where is your investigation focussed?	
When?	• When is the best time to carry out any fieldwork? • When must you complete your investigation?	
How?	• How will you carry out your investigation? • How will the investigation work be shared (if other students are involved)? • How (in what form) will the investigation be presented (booklet, charts, audio, etc.)?	
Why?	• Why is your topic worth investigating? • Why is this investigation geographically important?	
Who?	• Who is involved in your investigation? (You alone? You and another/others?) • Who is affected by the topic you are investigating? • Who might provide useful data? • Who might be interested in the findings of your investigation?	

3. Data

Data is the information that you gather and use in your geographical investigation.

The **gathering**, **analysis** (examination) and **presentation** of data will be a major part of your CBA 2.

You must also acknowledge the **sources** of your data.

Learn about this on pages 193-4.

4. Sustainability

Something is **sustainable** if it can continue without harming individuals, society or the environment now or in the future. Figure 1 shows how something can be **environmentally**, **socially** or **economically** sustainable.

Environment
- Minimise pollution
- Reduce waste
- Protect environment

Society
- Improve lifestyle
- Provide education
- Support community
- Promote equality

Economy
- Make profit
- Save money
- Grow economy
- Respect employees

SUSTAINABILITY

Figure 1: Environmental, social and economic sustainability

- Examine the **impacts** (effects/results) of the event that you have chosen.
- Examine how people have **responded**, or how they might **respond**, to these impacts.
- Explain how any of these impacts and/or responses are **sustainable** or **unsustainable**.

Reminder Box

*Write down any **sustainability issues** that you discover in your CBA 2 investigation.*

This will remind you to discuss these issues in your project.

_____
_____
_____
_____
_____
_____
_____
_____
_____
_____
_____
_____
_____
_____
_____
_____

5. Processes

A process is **a sequence of natural or human activities that changes our natural, human or economic environment**. For example:

- Sea or river **erosion** and **deposition** are natural processes that change our landscapes.
- Human **migration** is a human process that might increase the diversity of our population.
- **Cheesemaking** is a process that changes milk into cheese.

Reminder Box

*Name any **processes** that you discover in your geographical investigation.*

_____
_____
_____
_____
_____
_____

6. Patterns

Patterns occur when **geographical events happen or landscape features are arranged in a fairly regular manner**.

Patterns and processes are sometimes connected. For example, **nucleated (clustered) housing patterns** may result from roads being designed and constructed to meet at the bridging point of a river. On the other hand, **linear (ribboned) housing patterns** may result from many people building homes conveniently close to a single roadway.

Reminder Box

a. *If/when you discover **patterns** when preparing your geographical investigation, list those patterns here.*

_____
_____
_____
_____

b. *If/when you discover **connections** between geographical processes and patterns, mention those connections here.*

_____
_____
_____
_____

7. Systems

A system exists when **a number of things or processes work together to make a unified whole**. There are numerous examples of natural and human systems in our world. For example:

- **Bacteria** in the soil use a process called **humification** to changes leaves, twigs and other plant litter into **humus**. Humus helps to make **fertile soil** suitable for agriculture such as dairy farming.
- A **dairy farm** is a system that uses **fertile soil**, **cows** (and other inputs) to produce **milk** (and other outputs).
- A **creamery** is a factory that changes **milk** into **butter** and other outputs.

Notice that systems involve **inputs**, **processes** and **outputs**. Some processes that you identify under point 5 on page 185 might be part of a system.

Notice that different **systems** (such as the three examples given above) can be **interconnected**. An output of one system can be an input of another.

Reminder Box

a. Can you identify any physical or human **systems** connected to your CBA 2 topic?

_____
_____
_____
_____
_____
_____
_____

b. Are any of the **patterns** or **processes** that you identified earlier part of a **system**?

_____
_____
_____
_____
_____
_____
_____

c. Are any **systems** that you have identified **connected** with each other?

_____
_____
_____
_____
_____
_____
_____
_____

8. Conclusions

In this section, you will summarise the **main findings** of your investigation and the **data** that you examined. You will also discuss the **implications or relevance of your findings** for the local area.

For example, if you investigated a factory in your area, you might summarise your findings on **why** that location was chosen for the factory. You might also refer to positive and negative local **implications** of the factory. These implications might include direct employment, spin-off employment and other benefits. They might also include **sustainability issues** with regard to visual appearance, road traffic, demands on housing, noise and the treatment and discharge of waste water.

HELP! There's an awful lot of material in those features of quality. I might not be able to use it all!

That's OK! You won't have to use it all. The features are there to guide and help you. Just use as many as you can, as well as you can.

Junior Cycle Geography CYCLONE SKILLS BOOK

Getting started

Now that you have been introduced to the features of quality, you can start **planning your investigation**.

With the help of your teacher, try to decide on each of the following.

- Decide whether you prefer to produce **your own project** or to take part in a **shared project** with one or more other students. Think carefully about this. Your teacher will advise you.
- Decide also on the **format** of your geographical investigation. It could be presented in a (large-paged) booklet or scrapbook, or as a series of wall charts. It could be in the form of an illustrated recording, a PowerPoint presentation, a podcast or any other workable format. Choose the format that you think would suit you best and that you could produce most efficiently.

A possible step-by-step plan

Figure 2 on the following page shows the outline of a sample step-by-step plan for a CBA 2 project. This plan might be useful to you. It will be followed for the rest of this section.

Depending on the subject and format of your investigation, you might (having consulted with your teacher) choose to follow a different plan from the one in Figure 2. Feel free to do that; but remember to use as many as possible of the **features of quality** for CBA 2.

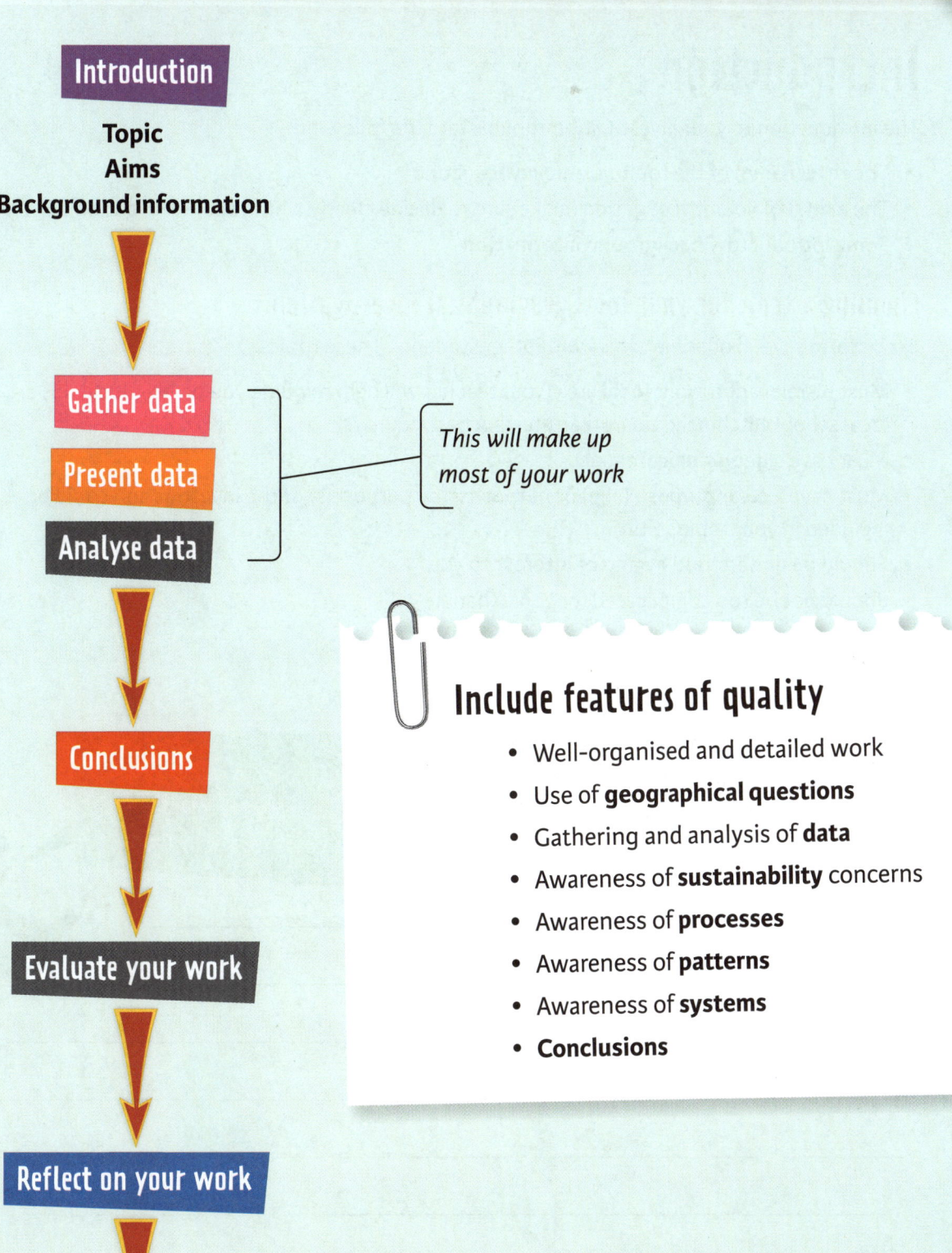

Figure 2: A possible step-by-step organisational plan for CBA 2

Introduction

The introduction to your investigation should state the following:

- The **title** (name) of the topic you are investigating
- The **aim(s)** of your investigation (make sure to decide clearly what your investigation will try to discover)
- Some introductory **background information**.

Choosing a topic for your local geographical investigation

Brainstorm several possible topics with other students. These topics:

- Must be relevant to your **local** area (your teacher will help to guide you on how local this area should be – local school catchment area/urban area(s)/coast/county?)
- Must have a **geographical** aspect
- Must have a **clear purpose** (aim) or **purposes**: the purpose(s) of the investigation could be included in your topic's title
- Should be **understood by** and **of interest to** you
- Should not be too complicated for you to handle.

Reminder Box

Write down a list of **possible topics**. Discuss these options with your teacher and with others to help you decide on the title of your chosen topic.

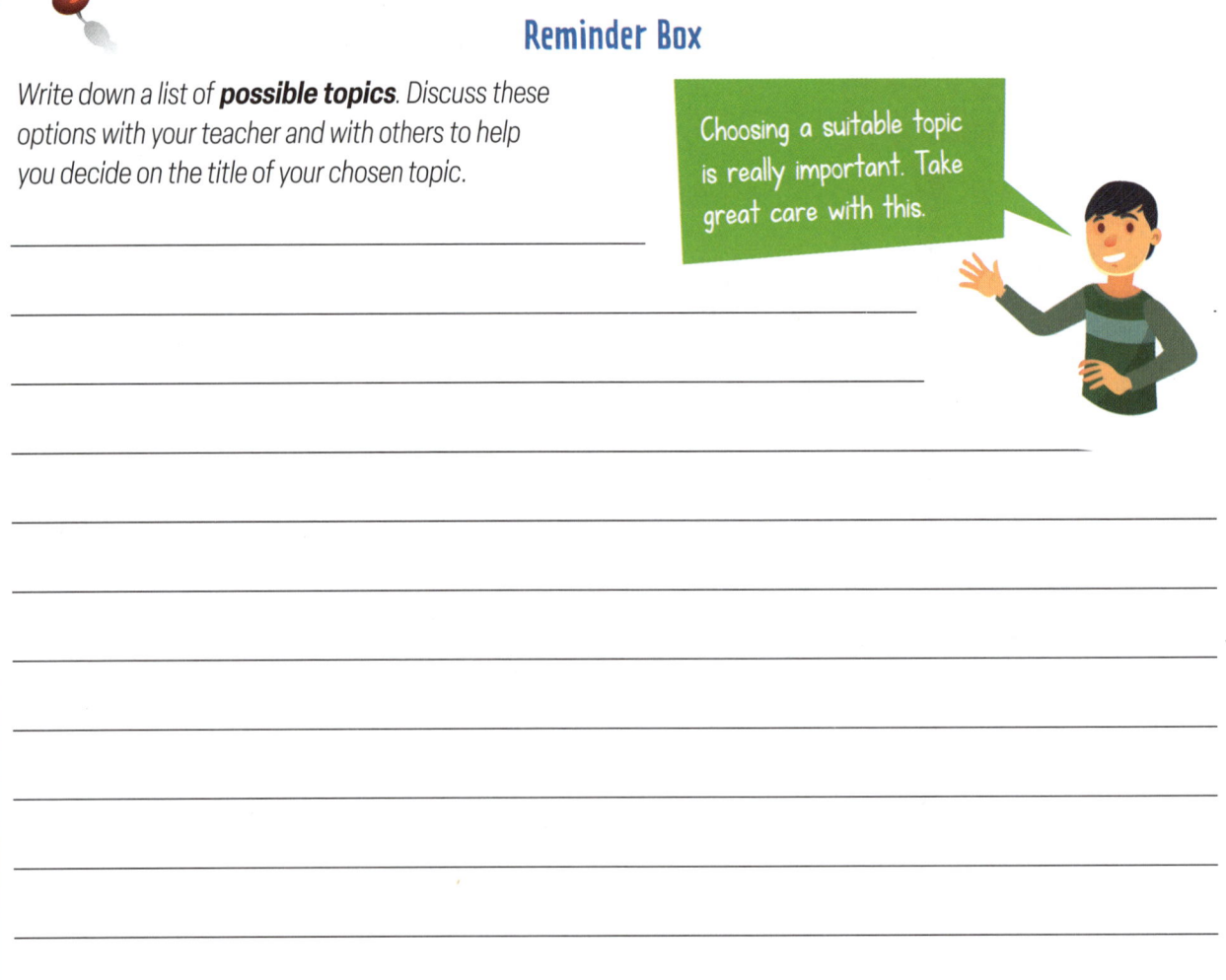

Choosing a suitable topic is really important. Take great care with this.

Deciding on the aim(s) of your investigation

When deciding on a topic, you should also decide on the aim or aims of your investigation. **An aim is a general statement of what you are trying to discover.** There are many possible types of aim, a few random examples of which are given below.

An aim might be to investigate **differences between places**. For example:

> *An investigation into how and why the shape, size and functions of my local town of … differs from those of … in the nearby town of …*

An aim might relate to **causes, effects or other connections** between things. For example:

> *An investigation into the causes and consequences of the opening of a large … factory in my local area.*

Some aims might be to investigate **change over time**. For example:

> *An investigation into how and why the shape, size and functions of … village have changed since the year …*

Some introductory background information

Name the **local area** of your investigation.

You could illustrate your introduction with relevant maps and/or photographs. A labelled **sketch map*** of this area, showing places or features relevant to your investigation, would be useful.

If you are investigating a problem or a controversy, state this briefly but clearly.

* *If you are drawing a sketch map, try to show a linear* **scale***. Also insert an* **arrow** *that indicates north.*

Reminder Box

a. The **title** of my investigation is:
   _____
   _____

b. The **geographical aspect(s)** of my topic relate(s) mainly to (tick one or more boxes below):
   - ☐ The physical world
   - ☐ How the physical world and people interact
   - ☐ Human activities
   - ☐ Places (change over time, etc.)
   - ☐ Others (specify):
   _____

c. The **aim(s)** of my investigation is/are:
_____
_____
_____

d. The **local area** of my investigation is:
_____
_____

e. **Reflection:** Why did I select this topic for a geographical investigation? Why/how is this topic relevant to my local area?

_____
_____
_____
_____
_____
_____
_____
_____
_____
_____
_____
_____
_____
_____

This will help you to reflect on your work.

Gathering, analysing and presenting data

Gathering data

Data is the **information** that you collect and use in your geographical investigation.

Sources are the people, places or things that **provide the data** that you use.

There are two general types of source that you should use: **secondary sources** and **primary sources**.

Secondary sources provide data that has been recorded already by other people. Some types of secondary sources are listed in the table below.

Examples of secondary sources that could be used	Tick (✓)
Print media: newspapers (local or national), magazines, books, local newsletters, posters, etc.	○
Broadcast media: News or current-affairs programmes on television, radio or social media	○
Digital media: Podcasts, Twitter, websites, etc.	○
Maps of the area being examined	○
Other: _____	○
Other: _____	○

1. Place tick marks next to the secondary sources listed that you will use for your research
2. Use the 'Other' space to name any type of secondary source that you might use but that is not listed in the table

Primary sources provide new and original information that you can get through **field research** (which is a fancy name for a field trip, interview, questionnaire, experiment, etc.).

The table on page 194 gives some examples of field research that you could use to collect data for your investigation. Tick any examples that you might use.

> You must use a **variety of sources** in your investigation. If the findings of some sources contradict each other, present and evaluate each point of view. Consider whether your sources are reliable in the information they provide.

Examples of primary sources (field research) that could be used	Tick (✓)
Experiments: local soil study, measuring weather, etc.	○
Field research: organised field trip to beach, farm, factory, etc.	○
Questionnaires/surveys of local people	○
Interviews	○
Witness statements	○
Other: _____	○

REMINDER

If you carry out field research, **decide** first what information you are seeking. Make sure that you use suitable **geographical questions**. **Collect** and **present** the information carefully and clearly and **use** the information to serve the aims of your investigation.

Analysing and presenting data

When you gather your data, you must **analyse** (examine) it carefully to find out what it reveals in relation to your investigation.

You must then **present** the data and its **findings** (what it reveals) in a clear and organised manner. Some data can be summed up in key points or paragraphs. Some can be presented in the form of well-explained tables of statistics, graphs and charts. The results of a questionnaire, for example, might be presented in the form of a table of statistics or by bar charts. The main findings of an interview could be summarised in your presentation, while the full record of each question and answer could be added to the end of the investigation in the form of an **appendix***.

Chapters 20 of the *Cyclone* textbook and Skills Book show how tables, graphs and charts can be used to present data. Chapter 13 of the Skills Book does the same in relation to weather data.

*An appendix is not an essential part of a written project. But, if there is information of exceptional interest or importance, it might appear as an add-on at the end of the project. An appendix should be clearly titled and referenced to a specific page of the project.

Acknowledging your sources

In any investigation, it is important to indicate the sources of your information. This could be done by including an **'Acknowledging sources'** section at the end of your project (see the step-by-step plan on page 189). To properly reference each source, use the guidelines in the table below.

Type of source	How to credit this source
Newspaper/magazine/local newsletter	Author, headline/title of article, name of newspaper/magazine, page number(s), date of publication
Book	Author, title, publisher, place and date of publication, page number(s)
Website/online newspaper	Web page title, URL, author (if known), date accessed/published
Other digital	Podcast title, Twitter address (e.g. twitter.com/JCGeography), etc.
Map	Title, scale, publisher, place and date of publication
Interview	Name of interviewee, date of interview
Visit	Location, date of visit

Conclusions

In this section, **summarise your main findings** and explain how they are **significant** or **interesting**. If, for example, your findings show that the number of young people in your local area is growing rapidly, you might stress the growing need for schools, play areas, sporting and social facilities, etc. It might also be useful to refer to **sustainability issues** in this section.

> **HINT**
> Findings and conclusions can sometimes contradict what you sought to prove in your investigation. Unexpected findings and conclusions can also be interesting and valuable. They might even prompt you to rethink or question your assumptions about things. This is a valuable learning process.

Evaluating your work

In this section, consider **how well your investigation went**. For example:

1. Which sources were best at providing the information that you needed, or at providing other interesting, relevant information?
2. Are you satisfied that all of your sources provided reliable and unbiased information? If not, comment on this.
3. Are there any significant gaps in the information that you tried to find? If so, why?
4. What worked well in your investigation?
5. What proved difficult or worked less well?
6. Name any valuable aspect(s) of your work.
7. Who (or what organisation) might find your work interesting or useful?

Evaluating and reflecting on your work might prove very useful when you complete the **Written Assessment Task**, which will be separate from but related to your CBA 2 investigation.

Be honest in evaluating and reflecting on your work. Not everything is likely to go smoothly in an investigation. Point out things that did not go well, as well as things that did. If you think you should do some things differently in any future investigation, explain what and why.

Reflecting on your work

In this section, discuss what you learned from doing your CBA 2. For example:

1. What interesting **knowledge or skills** did you learn?
2. How might you **use** some of this knowledge or these skills in the future?
3. If you were to do this investigation again, what would you do differently to improve it?
4. Do you have any useful **advice** for a student who will undertake a CBA 2 investigation next year?

Here are some important things to know about the Written Assessment Task.

The Written Assessment Task

- This will be separate from but based on your CBA 2.

- It will take place in Year Three, shortly after you complete your CBA 2 investigation.

- It will take place in school time, under examination conditions, over a total period of 80 minutes.

- It will be carried out under your teacher's supervision, but will be sent to the State Examination Commission for marking.

- It will make up 10% of your total Junior Cycle Geography mark/grade.

- It will come in the form of a paper or booklet.

A main purpose of the Written Assessment Task will be to seek from you a **focussed reflection** on various aspects of your CBA 2 investigation. Such aspects might include:

- ✓ **Skills** that you developed during your investigation

- ✓ The ability to **identify suitable aims** for your investigation

- ✓ The ability to **identify a variety of useful primary and/or secondary sources***

- ✓ The ability to **judge the value and reliability of individual sources**

- ✓ The ability to **examine, interpret and evaluate** the usefulness and reliability of various forms of **data****

- ✓ The ability to **draw appropriate conclusions**

- ✓ The ability to **recognise and describe** not only the **successful** parts of your investigation, but also any **difficulties or setbacks** encountered in your investigation

- ✓ The ability to **suggest how such setbacks might be avoided** or overcome in any future investigations

- ✓ The ability to **identify some possible value to others** of your investigation. How might your research and/or data be of interest to any person or organisation?

REMINDER
* A source is any person or thing that provides information. Sources could include written text, photographs, maps, tables of statistics, graphs, interviews, questionnaires, audio or video materials, etc.

REMINDER
** Data is any information provided by a source.

> On the following pages you will find a **SAMPLE REFLECTION AND EVALUATION TEMPLATE**. If you respond carefully to as many as possible of the questions asked, this could help greatly to prepare you for your Written Assessment Task. Fill in the sample reflection **after** you have completed your CBA 2 project.

Written Assessment Task

Sample reflection and evaluation template for CBA 2	
The questions marked with an asterisk (*) have been referred to **officially**, so make sure to respond to all of those fully. The other questions also raise useful reflection and evaluation topics. Respond fully also to as many of those as possible.	
CBA 2: *My Geography*	**Title/aspects of my chosen investigation***

What did I enjoy most about completing this CBA? *Give details.**

If you need more writing space, you could use some notebook pages and attach them to the end of this template.

Sample reflection and evaluation template for CBA 2

What were the main challenges/difficulties/setbacks that I faced? How could I overcome these in future investigations?*

Written Assessment Task

Sample reflection and evaluation template for CBA 2

What new geographical skills did I develop when completing this CBA? How could I use those skills in the future?*

Sample reflection and evaluation template for CBA 2
Which sources were most effective and which were least effective at providing the information/data that I needed? Most effective/why? _____ _____ _____ Least effective/why? _____ _____ _____ _____
Am I satisfied that the sources I used provided reliable and unbiased information? *If not, comment on this.* _____ _____ _____ _____ _____ _____
Were there any significant gaps in the data that I wished to obtain? If so, why? How could I prevent such gaps from occurring in any future investigation? _____ _____ _____ _____ _____ _____

Written Assessment Task

Sample reflection and evaluation template for CBA 2

Were some aspects of my investigation or its findings valuable or useful? *Explain.*

___

Who or what organisation might be interested in my research/data? Why?*

___

Sample reflection and evaluation template for CBA 2

What useful advice could I give to a student who will undertake a CBA 2 investigation next year?

_____
_____
_____
_____
_____
_____
_____
_____
_____
_____
_____

Is there any other suitable question that I think should be added to this template? *If so, write and respond to this question here.*

_____
_____
_____
_____
_____
_____
_____
_____
_____
_____
_____

Student: **Teacher:** **Date:**